W9-CZI-876

STUDIES IN THE LOGIC
of
CHARLES SANDERS PEIRCE

STUDIES IN THE LOGIC *of* CHARLES SANDERS PEIRCE

Nathan Houser,
Don D. Roberts, *and*
James Van Evra, *editors*

Indiana University Press

BLOOMINGTON AND INDIANAPOLIS

© 1997 by Indiana University Press

All rights reserved

No part of this book may be reproduced or utilized in any form or by any means, electronic or mechanical, including photocopying and recording, or any information storage and retrieval system, without permission in writing from the publisher. The Association of American University Presses' Resolution on Permissions constitutes the only exception to this prohibition.

The paper used in this publication meets the minimum requirements of American National Standard for Information Sciences—Permanence of Paper for Printed Library Materials, ANSI Z39.48-1984.

MANUFACTURED IN THE UNITED STATES OF AMERICA

Library of Congress Cataloging-in-Publication Data

Studies in the logic of Charles Sanders Peirce / Nathan Houser, Don D.
  Roberts, and James Van Evra, editors.
      p. cm.
    Includes bibliographical references and index.
    ISBN 0-253-33020-3 (alk. paper)
    1. Peirce, Charles S. (Charles Sanders), 1839–1914—Contributions
  in logic. 2. Logic—History—19th century. 3. Logic—History—20th
  century. I. Houser, Nathan. II. Roberts, Don D. III. Van Evra,
  James, date.
  B945.P44S78 1997
  160'.92—dc20 96-3853
1 2 3 4 5   01 00 99 98 97

B
945
.P44
S78
1997

091597- 4396 D2

All the men who are now called discoverers,
in every matter ruled by thought, have been
men versed in the minds of their predecessors,
and learned in what had been before them.

Augustus De Morgan
A Budget of Paradoxes, Vol. 1, p. 5.

CONTENTS

FOREWORD

Charles S. Peirce and Gottlob Frege form an interesting pair in the history of logic. They invented a notation for quantifiers and developed quantification theory almost simultaneously, independently of each other. They can therefore be regarded as the principal founders of modern logic. In philosophical logic, they stressed the importance of making a distinction between a speech act and its content, criticized psychologism, and characterized predicative expressions as incomplete or "unsaturated" signs. Frege and Peirce also shared a general neglect by the philosophical community of their time. Peirce held no academic position after 1884, and Rudolf Carnap tells that when he attended Frege's lecture course "Begriffsschrift II" in 1913, the entire class consisted of the lecturer, Carnap and a friend, and a retired military officer who was interested in new mathematical ideas.

Despite the similarities, Peirce and Frege are far apart philosophically. Jean van Heijenoort and Jaakko Hintikka have drawn attention to an important contrast between two different presuppositions underlying much of the twentieth-century philosophy of language and logic. It is a contrast between what may be characterized as the conception of language as the *universal medium* of communication and the view of language as a *calculus*. According to the former conception, a speaker is, so to speak, a prisoner of his or her language; it is impossible to "step outside" one's language and consider different interpretations (and reinterpretations) of it in a systematic way. According to the representatives of this tradition, the universality of language therefore makes a systematic, theoretical analysis of semantical questions impossible; semantical relations between language and reality are opaque or inaccessible. Frege's way of dealing with semantical questions indirectly, by means of "hints and clues," illustrates this way of thinking: in Frege's logical work semantical issues remained "hidden" behind axiomatic and deductive techniques. This view dominated the philosophy of language around the turn of the century; its main representatives were Frege, Bertrand Russell, and Ludwig Wittgenstein. The second main presupposition identified by van Heijenoort and Hintikka, the language as calculus view, is the cornerstone of the tradition of model theoretic semantics, represented by Ernst Schröder, Leopold Löwenheim, Alfred Tarski, and their followers. The central ideas of Peirce's work in logic and the theory of language belong to the latter tradition. In fact, some basic developments and ideas of this tradition have their origins in Peirce, whose "logic of relatives" was codified by Schröder, while Löwenheim's work was based on Schröder's.

The contrast between Frege and Peirce is particularly conspicuous in their approaches to semantic questions. Unlike Frege, Peirce tried to develop systematic theories of semantics, and his work contains the germs of many important contemporary developments in this field. He argued that the conditional statements of ordinary language cannot always be formalized as material conditionals, and explained the truth-conditions of such statements in terms of quantification over possible circumstances or courses of events. The gamma part of Peirce's theory of logical graphs (existential graphs) contains, in a rudimentary form, the basic ideas of the possible worlds semantics of modal logic, and his pragmatic analysis of complex sentences and quantifier phrases contains some of the main elements of game-theoretical semantics. Peirce investigated different logical languages and different logics: for example, he was interested in many-valued interpretations of logical connectives. He even argued that different methods of representation (different languages) and different logics are appropriate for different purposes. To follow the Greek poet Archilochus and Isaiah Berlin, we may say that in logic, Peirce was a fox, Frege a hedgehog.

The present volume attests to the multifariousness and the continuing fecundity of Peirce's work in logic. It includes papers that discuss some of the issues mentioned above, as well as papers in which Peirce's ideas are developed further and applied to contemporary issues. The papers are not restricted to logic in the narrow sense; some of them discuss questions of philosophical logic and semantics and the general theory of reasoning. The variety of topics represented in the present volume thus reflects the scope of Peirce's conception of logic as general semiotic.

Jaakko Hintikka
Risto Hilpinen

PREFACE

This collection of essays is an outgrowth of the principal logic symposium of the Charles S. Peirce Sesquicentennial International Congress held at Harvard University in 1989. That congress brought together over 450 scholars from 26 countries to commemorate Peirce's birth on 10 September 1839 and to celebrate his many contributions to the advancement of knowledge. The focus of the logic symposium was on Peirce's contributions to exact logic, with a special emphasis on linkages between Peirce's work and contemporary research in technical logic and foundations and in the history of logic. The majority of the papers fit this description, but the scope of the collection was broadened somewhat for the present book to include, for instance, issues in the philosophy of logic. We think the result is the most comprehensive account and exposition of Peirce's contributions to technical logic.

The primary purpose of this volume is to extend and promote the understanding of some of Peirce's most advanced work in logic and, in particular, to provide a useful resource for contemporary research by bringing his ideas to the attention of modern logicians and historians of logic. Many scholars who have studied his writings in logic and mathematics rank Peirce with Leibniz in the history of thought—some would rank him even higher. We believe there is important work to be done in settling Peirce's proper place within the larger contribution of the Algebraicists to the development of logic in the 19th century; but the rank Peirce has or should have is surely of less importance than the potential usefulness of his ideas for the present advancement of the science of logic. We believe that Peirce's researches in logic, only partly represented in these papers, are of continuing importance not only for their stimulating depth and originality but also for the technical achievements that may still be put to good use.

To facilitate the presentation of the papers that follow, a number of conventions have been followed throughout. All references have been standardized and keyed to a single list at the end of the volume. Multiple references in the text separated by slashes—e.g., "L 148/NEM 3: 785"—present different sources for the same text. References separated by commas or semicolons refer to different but related texts. Italicized citations—

"*Kempe 1885*"—are abbreviations for the complete information given in the references at the end of the book. If context makes clear who the author is, a citation may be further abbreviated to an italicized date standing alone. Works cited are generally listed by their original date of publication, with relevant later editions also mentioned. If a citation refers to a specific later edition, that date will appear in square brackets following the main reference date—"*Russell 1903 [1956]*: 24."

The following abbreviations are used for frequently cited sources:

Papers published during Peirce's lifetime: "P" followed by the number assigned in Ketner et al., *Comprehensive Bibliography*.

Peirce's unpublished writings: "MS" for manuscript or "L" for letter, followed by the folder number assigned in Richard Robin, *Annotated Catalogue*, or "*MS*" followed by the folder number assigned by the Peirce Edition Project for the new chronological edition. Page references to manuscripts are given as Peirce's own page numbers followed by the superscript "P"—e.g., "MS 12: 1^P"—or, more frequently, to page numbers assigned by the Institute for Studies in Pragmaticism with numbers corresponding to page placement in the Harvard microfilm edition of the Peirce papers. Manuscripts are quoted with permission from the Harvard University Department of Philosophy.

Collected Papers: volume and paragraph numbers—"3.365." The commonly used prefix "CP" is omitted here.

The New Elements of Mathematics: "NEM" followed by volume and page numbers.

Writings of Charles S. Peirce: A Chronological Edition: "W" followed by volume and page numbers.

Charles Sanders Peirce: Contributions to the Nation: "N" followed by volume and page numbers.

Semiotic and Significs: The Correspondence between Charles S. Peirce and Victoria Lady Welby: "SS" followed by page numbers.

Dates of Peirce's writings are occasionally given in square brackets in the citations to manuscripts or standard works.

The authors have for the most part been allowed to follow their individual preferences with respect to style and form. However, a few standardizations have been imposed on the papers by the editors for uniformity and general appearance. In a few cases, regularizing emendations were made to quotations: e.g., "De Morgan" is given a space between "e" and "M," and we use "MacColl" rather than "McColl" (in spite of Peirce's apparent preference for the latter). Such emendations were made silently.

We thank Debbie Dietrich, Aleta Houser, and Shea Zellweger for helping in various ways with the preparation of this volume. A grant from the

Social Sciences and Humanities Research Council of Canada made it possible for Robin Walsh to make the initial preparation of much of the artwork. We are grateful to Kenneth Laine Ketner, who was the principal organizer of the Peirce Sesquicentennial Congress, and to Harvard University for hosting the Congress on the site of Peirce's boyhood home.

It is hoped that the book is relatively free of errors, but each editor cheerfully blames the others—or the authors—for those which may be discovered.

Nathan Houser
Don D. Roberts
James Van Evra

STUDIES IN THE LOGIC
of
CHARLES SANDERS PEIRCE

1.

INTRODUCTION:
PEIRCE AS LOGICIAN

Nathan Houser

I want to offer a perspective on Peirce that may help unify the wide assortment of topics that are dealt with in this collection of essays. The subjects treated here range from Peirce's axiomatization of arithmetic and his set theory to artificial intelligence and his theory of proper names. No one will doubt that logic is useful for treating these subjects, yet one may wonder why they are brought together with more central questions about relations and inference in a book about Peirce's contributions to logic. The answer is that what Peirce counted as logic and what we count are not congruent. I will explain this below, but I will begin with some brief biographical remarks.

§1

Peirce was first and foremost a logician. It is now well known that he is (or at least was for a long time) the only person listed in *Who's Who* as a logician. Peirce claimed that from about the age of twelve, after reading his brother's copy of Whately's *Elements of Logic,* he could no longer think of *anything* except as an exercise in logic (*Fisch 1982b:* xviii). Charles's father, the great Harvard mathematician and astronomer, Benjamin Peirce, in a moment of deep paternal feeling, advised Charles to reconsider his dedication to logic, which he doubted could be a sustaining occupation from a practical standpoint. It is likely that Charles carefully weighed his father's advice and perhaps partly for that reason continued in his scientific work for the U.S. Coast Survey; but he could never give up logic. In 1878, when he applied to Daniel C. Gilman for the professorship of physics at Johns Hopkins, he wrote that it was as a logician that he sought to head that department. He had learned physics, he said, in connection with his study of logic. He explained that "the data for the generalizations of logic are the special methods of the different sciences" and that "to penetrate these methods the logician has to study various sciences rather profoundly."[1]

Peirce did not get the Johns Hopkins professorship in physics but he was hired the following year as part-time lecturer in logic. Although his official connection with Johns Hopkins was with the Philosophy Department, he quickly became a member of the Hopkins mathematical community, which included James Joseph Sylvester, for a time Arthur Cayley, and such brilliant young instructors and students as George B. Halsted, William E. Story, Christine Ladd, Fabian Franklin, Allan Marquand, Benjamin I. Gilman, and Oscar Howard Mitchell.[2]

For most of the preceding decade Peirce had exhibited a special interest in developing the area of overlap between mathematics and logic. Influenced by mathematicians who were striving to develop more abstract algebras and geometries and by logicians who sought to increase the purview of logic while making it more exact, following especially in the footsteps of Boole, Peirce worked hard to develop mathematical models that would inform logic. But in Baltimore in the early 1880s, without giving up his interest in the mathematics (primarily algebra) of logic, he turned more to the foundations of mathematics. Peirce had already contributed to this field—see, for example, his 1867 "Upon the Logic of Mathematics" (W2: 59–69)—but it was under the inspiration and probably the influence of the mathematicians of Johns Hopkins that he came to focus for a period on foundational questions. From his work at this time one might be led to think that Peirce had definite logicist sympathies, for apparently he believed that it was in logic that the key to the foundations of mathematics would be found. (I am here using "logicist" to mean one who accepts *logicism,* the general doctrine that mathematics is dependent on an epistemically prior science of logic. Thus, for the logicist, the principal subject for the study of foundations is logic. For further discussion, see Section 4.) But there was an important difference between Peirce's views and those of Dedekind, the foremost logicist of that time (Frege's views would not be widely considered until after the turn of the century). For Peirce, logic, in particular his logic of relations, made possible the analysis that was essential for foundational theory. Yet he did not accept the logicist view that foundational theory is logic. He would emphatically reject Dedekind's claim (*Dedekind 1888*) that mathematics is a branch of logic.

Peirce had worked in close consultation with his father ten years earlier when the elder Peirce was writing his *Linear Associative Algebra* (LAA, first published privately in 1870) and when Charles was himself writing his "Description of a Notation for the Logic of Relatives" (W2: 359–429). In the intervening years Peirce continued to investigate abstract algebras, and after his father's death in 1880 he prepared an annotated edition of LAA for J. J. Sylvester's *American Journal of Mathematics*. As addenda, Peirce included his proof that any linear associative algebra can be put into a relative (matrix) form (W4: 319–322) and his proof that there are only three linear associative algebras in which division is unambiguous (W4: 322–327). Peirce's work at Hopkins from 1879 through 1884 led to many notable results for theory of foundations and for symbolic logic. During these years Peirce made his contribution to the foundations of lattice theory—and caused a stir over his alleged distribution proof (W4: 163–209), worked out the first account of a single-connective logic more than thirty years before the better-known work of Sheffer (W4: 218–221), worked out his successful axiomatization of natural numbers (W4: 222–223, 299–309), succeeded in defining a finite set (W4: 267–268, 299–309), made one of his most substantial contributions to the theory of probabilities and to the general theory of induction (W4: 408–450), made some of his principal contributions to the logic of relatives (W4: 328–333, 453–466), developed an early form of truth-function analysis, discovered what we now know as Peirce's Law, and, as Hilary Putnam says, effectively introduced quantifiers as we know them today (W5: 162–190).[3]

After Peirce left Johns Hopkins, his research turned more decidedly toward logic with a graphical syntax, although this interest had begun earlier, perhaps with a fascination with geometrical constructions or with Euler's diagrams. To some extent, his interest stemmed from a Gaussian belief that algebra is a science of observation. For Peirce, this belief extended to all deductive reasoning, and it was the key to his lifelong interest in notation design and mathematical problems involving spatial elements (the four-color problem, for example). Probably the main influences for Peirce's interest in graphical logics were Clifford and Sylvester, with their use of chemical diagrams for representing algebraic invariants; Kempe, with his theory of logical form; and Clifford, with his graphical method. By 1882 (three years after Frege's *Begriffsschrift*) Peirce was specifically attempting to develop what he called spread formulas to represent relations, and he produced some graphs resembling switching diagrams (W4: 391–399). In 1886 he wrote to his former student Allan Marquand, recommending the use of electrical switching circuits for Marquand's logic machine, and he included in his letter the first known diagrams for that purpose (W5: 421–422). By 1897 Peirce had developed a complete graphical system called the Entitative Graphs (*Peirce 1897*), which was succeeded in less than a year by his final and most famous system, the Existential Graphs (EG). In 1903 he presented this system in a series of lectures at the Lowell Institute, with a printed brochure to guide his auditors (P 1035). By this time EG had three distinct sub-theories corresponding to propositional logic, predicate logic, and modal logic. The third part, which he continued to develop but never finished, had by 1906 been transformed into a theory involving possible worlds and different kinds of possibility. Peirce published an account of this "work in progress" in the *Monist* (*1906*), and it is thoroughly discussed in *Roberts 1973*.

In his day Peirce was widely regarded as a leading logician; there has probably not been another American who has been more widely acclaimed for his logic. Readers of this volume who are not convinced already that Peirce was a logician of the first rank are more likely to be convinced of Peirce's greatness by the evidence contained in the contributed essays than by the testimony of Peirce admirers. Even so, it may be helpful to review some of the estimates that have been made of Peirce's importance. This will help orient us in the problematic field of the history of logic by enabling us to see Peirce through the eyes of others and to get a sense of how the history of logic looks from different perspectives.

In 1877 Edward L. Youmans, editor of *Popular Science Monthly*, reported that the distinguished English mathematician William Kingdon Clifford regarded Peirce as "the greatest living logician, and the second man since Aristotle who has added to the subject something material, the other man being George Boole . . . " (*Fisch 1986:* 129).

In 1896 Ernst Schröder assured Paul Carus, editor at Open Court, that "however ungrateful [his] countrymen and contemporaneans might prove, [Peirce's] fame would shine like that of Leibniz or Aristoteles into all the thousands of years to come . . . " (L 392). Years later this comparison of Peirce with Aristotle would be echoed by no less a figure than Alfred North Whitehead, who wrote to Charles Hartshorne that America was becoming "the centre of worthwhile philosophy" and that, in his opinion, "the effective founders of the American Renaissance are Charles Peirce and William James. Of these men," Whitehead said, "W.J. is the analogue to Plato, and C.P. to Aristotle. . . . "[4]

In 1902 William James wrote that, when published, Peirce's logic "will unquestionably (in spite of certain probably obvious oddities) be recognized all over the world as an epoch-making work" (L 75c). James was referring to a book on logic that never appeared, but a similar view about Peirce's work is still frequently held: when Peirce's unpublished work on logic finally appears, won't everyone be surprised, even astonished, by his greatness! Certainly there are some important unpublished manuscripts; a few of them are discussed in this volume. But the extent and importance of the logical papers that Peirce published during his lifetime should not be discounted, as anyone must realize who examines the collection of published papers (said to have been first assembled by C. I. Lewis) in vol. 3 of the Harvard Press edition of Peirce's *Collected Papers*. Peirce's importance, so far, is due mainly to work published before James's remark.

As we move farther into the twentieth century, after the study of exact logic[5] had become widespread and the number of contributors had grown, estimates of Peirce's importance generally remained high. In 1922, Jan Łukasiewicz, rector of Warsaw University, said during an address:

> I belong, with a few fellow workers, to a still tiny group of philosophers and mathematicians who have chosen mathematical logic as the subject or the basis of their investigations. This discipline was initiated by Leibniz, the great mathematician and philosopher, but his efforts had fallen into oblivion when, about the middle of the nineteenth century, George Boole became its second founder. Gottlob Frege in Germany, Charles Peirce in the United States, and Bertrand Russell in England have been the most prominent representatives of mathematical logic in our own times. (*Łukasiewicz 1961 [1970]:* 111)

Karl Menger, in a paper entitled "The New Logic" (*Menger 1937*), examined the reconstruction that followed the failure of syllogistic logic, which had sufficed for some two thousand years. As the first steps in this reconstruction, Menger gives, first, the development of the algebra of logic (as a calculus of classes), which he attributes primarily to Boole, Peirce, and Schröder; second, the development of a calculus of propositions, which he attributes principally to Peirce and Schröder; and, third, the development of quantification theory (or calculus of functions), which he attributes to Peirce and Frege. Tarski (*1941*) credits Peirce with inventing the theory of relations, although Peirce never failed to acclaim De Morgan for his pioneering work.

In 1982, in a paper entitled "Peirce the Logician," Hilary Putnam expressed his surprise upon discovering "how much that is quite familiar in modern logic actually became known to the logical world through the efforts of Peirce and his students" (*Putnam 1982:* 295). Putnam pointed out that Whitehead had come to his knowledge of quantification theory through Peirce (Putnam could not find out where Russell had learned about quantification), and although Frege was *historically* first, Putnam attributes what he calls the *effective* discovery of the quantifier to Peirce and his students, by which Putnam means that it is to Peirce's work rather than to Frege's that modern quantification theory can trace its origin (*Putnam 1982:* 297).

Finally we come to Quine, from whom we might expect a cautious appraisal of Peirce. Indeed, in his 1985 review of Desmond MacHale's book on George Boole, Quine gave his principal endorsement to Frege, averring that "the whole of quantification theory . . . sprang full grown from Frege's brow." But he then went on to give this brief history of the origins of modern logic:

> General quantification theory is the full technique of "all," "some" and pronominal variables, and it is what distinguishes logic's modern estate.

Charles Sanders Peirce arrived at it independently four years after Frege. Peirce's work did indeed take off from that of Boole, De Morgan and Jevons. Ernst Schröder and Giuseppe Peano built in turn on Peirce's work, while Frege's continued independently and unheeded.

The avenue from Boole through Peirce to the present is one of continuous development, and this, if anything, is the justification for dating modern logic from Boole; for there had been no comparable influence on Boole from his more primitive antecedents. But logic became a substantial branch of mathematics only with the emergence of general quantification theory at the hands of Frege and Peirce. I date modern logic from there. (*Quine 1985:* 767)

Of course one can also find less laudatory, even dismissive, assessments of Peirce's contributions to logic. For example, consider the following remark from Eric Bell's *Development of Mathematics*, which, while recognizing Peirce's genius and potential, appears to discount his actual importance:

After De Morgan, the next outstanding advance toward mathematical logic was made in the United States by C. S. Peirce. For reasons into which we need not enter, Peirce's work failed to make the immediate mark its penetrating quality should have made. Peirce's own explanation for his lack of adequate recognition is contained in the unacademic remark, attributed to him on good authority, "My damned brain has a kink in it that prevents me from thinking as other people think." (*Bell 1945:* 556)

Among the more dismissive assessments of Peirce is one made indirectly by Jean van Heijenoort, who left Peirce completely out of his *Source Book in Mathematical Logic*. If we accept wholeheartedly the view promoted by Ivor Grattan-Guinness (see the second paper in this volume), that mathematical logic is a fairly restricted sub-domain—namely the logistic tradition (more or less)—within the broader tradition of exact logic, van Heijenoort's selection policy may seem well motivated. But van Heijenoort did not hold such a restricted view of mathematical logic, so his exclusion of Peirce can only be regarded, most charitably, as an unfortunate oversight.[6]

§2

Quine, as quoted above from his brief history of modern logic, remarked that "logic became a substantial branch of mathematics only with the emergence of general quantification theory." The relation of logic to mathematics is a point I shall return to in Section 4, but before that I want to address this question: Why didn't Peirce make a greater effort to connect his work with that of Frege, Whitehead, Russell, Peano, and others, who he must have realized were beginning to dominate the field? We know that he was urged by some of his friends to do just that. E. H. Moore of Chicago, an editor of *Transactions of the American Mathematical Society*, wrote to Peirce in July 1902 about "the foundations of the theories of natural numbers" and, after giving a set of Peano's axioms, remarked: "These relations are very closely like those you use. . . . Have you published your set?"[7] About three months later Moore wrote again:

> I requested Mr. Morley to send you a copy of the October number of the American Journal of Mathematics. It contains an article by Whitehead on Cantor's Cardinal Numbers. This is written in Peano's system of symbols, and makes use of Russell's additions on the algebra of relations in general. The Italian school believe in the new symbolism as a calculus in terms of which it is highly convenient to work, and not merely as an algebra of mathematical logic. It would give me much pleasure if you would let me know what you think concerning Russell's work, especially in comparison with your own. . . . [8]

Christine Ladd-Franklin also urged Peirce to respond to the new developments. She wrote to Peirce from Baltimore on 24 July 1904: "Do tell me how it strikes you—all this recent work of Bertrand Russell, Peano, Couturat & their school, which they make so much of. Don't you think they exaggerate both its originality & its importance? Are you not going to write something on the subject?"[9] In retrospect, knowing as we do the tremendous importance of the work of these "newcomers," it seems appropriate to ask why Peirce did not make a greater effort to participate in, or at least address, the supplanting paradigm. Yet the question could as well be reversed and, I submit, with more justice. Why didn't the "newcomers" make a greater effort to connect their work with that of Peirce, and in general, with the Boole-Peirce-Schröder school?

Consider how things looked at the turn of the century. Peirce and Schröder and their followers were in command of a powerful system of logic comparable, if not coequal, with the Frege-Russell-Peano logic. In some respects, especially with regard to the theory of individuals and the theory of relations, the logics of Schröder and especially of Peirce were advanced beyond anything by then developed in the new movement. It seems apparent that at least Peirce was aware of the tremendous power of his logics. (I speak of Peirce's logics in the plural because to understand how much he had within his "logical grasp" we must include his existential graphs and his iconic algebraic systems along with his general (universal) logic, which encompasses his quantification theory, his logic of relations, and his Boolean algebra of logic.) As late as 1905, in their article on symbolic logic for Vol. 9 of *The Encyclopedia Americana,* Edward V. Huntington and Christine Ladd-Franklin placed Peirce and Schröder at the core of the then coalescing lines of research that would evolve into the logic that has dominated work for most of this century. Russell is mentioned as having only recently (1903) announced "the surprising thesis that logic and mathemat-

ics are in reality the same science; that pure mathematics requires no material beyond that which is furnished by the necessary presuppositions of
any logical thought; and that formal logic, if it is to be distinguished as a
separate science at all, is simply the elementary, or earlier, part of mathematics." Huntington and Ladd-Franklin add that "it is too early to predict
what the final outcome of this new movement will be"; but of course we
know what happened. The logistic movement grew apace and the work of
hitherto recognized Boolean logicians (especially Peirce and Schröder) was
eclipsed by that of hitherto recognized mathematicians (especially, perhaps, Frege, Hilbert, Peano, and Cantor). I do not say that the achievements
of the Boolean logicians were lost or ignored, at least not all of them. As
Hilary Putnam has pointed out, more of the results of the Boole-Peirce-
Schröder school entered the canon than have generally been acknowledged
(*Putnam 1982*). But once the logistic movement got underway, the work
of "the old school" was relegated to the past and regarded, if at all, as part
of a superseded tradition. Except for Tarski, perhaps no major logician of
the twentieth century—at least not until quite recently—has deliberately
charted out a course of research based on the programs of the "old school."

So I return to the questions, why didn't Peirce pay more attention to
the work of Frege, Russell, Whitehead, Peano, and, in general, the new
school? And why, on the other hand, didn't *they* work harder to connect
their work with Peirce's and with Schröder's?

The record shows that to some extent Peirce did try to connect his work
with what he must have come to realize was the overwhelming advance of
the logicist program. There are a number of manuscripts and letter drafts,
many of them unpublished, that deal with the work of Cantor, Peano, and,
to a lesser extent, Whitehead and Russell. There is also a letter in the Houghton Library, dated 7 February 1912 from Cambridge University Press, in-
forming Peirce that his inquiry about *Principia Mathematica* is being for-
warded to the press's U.S. agent, G. P. Putnam and sons, and that the book
will cost $8.00. Apparently Peirce had decided that it was time to see what
the fuss was all about. Unfortunately, his health was about to take a serious
turn for the worse, and he would only live for two more painful years.

On the other hand, we know that Russell and Whitehead acknowledged
some debt to Schröder and even, in passing, to Peirce. Peano more emphatically recognized the influence of Peirce and Schröder. But while Peirce was
not about to give up a whole tradition to embrace a system and program
that he probably thought inferior to his own, the logicists were so caught
up in the excitement of their discoveries that there was little time and less
interest in examining the logical roots of their own movement.[10] Furthermore, because the purposes of the logicists were far more limited than, and
quite divergent from, those of Peirce, the idea of connecting their work in

a systematic way with that of Peirce (or of Schröder) might well have seemed pointless.

If Peirce may be thought stubborn for yielding so little (and so slowly), then his stubbornness is matched by the arrogance of Frege, who wrote in 1882 that "in the twenty years and more since it was devised the Boolean formula-language has by no means achieved such a signal success that to abandon henceforth the foundations it laid must be regarded as foolish, or that anything except a further development is out of the question" (*Frege 1882–1883 [1968]: 89*). This attitude, perpetuated by Frege's disciple Russell, helped create the gulf that obscured the fact that between the developments that Boole inspired and those that Frege inspired there was much in common. It now seems clear that Peirce occupied more of the common ground than did anyone else.

§3

Even though I have claimed that Peirce was not a logicist—that he did not think of logic as the foundation for mathematics—what I have said so far might suggest that Peirce's work in logic was primarily focused on the same questions that concern logicists. That would be a serious misconstrual. Although foundational questions were of great interest for Peirce, they were by no means his greatest concern as a logician. As a matter of fact, Peirce would probably not even have regarded his work on foundations of mathematics as work in logic, a point I shall expand on in the following section.

According to Peirce, logic, in its most general sense, is the formal science of representation or, as he sometimes said, the "objective" study of thought. It is the general study of signs, a normative science, and is coextensive with semiotic (for Peirce, *semeiotic*). Logic as semiotic divides into three branches: speculative grammar, critic, and speculative rhetoric.[11]

Speculative grammar, according to Peirce, is the study of the "general conditions of signs being signs" (1.444). In his 1901 article on logic (written with Christine Ladd-Franklin) for James Mark Baldwin's *Dictionary of Philosophy and Psychology*, Peirce wrote that this branch of logic considers "in what sense and how there can be any true proposition and false proposition, and what are the general conditions to which thought or signs of any kind must conform in order to assert anything" (2.206). The logician who concentrates on speculative grammar investigates representation relations (signs), seeks to work out the necessary and sufficient conditions for representing, and classifies the different possible kinds of representation. Speculative grammar is often presented as though it were the whole of Peirce's semiotic, and it is there that we first encounter some of Peirce's best-known trichotomies, including his famous division of signs into icons, indexes, and symbols.

Critic, according to Peirce, "is the science of the necessary conditions of the attainment of truth" (1.445). In his Baldwin article Peirce wrote that critic is "that part of logic . . . which, setting out with such assumptions as that every assertion is either true or false, and not both, and that some propositions may be recognized to be true, studies the constituent parts of arguments and produces a classification of arguments . . . " (2.205). Peirce described the sort of classification he had in mind:

> [Critic's] central problem is the classification of arguments, so that all those that are bad are thrown into one division, and those which are good into another, these divisions being defined by marks recognizable even if it be not known whether the arguments are good or bad. Furthermore, [critic] has to divide good arguments by recognizable marks into those which have different orders of validity, and has to afford means for measuring the strength of arguments. (2.203)

Thus, in addition to investigating truth conditions, the logician who concentrates on critic will investigate Peirce's well-known division of reasoning into abduction, induction, and deduction and the corresponding theories: abductive logic, inductive logic, and deductive logic. Much of what made up the traditional logic curriculum belongs in critic, as does much that is dealt with in philosophical logic, especially topics having to do with truth and reference.

Speculative rhetoric, according to Peirce, "is the study of the necessary conditions of the transmission of meaning by signs from mind to mind, and from one state of mind to another" (1.445). A focus for the logician who concentrates on speculative rhetoric is the relation between representations and interpreting thoughts (or interpretations). While Peirce usually said of critic that it is the science of the *necessary* conditions for the attainment of truth, in his Baldwin article he described speculative rhetoric as dealing with the *general* conditions for the attainment of truth. Peirce often emphasized the study of *methods* of reasoning as a main concern of speculative rhetoric and in Baldwin he suggested that this branch of logic might be better named "methodeutic" (2.207). It is not clear what all Peirce would have included as proper subjects for speculative rhetoric and one can easily imagine some overlap with the preceding divisions. Insofar as meanings are interpretants, the study of the relation between a sign and its meaning would seem to belong to speculative rhetoric even though the study of what we might call "meaning structures" would seem to belong to speculative grammar. And insofar as the conclusions of arguments may be regarded as interpretants (*interpretations* of conjoined premisses), the study of the different ways in which premisses and conclusions can be legitimately related might also be said to belong here. Whether Peirce intended that the analysis of meaning and the classification of arguments should somehow be subjects for speculative rhetoric—as well as for other branches of logic—is not altogether clear. I would also locate here Peirce's doctrine of pragmatism (pragmaticism), which is sometimes thought of as a theory of meaning or interpretation but which Peirce preferred to consider a doctrine of method.[12]

Around 1902, in the first chapter of his never completed "Minute Logic," Peirce summarized his views on the structure of logic (I have reconfigured this quotation to follow the order of the divisions of logic as we have been considering them):

> Logic is the science of the general necessary laws of Signs and especially of Symbols. As such, it has three departments. Originalian logic, or Speculative Grammar, is the doctrine of the general conditions of symbols and other signs having the significant character. . . . Obsistent logic, logic in the narrow sense, or Critical Logic, is the theory of the general conditions of the reference of Symbols and other Signs to their professed Objects, that is, it is the theory of the conditions of truth. Transuasional logic, which I term Speculative Rhetoric, is substantially what goes by the name of methodology, or better, of methodeutic. It is the doctrine of the general conditions of the reference of Symbols and other Signs to the Interpretants which they aim to determine. . . . (2.93)

This division of labor in logic is obviously related to Peirce's triadic theory of representation in which any sign is said to be in a triadic relation with an object and an interpretant.[13] The first division of logic focuses on the sign as such. The second division, building on the results of the first, focuses on the reference of signs to objects. The third division, building on the results of the two preceding divisions, focuses on the reference of signs to interpretants. There is an interesting resemblance between these divisions of logic and the Morris-Carnap triad: syntax, semantics, and pragmatics—but there are also important differences, most notably the greater breadth of Peirce's theory and its careful and systematic adherence to the fundamental triadicity of the sign relation. From the standpoint of philosophy, Peirce's semiotic logic encompasses much, if not all, of epistemology, theory of inference and philosophical logic, and theory of interpretation and scientific method.

§4

It will now be easy to see that some of the more "philosophical" papers, which may at first seem out of place in this book, indeed fall within the broad range of concerns that constitutes the work of Peircean logicians.[14] But an opposite difficulty confronts us. Many of the papers which may seem to fit perfectly in this volume, especially those that deal with the foundations of mathematics and with formal questions from the algebra of logic, are not strictly about logic as Peirce defines it. Where does such work fit within the Peircean framework? As a prelude to answering this question, I want to consider briefly what Peirce understood mathematics to be and the distinction he made between mathematicians and logicians. These matters are further addressed in several other chapters in this volume.

Peirce followed his father in defining mathematics as the science which draws necessary conclusions; it is the study of what is true of hypothetical states of things and thus it is exempt from the contingency characteristic of matters of fact. It is probably this view of mathematics that Russell had in mind when he said that "Mathematics [is] the subject in which we never know what we are talking about, nor whether what we are saying is true" (*Russell 1901b [1993]*: 366). Of course mathematicians must reason in order to draw conclusions at all. This leads some to suppose that a science of reasoning, or of inference, must precede mathematics and, therefore, that Peirce was mistaken in placing mathematics at the base of the ladder of sciences. However, such an argument is flawed, because if there has to be a science of reasoning *before* reasoning can be legitimately employed, then there can be no science at all. For even a science of reasoning must be developed by reasoning (4.242). So Peirce held that mathematics, the science which deduces what follows from suppositions, is more basic than logic, the science which establishes norms for evaluating reasoning.

Peirce sometimes tried to highlight the difference between mathematics and logic by contrasting mathematicians and logicians, as he did on 2 April 1908 in a letter to Cassius J. Keyser of Columbia University:[15]

> As for the difference between the mathematician and the logician—and the two kinds of thought have nothing in common except that both are exact—it is that the mathematician seeks the solution of a problem & has but a subsidiary interest in anything else, while the logician, not caring a snap what the solution may be, desires to analyze the form of the process by which it is reached, in order to get a general theory of the form of intellectual procedure.

Peirce had discussed this several years earlier in a chapter of his "Minute Logic," and had pointed out an interesting difference in how mathematicians and logicians conceive of the algebra of logic:

> The different aspects which the algebra of logic will assume for the two men is instructive. . . . The mathematician asks what value this algebra has as a calculus. Can it be applied to unraveling a complicated question? Will it, at one stroke, produce a remote consequence? The logician does not wish the algebra to have that character. On the contrary, the greater number of distinct logical steps, into which the algebra breaks up an inference,

will for him constitute a superiority of it over another which moves more swiftly to its conclusions. He demands that the algebra shall analyze a reasoning into its last elementary steps. Thus, that which is a merit in a logical algebra for one of these students is a demerit in the eyes of the other. The one studies the science of drawing conclusions, the other the science which draws necessary conclusions. (4.239)

So far it may seem that Peirce made an astute and cogent distinction between mathematics and logic, as I believe he did to some extent. Yet it is difficult to sustain one's confidence as one probes deeper into the areas of overlap between mathematics and logic. Consider, for example, the nature of mathematical objects—whatever it is that mathematicians reason about. Is a typical proposition of mathematics about abstract objects? Sets? Relations? Ideal structures? Possible worlds? According to Maxime Bôcher, modifying an earlier definition of A. B. Kempe (*Kempe 1894*), "If we have a certain class of objects and a certain class of relations, and if the only questions which we investigate are whether ordered groups of those objects do or do not satisfy the relations, the results of the investigation are called mathematics" (*Bôcher 1904*). Supposing that sets and relations and other equally abstract entities, or suppositions about them, are legitimate objects of mathematical inquiry, is mathematics really as neutral with respect to categorical truth as Peirce (and as Russell) claimed?

One's answer may depend on whether or not one is a realist with respect to "theoretical" entities. If one holds that the relations treated in mathematics are *real* relations—real in the sense that after sufficient investigation our understanding of these relations will be more molded by them than they by our will—then what we say about them is true or false in a quite robust sense. Such relations can be *represented* in systems of notation which, in turn, provide us with a means for investigation and discovery.[16] But as soon as we begin to deal with notational systems which we regard as representational systems, even if all that is represented are abstract entities such as sets and relations, it would seem that logic as the science of representation should come into the picture. It thus begins to appear that, for the realist, mathematics is a science of discovery almost in the sense of the natural sciences, and that one might be hard pressed to make a case for its priority over logic. One might be on safer ground as a nominalist, by declining to suppose that mathematical terms refer at all.

But are we not being led astray by this line of thought? Can we not use representations without having a science of representations—just as we can infer without a science of inference? And does not the priority of mathematics over logic and other positive sciences derive more from the generality and ideality of mathematical objects than from the hypothetical character of mathematical statements? The theoretical entities of mathematics, as ideal objects (*ens rationis*), are as distinct from matters of fact as are merely hypothesized states of affairs, yet what is concluded (discovered) from the study of abstract mathematical entities is not necessarily suppositional. What seems to be most important is that no matter whether the object to be considered is an abstract structure or a supposed fact, the mathematician's purview does not have to extend beyond his or her own imagination.

This is a complicated business, one which I will not pretend to have worked out completely. The study of Peirce's philosophy of mathematics and logic inevitably leads one into such conceptual thickets, and the paths out are far from well-trodden. (However, "parallel" paths in the logicist camp have been somewhat more beaten down.) Peirce's conception of the *internal* structure (or architecture) of mathematics seems equally recondite. Peirce wrote about many basic theoretical entities, including sets and collections of various sorts, and a variety of abstract algebraic structures. But it is not clear what importance he attributed to each for the development of mathematics and for subsequent sciences. What is most evident in his work is the importance Peirce attached to his basic analysis of relations.

According to Peirce there are only three fundamental classes of relations: monads, dyads, and triads. Of course he recognized that there are relations of greater adicity than triads, but he contended that they can all be reduced (with respect to their basic structures) to complexes of triads. Peirce further claimed that triads and dyads are not reducible, which leaves monads, dyads, and triads as the only relations fundamental to mathematics and the subsequent sciences. This reduction thesis has been called Peirce's remarkable theorem (*Herzberger 1981*) and is discussed by Anellis, Brunning, and Burch in their contributions to the present book (and in *Burch 1991*).

Did Peirce believe that it is the role of mathematicians or of logicians to discover and to investigate these fundamental relations? Perhaps the answer to this question is a key to deciding whether he was a logicist. Initially it may seem that the study of relations is too positive, too descriptive, to be mathematical in Peirce's sense. Yet Peirce often stressed the importance of observation for mathematics. The mathematician constructs models, or icons, of the objects to be investigated, taking care that the icons faithfully represent the relations involved, and then manipulates the icons and observes the results. Every mathematician and formal logician knows that what is discovered is general, that what is under scrutiny are not just the tokens actually at hand, but the forms of relations they embody. So it is not out of the question that the discovery and investigation of relations should fall to mathematicians. This view seems all the more compelling when it is noticed that the study of relations does not seem to be straightforwardly concerned with principles of intellectual procedure, as logic is on Peirce's view. It might be argued (by some philosophers), however, that what is really called for in the study of relations is conceptual analysis, and surely *that* is part of the work of logic. So isn't the theory of logic more fundamental than mathematics after all? Undoubtedly there is a logic of relations, i.e., a theory of reasoning with relational terms. (Sometimes Peirce called this the logic of *relatives*, to emphasize the related objects [correlates] rather than the abstracted relational structures.) But there is also a study of relations that is not concerned with reasoning as such, and thus it is not logic; it is concerned rather with the relations themselves and with the investigation and analysis of the structure of relations. This might best be called the mathematics of relations, or relation theory.[17] Even though Peirce sometimes spoke of this study as the logic of relations, insofar as what we have in mind is theory of relations per se rather than theory of inferences that involve relational terms, we are more properly dealing with mathematics than with logic. Peirce thus seems to escape the charge that he is a logicist, for one can argue that the "truths" that ground mathematics are themselves mathematical. Of course it is indisputable that Peirce recognized a logic of mathematics, but its function seems to be the analysis and evaluation of conceptions and inferences that are central to mathematics. Logic, in this sense, concerns mathematical reasoning, not the objects, whatever they are, that mathematicians investigate. Thus the logic of mathematics is chiefly critical and evaluative and cannot be foundational, or so it seems to me.

But Peirce recognized, just as Boole had, that it is possible to isolate mathematical structures to serve as the skeletons for our systems of logic. This insight hinges on the realization that inherent in reasoning are abstractable formal structures subject to mathematical representation. Peirce's labors to isolate and hone suitable mathematics for his logics are elaborated in several of the chapters in this volume. Here we may say that Peirce was working on the mathematics of logic rather than the logic of mathematics. Pure Boolean algebra is thus a mathematics of logic, as is Peirce's algebra of logic. In his mature years Peirce spoke of *dichotomic mathematics* as *the simplest mathematics* and the most relevant for logic. It is not clear whether he believed that the simplest mathematics, the mathematics of logic, is the basic mathematical system. If so, he would in a way be close to the logicist point of view, although he would refrain from identifying the basic mathematical structures with logic.

This is an unsettled issue in Peirce scholarship and one I can say little more about at present.[18] It does serve to highlight the complexity of the question of how mathematics and logic are related. What seems crucial for determining the relation of logic to mathematics, from Peirce's standpoint, is a thorough examination of his classification of the sciences. If one takes Peirce's classification seriously it seems clear enough that there is an interesting and important distinction to be made between mathematics and logic.

Peirce's classification seems to yield the following general picture: mathematics is at the foundation of the sciences and is the most fundamental and least dependent science; and only after phenomenology and the first two normative sciences—aesthetics, and ethics—do we come to logic, the third normative science. Logic is semiotic, which I have already characterized. Within mathematics, apparently at its own foundation, we find the theory of relations, where, presumably, Peirce's universal categories are first worked out and where his reduction thesis is established. With the theory of relations in hand, we can then go on to develop the other mathematical tools needed to achieve many of the results that I mentioned above.

It should also be noted that when Peirce says that a science is fundamental, or prior, to another science, he *does not mean* that the second can be *reduced* to the first. He means that the theory of the second depends essentially on the results of the more fundamental, or epistemically prior, science, but not the reverse. He would not deny that the dependent science may be useful for, and may even give direction to, the more fundamental science. Maryann Ayim (*1972:* 21–22), following a suggestion of Don D. Roberts, explains the dependency relations between the sciences by distinguishing between "principle-dependence" (science B depends on science A in that B takes principles from A) and "data-dependence" (science A depends on science B in that A uses data or facts from B). Only principle-dependence appears to involve epistemic dependency.

We must be careful not to surrender too much to our words. Peirce dealt with, and made contributions to, many of the foundational questions that have occupied the greatest mathematical logicians. He frequently did this work for the sake of mathematics, just as mathematical logicians do. He even sometimes called this work logic, although strictly speaking his system seems to require that we regard his foundational work as mathematics, not logic. But whether mathematics or logic, he did the work all the same, which surely is what *ought* to matter. Sometimes terminology counts more than it should. In 1885, Peirce submitted to the *American Journal of Mathematics* a second part of his famous Algebra of Logic paper, in which he had introduced quantifiers in their modern garb. In the second part, so far as it can be determined from the remaining manuscript pages, Peirce extended his theory of quantification and developed what he called his general algebra of logic. Simon Newcomb, the new editor of *AJM*, following J. J. Sylvester's departure for Oxford, hesitated to publish Peirce's work because *he wasn't satisfied that it was mathematics.* He told Peirce that if he would only affirm that the work was mathematics it would appear, but that *AJM* would not publish logic. Peirce stubbornly insisted that the work was logic, and Newcomb returned the manuscript. This is a case where less rigidity in the selection process might have significantly advanced knowledge, and it lends some support for the decision of the editors of the present volume to suspend the strict Peircean interpretation of "logic" and to extend its scope to include the more mathematical contributions.

This brings me back to the beginning. Much of the work mentioned above that Peirce would not have counted as logic, in particular, work on strictly formal issues in the mathematics of logic and on foundational questions, we (contemporary scholars) do. Notice, for example, that accounts of formal logic in most contemporary encyclopedias include substantial discussions of set theory and natural-number arithmetic. On the other hand, though we may not count the whole of formal semiotic as logic, Peirce did. Therefore, any issue that arises in these vast areas of thought was grist for the mill of the logic symposium at the Peirce Sesquicentennial Congress at Harvard University and, consequently, a candidate for discussion in this volume.

§5

While the themes and topics surveyed in the above sections go a long way toward characterizing Peirce's work as a logician, it is important to remember that the focus of this book is considerably narrower, and that many of the chapters deal with formal logic (and the mathematics of logic) at a quite technical level.

Some chapters address foundational questions, especially those by Shields, who examines Peirce's successful 1881 axiomatization of arithmetic, and Dipert, who discusses Peirce's neglected investigations of various sorts of sets and collections. Kerr-Lawson considers Peirce's "relaxed attitude toward epistemological questions" and finds that Peirce's philosophy of mathematics is compatible with the views of many practicing mathematicians.

Several chapters, including that of Grattan-Guinness and this introduction, discuss Peirce's view of the relation between mathematics and logic. Levy and Iliff give sustained discussions of this topic, and Van Evra looks at Peirce's early heavy dependence on mathematical analogies for logic, especially in his important 1870 paper on the logic of relations (DNLR). Van Evra raises a number of related historical and philosophical questions.

Merrill, who also focuses on Peirce's DNLR, considers the richness of Peirce's 1870 logic of relations and theory of quantification, and contrasts this early algebraic work with Peirce's better-known 1885 quantificational logic of relations. Brady takes a close look at some of the advances and discoveries that led Peirce to his 1885 model-theoretic quantificational logic; in particular, she shows how Peirce was influenced by his student O. H. Mitchell. Iliff's chapter complements Brady's by discussing the importance of Peirce's work in abstract algebra, especially matrix algebra, for his discovery of the quantifiers.

Four chapters, those by Burch (two papers), Brunning, and Anellis, examine aspects of Peirce's theory of relations, especially his unique notion of teridentity and his well-known but often criticized reduction thesis. Burch and Brunning discuss the special importance of a graphical syntax, and Anellis raises a number of questions about "valency proofs" and about the relation of Peirce's algebraic logic to the later work of Tarski. In his second chapter, Burch outlines his recent proof of the reduction thesis, which is the most ambitious attempt so far to establish the thesis that undergirds Peirce's whole system of thought.

Five chapters deal with Peirce's advanced "iconic" logic systems. Clark and Zellweger thoroughly examine for the first time Peirce's turn-of-the-century algebraic logic, which employs carefully designed signs for all sixteen binary connectives. Clark discusses the extent of Peirce's own work with this system, and Zellweger extends Peirce's ideas to take better advantage of group properties and symmetry relations. The chapters by Clark and Zellweger open a new field of study for Peircean logicians but, more importantly, they constitute the most advanced work yet produced in this field of "three-dimensional symmetry logic." The other three chapters of this group deal with Peirce's Existential Graphs (EG), his most developed "iconic" logic. Roberts discusses Peirce's efforts to provide a decision procedure for EG. Zeman shows how EG, with its iconic approach to modality, provided a notation which allowed Peirce to represent the conditional function in a new way, more in line with his long-held realist views. Sowa uses Peirce's EG as the foundation for a system of *conceptual graphs,* which provides the best available logic for representing the semantic structure of natural language and which, therefore, has far-reaching implications for cognitive science.

The rest of the chapters are somewhat less technical. The chapters by Grattan-Guinness, Hawkins, and Hiż are mainly historical. Grattan-Guinness paints a picture of the historical genesis and of the interrelations of the main logical trends, and provides a historical context for Peirce's contributions. Grattan-Guinness's chapter, which grew out of his general remarks on the logic symposium at the Harvard Congress, might well be considered a second part of the introduction to this book (although it should be noted that he did not see the chapters in their written form, and did not hear the Harvard presentation of a few of them). The chapter by Hawkins provides a carefully documented account of the relations (and non-relations) between Peirce and Russell, and he compares aspects of their logics. Hiż outlines Peirce's influence on the development of logic in Poland. Because of the acknowledged contributions of the Polish logicians to modern logic, Hiż's account provides indirect evidence of the continuity of development from Peirce to the present day in several important areas of logic. These are the most straightforwardly historical contributions, but almost all the chapters in this book contain historical themes.

The remaining seven chapters extend the scope of this collection beyond the exact and primarily deductive part of logic addressed in the rest of the book. Kent brings together many of the themes of other contributors under the guise of Peirce's pragmatism (and his left-handedness). Crombie, although focusing on Peirce's theory of deductive reasoning, finds that it is necessary to appeal to abductive and inductive processes. Kapitan explores more fully Peirce's theory of abduction. Burks examines Peirce's theory (logic) of evolution and compares it with his own views. He gives a full defense of his reductionist mechanistic theory of the universe (and of mind) and presents a real challenge to Peirceans who are committed to the ontological primacy of thirdness. Marostica examines Peirce's tychist (evolutionary) logic and finds that it addresses nonmonotonic reasoning of the sort studied in today's cognitive science. The final two chapters, those by Brock and DiLeo, examine Peirce's theory of proper names, Brock focusing mainly on Peirce's early *descriptive* theory, while DiLeo elaborates Peirce's mature *indexical* view. DiLeo's account is based on a fairly detailed consideration of Peirce's formal theory of signs and, thereby, extends the scope of this book to cover Peirce's broad conception of logic as semiotic.

While a wide variety and range of topics are discussed in this book, probably more than can be found in any other single work about Peirce's logic, a great deal has had to be left out. Inductive logic (including abduction), in particular, is underrepresented, given its centrality and importance for Peirce. This applies, as well, to Peirce's pioneering work on continuity and his studies in countability. His advanced quantification theory is hardly mentioned, and his work in modal logic is not thoroughly examined. The list could go on and on. It might hardly appear from the present collection of papers that Peirce's primary concern as a logician was the logic of science, although indeed it was.

Peirce also produced interesting work in the history and philosophy of logic that is not represented here. He was, in fact, one of the most informed historians of logic of his time, and he wrote a good deal on that subject. He was familiar with the work of almost all important logicians from antiquity on, including the writings of the medieval and renaissance logicians. One of his interesting ideas was that the difference between the ancient Greek view of logic as λογος and the Roman conception of logic as *ratio* marks a gulf between two different general outlooks, one committed to language and the other to calculation. This is a distinction that bears examination, for it might prove fruitful for organizing certain movements and predilections in the history of logic.[19] Peirce was a diligent student of different logical methods and programs, and he was a strong opponent of psychologism.

Probably the most ironic shortcoming of this book as a representation of Peirce as logician is that so few semiotic-related issues are treated. Given that logic *is* semiotic, in Peirce's opinion, it would seem that semiotic topics of the sort discussed in Section 3 would dominate a book about his logic. However, there is only so much that can go into a single book, and most of the Harvard Congress papers on Peirce's semiotic have been collected in a separate volume.

Even though this collection does not present the full extent of Peirce's contributions to logic, it does contain substantial discussions of many of his most important contributions, and it is hoped that it reveals something of the style and spirit of his exact thought. It is especially hoped that these papers will stimulate interest in Peirce's foundational studies and his somewhat renegade brand of algebraic logic, spanning as it does so much of the territory between Boolean Algebra and modern logic, and help build new connections between Peirce's work and contemporary logical investigations.

NOTES

1. Peirce to Gilman, 13 Jan. 1878. *Fisch and Cope 1952,* apx. 2, pp. 365–368.
2. For an account of Peirce's Johns Hopkins period, see *Fisch and Cope 1952, Houser 1989b,* and *Brent 1993,* ch. 3.
3. The titles of the papers referred to in this sentence are as follows: W4: 163–209: "On the Algebra of Logic"; 218–221: "A Boolian Algebra with One Constant"; 222–223: "The Axioms of Number"; 267–268: "Proof of the Fundamental Proposition of Arithmetic"; 299–309: "On the Logic of Number"; 328–333: "Brief Description of the Algebra of Relatives"; 408–450: "A Theory of Probable Inference"; 453–466: "The Logic of Relatives"; W5: 372–378: "On the Algebra of Logic: A Contribution to the Philosophy of Notation." These papers by no means exhaust Peirce's contributions for the period in question. W4 and W5 should be consulted for related work from the period.
   For accounts of Peirce's contribution to lattice theory and the problem of distribution see *Birkhoff 1948, Roberts and Crapo 1969, Dipert 1978, Salii 1988,* and *Houser 1985.*
4. Whitehead to Hartshorne, 2 Jan. 1936. *Lowe 1990: 345.*
5. I use the term *exact logic,* instead of the more common expression, used by many of the authors quoted in this paper (including Łukasiewicz in the following quotation), *mathematical logic,* to mean any symbolic logic systematically built up in an exact way after the practice of mathematics. So understood, nearly all of modern logic may be said to be mathematical; certainly algebraic logic, due principally to Boole, Peirce, Schröder, and more recently Tarski, will be included, along with the

more widespread system due principally to Frege, Peano, and Russell. Note that it is the use of mathematical methods and analogies in logic rather than the use of logic to inform and develop mathematics that determines whether a logic is mathematical under this view. However, it must be pointed out that according to Ivor Grattan-Guinness this view is quite mistaken. In his opinion, mathematical logic is only the logistic system that derives from Cauchy, Weierstrass, Cantor, Peano, and Russell. (See Figure 2.1, p. 25 below.)

The meaning I give to the expression "exact logic" (which has been used by a number of logicians, including Schröder) derives principally from Peirce, who defined the term for Baldwin's *Dictionary of Philosophy and Psychology*:

> The doctrine that the theory of validity and strength of reasoning ought to be made one of the "exact sciences," that is, that generalizations from ordinary experience ought, at an early point in its exposition, to be stated in a form from which by mathematical, or expository, reasoning, the rest of the theory can be strictly deduced; together with the attempt to carry this doctrine into practice.
>
> This method was pursued, in the past, by Pascal (1623), Nicolas Bernoulli (1687–1759), Euler (1708–1783), Ploucquet (1716–1790), Lambert (1728–1777), La Place (1749–1827), De Morgan (1806–1871), Boole (1815–1864), and many others; and a few men in different countries continue the study of the problems opened by the last two named logicians, as well as those of the proper foundations of the doctrine and of its application to inductive reasoning. The results of this method, thus far, have comprised the development of the theory of probabilities, the logic of relatives, advances in the theory of inductive reasoning (as it is claimed), the syllogism of transposed quantity, the theory of the Fermatian inference, considerable steps towards an analysis of the logic of continuity and towards a method of reasoning in topical geometry, contributions towards several branches of mathematics by applications of "exact" logic, the logical graphs called after Euler and other systems for representing in intuitional form the relations of premises to conclusions, and other things of the same general nature.
>
> There are those . . . who seem to suppose that the aim is to produce a calculus, or semi-mechanical method, for performing all reasoning, or all deductive inquiry; but there is no reason to suppose that such a project . . . can ever be realized. The real aim is to find an indisputable theory of reasoning by the aid of mathematics. (3.616–618)

6. With reference to his famous 1967 anthology, van Heijenoort wrote:

> "Mathematical" here means that derivations in logic proceed according to definite and explicit rules, the way we do sums (Frege's system, as he says, is "modeled upon the language of arithmetic"). Thus "mathematical" perhaps expresses a deeper feature of present-day logic than does "symbolic." Logic today is symbolic indeed, but the use of symbols, though convenient and perhaps even indispensable, is possible because logic is mathematical in the sense just mentioned. "Formal logic," too, would still be a perfectly appropriate name. But there is no point in searching for an epithet. There are no two logics. Mathematical logic is what logic, through twenty-five centuries and a few transformations, has become today. (*van Heijenoort 1967*: vii)

7. Eliakim H. Moore to C. S. Peirce, 22 July 1902 (L 299).
8. Eliakim H. Moore to C. S. Peirce, 14 Oct 1902 (L 299): 15.
9. Christine Ladd-Franklin to C. S. Peirce, 24 July 1904 (L 237): 201–202.

10. Although I believe that logicism (as a system of logic, not as a philosophy of logic) has significant roots in the logic of the Boole-Peirce-Schröder tradition, I wholeheartedly agree with Ivor Grattan-Guinness that there are roots in alternate traditions of mathematics that distinguish the logics of Peirce and Schröder from that of Russell. (See Figure 2.1, p. 25 below.)

11. Peirce sometimes uses other names. For a good discussion of the three divisions of semiotic, see *Savan 1988*.

12. It may be that something like what is today called "hearer's meaning" as opposed to "speaker's meaning" would be a subject for speculative rhetoric, in addition to a general theory of interpretation and, possibly, something akin to a theory of knowledge (belief) acquisition. To say, however, that something is a subject for speculative rhetoric does not preclude that it is also the subject, in a different sense, for speculative grammar or critic. In fact, speculative rhetoric presupposes results from the other branches of logic.

13. The theory is, of course, much more complex, but I can not go into it here. One good introduction to the complexities of Peirce's theory of signs is *Savan 1976*, especially in the second edition, *1988*.

14. Many papers presented at the 1989 Peirce Sesquicentennial Congress which are not included in this book fall under Peirce's broad definition of logic. Most of these papers are included in other congress collections.

15. This letter is from the Cassius Keyser Papers in the Rare Book and Manuscript Library of Columbia University.

16. To what extent notations actually constitute rather than simply reveal the relations we investigate is a question I shall leave for others to consider.

17. Perhaps, by extension, Peirce would have included his versions of set theory, continuum theory, and recursive function theory. I realize that much of this section is opposed to standard foundational views, and that it may be rather unintelligible to a reader steeped in logicist-inspired foundational theory. Perhaps the main thing to notice is that I am taking a generally realist view of abstract mathematical objects and am denying that they are "built up from" or are "reducible to" logical conceptions. On this view, for example, a set is a mathematical object even though our conception of sets may owe a great deal to the logic of mathematical reasoning.

18. For further discussion of this issue, see *Haack 1993, Houser 1993*, and several chapters in this volume.

19. A comparison of Peirce's distinction with that of van Heijenoort and Hintikka, as expressed in the Foreword to this book, might be worthwhile.

2.

PEIRCE BETWEEN LOGIC AND MATHEMATICS

Ivor Grattan-Guinness

§1. Introduction

[The utility of Boole's logic] consists in its giving one clear conception of logical relations, both of simple and of intricate kinds. . . . It was the application by Boole of his logical algebra to the doctrine of probabilities that proved that it was not a mere brilliant idea but was of real and great importance for logic. . . . It was before the epoch of Weierstrass; that is to say, before the present styles of mathematical reasoning had come in. . . .

At the same time as Boole, another mathematician, Augustus De Morgan began an important series of works on logic. . . . As a mathematician De Morgan lacked the genius of Boole, but his ideas were far more lucid. He opened up the logic of relations, which is of high importance; and he gave a complete formulation of an important kind of necessary reasoning which had been entirely overlooked. . . .

We now come to living men. Decidedly, the most profound among those whom I am to mention is, incomparably, Alfred Bray Kempe. . . . Another set of men who have done still more for logic by much profounder thoughts are those who have studied infinite numbers, both ordinal and those that express multitude. Dr. George Cantor has been the chief of them. . . . (C. S. Peirce NEM 4: 123–125)

Long have I been a fascinated spectator of the achievements of Peirce in logic and philosophy, but I cannot claim to be a Peirce specialist: for example, I have seen the manuscripts only from a respectful distance, while watching Max Fisch and his helpers in 1972 sort them into proper order for the first time.[1] So it is a special pleasure to be a commentator on his logic.[2] To keep the considerations within bounds, by 'logic' I shall always intend only the technical sense of the term and the mathematics to which it was related: I shall not comment on Peirce's broader logical concerns (where, for example, questions such as mathematical creativity might be posed in the context of his abduction), or the various areas on mathematics within which Peirce worked at one time or the other.

My own main areas of historical research include the history of mathematics and mathematical logic, and I draw upon this background to furnish the majority of the comments below: I shall lay emphasis on the context and the prehistory of Peirce's work. In the next section I shall distinguish mathematical logic from its algebraic brother, to which tradition Peirce's own work belonged. In Section 3 I shall effect comparisons between the two traditions, with some emphasis on mathematical matters; then in Section 4 I address, nervously, the question of the relationship between mathematics and logic for Peirce. The paper ends with a charming appraisal of Peirce by his father.

Throughout, the development of Peirce's thought is treated fairly lightly; for most of the matters discussed Peirce does not seem to have effected any major changes of strategy, although his standpoints and knowledge were enriched and expanded over the years. References include a few principal items of primary literature, but as a whole these sources are too numerous and complicated to gain practical treatment here: mainly I cite secondary items, and largely on topics around Peirce rather than on him.

§2. Logic, Algebraic and Mathematical

Logic, Symbolic. Symbolic Logic, or Mathematical Logic, or the Calculus of Logic,—called also the Algebra of Logic (Peirce), Exact Logic (Schröder), and Algorithmic Logic or Logistic (Couturat),—covers exactly the same field as Formal Logic in general, but differs from Formal Logic (in the ordinary acceptation of that term) in the fact that greater use is made of a compact symbolism—the device to which mathematics owes so largely its immense development. (Opening of *Huntington and Ladd-Franklin 1905*)

Figure 2.1 summarizes this section in, I trust, a semiotic-friendly manner. It is adopted from my survey (*1988*) of the interactions between logic and mathematics from 1800 to 1914, to which I refer for a more detailed account. The figure is not oriented around the contributions of Peirce, or indeed of any one logician.

I shall distinguish 'algebraic' from 'mathematical' logic, using both terms in their modern senses.[3] They *were* used during the nineteenth century and after, but with other senses intended, and other adjectives such as 'formal' and 'symbolic' were thrown around. In fact, the terminology got into a messy state, as is indicated by the hopelessly incoherent statement made by two highly competent practitioners of the art(/science), quoted at the head of this section. The principal lesson is that during the nineteenth century two different traditions emerged in logics which used

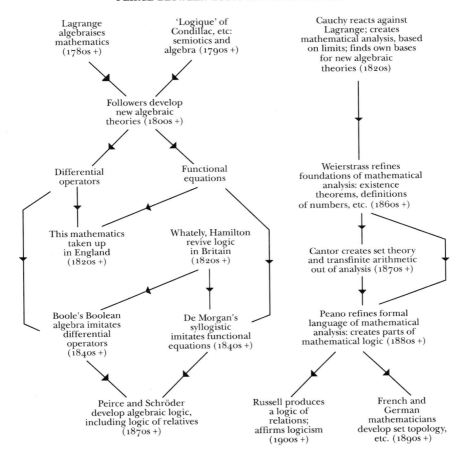

Figure 2.1

mathematics—a deliberately ambiguous phrase, used to exclude other traditions such as Kantian and (neo-)Hegelian logics, where mathematics was considered but not deployed.

2.1. *Algebra and Logic.* The pattern on the left in Figure 2.1 indicates some interest in connections between logic and algebra in France after the Revolution: I shall not pursue them here, but it is worth noting that the 'logique' was semiotic in character in Peirce's later jargon: it was concerned with the influence of signs upon knowledge and language (with algebra as an apotheosis), and led some French mathematicians to speak of 'rigor and clarity' in their work.

The algebras of functional equations and differential operators (but not the 'logique') were exported to England, where they became quite a fetish.[4] However, logic was revived to a remarkable and rapid degree in Britain after the publication of Whately's *Elements of Logic* in book form *(1826)*: new editions of this work and commentaries upon it soon appeared. (Peirce himself was to take up logic by reading an edition of Whateley (NEM 4: vii).) In particular, logic and algebra were to interact in substantial ways in two important English mathematicians.

2.2. *Augustus De Morgan.* De Morgan (1806–1871) took up functional equations, and indeed in *1836* wrote the first general account of their properties.[5] A simple example of such equations is

$$F(m + n) = F(m) + F(n) \text{ for all values of } m \text{ and } n \tag{1}$$

where the task was to solve for F, not m or n. This required special techniques, and indeed the topic was a challenging one for the time. It led not only to results of its own but also to two durable terms in algebra: 'distributive' and 'commutative,' proposed by the French mathematician Servois (*1814*) precisely in this context, Distributivity for functions is in fact defined by (1) extendable to any finite number of argument variables $m, n, \ldots$; commutativity is given by

$$F(G(m)) = G(F(m)) \text{ for all values of } m. \tag{2}$$

The point of interest here is that the structure of the theory was similar to the logic of relations which De Morgan was to introduce in his *1860a*; a commutativity (or not), the search for the inverse relations(s), and so on. While he did not discuss the links between these two theories, they are evident enough, and continue throughout the development of algebraic logic.

2.3. *George Boole.* Boole (1815–1864) was one of the principal English mathematicians who pursued differential operator methods, especially for solving differential equations. The theory of differential operators (which goes back to the eighteenth century) is based on reinterpreting the quotient of differentials dy/dx as $(d/dx)y$, the operation of differentiation d/dx on the function y: to improve the notation, the letter D denoted the operator. Boole (*1844*) put forward basic principles for functions of these operators to obey: two of them were distributivity and commutativity and the third was the 'index law'

$$D^m D^n = D^{m+n}, \text{ with } m \text{ and } n \text{ any positive integers.} \tag{3}$$

Three years later, in his first publication on logic, Boole (*1847*) proposed three basic laws for the objects and subjects $x, y, \ldots$ (mental operations, collections, or truth-conditions on propositions, in the various interpretations) which his Boolean algebra obeyed. Two of them were distributivity over subjects and commutativity; the third was the 'index law,' initially in the general form

$$x^n = x, \text{ with } n \text{ an integer} \geq 2 \qquad (4)$$

and later in the special case $x^2 = x$. The similarity between the mathematical and the logical theories could not have been more clearly stated: and, as with De Morgan, the analogy was to remain important in the tradition of algebraic logic. (For a general survey of De Morgan's contributions, see P. Heath in *De Morgan 1966*; and for Boole's achievements, see *MacHale 1985*.) Indeed, such similarities will arise in several of the points made in Section 3 below.

The next important stages of the development of algebraic logic included important modifications to Boole's system proposed from 1863 onwards by W. S. Jevons (1835–1882) (see my *1991*), and the union of both De Morgan's and Boole's theories by Peirce and E. Schröder (1841–1902), especially from Peirce's great 1870 paper on relations (3.45–149/W2: 359–42). This and other of his works are dealt with in detail in this volume (see also *Merrill 1978*): on Schröder—still a rather neglected figure—see especially *Barone 1966* and *Dipert 1980*.

De Morgan was still alive when Peirce started working in logic, and contacts were established late in 1867, when Peirce sent over his three logic essays published during the previous year, with handwritten corrections: these were incorporated in W2: 12–59 (see also p. 501). De Morgan's reply of 14 April 1868 (which is kept in the Peirce *Nachlass* at Harvard University) promised to send him offprints in return. For the next stage of their relationship see Section 5 below.

2.4 *Mathematical Logic*. Meanwhile, other French developments were taking mathematics in different directions. In (partial) reaction against the algebras, A. L. Cauchy (1789–1857) advocated a discipline of 'mathematical analysis,' in which the calculus, the theory of functions, and the convergence of series were brought under the umbrella of limits.[6] The main aim was to improve the level of rigor in these subjects, and one aspect is worth noting here: Cauchy greatly improved the logic of specifying necessary and/or sufficient conditions under which theorems were held to be true.

Peirce was aware of such aspects of Cauchy's work. "Gradually, beginning we may say with Cauchy early in the XIXth century," he wrote on one occasion, "a vast reform has been effected in the logic of mathematics which even yet is not completed" NEM 4: 157). Elsewhere, in an essay of 1893, he judged Cauchy's strengths and limitations as a proto-logician as follows: "Cauchy is often called a bad logician because he made many logical blunders. It is true that he did so, but nevertheless he knew how to reason better than others, who grope and never make a misstep because they only shuffle along" (4.151).

Gradually Cauchy's approach gained favor over the competing alternatives. In particular, it was adopted by K. Weierstrass (1815–1897) in his lectures at Berlin University from the late 1860s onwards, and exercised an enormous influence. Three strands need noting here.

First, out of some problems concerning Fourier series, G. Cantor (1845–1918) developed his set theory, initially for these mathematical purposes but then, in a much broader frame, to encompass basic parts of mathematics such as arithmetic and thereby mathematical analysis itself. Secondly, G. Peano (1958–1932) brought a more refined formal language into analysis, together with his followers he created essential elements of mathematical logic, including predicate calculus with quantification, with collections handled by Cantorian means (*Bottazzini 1985;* my *1986; Rodríguez-Consuegra 1991,* ch. 3). Thirdly, B. Russell (1872–1970) unified these two themes by extending Peano's logic with his own logic of relations (something the 'Peanists' failed to do) and then went a stage beyond Cantor in claiming that set theory had its own ground there. This was his 'logicism', the thesis that mathematics was contained within this form of logic.[7]

§3. Similarities and Differences Between the Two Traditions

The current phrase *mathematical logic* is ambiguous inasmuch as it may be understood to mean either the mathematics of logic or the logic of mathematics. (*Johnson 1922:* 137)

In this section I shall survey about a dozen aspects of algebraic and mathematical logic in order especially to suggest some perspectives of the former which could be worthy of further study. I take each tradition in its "settled" form, as represented respectively by Peirce's papers by 1890 and Schröder's *Vorlesungen* (which began to appear at that time) and the Whitehead/Russell *Principia Mathematica* of 1910–1913; I do not consider the historical origins of the various ideas and notions. In each case of similarity in an aspect, differences are also evident, and vice versa: I have made no effort to check the balance in fine detail. However, some aspects are more profound than others, and I begin with the three most significant ones.

3.1. *Links with Mathematics.* The inspiration of these two traditions in logic from different areas or branches of mathematics generated many differences. In algebraic logic, *laws* (such as distributivity and commutativity, and also associativity) were stressed, and (more with Schröder than Peirce) duality properties of theorem(-pairs) used. Analogies were deployed: + had properties like 'or,' × like 'and,' zero was like a contradiction, and so on. The expanding world of algebras of that time, with systems using more than one connective, hyper-complex numbers, and so on (*Novy 1973*) made the connections quite close. The influence on Peirce of his father was very fruitful here (as is emphasized in Section 4.4).

A particularly striking part of Peirce's logic, of which I know no analogy in mathematical logic, is the iconic system of 1902 for the sixteen binary connectives (see Clark and Zellweger in this volume).[8] Normally an algebra is a system of objects with connectives obeying certain laws of composition; but here the connectives themselves are the objects with iconic principles serving as, one might say, connectives of the second order. We see, of course, Peirce allying his logic with his semiotics in this passage; and it is worth recording that at its start he cited De Morgan, who himself indulged in sign language when classifying various kinds of syllogism.[9]

In mathematical logic, *axioms* were emphasized, and properties of completeness aimed at. Such hopes were only implicit in the sense that theory/metatheory distinctions were absent from virtually all logics of this time (a point made in Section 3.5 below); but still the hope was there. The importance of axioms is worth underling in, because only during the late nineteenth century did axiomatics begin to emerge as a major movement in mathematics. Furthermore, the principal stimuli came not from logics but from (Euclidean and non-Euclidean) geometries and abstract algebras (*Cavaillès 1938*)—topics which interested Peirce, incidentally, although they were not among his major concerns.

3.2. *Relationship with Mathematics.* This matter will be examined in more detail in the next section in the case of Peirce himself, but one general point should be made now. Mathematical logic, especially in the logicist version of Russell, was held to *contain* "all" mathematics;[10] algebraic logic maintained some *relationship* with it. For example, for Russell and G. Frege (1848–1925) integers were defined as sets of sets; for the algebraic logicians the number of members of a set was an auxiliary concept. The caution of the algebraic logicians is to be praised; for one of the most obvious points to make about Russell's logicism is that most mathematics of his time was not even sketched out (for example, there was not a line written on the calculus!), and it was not clear how such reductions were to be achieved.[11]

3.3. *Theories of Collections.* I use the word collections as a neutral term, to cover sets, classes, assemblages in an umbrella way. Like most logics of the nineteenth century, algebraic logic deployed versions of what has become known as 'part-whole theory,' a basically extensionalist conception of a collection which drew upon one subset relation. It corresponds to 'improper inclusion' in mathematical logic, where the theory of collections was based on Cantor's *Mengenlehre*. There, improper and proper inclusion were distinguished, and each was further distinguished from membership of an individual to a set, of a set to a set of sets, and so on.

A variety of attendant distinctions also emerged: I would mention especially that between the empty set, the number zero and nothing, which remains rather obscure in algebraic logic. (In particular, the status of the empty class there is unclear: if the collections really are extensional, then there should really be nothing.) The difference here is not just a technical one: it leads to significant differences in interpreting many logical ideas, as will be noted in Section 3.4. The eighteen-year-old N. Wiener (1894–1964) rather missed these points when he compared the logics of relations in Schröder and in Whitehead/Russell, and (otherwise correctly) showed a much structural similarity between them; Russell severely criticized him on these grounds (see my *1975*; and also Burch in this volume, for aspects of Peirce's logic of relations which were unique to him).

There was, of course, a penalty to pay for this extra power: the well-known paradoxes, which had to be a major concern for any enterprise of a mathematical-logical kind. By contrast, the algebraic logicians could set them aside, as most of them were not constructible within their systems.[12]

The uses made of set theory also differed. To take an important case, both Cantor and Peirce made significant contributions to the under-standing of continuity, but their views on its relationship to infinitesimals were quite different. Cantor's conception of the continuum comprised the standard rational and irrational numbers, and excluded infinitesimals ex-plicitly: Cantor even claimed to have a proof that their existence was im-possible. In expressing this hope he extended the view upheld by most fel-low Weierstrassians that infinitesimals were unnecessary (a line which Russell also adopted). By contrast, Peirce held infinitesimals to be an indis-pensable adjunct to continuity, and by means of the power-set construction (which he appears to have discovered independently of Cantor) he outlined means of constructing 'supermultitudinous collections' within which they were contained.[13]

While Peirce became much interested in Cantorian set theory, and made some significant contributions to it and to associated theories (such as axiom systems for integers), his logic still works with the part-whole the-ory just mentioned. He seems to have seen the *Mengenlehre* as part of mathe-matics, of course at its foundational end (compare his remarks of 1908 to P. E. B. Jourdain, NEM 3:879–888); by contrast, for Russell *Mengenlehre* was an essential component of mathematical logic and indeed received more detailed treatment in *Principia Mathematica* than any other branch of mathe-matics. In other words, Peirce would tackle Cantor's work; Russell had to eat it.

3.4. *Basic logics.* Both systems developed propositional calculi and predi-cate calculi of functions of one and of several variables with quantification. The similarities are of course considerable and even striking; but the dif-ferent approaches to collections affected the character of the logical con-cepts involved: in algebraic logic they were often extensional (although intensional interpretations were also admitted), whereas in mathematical logic intensionality was usually dominant. I am not considering the developments in logicism after World War I, when extensionality was given much favor.

Perhaps the most striking difference concerns quantification, where the respective interpretation of the universal and existential cases as infinite conjunctions and disjunctions, with the algebraic analogies of infinite products and sums and the symbols Π and Σ, is very typical of algebraic influences mentioned in Sections 2.1 and 3.1. Duality (mentioned in Section 3.1) was deployed also in arrangements of quantifiers. By contrast, the mathematical logicians rarely made any special use of duality. Another difference is that in algebraic logic equality was a prime connective, although after *1870*, Peirce preferred his symbol $\prec$ for the copula, with its interpretation as implication (3.47, 165–170/W3: 360, 4: 166–170). By contrast, in mathematical logic equivalence *never* enjoyed the same status (although definitions of identity were also used). This point will recur in Section 4.4.

Both logical traditions worked normally with bivalent classical systems; Peirce stands out for his sympathy for other types of logic, both modal and many-valued. The second case exemplifies Peirce's algebraic approach very clearly: possibly picking up a hint from De Morgan and Boole, Peirce extended Boole's law (4) in the factored form

$$x(1 - x) = 0 \text{ to the version } (x - \mathrm{f})(x - \mathrm{v})(x - z) = 0, \tag{5}$$

where 'v' and 'f' were true and false respectively, and 'z' was a new available case (see, for example, NEM 4: 118–123). Russell had concepts of necessity and possibility, but they were of minor significance in his logic, and he was cool to MacColl's proposal of modal logic in the 1900s (to which Peirce himself did not seem to respond).

Another difference concerns uni-connectival logic. Under the general influence of *Principia Mathematica*, Sheffer (*1913*) introduced his version of the propositional calculus based upon 'nor.' In 1880 Peirce had achieved the same aim with 'nand,' in an unpublished paper (now in 4: 12–20/W4: 218–221). Further, in an undated study of the 1880s he gave a symbolical presentation of the implicational propositional calculus NEM 4: 106–114) which bears an astonishing resemblance to the system employed in Spencer Brown's *Laws of Form* (*1969*), which is in effect a Sheffer calculus without brackets. It is not likely that Brown knew of this latter work of Peirce's; neither did the mathematical logicians of Peirce's day, nor did they attempt any comparable procedures of their own.

Finally, both traditions were much concerned with names; but whereas the mathematical logicians laid stress on the difference between names and definite descriptions (partly for motivations in mathematics itself, and for

Russell also because of the paradoxes), Peirce set off toward semiotics (which contains a theory of proper names), as we can read in DiLeo and Brock in this volume. No such concern excited mathematical logicians, although the need for good notation in mathematics was emphasized, and an excellent linear system was built up by Peano and Russell. Frege's tree-theoretic notation is a good planar design, too; it has received an absurdly critical press.

3.5. *Distinction of Theory from Metatheory.* Among the logicians of this period, only Frege took this distinction seriously, but even he did not observe it all the time. Peano and Russell not only lack it: their conception of logic is of a scale of generality that would *forbid* metatheory from operating. As a result, much mathematical logic is incoherent in content: implication is mixed up with inference, pieces of language conflate with denoted objects, and so on. (Of course Russell knew very well the distinction between a phrase and its meaning as he—unfortunately—called it.) There must be such conflations in Peirce also, but they appear not to attract the attention of his historians. Sometimes the distinction is built in without explicit discussion: for example, decision procedures based upon existential graphs (of which no analogue ever struck the minds of the mathematical logicians, incidentally) are in effect meta-theoretical.

3.6. *Popularity.* Overall two different worlds are revealed, despite the substantial overlap in content and purpose. By the late 1900s both areas were active, but the mathematical line was beginning to eclipse its algebraic brother as a logic in which mathematics played some major role. (On this point vis-à-vis Peirce, see *Putnam 1982.*) I think that the differences mentioned in Sections 3.2 and 3.3 are particularly important for the eclipse: that the question addressed in mathematical logic were more *specific* than those posed by the algebraist. It is not hard to imagine the reader of Peirce or Schröder at that time (or indeed long after), admiring the cleverness of their systems but wondering as to the purpose of it all. Here to end this section is a typical example of the kind of perception that I have in mind, taken from an excellent scholarly source, and an American one at that!

> The contemporary view of relations is due mainly to the development of the logic of relations by Charles Peirce, Schröder, and more especially Frege [*sic*] and Russell. We owe to Russell, more than to any other single philosopher, a clear understanding of the nature and importance of relations. (*Weinberg 1965:* 117)

Thus Frege gained an accolade (presumably for his notion of the ancestral—an important contribution to the logic of relations, but his only one), while De Morgan did not even gain a mention. Yet he has valuable things to tell us, as we shall now see once again.

§4. Peirce on Logic and Mathematics, and on Mathematics and Logic

We know that mathematicians care no more for logic than logicians for mathematics. The two eyes of exact science are mathematics and logic; the mathematical sect puts out the logical eye, the logical sect puts out the mathematical eye; each believing that it sees better with one eye than with two. (*De Morgan 1868*: 71)

This brilliant remark by an eminent one-eyed mathematician and logician sets the scene for our considerations. As was mentioned in Section 3.2, the logicism of Russell was intended to contain "all" mathematics. The fact that it strikingly failed to do so is not our problem here; but the question of the relationship between Peirce's algebraic logic and (parts of) mathematics is still to be posed. The relationship can be treated in two parts: mathematics applied to logic, and logic applied to mathematics (where the two senses of 'applied' may not be the same).[14]

4.1. *Mathematics Applied to Logic.* This position comes from Boole, who carried out "A Mathematical Analysis of Logic," to quote the title of his first text (*1847*); and this view continued in his successors, Peirce included. In De Morgan the situation was similar, although different in content: he used mathematical notations, or at least symbolic systems, in his logic (some examples were mentioned in Section 3.1), deployed analogies, noted that certain logical systems were commutative (or whatever), and so on. Peirce, in inheriting views from both men, followed and extended both kinds of line, and added mathematical techniques of his own: in particular, his existential graphs drew on ideas in (proto-)topology. For him "all formal logic is merely mathematics applied to logic" (4.228 [1902]).

But there are (to us, meta-)logical difficulties in this application. In particular, those mathematical theories had to be (assumed to be) consistent, in order that the application would be worthwhile; but there is the danger of a loop, for *a logical notion is assumed in an algebra used in logic.* In Boole, for example, consistency was taken for granted, possibly being related by him to his assumption that $0 \neq 1$ (and further properties to distinguish 0 from 1). In De Morgan the situation was still less clear: he applied algebras to his logic, but he did not discuss the consequences of this consistency condition. Further, remember from Section 3.5 that the distinction of theory from metatheory, which could convert the loop into a spiral, was not then known.

4.2. *Logic Applied to Mathematics.* In individual cases this can be appreciated: Peirce's 1870 application of the logic of relatives to errors of observation (W3: 114–118), for example, or to Cayley's abstract geometry (NEM 2: 368–640). But this relationship is hard to sort out in a general way. It involves not only the use of modes of reasoning in mathematical theories but also the place (if any) of mathematical objects in that logic.

First, the comparison with mathematical logic is instructive, even though the logicist program did not work. In particular, set theory was (supposed to be) part of that logic; how about the collections of Peirce's logic, for example? Do parthood and collectionhood belong to mathematics, or to logic, or both?

Again, Peirce thought that the premises of mathematics themselves were hypothetical, and this led him to distinguish mathematics from logic on the grounds that logic is categorical (4.233 [1902]/NEM 2: 595); but surely this distinction is itself one *within* logic, not between it and another subject. We may have a case here where the melding of theory and meta-theory mentioned in Section 3.5 may be evident. Elsewhere he took the more promising line that a task for logic was to analyze necessary conclusions as such (4.239 [1902]); but can mathematical theory be used in that process, and if so, how?

4.3. *Equality and Identity.* Naturally algebra played a major role in these questions for Peirce. One passage, of 1873, shows it very clearly; where "Logic may be considered as the science of identity" (of indiscernibles, in the tradition from Leibniz) alongside "a parallel science of equality, which is *mathematics* or the science of quantity" (W3: 92). Here analogizing, as mentioned in Section 3.1, is marked: equality of quantities in algebra, identity in logic (despite the primacy already assigned to $\prec$ that was noted in Section 3.4, incidentally). His interest in semiotics increases the links; but what sort of links were being forged? Naturally, he was well aware that algebra, and a fortiori mathematics, comprised more than the study of quantity (W3: 83–84, also of 1873), and so took up another definition.

4.4. *Necessity and . . . ?* Peirce's inspiration came from his father's monograph on linear associative algebra (*B. Peirce 1870*: on it, see *Pycior 1979*). Benjamin Peirce (1809–1880) began his main text by defining mathematics as "the science which draws necessary conclusions," and immediately remarked that this definition was wider than the normal ones, which emphasized "quantitative relations," and that "qualitative" ones, "of which a fine example is given by Boole," had independent status. Charles was to accept this philosophy of mathematics with enthusiasm (for example, 3.558 [1898], and 4.229 [1902], immediately following the remark quoted in Section 4.2): indeed, in 1881 he had the lithograph printed in a mathematical journal with his own appended notes (some now reprinted in W4: 312–327). In addition to the obvious resonances concerning algebras and relations, two points immediately arise.

First, to us the word 'necessary' links with possibility; but at that time the closer association would be with probability, which itself treated (among other things) belief and so melded with psychology, a descriptive science to which logic was a normative counterpart (hence a use of 'necessary'). Such links were quite clearly made in Boole and De Morgan, who both wrote on probability as well as logic: indeed, De Morgan's book (*1847*) was called *Formal logic, or the Calculus on Inference, Necessary and Probable*, and Boole's second one (*1854*) included the phrase "the laws of thought, on which are founded the mathematical theories of logic and probability."[15] In one of his first papers on logic, Peirce even stated in 1867 that "The principal use of Boole's Calculus of Logic lies in its application to probability" (3.1/W2: 12).

One consequence for logic is as follows: probability handles values between 0 and 1, and logic is a sort of extreme case; then in logic are the values 0 and 1 pukka *quantities*, and not just distinguishing states? Is quantity, including integers, acceptable within algebraic logic? Or are we again in the loop of mathematics applied to logic?

Second, Benjamin Peirce's tract was written within a tradition of algebra in which De Morgan played a major role and in which a distinction was made between the form and the matter of an algebraic theory (*Pycior 1983, 1987*). One consequence of this philosophy was that necessary conclusions (or deductions) were made from form alone.[16] So to understand Peirce's position, we need to examine this category more closely.

4.5. *Form*. The role of form exemplifies another analogy with algebra which has been mentioned earlier, for logic was often held to be concerned with the form of reasoning independent of the language within which it was expressed. However, this analogy had been weakened, in that the place of language within logic had been raised from Whately onwards; and the introduction of predicate calculi (of various kinds) into logics had increased this tendency further. On the other hand, the emphasis on form doubtless encouraged interest in the logic of relations: it may be no coincidence that three of its four major founders belonged to the tradition of algebraic logic.

In addition, the importance of form in mathematics was strongly argued by the amateur British mathematician and philosophy A. B. Kempe (1849–1922), in a remarkable paper (*1886*) published with the Royal Society "On the theory of mathematical form." It aroused no particular interest among the mathematical logicians, but Peirce quickly responded with a letter to Kempe, which motivated Kempe to publish a short "Note" (*1887*) containing modifications.[17]

Kempe 1886 contained a wide range of considerations of great generality concerning mathematical objects: for example, he worked with "heaps" (*1886*: 7), which are not called "multi-sets" and are more general than Cantorian set theory in allowing an object to belong to them more than once. (*Hailperin 1976 [1986]* interprets parts of Boole's version of Boolean algebra in terms of signed multi-sets, where negative measures of membership are allowed.) This was an important extension, for his general considerations often involved "distinguished and undistinguished units" (but with no underlying theory of identity supplied), and the types of "plurality" that they could constitute. The part on logic followed Boolean lines for "classes" composing the "full set" (*1886*: 63–70). Form, to quote a rather clearer statement from a later paper, was what "remains" when the "special peculiarities and characteristics of the individuals and relations" were set aside.[18]

This process of setting aside is similar to Cantor's idealist processes of abstraction of nature and order-type from sets to "define" cardinal numbers from general sets: indeed, Kempe's whole approach is unusual for the time (outside Cantor) in the prominent place given to collections of various kinds (and of which I have not given a full account here). In (*1894*: 5) it led Kempe to this "provisional" definition of mathematics: "*the science by which we investigate those characteristics of any subject-matter of thought which are due to the conception that it consists of a number of differing and non-differing individuals and pluralities.*"

Kempe developed his approach in terms of algebraic laws and their defining properties: in fact, he (*1886*: 56) cited B. Peirce (*1870* in the edition of 1881 edited by Charles) as a source for some of these properties. But he also dressed his ideas in geometrical and (proto-) topological clothing. This excited the interest of Peirce, who saw in the work a means of developing the logic of relations (which Kempe himself had not done) and related it to his own graphical representations mentioned in Section 4.1.[19] Peirce was further interested in Kempe's work on topology, by his work on linkages and his unsuccessful attempt to prove (De Morgan's) four-color problem (see, for example, NEM 3: 449–463). Elsewhere he noted the work of J. J. Sylvester (1814–1987) on the graph theory of chemical notation, of which he also approved (it was akin in spirit, of course, to his own studies of graphs).[20]

If Kempe's views on form are consonant with Peirce's, at least in part, then we might conclude that Peirce's philosophy of necessary conclusions was an extension of his father's position, which was oriented around algebra, to a general assertion of mathematics as matterless. The closest approximation to this position that I have found occurs in a paper published in 1908, where Peirce stressed again that mathematics required necessary reasoning, allowed that the profoundest invention, the most athletic imagination, was nevertheless required by it, and that "originality is not an attribute of the *matter* of life . . . but is an affair of *form*" (4.611); but even here some essential relationships were not clearly indicated.

In addition, there is also the interpretation of quantification (Section 3.4), as to when it is applicable to mathematics; for it seems to require the existence of the objects over which it takes place—as in mathematical logic, where in fact the ontological status of Cantorian sets (and sets of sets, and sets of . . .) constitutes a considerable problem for logicism. Peirce's own writings on quantification in *logic* do not seem fully to penetrate this issue (*Dipert 1984*), never mind its bearing upon the relationship with mathematics. Perhaps some pertinent Peircean manuscript awaits the light of day, to help clarify an important but hitherto unclear aspect of his philosophy of mathematics.

4.6. *Final Remarks.* There are of course other pertinent passages in Peirce's published works, but I have not perceived a clear stand taken on the issues raised here. Doubtless many more passages are to be found in his hitherto unpublished manuscripts; my comments are offered as a source of perspectives from which all these sources may be read. I conclude meta-theoretically with a few remarks on the types of answer that may be found.

First, I recall from Section 1 that the question posed in this paper concerns the relationship between mathematics (and mathematical reasoning) and his technical logic, not to logic in broader senses. Second, Peirce may never have found a definitive position—a likely possibility in someone who started so much and finished so little! Finally, a main issue is *how much* mathematics is to be related to logic. Here I suspect that the influence of algebra, while beneficial in stressing laws and forms (and doubtless stimulating the development of algebraic logics of relations) became *too* dominant and restricted the scope of his position. If so, then a similarity with mathematical logic reveals itself; for, as was noted in Section 2.2, Russell's position of containing mathematics within his logic was too closely linked to mathematical analysis and set theory to be convincing.

§5. Strands in Peirce's Life

Three main influences in Peirce's logic, especially its mathematical aspects, may be detected; and each of them involves algebras of his time. First and possibly the most important, came Boole's algebra, with its attendant concern with operators. Second was De Morgan's contributions, especially the logic of relations; and Peirce deployed some traces of functional equations (see, for example, 4.301–302 [1902]).[21] Third lay the panoply of "common" algebras and the laws which they did (not) satisfy: De Morgan was again important, but the dominant predecessor was Peirce's father, especially with the lithograph.

In this essay I have stressed aspects of the context and prehistory of Peirce: one must avoid Peirce-myopia, for Peirce himself did not start with Peirce, and so his students must not do so either. I end with a nice detail exemplifying this point. It dates from a crucial year in his development, 1870, his 32nd year. He published his first major contribution to logic (*1870*), in which Boole's and De Morgan's ideas were brought together; his father put out in lithograph form the tract on algebra (*B. Peirce 1870*), which was to influence Charles both technically and philosophically; and the expedition to Europe under his father's leadership to see an eclipse enhanced his scientific career. During that trip he came to London, and visited the aging De Morgan in order to leave both his own and his father's new works, and the following charming letter "To Professor A. De Morgan &c. &c. &c. From Professor Benjamin Peirce of the U.S. of America to introduce his son Charles S. Peirce Esq":[22]

> To   Professor Augustus De Morgan
>   &c. &c. &c.
>
>   U.S. Coast Survey Office[23]
>   Washington, $18\frac{6}{17}70$
>
> My dear Sir-
>   I presume upon the universal brotherhood of science to introduce to you my son Charles S. Peirce Esq. who is a devoted student of Logic and I think that he has original thoughts which you may regard as deserving your consideration. He carries with him a memoir which he has written upon one of the subjects of your own learned investigations and also one which his father has written upon Linear Algebra and which was not printed without a careful perusal of this [?] and [of a] previous treatise upon the same subject which was printed by yourself in the Memoirs of the Cambridge Philosophical Society.[24] My son is now my private and confidential aid in my most important investigations in the U.S. Coast Survey, and I sincerely commend him to your favorable notice.
>   Very respectfully
>   and with great admiration
>   Yours truly
>   Benjamin Peirce
>   Superintendent U.S.
>   Coast Survey

In 1898 Peirce recalled his discussion with his senior colleague in a way which expresses my main theme rather well.[25]

> I was not the first discoverer [of the logic of relatives]; but I thought I was, and had complemented Boole's algebra so far as to render it adequate to all reasoning about dyadic relations, before Professor De Morgan sent me his epoch-making memoir in which he attacked the logic of relatives by another method in harmony with his own logical system. But the immense superiority of the Boolian method was apparent enough, and I shall never forget all there was of manliness and pathos in De Morgan's face when I pointed it out to him in 1870. I wondered whether when I was in my last days some young man would come and point out to me how much of my work must be superseded, and whether I should be able to take it with the same genuine candor. . . .

A C K N O W L E D G M E N T S

For comments on a draft I am grateful to T. Hailperin and N. Houser. Benjamin Peirce's letter to De Morgan was transcribed with the kind permission of the University of London Library.

N O T E S

1. Let us not forget the late Victor Lenzen, who brought the manuscripts (and library) to Harvard from Peirce's home in 1914. I met him (also in 1972), and I remember the long account which he gave me of that occasion: to fulfill his charge, he took the materials from the house in winter on a horse-drawn sledge (compare his *1965*: 8)

2. At the symposium on "Peirce's contributions to logic," of which the written (and expanded) versions of the lectures form a substantial part of this volume, I played the role of discussant. This article takes a similar role, though I shall not comment on the papers as such. At the symposium I used those lectures to build up my own piece, drawing heavily upon them for my presentation, especially the material in Section 3 here.

3. For example, the phrase "mathematical logic" was introduced by De Morgan in 1858 (*1966*: 78); but it served to distinguish logics using mathematics from purely prosodic "philosophical logic" (his expression again). As we shall see in Section 2.1, his own logic was part of the algebraic tradition. "Boolean algebra" was not then a common phrase; although Peirce sometimes used it (in the spelling "Boolian") and also the more common phrase "the algebra of logic." Mathematical logic was often called "logistic" in the early part of this century, when Russell's position was still emerging and also the writings of Peano and his school (who came close to logicism but did not espouse it).

Some indications of the differences between the two traditions can be gained from Styazhkin (*1969*), who however speaks only of "mathematical logic" in his title. Certain other lines from logic and mathematics could be added to the diagram, but they do not seem to me to be sufficiently important to justify the resulting complications: the contributions of H. Grassmann are the most meritorious of this genre (compare Note 7).

4. A slight exception can be made for Babbage, who studied functional equations with great enthusiasm and also took some note of "logique." On the algorithmic character of his thinking, see my *1990b* and *1992*. My student M. Panteki completed a doctorate (*1992*) on the links between algebras and logics in England from Babbage's time to Boole and De Morgan.

5. On these two algebras of functional equations and differential operators see respectively *Dhombres 1986* (where, however, De Morgan's essay is only mentioned) and *Koppelman 1971*; on the French prehistory of both, see my *1990a*, chs. 3, 4, and 10.

6. On this history, see, for example, my *1990a*, esp. chs. 10 and 11; and *Bot-*

tazzini 1986. Both of these histories also treat the contemporary development of complex-variable analysis, but I shall not consider it here.

7. Naturally, I have passed over some significant factors: some place for Grassmann in Peano's thought; the role of Dedekind, especially in Cantor; Frege and his neglect; and the place of Whitehead, who started off in algebraic logic in the 1890s (see Note 15) but switched into mathematical logic under the influence of Russell. For details of this history, see, for example, my *1977, Vuillemin 1968, Lowe 1985, Winchester and Blackwell, eds. 1988,* and *Rodrìguez-Consuegra 1991.* Recent research on Russell's work in logic and philosophy is surveyed in my *1990c.*

8. NEM 3: 272–275 contains a passage in which the sixteen connectives appear, although not in a full boxed layout. This text was permitted only six lines of footnote in the *Collected Papers* at 4.261. Note Peirce's other layouts at 4: 268–273.

9. These notational patters are evident in De Morgan's book (*1847,* chs. 5 and 7), but are best seen in his series of articles: see *1966,* esp. Pp. 12–20, 41, 158–163 (including a logical "zodiac"), 245, 316–322. Presumably Peirce saw at least some of these texts.

10. In fact Russell claimed that pure mathematics was contained in his logic; but his definition of pure mathematics (propositions of the form $p \supset q$) is so bizarre that it does not help. On Russell's conception of logic, see *Griffin 1980.* Frege's logicism was confined to arithmetic and parts of mathematical analysis (although it is not easy to say how much: *Simons 1987* is the best guide).

11. To avoid misunderstandings, which always arise among philosophers when I make this remark, I stress that it is intended *historically*: I make no appeal to Gödel's theorem of 1931 or to the redesigning of logico-mathematics which has happened since. An important problem for Russell was the axioms of choice, which emerged during the 1900s (my *1977; Moore 1982*) and appeared quite frequently in the bits of mathematics that he handled. Russell was an important contributor to their emergence: but their legitimacy as logicistic objects was suspect, since the specification of the choice set required an infinite list of instructions whereas his logic was finitary.

These axioms were developed during the last decade of Peirce's life, and they excited much philosophical as well as mathematical interest. Did Peirce himself react?

12. However, the liar paradox would be constructible if truth-values were regarded as part of the system. Early on Peirce preferred to conclude from the paradox that "*every proposition asserts its own truth*" (*1868; 5.340.* Compare *1901*; 2.618, an entry for Baldwin's *Dictionary*). Schröder's own approach, via "consistent" manifolds, and the associated type theory, is not very satisfactory (see my *1975*: 124–126).

13. Peirce sketched out his theory of infinitesimals at different times and with varying degrees of detail. His paper on the logic of relatives (*1870*) already contained a part on "infinitesimal" cases (3.110–111/W2: 395–408). Among later published and unpublished writings, see especially 3.563–570 [1900] and NEM 3.87–89, 120–126; 4: 55–56. For commentary, see *Dauben 1982.*

14. Since writing this paper, I have had the chance to study more Peirce manuscripts and correspondence at the Peirce Project at Indiana University. The idea of mathematics interacting with logic is made by Peirce in exactly this way in, for example, the first draft letter of 19 January 1902 to J. Royce (L385): "I propose to confine all I have to say of the two subjects of the Logic of Mathematics & the Mathematics of Logic to a single long chapter, which I am now engaged upon"; see 4.239–244 and the succeeding unpublished chapter, MS 430.

15. In Boole's case there were other links between logic and probability, such

as compound events and Boolean combinations of simple ones. The links were weakened by (among others) Venn, especially in his treatise (*1866*) on probability. On the history of probability logic (that is, logical consequence in a probabilistic setting), including contributions made by Peirce, see *Hailperin 1988*.

16. In his so-called *Universal Algebra* Whitehead (*1898*) took a similar position about necessary conclusion, though again he offered little explanation. But he did not praise Peirce much in his book, and on p. 155 he even rejected the need for a logic of relations. Recall from Note 7 that he was then in his pre-Russellian phase.

17. I examined the *Nachlass* of Kempe in 1973 at the Historical Manuscripts Commission in London, when it was being sorted; it is now preserved in the West Sussex Record Office, Chichester (MSS NRA 17595). It contains some interesting letters from Peirce of 1905, but not the letter mentioned in the text. Kempe referred to it again, giving the date 17 January 1887, in a reply (*1897*) to a paper of the previous year by Peirce on relatives (3.446–552: Kempe's reply is not mentioned by the editors).

There is also a letter of 1885 from G. G. Stokes (as a Secretary of the Royal Society), mentioning referees' comments and suggesting a change of title from that used in the published version (*1886*); this differs slightly from that of his summary (*1885*), which apparently already published. Kempe (*1886*) alluded to these reports in footnote 1. They are kept in the Archives of the Royal Society (MSS. R. R. 9.287–288): the authors are Cayley (keener on the algebraic parts than the others, suggesting the title and the omission of a part on geometry, puzzled about the logic, complains about lack of clarity) and Sylvester (praises the philosophy but is also unhappy over clarity). It is worth remarking that there are a number of letters from Sylvester in Kempe's *Nachlass*.

18. *Kempe 1894*: 8–9 (on p. 6, B. Peirce's definition was quoted); also in *1897*: 455. Kempe's contributions are noted in the early pages of *Vercelloni 1989*. In some respects his ideas are astonishingly similar to Dedekind's views on the nature of numbers, which appeared little more than a year later (*1888*). For example,

Dedekind, art. 2	*Kempe*, art. 25
It very frequently happens that different things *a*, *b*, *c*, . . . can for some reason be considered from a common point of view, can be put together in the mind, and one then says, that they form a *system*. . . .	If every component unit of a collection is distinguished from every unit which is detached from the collection, the collection will be termed a system.

It is very likely that both authors took the word system from its use by French authors in the nineteenth century (for example, Cauchy) to refer, in a informal way, to collections of things in general. My own use and quotation of the word elsewhere in this paper belongs to this tradition.

19. See especially 3.423, 3.468–469; 4.325, 335; 5.505; 6.174; and above all NEM 3: 431–440, where on p.431 Kempe's "sad want of training" in logic was lamented in a reference to Kempe's essay (*1890*) comparing "the logical theory of classes and the geometrical theory of points."

20. *Kempe 1897*: 457; Peirce at W4: 211 and 3.470. Peirce's relations with Sylvester were difficult, especially when they were both at Johns Hopkins University: see, for example, 3.320, 646 and 4.305, 611 (which also contains a great compliment to Sylvester's mathematical ability); or W4: 675–677.

I wonder if Peirce knew of two other links: the connection between the

'Logique' of Condillac and the new chemical notation of equations proposed by Lavoisier and others; and the attempt made by the English chemist B. Brodie in the 1860s to create a Boolean algebra for chemistry (see *Brock ed. 1967*, especially the letters to Brodie from mathematicians).

21. The influence on Peirce from Venn to Jevons seems to have been slight relative to those of Boole and De Morgan (for evidence, compare the number of entries in the indexes of volumes of the *Collected Papers* and the *Writings*), and it often consisted of messages from the earlier two figures. For example, in 1867 Peirce modified Boole's definition of logical addition to allow for intersection in the components (3.3/W2: 12), but he did so independently of Jevons's and De Morgan's definitions (the former being offered explicitly as a modification of Boole: see my *1991*). Grassmann and MacColl were still further in the background. See Peirce's interesting footnote of 1880 in 3.199/W4:182.

22. I quote from the envelope of the letter; both documents are stuck on to De Morgan's copy of *Benjamin Peirce 1870* (which is kept in the University of London Library). To ease the difficulties caused by his vile handwriting, two apparently missing words were added in square brackets.

23. The date on the next line is rendered in Benjamin Peirce's manner. It was 17 June 1870, the day before Charles sailed from New York (according to W2: xxxiii). On the last page of his copy of the lithograph for De Morgan, Benjamin wrote by hand the date 15 June, and numbered the copy 12 (out of 100).

24. *Sic:* the journal was a *Transactions*. The allusion is probably to *De Morgan 1842–1849*.

25. 4.4 (an important passage, *not* indexed under De Morgan at the end of the volume). On this episode, compare Peirce's footnote of 1902 at 4.301, and his recollection of 1908 to P.E.B. Jourdain in NEM 3: 882–883.

3.

PEIRCE'S AXIOMATIZATION OF ARITHMETIC

Paul Shields

§1. Introduction

Peirce's paper "On the Logic of Number" first appeared in *American Journal of Mathematics* in 1881 (2.252–288). Among its other accomplishments, this remarkable paper provided what was probably the first successful axiom system for the natural numbers. Since Peirce's paper is also included in the most recent volume of the *Writings of Charles S. Peirce*, there is some timeliness in taking a good look at what Peirce achieved (W4: 299–309).

My purpose is primarily historical and explicative, so my procedure will be simply to set forth the main technical achievements of Peirce's 1881 paper as I see them. Let me preface this undertaking, however, with two brief remarks on the context of this paper—remarks which also bear upon the interpretation of Peirce's philosophy of mathematics.

The first thing to notice is that Peirce's paper appeared the year after the death of his father, Benjamin Peirce, in 1880. It immediately preceded, in the same issue of the *American Journal of Mathematics*, the posthumous publication of the elder Peirce's "Linear Associative Algebra," which began with the famous definition of mathematics as the science which draws necessary conclusions (*B. Peirce 1870 [1881]*). I think that Peirce's 1881 paper was intended to reinforce his father's definition of mathematics, and that one can detect this support in the introduction:

> Nobody can doubt the elementary propositions concerning number: those that are not at first sight manifestly true are rendered so by the usual demonstrations. But although we see they *are* true, we do not so easily see precisely *why* they are true; so that a renowned English logician has entertained a doubt as to whether they were true in all parts of the universe. The object of this paper is to show that they are strictly syllogistic consequences from a few primary propositions. The question of the logical origin of these latter, which I here regard as definitions, would require a separate discussion. (3.252)

Peirce's immediate project was to show that arithmetic draws necessary conclusions. But since Peirce was aware of the extent to which analysis itself had been arithmetized, his larger purpose was to provide the missing link in the argument that mathematics as a whole draws necessary conclusions.[1] Now it is important not to confuse this undertaking, ambitious as it is, with the even more ambitious program of logicism.[2] In fact, Peirce interpreted his father's definition to mean that mathematics, the science *which draws* necessary conclusions, is essentially different from deductive logic, the science *of drawing* necessary conclusions. And he distinguished his own conception of mathematics from the logicism of Dedekind by observing that the latter "would *not* result from my father's definition."[3] This is why Peirce cautioned us that, "the question of the logical origin of [these few primary propositions] . . . would require a separate discussion."

The second point I want to make about the context of Peirce's paper concerns his reference to "a renowned English logician"—John Stuart Mill. I believe it helps to clarify Peirce's intentions to view Peirce as responding to Mill's empiricism. But I also think that there is a larger historical lesson to be learned from this reference—a lesson which can also be inferred from Frege's extensive polemic against Mill in his *Grundlagen* several years later (*Frege 1884*: 9–17, 29–33). We are dealing with an intellectual environment in which John Stuart Mill's empiricism was a force to be reckoned with—a widely defended, if not defensible, alternative in the philosophy of mathematics. Recognizing this, I think, can give us a new perspective on the tremendous changes that took place in the last decades of the nineteenth century—changes in which Peirce was a significant factor.

§2. Peirce's Axioms

So what do we find in Peirce's paper? The first thing Peirce did was to spell out the few primary propositions from which the elementary propositions concerning number can be derived, i.e., he provided an axiom system for the natural numbers.

Let N be a set, R a relation in N, and 1 an element of N; assume definitions of "minimum," "maximum," and "predecessor" with regard to R and N.

1)	N is partially ordered by R	(3.253)
2)	N is connected by R	(3.254)
3)	N is closed with respect to predecessors	(3.256)
4)	1 is the minimum in N; N has no maximum	(3.257)
5)	Mathematical induction holds for N	(3.258)

I should hasten to point out that this is *not* what the actual text of Peirce's 1881 paper looks like. So let me go through and briefly justify each axiom, explaining its origin in the text and clarifying some of Peirce's apparently peculiar terminology:

1) Peirce's first paragraph describes a relation—think of it as the relation greater than or equal to—which is transitive, antisymmetric and reflexive in a given system of objects. He calls this system—what we would today call a partially ordered set—a system of quantity.

2) Peirce's second paragraph adds the requirement that a system of quantity be connected, and he calls the result a system of simple quantity. Such a set—both partially ordered and connected by a given relation— is variously referred to today as being simply, totally, or linearly ordered. So one could combine Peirce's first two axioms into the single requirement that N be simply ordered by R. To the best of my knowledge, these two axioms are among the earliest abstract definitions of partially and simply ordered sets.[4]

3) Peirce's third paragraph describes a system in which every element, except the minimum, has an immediate predecessor. Unfortunately, Peirce refers to such systems as being "discrete"—as opposed to "continuous" or "mixed" systems for which the property does not hold. This terminology reflects Peirce's early struggles with the notion of continuity (*Potter and Shields 1977*). It is unfortunate because it is likely to distract a modern reader from Peirce's otherwise clear statement of his third axiom—as requiring closure with respect to immediate predecessors.

4) Peirce's fourth axiom is straight forward, requiring what he calls a semi-limited system—having a minimum element but no maximum.

5) Peirce's final paragraph again invites confusion since he starts out by stipulating that his system be "infinite." But further on it becomes evident that the final axiom has nothing to do with being infinite in the ordinary sense of the word, and that Peirce is simply describing the principle of mathematical induction—the principle he later referred to as Fermatian inference.[5] For example, he says "an infinite system may be defined as one in which from the fact that a certain proposition, if true of any number, is true of the next greater, it may be inferred that that proposition if true of any number is true of every greater." Note that Peirce's formulation here is sometimes call "mathematical induction starting with k." For a semi-limited system it is equivalent to ordinary mathematical induction, which is why I have only listed the ordinary version. Peirce chose the "starting with k" formulation in his 1881 paper because he also wanted to discuss systems of integers. In general, Peirce shows a remarkable sophistication about the axiomatic

function of mathematical induction. He describes it as guaranteeing that every quantity can be reached by a series of successive steps, and explains that he calls such systems infinite. In order to distinguish them from systems which are "super-infinite"—apparently recognizing the role of mathematical induction in minimalizing ordinal models.

These five axioms constitute what Peirce called "ordinary number"—a system of infinite, semi-limited, discrete, simple quantity (3.260).

3. The Peano Axioms

Scholarship has traditionally attributed the first successful axiom systems for the natural numbers to Richard Dedekind and Giuseppe Peano (*Dedekind 1988, Peano 1889*). Now it is perhaps more instructive to compare Peirce's paper with Dedekind's 1888 booklet, *Was sind und was sollen die Zahlen*, especially given Peirce's later claim that his 1881 paper influenced Dedekind (MS 316a). But since Dedekind's axiom system would require more machinery to present, and since his approach is essentially similar to that of Peano, we will contrast Peirce's system instead with a version of Peano axioms:

Let N be a set, "successor" a function on N, and 1 an element of N.
  1) 1 is an element of N
  2) N is closed with respect to successors
  3) No two elements of N have the same successor
  4) 1 is not the successor of any element of N
  5) Mathematical induction holds for N

I have made no attempt to be historical.[6] I believe most readers will recognize the Peano postulates, and my purpose is merely to facilitate a comparison with Peirce's axioms. Note that there is a redundancy between the specification of notation and the first two axioms.

§4. Comparison

In general, Peirce's style of presentation was clearly less formal and more dependent upon ordinary language than the axiom systems of Dedekind and Peano. This is partly the difference of a decade, and I don't think Peirce was much, if any, less precise. But I also acknowledge that the works of Dedekind and Peano were pioneering in many respects, apart from their axiom systems. Dedekind's informal set theory and Peano's advances in formal notation were important accomplishments, and I do no mean to detract from them.

The most striking difference in the content of these axiom systems is that Peirce chose a transitive relation as primitive, while Dedekind and Peano chose a non-transitive successor function. This difference is worth a brief historical-philosophical digression.

In *Principles of Mathematics*, Bertrand Russell defended at some length the philosophical priority of the transitive approach, claiming that "all order depends upon transitive asymmetric relations" (*Russell 1903*: 218); and Peirce himself described transitive relations as the "basis of all quantitative thought" (4.94). As it happens there is an actual historical thread linking Peirce and Russell on this topic. In his 1901 paper "On the Notion of Order," for instance, Russell said that partially ordered series "have great philosophical importance" and that their theory "is one of the most essential parts of logic" (*Russell 1901*). But at the beginning of this same paper Russell had acknowledge that "the following account of the genesis of order is virtually identical with that of Mr. B. I. Gilman" (*Russell 1901*: 30), the reference being to the paper, "On the Properties of a One-Dimensional Manifold," by Gilman, which had appeared in the first issue of *Mind*. Gilman's paper, also cited in *Principles of Mathematics*, had claimed that partial ordering exhibited "the ultimate constituents of the notion of one-dimensionality" (*Gilman 1892*). Gilman was a student of Peirce's at Johns Hopkins.[7]

In any case the argument is simply that many natural species of order—fractions, real numbers, moments in time—can be described by transitive relations but not by successor functions. Thus Peirce's approach has the advantage of allowing one to describe the natural numbers *genus et differentia*, as one particular species of order. We can detect this motivation in Peirce's statement of his third axiom, and it is not difficult to see how transitive relations would be far more attractive to one as intoxicated with the notion of continuity as Peirce.

The other side of this issue, of course, is that Peirce's approach cannot hope to compete with the simplicity and elegance made possible by Peano's choice of a successor function. There is little danger, I think, that Peirce's axioms will supplant the Peano postulates in elementary textbooks.

On more technical questions, it is important to note that the primitives of Peirce and Peano can be cross-defined. Peirce implicitly defined a successor function in the course of expressing his third axiom, while Dedekind and Peano defined Peirce's transitive relation greater than or equal to in terms of their own functional primitives.[8] This latter task was the more difficult, and it followed the procedure originated by Frege, in 1879, to characterize what was later called the "ancestral"—of a function (*Frege 1879*: 69).[9]

Given these normal definitions of primitives, Peirce's 1881 axiom system is equivalent to the Peano postulates. In my dissertation I took considerable pains to prove, formally, the equivalence of the Peirce and Dedekind axioms (*Shields 1981:* 103–119). I will sketch out enough of an argument to suggest how a more formal demonstration might be constructed.

5. Equivalence

The outlines of the equivalence are not that difficult. The mathematical induction axiom is the same in both systems. Moving from Peano to Peirce is cumbersome, since most of the individual proofs require the definition of Peirce's transitive relation using ancestrals or chains. But in this direction I can appeal to your confidence in the Peano system and merely observe that the properties set forth by Peirce certainly *ought* to be true of the natural numbers.

Moving the other direction, from Peirce to Peano, it is clear that Peano's first and fourth axioms are no problem. Similarly, Peano's third axiom, the uniqueness of successors, follows from the definition of successor and the antisymmetry and connection clauses in Peirce's first two axioms. In fact, the only real difficulty in the entire argument seems to be the derivation of Peano's second axiom—closure with respect to successors—from the Peirce axioms. In visual terms, we have to rule out models for Peirce's system which have embedded within them infinite descending sequences; i.e., we have to show that Peirce's system is well-ordered. Fortunately, mathematical induction is sufficient:

> Choose an arbitrary element of Peirce's system and assume that this element does not have a successor. Use this as a basic step, and consider for induction the property of not having a successor. The inductive step turns out to be vacuously true since an element will never have a successor when it is itself the successor of an element which has no successor. So, whenever an element has no successor, it follows by mathematical induction that all elements greater than the element must also lack successors. But this would contradict the Peirce axioms requiring closure with respect to predecessors and no maximum—which is the second Peano axiom.

There is a certain elegance in the way Peirce's mathematical induction axiom functions simultaneously to minimalize and well-order his system.

To be strict there remains one final step, the derivation of the primitive definitions which must be added to each system (see *Shields 1981*).

If Peirce's axiom system is equivalent to the later systems of Peano and Dedekind, this means that Peirce's 1881 paper contains what is probably the first successful axiom system for the natural numbers in history. This is not at all the conventional wisdom on this topic, and I think it is about time that we adjust our footnotes accordingly. There is a sense in which, to paraphrase Russell, the logical "weight" of Peirce's few primary propositions is equal to the entire body of traditional pure mathematics. It seems only fair to balance such logical weight with at least a few ounces of historical weight.[10]

§6. Arithmetic, Cardinals, and the Infinite

Let us leave Peirce's axiom system and turn to the remainder of his 1881 paper. The central portion described the arithmetic of natural numbers. Peirce first gave recursive definitions of addition and multiplication and then proceeded to prove associative, commutative, and distributive laws for these operations. Fraenkel and Bar-Hillel, in their *Foundations of Set Theory*, claim that this is one of the earliest-known examples of definition by recursion.[11] Peirce's definitions were quite similar to those of Dedekind's work at the end of the decade.[12]

After a short treatment of integers—which Peirce called an unlimited system—two passages near the end of his paper break entirely new ground. The first introduced the notion of the cardinality of a set—what Peirce called "the number of its count." In essence, he proposed that the cardinality of a set be determined by placing it in one-to-one correspondence with an initial segment of natural numbers (3.280–281). This is probably the first instance of what has become the dominant modern method of defining cardinals.[13] And it was again paralleled, down to the analogy with counting, by Dedekind's approach in *Was sind und was sollen die Zahlen* (*Dedekind 1888*: 109,110). This first passage ended with the claim that a set is finite just in case its corresponding initial segment has a maximum.

The second passage provided a contrasting, purely cardinal definition of what it means to be a finite set, based upon De Morgan's syllogism of transposed quantity. According to Peirce (3.288):

> If every S is a P, and if the P's are a finite lot counting up to a number as small as the number of S's, then every P is an S. For if, in counting the P's, we begin with the S's (which are a part of them), and having counted all the S's arrive at the number n, there will remain over no P's not S's. For if there were any, the number of P's would count up to more than n. From this we deduce the validity of the following mode of inference:
>
> > Every Texan kills a Texan,
> > Nobody is killed by but one person,
> > Hence, every Texan is killed by a Texan,
>
> supposing Texans to be a finite lot. . . .

This was the first appearance in Peirce's published work of this famous definition, over which arose the controversy with Dedekind that prompted Ernst Schöder to write the essay "Über zwei Definitionen der Endlichkeit" (*Schröder 1898b*).

Note that both of the above passages imply a corresponding definition of the infinite—an infinite set is one that is *not* finite. It should be remarked, however, that these definitions are *not* equivalent. The first depends upon Peirce's system of ordinary number, while the second is purely cardinal in that it can be stated without reference to such a number system. The two properties are often referred to today as being "finite" and "Dedekind-finite" respectively. In the above passage, Peirce derived the latter from the former. It is known today that to prove their full equivalence by making the reverse derivation, as Dedekind attempted, requires the axiom of choice.[14]

§7. Conclusion

In conclusion, I shall make two modest suggestions. The first is that Peirce scholars should pay some attention to the role that axiomatization plays in Peirce's philosophy of mathematics. In volumes 3 and 4 alone of the *Collected Papers*, there are at least ten separate axiom systems for the natural numbers, presented within a variety of contexts and using different formalizations.[15] Peirce clearly thought this was important.

My second suggestion addresses the question of why Peirce's 1881 paper, clearly a landmark in the history of mathematical thought, has not been more widely acknowledged in recent scholarship. Could it be that our tendency to split the nineteenth century into two separate traditions—a tradition of algebraic logic which includes Boole, De Morgan, Jevons, Peirce, and Schröder, and a tradition of mathematical logic which includes Frege, Dedekind, Cantor, and Peano—has produced a kind of blindness when it comes to individual contributions which do not fit neatly into these traditions? Perhaps these categories have outlived their usefulness. At the very least, it seems that Peirce's work would be more accurately viewed as spanning both traditions.

N O T E S

1. Peirce's discussion of the doctrine of limits at the end of the decade shows his familiarity with the outlines of arithmatization. See (4.118n.1), and *Eisele 1957*. Although it is difficult to say just what Peirce read, he explicitly cites Lagrange's *Theorie des Fonctions* as well as works by Cauchy and Duhamel on the method of limits (6.125).

2. Such a conflation of axiomatization with logicist reduction seems to underlie the treatment by Thomas Goudge of Peirce's earlier approach to axiomatization using Boolean algebra, in his 1867 paper "Upon the Logic of Mathematics" (3.20–44). This misapprehension (*Goudge 1950:* 55–67) is discussed further in *Shields 1981:* 153–158.

3. 4.239, emphasis added. Historically there have been various other interpretations of Benjamin Peirce's definition. Hermann Weyl, for instance, understood it to mean that any distinction between logic and mathematics had been "obliterated" (*Weyl 1963:* 63). But there is no evidence that Peirce himself ever interpreted it in other than the manner indicated. He even claimed that this was the interpretation his father had originally intended: "It is evident, and I know as a fact, that he had this distinction [between logic and mathematics] in view. At the time when he thought out this definition, he, a mathematician, and I, a logician, held daily discussions about a large subject which interested us both; and he was struck, as I was, with the contrary nature of his interest and mine in the same propositions" (4.239).

4. In symbolic notation:

> simply ordered = $_{df}$ partially ordered + connected
> partially ordered = $_{df}$ transitive + antisymmetric + reflexive
> transitive = $_{df}$ $(x)(y)(z)(xRy$ & $yRz \rightarrow xRz)$
> antisymmetric = $_{df}$ $(x)(y)(xRy$ & $yRx \rightarrow x = y)$
> reflexive = $_{df}$ $(x)(xRx)$
> connected = $_{df}$ $(x)(y)(xRy \vee yRx)$

5. In the *Collected Papers* this paragraph (3.258) is confusingly cross-referenced with 3.564 which refers to the controversy with Dedekind over the cardinal definition of the infinite.

6. Peano's actual presentation was highly formal and included nine axioms, four of which are today generally subsumed by the underlying logic (*Peano 1889:* 113).

7. Gilman cites Peirce, and there is little doubt that his work was strongly influenced by Peirce. It is interesting to note that Russell directly acknowledges Peirce's discovery of irreflexive (aliorelative) relations (*Russell 1903:* 203; *1919:* 32).

8. For Dedekind, *x* is greater than or equal to *y* just in case *x* belongs to every chain to which *y* belongs (*Dedekind 1888:* 74). Peano first defines arithmetical operations (*Peano 1889:*116).

9. In an 1890 letter to H. Keferstein, Dedekind remarked that "Frege's *Begriffsschrift* and *Grundlagen der Arithmetik* came into my possession for the first time for a brief period last summer (1889), and I noted with pleasure that his way of defining the non-immediate succession of an element upon another in a sequence agrees in essence with my notion of chain . . . " (*van Heijenoort 1967:* 101).

10. It should be noted that the priority of Peirce's axiomatization was recognized in Peirce's own lifetime by the Dutch mathematician Gerrit Mannoury, in his *Methodologisches und Philosophisches zur elementar-Mathematik.* (I am indebted to Max Fisch for bringing this work to my attention.) Mannoury remarks that "A rigorous foundation for the theory of finite numbers, however, was successfully given for the first time by the American C. S. Peirce. This was reproduced in another form or independently rediscovered by others (principally Dedekind, Frege, and Peano)" (*Mannoury 1909:* 51). Evert Beth also follows the lead of his countryman (*Beth 1959:* 113, 360) but Quine, while apparently recognizing Peirce's accomplishment in his review of the *Collected Papers* for *Isis* (*Quine 1935:* 294), seems to have forgotten it in his later work (*Quine 1969:* 92), citing only the standard treatment by Wang (*Wang 1957*) which does not mention Peirce. Against these three stands a long list of scholars who have overlooked, or misunderstood, Peirce's axiomatization.

11. Fraenkel and Bar-Hillel also claim that Peirce's use was not known to either

Dedekind or Peano (*Fraenkel and Bar-Hillel 1958:* 293). The technique probably originated with Hermann Grassmann (*Grassmann 1861*).

12. Cf. 3.262 and *Dedekind 1888:* 97, 101.

13. Although it would hardly be fair to credit Peirce, in 1881, with understanding the full implications of this approach, he does return to this view later in his life (4.674) with considerably more sophistication.

14. The first complete survey of definitions of finitude in relation to the axiom of choice was *Tarski 1924.*

15. 3.20ff., 253, 562A-I; 4.110, 160ff., 188, 335, 336, 341, 606, 664ff.

4.

PEIRCE'S PHILOSOPHICAL CONCEPTION OF SETS

Randall R. Dipert

In this essay I propose to examine some of Peirce's reflections on the notion of a set. My main interest is "philosophical"—in coming to identify what notions and motivations stood at the basis of his theory of collective entities—rather than "technical" (in studying his theorems about collections, infinities, and so on). The very first thing I have to do however is to disown the title: Peirce rarely used the term 'set,' instead using 'collection' as the term for a basic collective entity (including as a translation of Cantor's '*Menge*').[1]

§1. Trouble in Paradise, and Not Only Paradoxes

Peirce was, together with Schröder, one of the major proponents of Boolean approaches to logic in the last quarter of the nineteenth century. Consequently, he could already be credited with a development of theories to deal with one kind of collective entity, "classes."[2] In the middle portion of his career (roughly, the 1880s), and quite independently of work by Frege, Dedekind, and Cantor, Peirce also developed theories of infinity, of the continuum, and of the nature of numbers, in addition to strictly logical work, such as the refinement of Boolean theories and their extension of Boolean algebras to relations and propositional logic. In the latter portion of his career (after 1890), after his retirement to Milford, Peirce became familiar with and studied carefully the works of Dedekind, Cantor, Borel, and others. He had considerable difficulty in receiving these papers in a timely manner, apparently seeking them out on trips to New York City. It is however in reaction to these papers that Peirce began to develop further his own, original theory of sets, and to criticize what was rapidly becoming the standard account of collective entities. So far as I am aware, Peirce's views have never played a major role in any major discussions of set theory as we now know it. His views on the continuum have been until recently poorly understood, and his definitions of infinity have been regarded as a curiosity.

However, except for some lucid but difficult writings on the continuum (whose perhaps clearest exposition can be found in *Peirce 1992*, the 1898 Cambridge Conference Lectures), and his letter to *Science* of 1900 on Dedekind's definition of infinity, it is the state of Peirce's available writings on the topic, and not lack of receptivity of the logical community, that explains his minute impact. Why anyone should be interested in Peirce's especially diffuse and difficult writings on this topic requires an explanation. Peirce's treatment on the topic stretches over dozens, even hundreds of places in his writing—many in fragmentary essays and notes still only in manuscript, with no single, extensive summary of his views. They require an understanding not only of set theory as we know it, but of early formulations and terminology, especially Cantor's own, and Peirce makes frequent use of his own idiosyncratic philosophical notions (e.g., secondness), medieval and distinctively Peircean notions of individuality and reality, and applies his own logical tools, now rarely encountered (e.g., the "syllogism of transposed quantity" and the notation of the theory of relatives).[3] One defense "internal" to Peirce scholarship would be that to understand his logic and later metaphysics, it is important first to understand his work on collections and especially on the continuum. This is essentially the path first taken by Murray Murphey in his still quite valuable *The Development of Peirce's Philosophy* (1961). (A more recent, and I think more successful, reconstruction using Robinson's non-standard analysis of Peirce's difficult writing on the continuum is to be found in *Putnam 1995*.) My main interest will rather be motivated by certain issues in the history of logic, set theory, and mathematics.

In spite of its formidable development in this century, its extensive use in our schools, and the widespread presumption of great usefulness, I think modern set theory remains philosophically suspect—either as a foundation for mathematics or as a theory of anything conceptually vital. One bit of evidence for this reactionary thesis is the plurality of theories; we should properly speak of set theor*ies* (Zermelo-Fränkel [ZF], with or without *Urelementen;* von Neumann-Bernays-Gödel [NBG]; Quine's New Foundations [NF]; and so on). The substantial differences among them (and with Cantor's own theory) bespeak substantive differences of what they are theories *about.* Furthermore, certain results in the twentieth century, notably the independence results for the Axiom of Choice and for the Continuum Hypothesis, as well as a variety of proposed axioms of infinity, indicate a "looseness" about what it is we take as the basic notion of set. The independence of the Axiom of Choice is I think an especially serious problem.[4] Both developments—the plurality of set theories, even within an iterative framework, and the independence results—tell us that the subject matter of these theories, ostensibly "sets," is not so clear and well-defined as it is often taken to be—at least since the hurried evacuation from the troubled paradise of naive set theory.

By an 'iterative' conception of set, I mean a theory that first stipulates the existence of certain non-problematic primitive sets (typically, only the empty set), and then stipulates the existence of sets that are the results of applying certain well-understood operations (forming a power set, union, intersection, "Separation," and so on).[5] These are the devices most conspicuously at work in ZF set theories. It is only the power-set axiom that generates the hierarchy of *types* of sets—$\{\wedge\}$, $\{\{\wedge\}\}$, and so on—that distinguish it from earlier, single-type theories of classes, such as those used by Boole and the later Booleans. Certain other axioms, such as infinity, choice, and so on, fall outside of the motivations of the iterative conception itself, but are necessary to ensure the adequacy of set theory as a foundation for mathematics. Consequently, they are, seen from the modest constructive/iterative impulses themselves, *ad hoc.*[6]

In this historico-philosophical criticism of set theory, I partly echo the views of Michael Hallett (*1984*)[7] as well as all those who have sought alternative foundations for mathematics (such as Curry, or those inspired by intuitionist, higher-order, or intensional logics)—although the motivations for these efforts are diverse and often complementary, sharing only a dissatisfaction with traditional set-theoretic foundationalism. Critiques of the very bases of set theories have of course met with resistance from non-philosophically inclined practicing mathematicians and logicians, who reply that these are details, and the basic conception must be sound, since it "works." Critics then find themselves in an uncomfortable position similar to that of Berkeley's quixotic attacks on the infinitesimal calculus.

My own view is that there were a large number of metaphysical, phenomenological, and epistemological problems with collective entities that were inadequately discussed at the time of the crucial turning point in history (roughly 1880 to 1920). It is not just the conception of infinite sets that is problematic, but the conception of any set: the notion of a set itself is fundamentally mysterious. Max Black (*1971*) offers an especially careful account, both of historical resistance to sets and of their conceptual difficulties. Cantor (*1895*, quoted and translated in *Fraenkel 1968*: 9) had written, "A set is a collection into a whole of definite, distinct objects of our intuition or of our thought" ([eine] Zusammenfassung von bestimmten wohlunterschiedenen Objecten unsrer Anschaung oder unsreres Denkens zu einem Ganzen). Both Frege and Dedekind complained that they could not decipher from this what a set was (*Hallett 1984*: 34). I will document below that

Peirce's objections to this early formulation of the concept of set, a formulation never really improved upon in our century, were extensive, and they were more detailed and subtle than Frege's and Dedekind's.

But the inadequacy of most characterizations of what precisely a set *is* deserves to be hammered home. This quotation from Felix Hausdorff's *Grundzüge der Mengenlehre* (*1914*: 1) is especially illustrative of the extraordinary lack of clarity in early descriptions of the notion of "set": "A set is a gathering together of things into a whole, that is, into a new thing." (Eine Menge ist eine Zusammenfassung von Dingen zu einem Ganzen, d.h. zu einem neuen Ding.) The definition is multiply interesting. There is a use of "*Ding*" probably derived from Dedekind, a use of whole (but Cantor's early *Einheit*, "unity," seems more perspicuous than his later term, '*Ganze*,' used by Hausdorff), a carefully chosen word for the activity of putting things together ('*Zusammenfassung*' from *Cantor 1895*) in an abstract or mental way rather than, say, the spatial organization that might be suggested by *Zusammensetzung* or *Zusammenstellung* (a connotation that is unfortunately more strongly suggested by the English 'collection,' as in coin- or stamp-collection), and a hint of a constructive/iterative conception in the view that sets are not pre-existing, but rather, we *make* them out of other things ("a *new* thing"). Hausdorff seems to have realized its inadequacy, for in the third edition (1937) he gives 'set' another equally problematic but at least more Cantorian definition: "A set is a plurality thought of as a unit," and he then comments: "If these or similar statements were set down as definitions, then it could be objected with good reason that they define *idem per idem* or even *obscurum per obscurius*. . . . In our opinion, it does not detract from the merit of Cantor's ideas that some antimonies . . . still await complete elucidation and removal."[8] Hilbert seemed to have believed that set theory especially needed a "formal" axiomatic treatment precisely because the nature of sets was so unclear: "Here (in the spirit of the axiomatic method) one understands by 'set' nothing but an object of which one knows and wants to know no more than what follows about it from the postulates," (*1925*, quoted in *Hallett 1984*: 38). For a careful discussion of Frege's and Russell's difficulties, consult *Black 1971*.

Instead of first achieving clarity and agreement about the basic *concept* of a set, logicians and mathematicians have preferred—maybe properly so—a rapid refinement and correction of theories formally as close as possible to those of Cantor, politely ignoring Cantor's own metaphysical agenda or any other broadly philosophical concerns. Where Frege, Schröder, Russell, and even Cantor and Hausdorff refused to tread, or trod only with hesitation, we have cheerfully rushed in, emboldened only by a growing tradition of lack of foundational criticism.

Returning to the late nineteenth century allows us to rethink our options, and to examine the issues at a point where discussion of "conceptual" and foundational issues withered away. Some of these issues have re-turned—stimulated by Boolos' and others' work on the iterative conception of set, for example—while others have yet to be carefully considered. Peirce's ideas are especially useful, I think, because they are the work at this important historical juncture of an extremely sophisticated mind, and one who knew the troubled historical baggage associated with collective and especially but not exclusively infinite entities.[9] Peirce also was distant enough from the site of the European discussions, and mature enough to withstand immediate acquiescence to the views of others, that he gives us a chance to look at the issues with an independence of judgment that is difficult for us now to muster.

One charge that could be made against my own unease with set theory is that I and others are being quite naive. Namely, I am presupposing that there *is* a single, best set theory, or that there is a single conception of sets in our heads, or in the world, when in fact there isn't. I am however not presupposing that there is a single true, inborn or *a priori* notion of set. I am happy for the purposes of this essay to feign total "pragmatism"—that the set theory is best, or even "true," which works the best. But, with Peirce, I would insist that this "fitness" must be conceived in a very robust way and not, say, simply for ease or neatness in organizing one branch of mathematics alone (e.g., arithmetic). The function determining fitness should include the usability of the notion in *all* of our thought, or we should clearly specify what our narrow purpose or goal is. The goal of global fitness might return us to the issues involved in Cantor's theological and idealistic views, and to the still more complicated motivations we see in Peirce.

One of the problems with the narrowly pragmatic approach to set theory is as follows. Let us just take as the foundation for set theory whatever conception of set is neatest and least problematic.[10] But then, why would we want such a set-theoretical foundation for mathematics (and the other sciences) at all? The original motivation for set theory was either one of metaphysical or epistemological foundationalism—of, for example, explaining the abstract and opaque (natural) numbers in terms of the metaphysically primitive, or in terms of the conceptually clearer and nearer. (In Peano, we see a third possible motivation. Rejecting foundationalism of a Fregean sort, he seems to have pursued set theory for its role in organizing mathematics, consolidating its notation, and facilitating future research.) Things would have been fine had naive set theory worked out. But without some sort of non-technical credentials in our common thought, or in a metaphyscial theory, a modern notion of set cannot provide a solid "foundation."

These themes are surprisingly close to methodological issues that Peirce in fact began to address. When he encountered logicism—the attempt to reduce mathematics to logic including set theory—he counterproposed with what we might term "reverse logicism": to treat logic as "rooted in" mathematics. His simple argument was that the results and techniques of mathematics (understood however as a general, rigorous theory of representations and diagrams of all sorts for necessary reasoning) were far better known and highly developed than logic, therefore it was logic which needed help from mathematics, not mathematics from logic. (See *Dipert 1995*.)

Reflecting on this view, and on the "foundational" views close to Frege's, are distinctions among the numerous dimensions—pedagogic, prudential (e.g., facilitating development), conceptual, semantics for natural language, epistemology, metaphysical, organizational, and so on—along which we might think a logic of set-theoretic reduction, foundation, or analysis is useful. These senses of "foundation" have generally been poorly distinguished. More common is the attempt to try to achieve two or more distinct such purposes with a single logic or set theory when, for example, it is not obvious that the same theory could optimally display clearly one's metaphysics, serve as a foundation for mathematics, and be an iconic and facilitating notational system for future research—roughly, Frege's perhaps not-jointly-realizable concerns.[11]

In the early 1890s, and accelerating rapidly with his slow digesting of Cantor's work in the mid-1890s, Peirce both advanced his own technical theory of collections, multitudes, infinities and the continuum[12] and integrated these discoveries into his formidable metaphysics of that period. By the time of his contribution "Symbolic Logic" to Baldwin's *Dictionary (1902)* we find him saying:

> It is, therefore, necessary to make a special study of the logical relatives "is a member of the collection _____," and "_____ is in the relation _____ to _____." The key to all that amounts to much in symbolical logic lies in the symbolization of these relations. (4.390)

§2. Peirce and Boolean Collections/Classes

Although we could treat Peirce's writing on collective entities strictly from the point of view of his own involved philosophical considerations, I think it would be most useful for us to look at Peirce's theory in the context of his reactions to the works of Dedekind and Cantor. Even before beginning this investigation, it is important to make some general remarks about later Boolean conceptions of collections, such as those we find in Peirce's writing before the influence of Cantor and Dedekind, and in Schröder *Vorlesungen (1890–1895)*. Some representations for classes had of course been used in pre-Boolean symbolic logics (such as those of Leibniz, Ploucquet, and Lambert). In the then-predominating intensional logics (to which Jevons's and Frege's logics are throwbacks), it was of no great moment whether the referent of a term was treated as a class of properties or as a conjunction of all of those properties. No one seems to have seriously contemplated—although there were problems with proper names—the possibility that a thinkable complex property was, for example, composed of an infinite number of primitive properties. And, consequently, no one seems to have sensed danger in being rather loose about whether these referents were classes, or about precisely how to describe what a class" was. (Earlier medieval, non-symbolic theories avoided classes altogether by letting terms extensionally refer [*suppositio*] individually but indiscriminately to each among possibly infinite such entities.) It is with the extensional, Boolean conceptions of classes as referents for terms, and the application of De Morgan's idea of a "universe of discourse" when classes, possibly infinite ones, first start being considered seriously in logic. There were of course parallel discussions of infinite collections, totalities, and so on occurring in geometry, the foundations of the calculus, and, later, what came to be known as analysis. These arose out of late-medieval, and even much earlier Greek discussions of space, time, and motion.

In the works of Boole and his followers through Venn, there is no evidence of any elaborate consideration of what precisely a class is. Theirs was a theory in which one talks informally about a kind of *Urelementen*, and formally about classes of these, but no further (e.g., going to classes of classes). The question of what kinds of things could constitute the members of the class, what kind of thing a class was (other than implicit definitions stipulating rules governing the class operations of multiplication, taking the complement, and so on), and rules governing what it is for an entity to "be a member of" another entity (class), were all left unaddressed. This state of affairs later brought the accusations that the Booleans had "confused" the subset relation and the element relation. (The charge originated in Padoa and Frege, and was most forcefully expressed in *Russell 1903*: 19. The hasty charge of confusion is a bit peculiar, since explicit notation for the difference was slow in coming even to Peano, Cantor, and Frege.)

Given their interest only in classes of one level, namely, what we would call sets of individuals, it is quite natural that they would not have seen the need for any elaborate machinery to describe the element-of relation. More charitably still, we could see the Boolean class logic as an anticipation of the part-whole calculus of the mid-twentieth century, a metaphysically cautious theory in which there is no distinct element-of relation, or even conceive it as anticipating the motivations for a theory such as Quine's "pragmatic" New Foundations, in which one deliberately conflates an entity and its singleton, and thereby (partly) erases a distinction, often thought to be useful, between element and subset. It is then hard to see whether one should charge the Booleans with being very old-fashioned and slow in failing to see the difference, or very modern and metaphysically circumspect in doing their best to ignore it! Norbert Wiener disputed the charge of confusion, citing Schröder's hierarchy of manifolds, but this concerns what I consider a later development, and is discussed below.[13]

In arguments on behalf of the restricted universe of discourse, De Morgan had argued that we could not use the notion of a "class of all things," primarily because he wished to use complementation, and the complements of well-behaved and defined classes might not be well-defined (*De Morgan 1850; cf. Hallett 1984*: 241). Similar reservations probably lie at the base of Peirce's and Schröder's hesitance to embrace "very large" collections and their desire to treat collections only within metaphysically homogenous and iteratively constructed domains.

Both Peirce and Schröder developed techniques within a Boolean framework, at first uninfluenced by Peano or Cantor, for dealing with some of the more complex kinds of entities evolved in "full" set theories—that is, with entities other than classes of individuals. Peirce's technique involves what he later called "logical dimension." The idea apparently first originated in O. H. Mitchell's contribution to the 1883 *Studies in Logic* (written under Peirce's tutelage), but is developed there only slightly, and was extravagantly praised by Peirce in 1883 (3.348), in 1911 (3.624), and at several other places.[14] This proposal is of "an element or respect of extension of a logical universe of such a nature that the same term which is individual in one such element of extension is not so in another" (3.624 [1911]). The idea is perhaps like the Cantor-Hausdorff notion of a plurality "thought of" or "taken" as a unity—except that Mitchell and Peirce provide further analysis of what this means. The language is hardly perspicuous, but Peirce seems to have in mind universes composed of the same basic stuff or even entities, but individuated in different ways. He gives as an example a universe containing an enduring person, as compared with one composed of what we would now call "time slices" of that person.[15]

It is not difficult to see how this can be turned into a rather clumsy and semantically based set theory: a set is construed as a single individual, in one "dimension"—i.e., under one interpretation—but as individuated by its members in another. The "element-of" relation is thus explicated in terms of a semantic relationship between what we would call two interpretations of the "same" domain. Peirce also explicitly allows that a dimension may include *possibilia* ("hypothetical states of things"). This remark seems to harbor the intention of examining modal statements, although Peirce cannot be credited with a "possible worlds semantics" since he did not develop this remark. His later language in describing a conception of sets more clearly inspired by Cantor seems to confirm this interpretation: sets are considered as "individuals" that are also considered (along a different dimension?) to be composed of individuals. Peirce's semantic theory of "element-of" relation is however not much developed beyond this, and certainly not symbolically, as far as I can see.[16]

Although Schröder rejected these techniques of Peirce and Mitchell (cf. *Schröder 1895 II*: 295, of which volume Peirce says "he is far below himself" [3.624]), he himself used a notion which also addresses collective entities composed of other than individuals (cf. *Schröder 1890 I*: 248f., and *1890–1895 II*, cf: 461), where he considers Peano's work. This feature was first noticed and discussed in *Church 1939* (although I think we can today better see Schröder's theory as prefiguring the iterative notion of a set). Schröder says that from a "pure" manifold (*reine Mannigfaltigkeit*), that is, one that contains only individuals, another manifold can be constructed that contains as its individuals all the subclasses of the first manifold—a "power set."[17] This process can be continued, and we then have a "hierarchy of the first manifold" with increasingly numerous elements. Schröder stresses, however, that we cannot mix these manifolds, and thus we cannot have a manifold whose individuals are of different "types"—a system of rigid stratification that gave Church's paper its title.[18]

One aspect of Peirce's later theory of collections, which firmly separates it from modern accounts, is his extraordinary concern first to clarify what an "individual" is. This concern is partly metaphysical (i.e., in establishing the entities which might be capable of separate "existence"), but also seems to stem from a natural need to establish the metaphysical status of entities before one addresses the status of *collections* of these entities. Schröder seems to have followed Peirce in this penchant—at least as far as any non-philosopher would care to.[19]

This obsession with specifying what counts as an individual comes as early as 1870 (3.93f.) and as late as the historical definition(s) at 3.611f. (1911). There are such a massive number of substantial discussions in all of his published and unpublished writings of what it is to be an individual that I can here only refer to several representative passages. The 1870 definition equates an individual with a logical atom, "a term not capable of logical division . . . of which every predicate may be universally affirmed or denied" (3.94 [1870]). This seems in turn to require that being an individual is relative to a set of predicates as given in a theory, or perhaps to what we would now call a model/interpretation.[20]

The first corresponds to Peirce's requirement of "being fully determined" (which, however, he distances himself from first in 1870, and more fully in 1911, moving to a definition in terms of secondness that is discussed below). The second accommodates the semantic notions of "respects" or dimensions. We can regard Peirce as dissatisfied with his own previous characterizations, since he admits in 1911 that "[A] new discussion of the matter, on a level with modern mathematical thought and with exact logic, is a desideratum" (3.612).[21]

Schröder's definition states that an individual is a domain (*Gebiet*) which (in a given theory? applied to a given universe of discourse?) has no proper subdomain, i.e., *i* is individual iff for any *x*, $x \nleqq i$ implies $x = i$ and $i \nleqq y$ for some *y*—i.e., is a "logical atom," analogous to, and what probably directly inspired, *Urelementen*.

So far as I am aware, there has been little discussion of the issue—what allows something to qualify as a possible member of a set—since the time of Peirce and Schröder (other than, say, unhelpful stipulations of the existence of at least the empty set). The other issue, of the sense in which a collection is itself an individual (in Peirce's terminology), or what it is to "consider" objects as constituting a collection, has also remained relatively unexplored outside of the phenomenological literature. Early and not very philosophically enlightening talk of "mentally gathering together," regarding, imaging, and so on (*zusammenfassen, betrachten, ansehen, anschauen, abbilden*) seems still to be with us.

§3. Peirce and Dedekind

Peirce was extremely covetous of his apparently correct claim to have been the first to give a technically adequate definition of an infinite collection (in "On the Logic of Number," published in 1881).[22]

His competition was of course Dedekind's definition in *Was sind und was sollen die Zahlen (1888)*, and Peirce's claim was reiterated in a published letter to the editor of *Science* in 1900. For Peirce, a finite collection is one in which a syllogism of transposed quantity within it is always valid, and an infinite collection is one in which some such syllogisms are not valid. In other words, Peirce's definition distinguishes finite from infinite collections according to the arguments that are valid when these collections are taken as universes of discourse. (Schröder had already argued for their equivalence in 1898 [*Schröder 1898b*], to which Peirce makes no reference. See also *Miller 1904*.)

Peirce criticized Dedekind's definition,[23] and I will argue that his criticism is extremely subtle and significant for the concept of set. Peirce's criticism is nevertheless extremely "compactly" expressed, to put the matter charitably.

In response to Dedekind's definition, Peirce argues, first, that Dedekind's notion of *echter Theil* is not defined, and that correctly defining this notion of "proper part"—what we would now call a "proper subset"—would bring one to Peirce's conception. This is at first glance more than a little puzzling, but I think Peirce's point is substantially correct. To see this we must examine Peirce's notion of a syllogism of transposed quantity. An example of a syllogism of transposed quantity is:

Everyone loves someone.
No one is loved by more than one person.
Therefore, everyone is loved (by someone).

It is not difficult to see that this necessarily holds in finite domains but can fail in infinite ones. From the hindsight of the history of set theory, it might at first be hard to see why anyone would want to define infinity in this fashion. But knowing Peirce's framework explains his desire to do so. Every collection save the empty one seems to be viewed as a potential universe of discourse, which of course links Peirce's set theory with his (Boolean/De Morganian) logic. This thesis is confirmed when we find Peirce first speculating about very large class-like entities (or entities with indistinct elements) on the order of the continua, and then find him speculating about logics that take as their universe of discourse these continua.[24]

More important, I will soon argue, Peirce seems to suggest or presuppose that all sets must be formed according to what has come to be known as the "iterative" notion of a set. In his terminology, to know a collection is infinite we must know there is a relation holding among the individual members such that: (a) the premises of the syllogism of transposed quantity using this relation ("love" in the above example) are true; and (b) a conclusion of the correct form using the relation false. This seems to require more knowledge of the individuating properties of, and relations among, the members of the set than we typically demand.[25] Although we might know there *exists* such a relation without knowing what it is, Peirce seems to additionally require ("constructively") that, when we know the set is infinite, we have identified the relation. That is, we have to know not only the property that specifies the original set (e.g., that of being a human being) but also, on Peirce's account, some knowledge of the internal structure of this set (e.g., some formal specifications about who loves whom) that go beyond the constituting property of the set. We could call this a relational principle of individuation of its members (at least relative to *this* set).

These rather strong prerequisites on the determination that a collection is infinite puts us in a good position to see why Peirce would protest that Dedekind's notion of *echter Theil* (proper subset) is undefined, and that if it were specified, would amount to his own definition. We can here try to fill in Peirce's reasoning. How do we know a set *is* a proper subset of an infinite set? We cannot merely suppose that we can take some members away, for without specifying which members and how, we cannot be sure we have an *echter Teil.*

To know that P is a proper subset of W, we must know: (i) that all the elements of P are elements of W, and that (ii) some elements of W are not elements of *P*. Observe that, for (i), to know that *P*'s elements are elements of *W*, we must know enough about all of *P*'s elements such that we can tell when they are, and when they are not, elements of *W*. But this requires only those elements in *W* that are also in *P* to have sufficient individuality or distinctness. For (i) alone we are not required to know at all about the specifications of the elements of *W not* in *P*. They might be epistemologically, or even metaphysically, indistinct. However, we must know enough about at least one element of *W* not in *P* to know that it *is not* in *P*. This is condition (ii). So, some property or relation must hold on some of the members of the collection, *W*, but not others (i.e., those in *P*), and we must know this. In other words, we are back, if we choose Dedekind's method, of requiring certain kinds of knowledge about the internal structure of the original infinite set, *W*.[26]

If I have correctly read his reservations, Peirce is dealing with issues related to choice functions and the Axiom of Choice.[27] We will see similar issues, of the distinctness of individuality of members of sets, raised in his reaction to Cantor, discussed below. I see no other reason why he would have developed and persisted with his definition of infinity, and why he would have objected to Dedekind's notion of proper subset. If I am right, he is in fact weighing into the debate on issues of enormous sophistication. Peirce's "constructivist" leanings are almost surely bound up with his philosophical conception of a collection. Namely, a collection is an *ens rationis* that is itself treated as an individual (2.324: "A collection is logically an individual"). It is composed of individuals. These individuals, *qua* individuals, must be distinguishable from one another. For Peirce, unlike his strictly mathematical colleagues, declaring something to be an individual was no lighthearted affair, but rather a matter fraught with metaphysical significance.

We can also see this conception of individual at work in his rather remarkable discussions of continua. Namely, he argues (in opposition to one of Cantor's first major papers, "Über unendliche, lineare Punktmannigfaltigkeiten") that all the determinable points on a line (a continuum) do *not* constitute a collection, and hence do not have a multitude (=cardinality).

> The explanation of their not forming a collection is that all the determinable points are not individuals, distinct, each from the rest. For individuals can only be distinct from one another in three ways: First, by acts of reaction, immediate or mediate upon one another; second, by having *per se* different qualities; and third by being in one-to-one correspondence to individuals that are distinct from one another in one of the first two ways. Now the points on a line not yet actually determined [the points not yet determined] are distinct from one another in one of the first two ways. Now the points on a line not yet actually determined [the points not yet determined] are mere potentialities, and, as such, cannot react upon one

another actually; and, *per se*, they are all exactly alike; and they cannot be in one-to-one correspondence to any collection, since the multitude of that collection would require to be [*sic*] a maximum multitude.[28] (3.568 [1900])

The invocation of "reaction" is a key element of Peirce's metaphysical/phenomenological category of "secondness." Certain non-trivial relations like "love" would apparently qualify. This is presumably in fact the only way we can conceive of infinite collections, since clearly conceiving of all of the individuals in an infinite collection as having distinct properties in a way that does not trivially derive from relations among the individuals seems highly problematic for a finite mind. I conjecture that it is for this reason—the desirability of contemplating infinite collections—that Peirce tries to reform his definition of an individual. His earlier, "monadic" definition (suggested in *1870*) would seem to require the contemplation of an infinite number of distinguishing properties (for individuals to be distinct). His definition in terms of secondness, or what we would call an ordering relation, requires only the contemplation of one relation. (Observe that he must be conceiving of the relation intensionally, as an algorithm or finite description, for we could not contemplate the relation extensionally as a class of ordered pairs; this would beg the question of the entertainability of infinite collections. This view is in turn consistent with his views about the "real" status of rules and laws.)

In this discussion of the "individuality" of members of a collection, there is enough information to guess about the way Peirce would have reacted to the Axiom of Choice. If one has a sufficiently detailed grasp of the members of a purportedly collective entity to form proper subsets, then the entity in question might be a collection.[29] If one cannot do so, then it is not a collection at all, since the members are not sufficiently discernible (by some internally discriminating relation or properties). One cannot *legislate* an ability to form proper subsets of a given purported collection. That is, of any purportedly collective entity, one cannot simply assert that one can, say, partition it. Rather, the ability to do so is a consequence of, and hence required for, its being a collection in the first place. A conception of the members of the collection as individuated entities comes first, and of the collections of such entities, the Axiom of Choice will in fact be *descriptively* true.

§4. Peirce and Cantor

A comparison of Peirce's theory and Cantor's requires first a rapid introduction to Cantor's views and terminology, with an emphasis on areas of philosophical interest and controversy. For Cantor a set (*Menge*) was a collection (*Vielheit*) that could be considered as a single entity (*Einheit*). For each collection, or at least each set, there was a measure of the "size" using the method of one-to-one relations introduced by Bolzano. This measure was called the cardinality (earlier: *Mächtigkeit*, "power") of the set, and of most interest are of course the transfinite cardinalities, denoted by the aleph series. Cantor makes the suggestion, consistent with, but not explicit in, his earliest writings, that some collections are too large to be considered as single entities, and hence that there are collections that are not sets. He describes these as "absolutely infinite" (*absolut unendliche*), or as "inconsistent" (*inkonsistente*) collections. Indeed, there is some reason to think that Cantor was aware of these unruly collective entities before the explicit discussions, begun by Russell, of what have come to be known as the "set-theoretic" paradoxes (see the hint of this in *Dauben 1988*); for this reason, perhaps, Cantor was a good deal less shaken by their appearance than was Frege. In fact, Cantor not only entertains the possibility of the existence of these very large collective entities (somewhat similar to the proper classes of NBG set theory) but also rejoices in the fact, equating them (or, if singular, it) with a Hegelian Absolute or a theological entity. The details of these philosophical issues—what few meager ones we have—are beyond the scope of this paper, and it is a good historical question whether for Cantor the theological-metaphysical notions (e.g., of a super-transfinite collective entity) came first, and then the technical details of what we know as set theory, or vice versa. (See *Hallett 1984*: xi.) Little careful work has been done on this "transcendental" aspect of Cantor as a philosopher.

Peirce's God affected his logic also, in holding him strictly to the Boolean view that every universe of discourse has at least one element, God—or even only one element. His presupposition seems to have been that there is only one necessarily existing entity. A zealous interpretation of his view, nowhere else consistently maintained that I can see, brings him to an amazing dispute with Schröder in which Peirce apparently maintains that logic can contain no expression for the fact that a class is midway in size between the empty class and the whole universe of discourse—in spite of his own existential and other quantifiers. (See *Dipert 1978*: 270ff., the Peirce-Schröder correspondence [*Houser 1990*], and 6.452f., 494.)

Although Peirce was slow in acquiring some of Cantor's papers, and in fact repeatedly and apparently seriously claimed not to understand all of them (especially material with the ordinals), he wrote fairly extensively about his understanding of Cantor's work, climaxing in elaborate drafts of letters to Cantor of 21 and 23 December 1900 (L73). The question of whether Cantor received and read a letter from Peirce, or even whether Peirce mailed them, remains a mystery.[30]

Some of their differences are merely terminological. What Cantor called a set, Peirce calls a collection. What Cantor calls the cardinality of a set, Peirce calls its "multitude." For measuring large sets, both use Bolzano's measuring tool of one-to-one functions, although Bolzano himself did not apply the technique to infinite collections. Like Cantor, and apparently also independently of knowledge of the paradoxes, Peirce argues that some purported collective entities can be verbally described but are not in fact collections. He has no generic name for such terms (unlike Cantor, for whom they are still *Vielheiten* even if they are not *Mengen*), instead indicating them by their cardinality.

Perhaps most idiosyncratically, Peirce argues that this cardinality—a greatest possible cardinality—is best described as a continuum. These noncollections, such as "all the points on a line," would have a "maximum" multitude, and since there is no maximum multitude for collections, they cannot constitute a collection. (His argument explicitly relies on a Power Set Axiom and theorem about multitudes, namely, the multitude of a collection is always exceeded by the power set of that collection.) On this issue, we have a similarity between Peirce and Cantor, and also a major difference of opinion. The similarity is that both theorize about infinite entities not subject to rules governing well-behaved collections or sets. The difference lies in the fact that for Cantor, this entity or these entities take on certain purely metaphysical or theological dimensions. (Also, Cantor does not speculate about these entities' cardinality.[31]) For Peirce, the "large" collective entity is to be identified with the continuum, that is, with what Cantorians take as a set of perfectly respectable transfinite cardinality, $\aleph_1$. For Peirce too, however, this super-transfinite entity occupies an extremely significant philosophical dimension (*Murphey 1961*). Peirce's views on the noncollective continuum and its maximum cardinality suggests a certain similarity with Julius König's attempt in 1904 to show that the power of the continuum was not an aleph (*König 1905*; see also *Moore 1982*: 86f.). Zermelo later discovered a mistake in König's proof, but the question of whether something like König's conclusion might be true endured—an effort König never abandoned.[32]

A record of Peirce's first extensive contact with Cantor's work survives in the form of his notes on Cantor's "Beitrage zur Begründung der transfiniten Mengenlehre" (*1895*). These notes are especially interesting in that his initial reaction to the definition of collection and to the individuality and distinctness of the members of collections in Cantor's theory prefigure almost all of the points he makes more extensively in the large number of writings on sets from 1900 to 1905:

> He defines a "Menge," that is, a Collection, as follows: "jede Zusammenfassung M von bestimmten wohlunterschiedenen Objecten *m* unsrer Anschauung oder unsreres Denkens zu einem Ganzen."
>   This is remarkably accurate. No need of the subjectivism of "unsrere Anschauung oder unseres Denkens." He notes that the *members* which he calls "Elemente" are *definite*, "bestimmten" but not that they are also *individual*, and that they are *independent*, "wohlunterschiedenen." But he does

not analyse that character, and say in what it consists." (MS 821 undated,
NEM 3: 1110–1113)

The most sustained remarks Peirce ever wrote on the philosophical foundations of the theory of collections are contained in the drafts of his letters
to Cantor of December 1900. The divergence there of his views on individuals and collections from earlier and later writings, as well as considerable differences among the drafts and draft fragments, indicate that these
were tentative and exploratory—even at this very late stage in Peirce's philosophical development. But the detail, care of the handwriting, and extent
of the drafts reveal, I think, the importance Peirce attached to the topic.

An individual "has no generality" and is fully determinate (*"bestimmt"*
Peirce adds for Cantor's benefit on p. 2). He later writes of a distinction
between a "perfect individual" and a "virtual individual." Realizing that our
conception of an entity is typically incomplete, a perfect individual has
properties that are completely fixed and determinate, but a virtual individual is partially undetermined in our thought—even though "something . . .
compels the facts of observation and the deductions from them to be such
as they would be if there were an individual there" (pp. 15–16). An imagined inkstand has no "self-identity" (no fixed characteristics), since one can
imagine it tall or short at will. An individual does have self-identity (some
fixed characteristics?). He then writes of a case, that of a triangle, in which
"individuality is still weaker" (than the inkstand); thus even virtual individuality comes in grades.

This discussion, and long discussions of reality, firstness, secondness,
and so on, are certainly not ideally directed toward the non-philosopher
Cantor and seem to have more to do with the "reality" of entities than with
their individuality; the two are connected, but the latter question emphasizes the way in which an entity is distinguished from others. If there is a
philosophical failure in the tardy reflections of Peirce on the nature of collections, it is that he seems oblivious to the niceties of metaphysical categories. He seems to assume—and this is explicit in the drafts to Cantor—that
the issue of the reality of an entity is precisely the same as the way it is
individuated from other entities. Reality, individuation, spatio-temporal (or
other) boundaries, and countability are I think instead distinct.[33] We can
know something exists without being able to say precisely where it ends and
another starts, or the way it is individuated from other things. We can know
it *is* individuated, without knowing what the criterion is, and hence without
being able to count the members of a set. Or, for example, we can know
there are five things in a box without knowing how to tell one from another.

What appears to be a second draft (pp. 20f.) distinguishes between
primitive and derivative individuals. An individual is a "subject of which
every predicate is either universally true or universally false." A primitive

individual is one "having characters which it might, without contradiction, be supposed to possess although there were no regularities among other things, except certain regularities presupposed to render knowledge possible." The idea is that a primitive individual is metaphysically independent of other individuals.[34]

A primitive individual had apparently been called in the first draft (p.1) an "independent" individual: subjects are independent "if it is logically possible for one to have any character while another has any character, provided these two characters are not themselves, either directly or indirectly, in any logical relation which prevents one from being affirmed (or denied) while the other is affirmed or while it is denied." Peirce here seems to have in mind relationships, e.g., "Nor does it prevent A and B from being independent that both cannot be the sole thing adored by C." Peirce does not seem to have specifically anticipated the possible dependence of an entity on the set containing it, but rather views the condition in terms of non-logical relations.

A derivative individual is an individual "which has no characters except such as truly consist in certain regularities among other things, beyond such regularities as are requisite to cognition. Thus, the German People is a[n] [derivative] individual of which nothing whatever is true except in virtue of facts concerning individual men, their country, etc." Similarly for a line regarded as an "aggregation" of points. A collection then is such a derivative individual. Independently of the issue of derivative and independent individuals, Peirce could here be accused of conflating aggregations and collections. But he does not say that a collection is defined simply as a derivative individual. While both are derivative individuals, the difference between a collection and an aggregation may still lie in our conception in the case of a collection of the individuality of its components. Primitiveness, like independence, is not absolute, since Peirce speaks later of "relatively primitive" individuals. Thus, apparently, $\{A\}$ is primitive relative to $\{\{A\}\}$, but not to A alone.

Two individuals are "coordinate" if they are of the same metaphysical type or kind. Two points are coordinate, but a point and a line are not; a material thing and a relation are not coordinate, nor is a movable object and its position, and so on. His definition of metaphysical/logical similarity is Aristotelian: "there is nothing in their essential nature to prevent one from taking any character, or predicate, which the other can take" (draft 1, p.2). In draft 2 he calls this relation of metaphysical type "subjects of the same category" and defines it similarly (p. 21). His theory of metaphysical types has great importance, since collections can be formed only with members of the same metaphysical type. In this, he was perhaps consciously or unconsciously echoing Schröder's type theory. Neither in Peirce nor in Schröder were there concerns about avoiding the paradoxes: their pre-paradoxical concerns seem to have been metaphysical or phenomenological. Although Peirce does not use the example here,[35] he clearly regarded an individual and a singleton containing this individual as not coordinate. Thus legitimate collections might be: {A, B}, {{A}, {B}}, and so on, but not {A, {A}}—i.e., collections with members of different metaphysical type. Although paradox avoidance is apparently not his explicit purpose, this anticipation of type theory is even clearer and more dramatic than Schröder's.

Collections are defined (draft 2, p. 21) as "a derivative individual which possesses no characters except such as consist in certain characters of certain individuals, distinct from, of the same category with, and more or less independent of, one another, and that in one or other or both of the following ways. . . . " He then allows that a collection may be formed by (i) specifying the characters of its "relatively primitive individual" members ("ineunts")—i.e., a kind of set abstraction—or, more remarkably, (ii) "the characters of the collection may consist in the possession of certain characters by others of the individuals of the same category as its ineunts, these being termed its exeunts." He seems to assume that the collection of all individuals of a given metaphysical category may always be formed, and that the characterization of subsets is then accomplished directly (by characterizing the ineunts) or indirectly (by characterizing the exeunts). The first idea suggests rather strongly then the notion of Separation in ZF set theory.[36]

In these drafts, properties of sets are then examined and some of Cantor's own results are duplicated by ordering relations on members of the set, as in the published letter about Dedekind from nine months earlier. These reflections on the notion of individual and collection were used extensively in letters and essays between 1900 and 1905.[37]

Although Peirce's work on the philosophical foundations of a theory of collections is neither complete nor in all places clear, I regard it as extremely sophisticated and as anticipating many of the foundational and philosophical issues that were later considered or still remain ahead of us. For Dedekind, Cantor, and the other set-theoretic extensionalists (see *Hallett 1984*: 299–300), the problem concerns what it is to think of a disparate collection as a unity (Cantor: *Einheit*; Dedekind: *Ding*). For the intensionalists, Bolzano and Frege, this is no problem; it is just to have a concept. But for the intensionalists, the problem is the one of what sense this entity is "collective"—in what sense one thereby conceives of (all?) the objects "falling under" the concept. For extensionalists, it is the conception of the collective entity that is problematic, for the intensionalists it is the conception of the "members."[38]

In a deeper and historically richer theory of "thing" and "individuality" than Frege and Cantor seemed willing or able to utilize—and in the *ad hoc* if at least honest manoeuvre of two "logical dimensions"—Peirce, I believe, offered us hints of a way to understand collective entities. At the very least, he saw the problem.

§5. Conclusion

I conclude by making two small observations, which indicate Peirce's sophistication. First, Peirce established, in an 1897 paper only slightly influenced by Cantor's work, that the multitude (cardinality) of the power set of a set always exceeds the multitude of that set. He uses this fact to show that there is no maximum multitude (for if there were, we could form its power set, which would have a larger multitude, contrary to our assumption). He assumes that the power set always is a set—i.e., the Power Set Axiom. In most places he assumes there is a denumerable hierarchy of transfinite multitudes, corresponding precisely to Cantor's $\aleph$ series (see 3.631) [1902]). But in one of his last papers, he voices this speculation:

> After it [the first infinite multitude, corresponding to Cantor's $\aleph_0$] follow one by one an endless series of abnumerable multitudes. Yet as far as I know (I am not acquainted with the work of Borel, of which I have only quite vaguely heard), it has never been exactly proved that there are no multitudes between two successive abnumerable multitudes, nor, which is more important, that there is no multitude greater than all the abnumerable multitudes. (4.656 [c. 1909])

The first issue is a clear statement of the Continuum Hypothesis. It is unclear whether Peirce was aware at this point of the discussions of Hilbert, Zermelo, and König on the topic. The second issue concerns "higher order" axioms of infinity.

In the same essay in (4.650 [c.1909]), we find a long argument against conflating the individual, Julius Caesar, and the collection whose sole member is Julius Caesar (see also MS 469, which discussed the same point):

> Some writers whose logical conceptions would seem to be in a state of disintegration have supposed the collection whose sole member is Gaius Julius Caesar to be identical with Gaius Julius Caesar himself—a strange confusion considering that the latter was a man of immense force of intellect who was brought into the world by a grossly unskilled operation of surgery, while the other is nothing but an *ens rationis* brought into being by the idea of that man without any surgery at all.

We can take this passage, alternatively, as a conclusive refutation of the view that all Booleans always confused the subset and member-of relation, or we can take it as the very first critique of Quine's New Foundations. Observe that there is a philosophical point to this passage, which might appear to be a joke except for Peirce's general lack of humor: individuals (Julius Caesar and the singleton Julius Caesar) are distinguished by having different properties.

I have merely dabbled in the work that remains to be done on Peirce's conception of set. Scattered throughout his large lectures, fragmentary essays, and correspondence from 1890 to 1905 are a striking number of reflections on sets and on the continuum. It also has to be admitted that Peirce's work on sets has none of the developed character of the work of Cantor, Frege, and their followers. No single line of thought is developed in a rigorous, symbolized manner over many pages. (Nor, I think, do Peirce's technical observations on transfinite sets depart in major ways from the Cantorian school; the exception here is Peirce's treatment of the continuum.) Instead, what one sees in Peirce's work, and especially in his criticism of Cantor and Dedekind, are sustained, historically informed, and philosophical probing examinations of notions of distinctness and individuality, and of the notion of a collective entity itself. Since remarkably little has been done in this area since the first useful discussions of collective entities began, it is remarkable and not a little unflattering that Peirce's observations fully retain the insightful character and depth that they had a hundred years ago. Much work of a historical nature remains to be done on figures such as Peirce and König, who early expressed strong reservations about set theory (quite independently of the razzle-dazzle of the paradoxes). Much philosophical work also remains to be done on what precisely a collective entity is, and on what characteristics its elements must have.[39]

N O T E S

1. For rare uses of the term 'set,' see MS 37b ("On the Numbers of Forms of Sets"); MS 246 (2 January 1889); and the letter to William James reprinted in NEM 3: 820, in which the expression 'primitive set' occurs. For the use of 'collection' as Peirce's translation of Cantor's *Menge*, see the draft letters to Cantor (L73/NEM 3: 767ff.) discussed below; the draft definition for *Baldwin's Dictionary* in MS 1147/NEM 3: 1116–1117; the etymology of 'collection' in MS 169/NEM 1: 124–126; the letter to Frankland of 8 May 1906 (L148/NEM 3:785): "I translate Cantor's *Menge* by 'collection' "; the discussion of terminology in the letter to Moore (L299/NEM 3: 915); and many other places in his writings.

2. Exactly how Boole, De Morgan, and others conceived of the referents of their (algebraic) terms is however a complex matter. In his *Mathematical Analysis of Logic* (*1847*), Boole seems to demur from calling them classes, instead referring to the referents of letters as "elective operations"—e.g., something like functions or operations on collective entities. A great deal of the lack of clarity in nineteenth century logic is attributable to the understandable (for the times) lack of a clear theory of semantics for the theory of models, interpretations, and so on which would have forced clarity on the exact nature of the objects in the domain of intended interpretation.

3. For the beginning of a formal, symbolized, and axiomatic account of Peircean set theory in the notation of his logic of relatives, see MS 26/NEM 3: 64ff.), "On Multitude," presumably from the late 1890s. It uses a definition of 'collection' (roughly as that which is the extension of a predicate) not later repeated in the more unified series of discussions on collections following the draft letters to Cantor of 1900.

4. See the remarks on A. Mostowski and J. Dieudonné quoted in *Moore 1982*: 4.

5. See M. Hallett's survey of the various iterative notions, from ZF itself, to works of Schoenfield, Boolos, Wang, and Scott in *Hallett 1984*: 214–223.

6. We could also distinguish between the "constructive" components of the iterative conception, such as the existence axioms I have cited, and the negative, prohibitory constraints imposed by an axiom such as that of *Fundierung*.

7. Cf. *Dauben 1988*, which led me to Hallet's book after most of this paper was completed.

8. "Eine Menge entsteht durch Zusammenfassung von Einzeldingen zu einem Ganzen. Eine Menge ist eine Vielheit, als Eihneit gedacht. Wenn diese oder ähnliche Sätze Definitionen sein wollten, so würde man mit Recht einwenden daß sie idem per idem oder gar obsurum per obscurius definieren."

9. I have in mind his familiarity with the medieval nominalist/realist disputes and his reservations about infinite entities in Berkeley, Gauss, and others. This is a historical perspective that I suspect Russell, Frege, Cantor, et. al. lacked.

10. But see *Dipert 1982* on the difficulty of doing this for a conceptual entity so important as that of "ordered pair."

11. The difficulty of many and conflicting purposes in modern symbolic logic is addressed in *Dipert 1991*.

12. For the importance and novelty of Peirce's research on the continuum and infinitesimals, see *Putnam 1995*.

13. See *Wiener 1913* and *Grattan-Guinness 1975*.

14. See MS 789, where 'universe of discourse' is defined in such a way as to permit propositions to refer to several such universes. For a discussion of Mitchell's work, see *Dipert 1994*.

15. But see 4.647f. [c.1909], where he seems to oppose precisely this technique of time slices; the entire topic of dimensions seems to be foreshadowed—thus my reluctance to give credit to Mitchell—as early as 1870 (3.93–94), where Peirce talks about what is "one in number from a particular point of view."

16. But see also MS 640 and MS 27 on "first and second intentional collections." I develop the symbolism a bit more in *Dipert 1978*: 327f. and perhaps, I suggest there, it could also be made to handle intensional aspects of "guises" of entities. See also *Dipert 1990*.

17. Observe that what counts as an individual does seem to shift, as in the Peirce-Mitchell system. Either Schröder did not understand Mitchell's point, or Peirce presented his and Mitchell's (in 1902) in a way that became precisely Schröder's view. It is interesting that, so far as I am aware, Peirce never shows an awareness of Schröder's notion of a hierarchy.

18. Peirce at this time (but see below) did not seem to impose any similar condition, and the motivations for Schröder's condition are mysterious. The matter is further discussed, although not always adequately I think now, in *Dipert 1978*: 224f., 246f.

19. Cf. Peirce's praise at 3.612 of Schröder's discussion in *Schröder 1890–1895 III*, lecture 10—which indeed is masterful, I think.

20. In note 8 of the 1870 paper, Peirce writes that an "absolute individual" can

not be realized in sense or thought, nor can it exist. He seems to have in mind an individual specified in all its properties (not just some contemplated ones); and that any individual is subject to change. It is perhaps less clear why every individual must be *thought of* as changing, although it may indeed change in virtue of "now being thought of" (but is at least absolute during the time it is thought of as fixed). He then formulates a nice paradox on the problem.

21. In 3.96 (1870), Peirce in fact gave a formal characterization of what it is to be an individual, but never referred to it again. He wrote (formula 96) that if X is an individual, then $y^X = yX$, which is apparently a second-order formula indicating that if X is individual then, for any dyadic relation Y, the class of things that are in the Y-relation to all X's is identical to the class of things in the Y-relation to [some/an] X.

22. Cf. *Moore 1982*: 25n.11, which says "This definition is found (in 1881, 3.288) in embryo, but not clearly formulated until 1885, 3.402." I do not see any unclarity in the 1881 formulation, or any substantive difference. I suspect Moore disqualified Peirce's 1881 prize because it was not there symbolized. See *Moore 1982*: 24–25 for a discussion of Dedekind's and Cantor's definitions.

23. Peirce misstates and possibly misunderstands Dedekind's definition— "every echter Theil is similar to the whole collection" (3.564), whereas the definition in modern formulation is only that *at least one* proper subset is equinumerous with the whole set.

24. 2.339 (c.1902): "The universe of a logical subject has always hitherto been assumed to be a discrete collection, so that the subject is an individual object or occasion. But in truth a universe may be continuous, so that there is no part of it of which every thing must be either wholly true or wholly false. . . . But the logic of continuous universes awaits investigation. . . . " This position is also related to Peirce's "logic of vagueness."

25. Compare Frege's similar view discussed in *Black 1971*.

26. It might at first seem that the required knowledge of the internal structure of the set is weaker for Dedekind's method than for Peirce's. But from the partition required for Dedekind's method my guess is that we can generate a relation of the sort Peirce needs; it is easier to see that from the knowledge required for Peirce's method, we can generate a clearly *echter Theil*. In the example above, either the beloved or the unbeloved will constitute a proper subset equinumerous with the whole infinite super-set.

27. But see his remarks at 4.99 (1893), where he states that "no contradiction can emerge" from the assumption that the members of an infinite set cannot be ordered. As *Moore 1982*: 49 notes, this view must have changed by 1897, when he offers a proof of Trichotomy. My explanation is that he had come in this later period to the conscious view that a Choice/Ordering principle was true of collections by their very nature.

28. The third way of establishing individuality is surely unnecessary, since we could not establish that there was a one-to-one correspondence with another collection if we could not distinguish the individuals by one of the first methods. This seems to be an instance of Peirce's notorious tendency to formulate alternatives in triads.

29. I say "might be" because the individuality and distinctness of the members with respect to one another is at most a necessary condition for their forming a well-defined collective entity. Another necessary condition would involve distinguishing this collective entity from others.

30. J. Dauben informs me that there is no record of these letters in the Cantor

Nachlass in Göttingen (although we know the correspondence to be incomplete). Peirce's drafts and false starts have an oddly tentative and unusually humble ring, and yet Peirce did not normally draft letters he did not send. Furthermore, there is independent evidence of some communication between Peirce and Cantor: in a letter of 31 december 1903 to E. H. Moore (L 299/NEM 3: 923), Peirce writes: "There is a paper of Cantor's [in the *Zeitschrift für philosophischen* [*sic*] *Kritik*] that I have never read (although he sent it to me). . . . "

31. This reluctance and a penchant toward using the singular suggest an affinity with the Kantian *Dinge[e] an sich,* although Cantor apparently has no difficulty in calling the entity a *Vielheit.*

32. Compare König's related views in these passages: That the word 'set' is being used indiscriminately for completely different notions and that this is the source of the apparent paradoxes. . . . I do not want to represent any of this as something new" and on the concluding page, "the second number class cannot be considered to be a complete set, that is, totality of *well-distinguished* elements that are altogether conceptually distinct" (emphasis mine) in his 1905 essay (*van Heijenoort 1967*: 145–149).

33. This view was once impressed on me by H. N. Castañeda.

34. This condition seems troublesome, since there are *always* entities of which an individual is not independent: namely, the collections having that "primitive" individual as a member! The ability to form such dependent collections from more independent elements may, however, be part of what it takes "to make knowledge possible" (in Peirce's words).

35. But he does at p. 22: "The collection which as *A* as (only) ineunt is no more to be confounded with *A* than with *B*," and see below as well.

36. Compare the slightly different definition of collection in draft 1, pp. 3f.

37. "A *collection* is an individual object concerning which no facts whatever are true except such the truth of which consists in the truth of facts concerning other individual objects independent of the collection and of each other; and these objects are called the members of the collection" ("The Theory of Multitude," MS 14/NEM 3: 1056 [dated by R. Robin as c. 1903, but which I would date as shortly after the 1900 draft letters to Cantor, because of the use of the terms 'ineunt' and 'exeunt']).

A collection is a species of abstraction. "A collection is a substance whose existence consists in the existence of certain other things called its members. ¶An abstraction is a substance whose existence consists in something being true of something else" (Lectures on Pragmatism II, [1903] MS 302/NEM 4: 164).

"A Collection is anything whose being consists in the existence of whatever there may exist that has any one qualtiy; and if such thing or thing exist, the collection is a single thing whose existence consists of all those very things. According to this definition a *collection* is an ens rationis" (Lowell Lecture III [1903], MS 459/NEM 3: 353).

Peirce complains that Cantor gave no definition of collection (*Menge*) but used a "false method of definition much affected by Kant"; and in a later letter ([1904], NEM 3: 923) he wrote: "Cantor thinks he defines a collection; but he (like most Germans) does not know what a definition is." (See also MS 27/NEM 3:1069–1071 criticizing Cantor's definition presumably from the same period. At the end of this short essay, Peirce notes "I have seldom met with a conception more difficult of logical analysis than that of a collection . . . [i]n a more general sense, a collection is simply an individual object whose being consists in the being of whatever objects there may be of a certain general description, these objects being called its *members,* so that every proposition concerning the collection as subject is equivalent to some relative proposition concerning the members as subject.") Peirce's complaint is that Cantor describes the psychological experiment by which we can come to think of collections, but does not give the criterion by which it can be truthfully predicated of entities. "In a wider sense, a collection is an individual object such that the truth of any proposition asserting any predicate of it consists in the truth of a relative predicate of certain individuals called its members, or, what is the same thing consists in the truth of a proposition all the blanks in whose rheme are filled by proper names or designations of certain individuals called the members of the collection." Later (NEM 3: 916) Peirce questions whether this is part of the *definition* of 'collection': "For it does not usually belong to a definition to state the conditions of the *existence* of the definitum . . . [on] the other hand, not merely the *word* collection, but the *thing itself* is a creation of thought" (Letter to E. H. Moore [2 January 1904]), L 299/NEM 3: 900–901).

38. See *Dipert 1990* for a discussion of individuals in extensional and intensional logic.

39. I thank Nathan Houser for renewing my interest in the topic of Peirce on sets, for focusing my attention on Peirce's letters to Cantor, and for numerous comments on sources and on my interpretation of them. For other helpful comments, I thank Ivor Grattan-Guinness, Gregory H. Moore, and my colleagues Morton S. Schagrin and Kenneth L. Lucey.

5.

PEIRCE'S PRE-LOGISTIC ACCOUNT OF MATHEMATICS

Angus Kerr-Lawson

§1

Charles Peirce's work on the philosophy of mathematics largely preceded the great foundational struggles of the early twentieth century. Of course, Peirce did comment on some of the issues which constituted these struggles, for instance, the claim of Dedekind and Frege that mathematics can be reduced to logic. These issues have been ably treated by Stephen Levy (*1982a*). The aim of this paper is rather different, and more speculative. I wish to bring out a number of similarities between Peirce's views and the philosophical position which has emerged among mathematicians following the failure of each of the three best known accounts of the foundations of mathematics.

Today there is impatience among many mathematicians with theoretical questions about knowledge and truth. Although these question were urgent in the era when paradoxes had introduced confusion into actual mathematical practice, they have since become dormant, if not entirely answered. The epistemology of mathematics is not seen to be a pressing problem among most of today's practicing mathematicians, for reasons which would probably be shared by Peirce. A comment: when I refer in general terms to today's mathematicians, I am not working from any empirical data other than that of experience; rather I express a personal view, which I believe is widely shared by other mathematicians.

One source of this similarity between Peirce and today's mathematicians is a certain dual view of both truth and existence, a view at odds with the prevailing opinion of philosophers. I believe that this shared position is a viable one and it offers reasonable answers to some questions often aired in philosophical debate about mathematics. I shall deal with one such question: if we assume that mathematical truth is to be construed in a realist sense as the truth about objects, and that mathematical knowledge is to be understood in a causal or partly causal sense, then it becomes difficult to explain mathematical truth. Mathematicians seem to many philosophers to embrace an untenable combination, of Platonism about a remote mathematical existence, with an axiomatic or combinatorial approach to mathematical knowledge. They want to have their cake and eat it too. I shall argue that mathematicians, however little they may be concerned with philosophical issues, have a valid response available to them, based upon a rather different understanding of existence and truth, and shared in large part by Peirce.

§2

Peirce was never happy with the logicist thesis of Dedekind. But the development of this thesis by Frege, Russell, and others had as its upshot not a reduction of mathematics to logic but a demonstration that the universe of sets is a rich enough structure to contain all (or almost all) areas of mathematics within it as suitably defined sub-parts. Peirce might have been a good deal more happy with this consequence. For him, mathematical theory describes hypothetical states of affairs which might or might not find actual realization in the world. He would not interpret a theory, after the current fashion, in terms of relations on a universe of sets. However, this set-theoretic interpretation of a language might be seen as an abstract version of the hypothetical states of affairs he describes.

Mathematicians commonly see themselves as providing a vast array of precisely defined structures, a resource from which scientists may draw any that prove useful as models of real situations. Which structures are in fact selected is not seen as their concern; applied mathematicians will deal with ideal mathematical models in which physical situations find a structural realization, while pure mathematicians tend to adopt an esthetic ideal wholly independent of applications. It is indeed true, traditionally, that most interesting mathematical notions have arisen from practical concerns. However, this is no longer so obvious today; and anyway, pure mathematics is seen to be entirely free-standing and autonomous, whatever its origins. Moreover, although they freely use terms like 'existence,' 'truth,' and 'knowledge,' mathematicians remain for the most part clearly aware that these terms are not being used in the sense proper to natural science. I shall focus upon this usage and this awareness, which were certainly shared by Peirce: "Mathematics studies what is and what is not logically possible, without making itself responsible for its actual existence. Philosophy is *positive science*, in the sense of discovering what really is true" (1.184).

Real truth and *actual* existence, then, come into play with the question of whether or not mathematical structures are actually realized, apart from the formal context; formal truths and the existence of the entities of pure mathematics have a secondary status. This position I shall refer to as a *bi-categorial* view, meaning by this a metaphysical stance which assigns a priority to the existence of actual things and to truths about these. Ideal objects, such as the logical possibilities of mathematics, do not thereby become non-entities; but they have a different ontological status, and truths about them are not sufficiently distinguished merely by characterizing them as *a priori*.

Certain further positions are characteristic of this bicategorial view of mathematics, in the sense that they become natural or inevitable in the light of the double level assumed. Various problems which might be keenly felt by members of the empiricist school are not seen as serious difficulties. One example is the problem of explaining in some plausible manner how mathematics can take on such importance in physical theory, a problem supposed to be especially difficult for those who grant to mathematical truth some sort of special non-empirical status. Given the bicategorial position, however, it is difficult to see this as a major problem: with the vast range of structural possibilities exposed through mathematics, there should be no surprise when physical events turn out to conform to certain of these. A second example, pertinent to the discussion below and heartily endorsed by most practicing mathematicians, is the relaxed attitude toward the epistemology of mathematics characteristic of the bicategorial view. The question whether or not some mathematical structure is realized as the form for actual things and events is a scientific one; and the reliability of answers to this question raises difficult epistemological issues. Such issues, however, do not arise in the same form for mathematical knowledge (if indeed one uses the word 'knowledge' in this context), which consists in a careful specification of this mathematical structure and a logical exploration of the properties determined by this specification. Another example, closely related to the last, is a liberal approach to the ontology of mathematics; given that no mathematical entities are existences in the fullest sense, there seems little point in restricting oneself to sets in formulating a mathematical ontology, or in adopting an invidious attitude toward any mathematical entities at all.

It is the bicategorial view, moreover, which permits these last two examples, a relaxed epistemology and a liberal ontology, to obtain simultaneously. We turn to this issue next.

§3

A standard epistemological problem is to reconcile the knowledge we have of the world with the realist view of what might be the truth about that world. Similar problems have received special attention in the case of mathematical knowledge, from Benacerraf and from several others after him (*Benacerraf 1983*). It is suggested that the situation is even more difficult with mathematics, when we construe knowledge as a relation tied to causality, because of the remoteness of mathematical objects from any sort of causal connections. The contention is that the "standard" realist account of truth is connected to a Platonist view of mathematical objects, and that there is a serious problem, given Platonism, of explaining mathematical knowledge: what causal relation can there be with eternal objects lying outside the space-time continuum?

The view that mathematics presents especially difficult problems for knowledge was certainly not shared by Peirce, who applied the term 'perfect' to mathematical knowledge. (He did not thereby make far-reaching claims that this knowledge is certain or *a priori*.) And I think that today's mathematicians would be puzzled by the suggestion that something about causality interferes with their knowledge of mathematical truths. Even though they often have Platonist leanings of various kinds, their attitude toward realism has little effect on the widely agreed method of confirming mathematical truths, which depends upon axiomatic formulation into a clearly defined theory. It is understood that objects under discussion will be completely specified by this formulation, along with possible refinements in the logic and mode of presentation being employed. This should lead to agreement on what is known about a single structure or to an understanding that two different structures are under consideration.

Consider Gödel, frequently cited as the classic case of a Platonist, in the light of his statement that the truth of the axioms of set theory "forces itself upon us," and similar well-known phrases. It is nevertheless clear that Gödel seeks only a suitable addition to the axioms of set theory, which would settle the continuum hypothesis in a generally acceptable manner. Had this occurred, he would have taken this enhanced theory of sets to be entirely satisfactory from an epistemological point of view. With Gödel, as certainly with others less inclined to Platonism, what is taken as mathematical knowledge has to do with finding a consistent axiomatic presentation which exposes the bare bones of some mathematical structure.

With set theory, the quest is for a fresh axiom. However, as Gödel himself has shown so forcefully, a mathematical structure need not be completely specifiable by a set of axioms. What is important, in general, is that some perfectly clear specification be given, in whatever form, so long as it accords with mathematical norms. How many of the properties of the structure become known depends on the deductions that can be made from these assumptions. This may remain a mystery, but there should be none about the way the structure is defined. With the investigations of physics, no such specification is possible, and the theories dealt with are always tentative descriptions of external events.

In posing this alleged difficulty in the philosophy of mathematics, Benacerraf makes quite explicit the assumption that both knowledge and truth receive a uniform treatment, covering both natural and formal science. He takes this assumption to be called for; yet the bicategorial view is precisely its denial. Mathematicians have been accused of being careless on foundational issues when they accept some measure of Platonism without any reckoning of the attendant epistemological problems. Such a critique, however, often imposes monocategorial presuppositions, which are justified in ways unacceptable to these mathematicians. We turn now to a consideration of this monocategorial view.

§4

According to Peirce, mathematics does not deal with matters of fact:

> For all modern mathematicians agree with Plato and Aristotle that mathematics deals exclusively with hypothetical states of things, and asserts no matters of fact whatever; and further, that it is thus alone that the necessity of its conclusions is to be explained. This is the true essence of mathematics. (4.232)

In many respects, this characterization resembles that of the logical positivists, who explain the *a priori* nature of mathematical truths by holding them to be analytic and empty of empirical content. In spite of these similarities, however, it is some of the marked differences which will concern us here: (1) Peirce does not take mathematical statements to be empty of content, even though they are hypothetical; mathematical truths are not true merely because of linguistic usage or convention. (2) The positivists, like Kant, were eager to explain the *a priori* nature of mathematical statements. For Peirce, the above characterization explains their necessity; but for the most part, he discusses the *a priori* only when criticizing improper utilization of *a priori* methods in science. (3) Peirce's position has its proper setting in the bicategorial view, something alien to logical empiricism, and to the entire empiricist tradition.

In Quine's empiricist attack on logical empiricism, interestingly, two of these differences Peirce has with empiricism disappear. Quine attacks the linguistic *a priori*, thereby siding with Peirce both in rejecting linguistic conventionalism and in downplaying the *a priori*. It is the third difference—the bicategorial—which legitimates Peirce's view that mathematical knowledge differs markedly from empirical knowledge; and it is Quine's strong rejection of any hint of the bicategorial which underlies his view that mathematical and even logical claims call for empirical justification. Peirce says:

> The certainty of pure mathematics . . . is due to the circumstance that it relates to objects which are the creations of our own minds. (5.166)

> The assurance of the mathematician is due to his reasoning only concern-

ing hypothetical conditions, so that his results have the generality of his
conditions. (5.8)

Led originally by Quine, the attack on analyticity and any account of
a priori truth grounded on linguistic convention is itself firmly based on a
unique ontological category of existence and on a linguistic criterion for
existence in this univocal sense. Unlike logical empiricism, where ontology
is little discussed, Quine's writings militantly advance a monotone ontology
of existence resting on the quantifications of language. Thus both the logical empiricists and their empiricist critics stand opposed to the bicategorial
view, and they tend to see this view as an aberration brought on by the lack
of philosophical training.

Quine expresses what might be called the classical objections to any
distinction between two ontological categories, such as being and existence
(*1963*). The argument rests on the belief, false as far as the mathematical
bicategorial view is concerned, that the source of the distinction lies in the
ancient problem of talking about non-existent things, and that Russell's descriptions offer the obvious solution. While this argument is special in its
assumptions, it is typical of a barrage of other empiricist arguments in its
method, which relies completely on a linguistic formulation of the problem
and its solution.

Consider Quine's argument, intended to refute the notion that observation of nature is relevant only to questions of existence in space-time
(*Quine 1963:* 3). The following example, he holds, is a refutation: consider
whether or not the ratio of the number of centaurs to the number of unicorns exists. As a ratio, it would be an abstract entity, if it existed. However,
we can only determine its failure to exist by means of a fruitless search for
centaurs and (especially) unicorns in space and time. In this treatment,
everything depends upon the linguistic description used to single out the
number. However, if we set aside this description as extraneous, what is at
stake is the purely formal question of the ratio $0/0$, a question about formal
existence having a non-empirical negative answer.

Similar considerations apply to the number 9, and Quine's designation
of it as "the number of planets." The bicategorial view would accept the
essentialist implications of a favoritism toward some of the traits of the number 9, but would not feel bound to carry this essentialism over to the objects
of space-time.

Those who embrace the bicategorial doctrine—most mathematicians,
in my view—are little influenced in their ontological judgments by questions of language. If we were to be guided by the language we use, we would
probably opt for a single existential category to represent the universe of
discourse and a single notion of truth applying to any and all propositions
in our language. With the focus on the philosophy of language, it is natural

to treat these uniformly; however, there is little reason for mathematicians to be swayed by this consideration.

§5

I believe, then, that a weak form of Platonism can be adopted for mathematical structure in the context of the bicategorial view. It must be made clear, however, that the Platonic structures are not posited to give intuitive knowledge of some mysterious realm, or indeed to serve any epistemic purpose. Rather they are posited to account for the shared and eternal nature of transparent mathematical objects, something not accomplished satisfactorily by literalistic accounts dealing with marks on paper or ideas in the head.

The kind of weak Platonism I am suggesting could be called structuralism. However, when it is said that mathematicians study structure, we need not take this to mean a study of structures in the set-theoretic sense, where a structure consists of a universal set and relations assigned to the non-logical constants of some language. Other theories entirely different from set theory, for instance, that of categories, are sufficiently rich to allow all of mathematics to be carried out within them. 'Structure' can equally well refer to the structure of categories, with no connection to sets. And with categories, there is not the temptation to treat them as concrete individuals, as there is with sets of concrete individuals. Mathematical categories are more obviously universals, belonging in a different *ontological* category from particles and forces.

Essential to the bicategorial view, as we have seen, is the uniform treatment given to any and all mathematical universals. The question of which set-theoretic definition of the integers is correct, Zermelo's or von Neumann's, will not be taken seriously by most mathematicians. It is with questions of mathematical existence, where the division becomes most acute between mathematicians and empiricist philosophers of mathematics. I know that a split ontology seems outlandish to many philosophers. But the tables can be turned, with the monocategorial view having consequences no less outlandish to mathematicians. What sense is there to the question "Which comes first, sets or categories?" Both after all give us the same mathematics. Parenthetically, the structures do turn out to be different, in some quite non-trivial ways—all of which encourages the liberal approach of contemporary mathematicians, who find and welcome more than one global approach to their subject.

Thus mathematicians tend to be mystified when asked by philosophers which mathematical entities exist and which do not. They do not comprehend the urge for excluding some in favor of others. This exclusive stance has its origins in empiricism, with its horror of any separation between real things and ideal objects, and with its imposition on mathematical objects of the parsimony proper to the granting of existence to physical objects.

Peirce's view—that mathematics deals with necessary reasonings about hypothetical objects and calls for a modal treatment—is not formulated in such a way as to command much support among mathematicians. On this point, his work is closer to that of some recent philosophers: I have in mind Hilary Putnam and his paper "Mathematics Without Foundations" (*1983*). As modal logic is understood today, questions of epistemology are shut off from questions of necessity and possibility, especially as formulated by Kripke. With a view such as this, mathematics is freed from the need for an epistemological foundation.

Most mathematicians, I think, achieve a similar freedom, but in a rather different way. They have been largely untouched by the latter-day surge of interest in modal logic, perhaps differing little from Russell's negative opinions on the subject; and the ontology of possible objects is not a congenial backdrop from which to view their subject. They may adopt a weak demythologized Platonism in terms of structure, rather than with possible objects. Perhaps conceptual structures better reflect the feeling of mathematicians of a complete detachment of their field from questions of existence in the world, something not felt by Peirce. Nevertheless, Peirce clearly held a more detached view of mathematics than others of his time. For example, consider Russell's worries, so antiquated today, about whether the world is infinite, which he considered crucial to his mathematical doctrines, and Peirce's attack on him on this point.

There remains, however, a similarity between accepting modal possibilities and accepting ideal structures, in the sense that both can inform concrete realizations in the realm of actuality. Both offer possible forms which may or may not arise in things; whether they do or not is then a question which touches epistemology.

6.

PEIRCE'S THEOREMIC/COROLLARIAL DISTINCTION AND THE INTERCONNECTIONS BETWEEN MATHEMATICS AND LOGIC

Stephen H. Levy

§1. *Introduction*

Peirce defined mathematics as "the science which draws necessary conclusions" and logic as "the science *of* drawing necessary conclusions." Later, he made his "first real discovery" in mathematical method of two types of necessary reasoning—theoremic[1] and corollarial. In this paper, I refine the distinction and show that it illuminates not only the nature of mathematical reasoning, but the relationship between mathematics and logic as well.

The next section of the paper considers Peirce's stated view on the interconnections between mathematics and logic, and shows that when Peirce focused on other logical/mathematical problems, he revealed another view more in the spirit of his overall philosophy and closer to the truth. The following section analyzes Peirce's theoremic/corollarial distinction and shows how, in accord with this implicit view, some mathematical theorems depend on logical axioms in an important (Peircean) sense of the term. In the last section, we see how Peirce's distinction illuminates the traditional distinction between analytic and synthetic propositions and corrects the view that all mathematical propositions are analytic. In addition, we see that Peirce's distinction supports the contemporary view that logic and mathematics are interdependent.

§2. *The Interdependencies of Mathematics and Logic*

Peirce, following his father's definition (*B. Peirce 1870*), often defined mathematics as "the science which draws necessary conclusions" (3.558). In contrast, Peirce defined logic, or more precisely *deductive* logic, as "the science of *drawing* necessary conclusions" (4.239, his emphasis). Peirce acknowledged the closeness of the definitions and pointed out much that is common in the two subjects. For instance, logicians and mathematicians both use observation, manipulate diagrams, and engage in theoremic and corollarial reasoning, the topic of our next section.[2]

Despite these common features, differences abound between the two subjects. Peirce emphasized that mathematics can draw necessary conclusions because it deals solely with "hypothetical states of things" (4.233). He wrote that

> a "hypothesis" is a proposition imagined to be strictly true of an ideal state of things. In this sense, it is only about hypotheses that necessary reasoning has any application; for, in regard to the real world, we have no right to presume that any given intelligible proposition is true in absolute strictness. (3.558)

Since mathematics requires hypotheses, it would seem that the mathematicians must create the hypotheses as well as reason from them, as indeed they do. Put another way, since mathematics would not exist without its hypotheses, it seems that hypothesis creation should be included in the definition of mathematics. Moreover, Peirce's account of theoremic reasoning, which we will discuss, reveals that hypothesis or axiom formation is central to mathematical reasoning. Yet Peirce often hesitated to include it in the definition and only rarely acknowledged that hypothesis creation belonged there at all. On one occasion when he did acknowledge it, however, he let the emphasis fall on the hypothesis formation, reworked his definition, and wrote:

> it is an error to make mathematics consist exclusively in the tracing out of necessary consequences. For the framing of the hypothesis of the two-way spread of imaginary quantity, and the hypothesis of Riemann surfaces were certainly mathematical achievements.
>
> Mathematics is, therefore, the study of the substance of hypotheses, or mental creations, with a view to the drawing of necessary conclusions. (NEM 4: 268)

Apparently, Peirce sometimes hesitated to acknowledge that hypothesis formation was an essential part of mathematics because, as I show later, it conflicted with one of his generally favored doctrines about the relationship between mathematics and logic, namely the total independence of mathematics from logic.

In contrast with mathematics, which deals with hypothetical states of things, logic deals with the "real world." It is an "experiential, or positive, science . . . [that] rests upon a part of our experience that is common to all men . . . " (7.524) and one that "is categorical in its assertions" (4.240). Some examples of such categorical assertions in logic would be statements that any argument of a particular form, say modus ponens:

$$(P \prec Q)$$
$$P$$

Hence, Q
is valid and that any statement of a particular form, say contraposition:

$$(S \prec P) \prec (\sim P \prec \sim S) \qquad (3.196)$$

is logically true.

Another difference between the two subjects is that whereas the mathematician draws inferences, the logician analyzes them into their most fundamental parts. So

> [t]he mathematician wants to reach the conclusion, and his interest in the process is merely as a means to reach similar conclusions; the logician wishes to make each smallest step of the process stand out distinctly so that its nature may be understood. (4.533)

To promote this understanding of reasoning's elementary steps, you must analyze the nature of certain concepts involved in these reasoning and from which they often spring. Peirce offers several such analyses, including those of set, number, and infinitesimal (cf. *Levy 1982a, 1982b, 1983, 1986, 1991*). Besides illuminating these concepts themselves, Peirce's analyses also reveal insights into his understanding of the relations between mathematics and logic, as will soon be apparent.

Peirce pointed out that logic depends on mathematics. This is indeed evident, for example, from the frequent use of mathematical induction—an essential mathematical technique that Peirce preferred to call Fermatian inference, in honor of its discoverer—to prove theorems in mathematical logic. However, the contrary, Peirce argued, was false: mathematics does not depend "in any way upon logic" (4.228). By this he meant that whatever logic mathematicians use is unconscious or intuitive, their *logica utens*. They need not appeal consciously to the results of the *science* of deductive logic, what he called their *logica docens*. So he wrote in 1894:

> The mathematician . . . needs no theory of reasoning, because no difficulties arise in mathematics which require a theory of reasoning for their resolution. . . . All the special sciences . . . repose on metaphysics, and therefore, . . . some of them require a theory of logic. But pure mathematics can postpone such a theory. (NEM 4: 98)

Peirce continually stressed this view. In a paper written in 1902 he wrote "Mathematical reasoning is so much more evident that it is possible to render any doctrine of logic proper—without just such reasoning—that an appeal in mathematics to logic could only embroil a situation" (4.243). In these passages and others, Peirce seems to reject *any* sense or thesis of dependence of mathematics on logic. Yet, despite such passages, Peirce's discussions on other topics in logic and mathematics reveal an entirely different position—that in many cases, mathematics depends quite essentially on logic in a variety of ways.

At issue here are several senses of dependence, logic, and mathematics, and several variants of the thesis corresponding to them. Peirce himself shifts between more than one of these. We'll enumerate the more important variants as we proceed.

There are numerous passages (more perhaps than those in which he states his official view) where Peirce suggests, or outright declares, that mathematics depends on logic. In large part, these occur where he analyzes the nature of sets (collections), number, multitude, and infinitesimals. Some passages suggest merely that logic is like a good pair of glasses which enhances your vision but without which you could still find your way. In this weak vein, Peirce wrote,

> such a subject as the theory of functions is simply sown with logical pitfalls.
> . . . [I]f Cauchy and others had just devoted a few days to the study of the
> logic of relatives as I would expound it . . . their minds would have been
> far clearer and they would have advanced more rapidly. (NEM 3: 949)

But eventually, culminating with the theory of limits as developed by Weierstrass, the theory of functions did reach a polished state even without the "study of the logic of relatives." So logic was not necessary for this line of mathematical development. This line of thought leads us to the

> *Weak ML Dependency Thesis:* Mathematics would advance more easily if
> logic came to its aid, but that without logic, mathematics would eventually
> get to where it is going, anyway.

Variants of this thesis turn on what results of mathematics we focus on—discovery of concepts, theorems, rules of inference, etc.

In other passages, Peirce suggests the stronger claim that mathematics most definitely requires logic for its corpus of results to be what it is and will be. Without it, mathematics would be very different from what it is. In this stronger vein, Peirce wrote to Cantor that

> a student of such subjects as yours is much more a logician than a mathe-
> matician; that is to say, however essential to your work it is to deduce the
> consequences of exact hypotheses (which is mathematical business) . . . ,
> yet still more it belongs to you to *frame* those exact hypotheses by the analy-
> sis of unclear notions; and that makes you a logician of our school of Exact
> Logic. (NEM 3: 769)

This passage plainly implies two things. First, that the analysis of unclear mathematical notions and the associated framing of hypotheses about these notions is the work of the logician. It is *logical* analysis. This is also what generates *logica docens*, the theory of logic—its results, hypotheses, products, and methods. Second, those very same hypotheses are the stuff from which the mathematicians draw their necessary conclusions (at least in this branch of mathematics), and without which the mathematicians' particular results would be impossible, or at least quite different from what they are. So in this realm—Cantor's theory of multitude, in Peirce's terminology— mathematics relies or depends quite directly on the results or products of "Exact logic"—i.e., *logica docens*, the science, not the instinct, of logic.

This line of thought embodies the

> *Strong ML Dependency Thesis*: The results of mathematics would be quite different from what they are, if the results of logic were different from what they are.

The passage cited suggests at least two variants of this thesis—namely, that mathematical *concepts* depend on some logical concepts and that mathematical *theorems* depend on some logical axioms. We can further refine these variants by replacing *depend on* with *definable in terms of* in the first case, and with *derivable from* in the second. We then have the variants that mathematical concepts are definable in terms of some logical concepts and that mathematical theorems are derivable from some logical axioms.

Note that these theses are close to, but very different from, the two central theses of Whitehead and Russell's logicism—that *all* mathematical concepts are definable in terms of logical concepts *exclusively*, and that *all* mathematical theorems are derivable from logical axioms *exclusively*. Peirce consistently argued that logicism was misguided (see *Levy 1982a, b*). No, what Peirce is suggesting in the previous and following passages are two less sweeping dependency theses:

> *Definability ML Dependence Thesis*: *Some* mathematical concepts are definable in terms of *some* logical concepts (perhaps in conjunction with some mathematical concepts as well).

> *Derivability or Theorem ML Dependency Thesis*: *Some* mathematical theorems are derivable from *some* logical axioms (perhaps in conjunction with some mathematical axioms as well).

In a paper he presented before a mathematical society, Peirce wrote in the same spirit "It appears to me that the great difficulty under which the doctrine of multitude labors is that its implicit hypotheses have never been explicitly stated; and there is great *logical* difficulty in stating them. Thus, Cantor has never defined a collection (Menge)" (NEM 3: 1069, emphasis added).

It is the task of *logic* to clarify and define the hypotheses or axioms of this branch of *mathematics*, the "doctrine of multitude" or set (collection, *Menge*) theory. To indicate that this is not simply a slip of the pen, a paragraph later in the same paper, he says, "I have seldom met with a conception more difficult of *logical* analysis than that of a collection" (NEM 3: 1070, emphasis added). So the problem of analyzing or defining a collection is a logical one. It consists in articulating the hypotheses or axioms of set theory (an enterprise to which Peirce contributed and which would accelerate in the years following his writing of these words), and analyzing—if not defining—the concepts comprising the axioms.

In these passages, however, it is not always clear whether Peirce thought that the concepts (such as set or collection) in these axioms were in fact *logical* concepts. It is conceivable that they might sometimes be *mathematical* concepts to which logical analysis must be applied. We'll analyze the distinction in the next section of the paper, and conclude that the notion of set is, indeed, in part a logical concept. However, we should still distinguish another dependency thesis conceptually, if not actually, distinct from the Definability and Derivability Theses above. This is the

> *Analysis ML Dependency Thesis:* Some mathematical results are the product of (depend on) the process of logical, not mathematical, analysis.

The strong version of this thesis would add the clause that the mathematical results in question could not be obtained otherwise—i.e., they would not exist if the logical analysis were not performed. Note that this is exactly what Peirce suggests in the next-to-last passage quoted: As logical analysis discovers new set theoretic axioms, new mathematical theorems that could not be proved otherwise will be discovered and proved. Historically, this is indeed what has occurred.

The notion of infinitesimals is another case in point where, Peirce noted, the development of mathematics depends on logic. The dependence here is both indirect and direct. Peirce's theory of infinitesimals depends on logic indirectly, because it depends on—is derivable from—his and Cantor's theory of infinite multitude (see *Levy 1991*). But as we noted a moment ago, these in turn depend on logic in that analyzing the fundamental notion of a set requires *logical* analysis (Analysis Thesis) and the notion of a set itself is analyzable in (some) logical terms (Definability Thesis). In addition, Peirce's theory of infinitesimals depends on logic directly in that, for instance, the nature of equality between infinitesimal and other quantities requires logical analysis (Analysis Thesis). In an article in the *Century Dictionary*, Peirce stated the point thus:

> If a is a finite quantity, and i is an infinitesimal, we always assume $a + i = a$, a fundamental property of the infinitesimal calculus; but whether this is so because the infinitesimal is a fictitious quantity strictly 0, or because equality is used in a generalized sense in which this is true, is a question of *logic*, concerning which mathematicians are not agreed. ("Infinitesimals," *Century Dictionary* [1889], emphasis added)

Incidentally, Peirce's logical analysis of this mathematical question foreshadowed Abraham Robinson's. Both thinkers took the second path suggested here—that "equality is used in a generalized sense." Whereas Peirce called the two quantities *perequal* or *adequal*, Robinson later described them as being *infinitely close.*

In extending his analysis of infinitesimals, Peirce was often seeking a definition of *continuity*, a central concept in his philosophic thought and one which "supposes infinitesimal quantities" (6.125). He maintained that a logical analysis of continuity would promote the development of the then fledgling branch of mathematics, topology, or mathematical topics: "In order to set it [topology] upon a proper basis, it seems first necessary to produce a workable definition of continuity, an extremely difficult task of *logic*" (NEM 4: xxiii, emphasis added). Or again, he wrote:

> If we define a continuum as that every part of which can be divided into any multitude of parts whatsoever—or if we replace this by an equivalent definition in purely *logical* terms—we find it lends itself at once to mathematical demonstrations, and enables us to work with ease in topical geometry. (3.569, emphasis added)

If a logical definition would enable us to "work with ease," could we work at all in topology without one? If the claim is that topology could still advance without a logical definition of continuity (though more slowly and with greater difficulty), then we have the weak versions of the Definability and Derivability Theses. If, on the other hand, the claim is that topology *required* a logical definition of continuity to develop as it has, then we have the strong versions. The use of "necessary" in the first passage here suggests the strong versions, but the "seems" suggests a cautious turn toward the weak versions.

At the turn of the century, the set-theoretic paradoxes were discovered in the span of a few short years. Peirce apparently became aware of Cantor's paradox of cardinal numbers in the late 1890s. In a paper of 1899, he wrote, "the subject of multitude requires great logical caution, and leads to a paradox which cannot be satisfactorily treated without a thorough logical preparation. For this reason, I have been obliged to begin with a section of logic" (3.569). So it is *logical* analysis that will find a way out of the paradoxes at the foundation of set theory and mathematics generally (Analysis Thesis). Peirce himself did such analysis and recognized that the notions of "all multitudes" (4.652) and "all possible collections" (4.253) were inherently absurd or self-contradictory. The first is so because the supposition that there is a set of all multitudes conflicts with Cantor's result that there is no maximal multitude. The second is so because it falsely presumes the universal set of all sets to be a member of itself, whereas Peirce's constructive notion of set requires all members of a collection (set) to exist before the collection itself can exist. Rejecting the two absurd ideas puts Peirce on the track of what became one common solution (that of NBG, the set theory of von

Neumann, Bernays, and Gödel) to Cantor's paradox as well as others, namely, that the cardinal numbers and the universal class are, in contemporary terminology, *proper* classes and so cannot be members of other classes in turn (see, for example, *Mendelson 1964: 170, 184*). It is this and allied logical analyses that enable set theory and mathematics to be built free of obvious contradiction—a clear dependence of mathematics on logic (Strong Definability and Derivability Theses).

Finally, Peirce's discussions of one of his favourite forms of inference—De Morgan's syllogism of transposed quantity—reflect the same view. One example of the inference Peirce offered was

Every Texan kills a Texan.
Nobody is killed by more than one Texan.
Hence, Every Texan is killed by a Texan. (3.288)

To generalize, we can express the inference as follows:

Everything R's something.
Nothing is R'd by more than one thing.
Hence, everything is R'd by something.

Now, some analyses of this form of inference are examples of logic. In fact, Peirce remarked in 1893 that De Morgan's discovery and analysis of it "constitute one of his claims to be considered the greatest of all formal logicians . . . " (4.103). Following the view Peirce professed, one might take this to be an inference form that mathematicians and others use instinctively (with their *logica utens*). If so, perhaps mathematics could advance without any logical analysis of the syllogism of transposed quantity. However, Peirce pointed out what De Morgan and others had missed—that this form of inference applies only to *finite* sets, not infinite ones: "it was not until my paper of 1881, 'On the Logic of Number,' that the limitation of its validity to finite collections was first noticed . . . " (NEM 3: 338).

In noting this *logical point* that *logica utens* overlooked, Peirce himself (through an instance of *logica docens*) promoted the development of *mathematics* by warning against the misguided application of the syllogism of transposed quantity to infinite sets. True, this does not appear to be an essential dependence of mathematics on logic because it seems only to have prevented mistaken developments and not necessarily to have promoted positive ones. Nonetheless, other rules of inference, proof techniques, or methods have promoted genuine positive developments in mathematics. We have previously mentioned proof by mathematical induction (Peirce's Fermatian inference) as an example of the way, to the contrary, logic depends on mathematics. But, in the next section, we shall note how refinements of mathematical induction illustrate as well the dependence of mathematical proofs on *logica docens*.

This line of thought leads us to another dependency thesis:

Proof Technique Dependency Thesis: Some mathematical theorems have been proved using a proof technique, rule of inference, or other method that has been discovered, developed, or refined in logic.

In this section, we have seen several passages in Peirce's work that establish that, despite his oft-stated position that mathematics does not depend on logic, on another level he maintained to the contrary that mathematics does indeed depend on insights and results of logic. To reconcile Peirce's views on this matter, it is natural to suppose either (i) that Peirce changed his view on the question or (ii) that he consistently denied some dependency theses, but not others: When he denied that mathematics depended on logic, he was thinking of the theses he denied; when he suggested otherwise, he was thinking of the different dependency theses he embraced, or at least had not rejected.

Attractive as these suppositions may be, they do not hold up to scrutiny. As for (i), if there had been a change in view, the writing espousing the one view would (in great measure, if not completely) precede those espousing the other. But the dates associated with both points of view are intermixed—the writings in the 1890s and the first decade of the twentieth century reveal both positions. Further, if there were a change, there seems to be no word as to when it might have occurred, or why. Presumably, if Peirce had changed his view, he would have noted the fact since he had emphasized the view of mathematical *in*dependence of logic so vigorously on several occasions and since he often commented on his previous writings.

As for (ii), in most cases, Peirce simply denies *all* dependency theses at once. Earlier, we cited two passages with the sweeping denial of any form of dependency of mathematics on logic (*logica docens*), namely, 4.243 and NEM 4: 98. At the start of the latter paper, Peirce declared, "It does not seem to me that mathematics depends in *any* way upon logic. It reasons, of course. But if the mathematician ever hesitates or errs in his reasoning, logic cannot come to his aid" (4.228, my emphasis). So it does not appear that Peirce intends to deny only some dependency theses, while accepting or remaining neutral on others. He denies them all.

In a rare instance, however, Peirce admits that there is a dependency lurking in mathematics. To make it out, we must note—as Peirce does—the familiar but often forgotten distinction between a *process* and its *results*. The process of mathematics consists of, among other things, the investigation, reasoning, or research *leading* to the invention or discovery of an axiom, theorem, or proof (sample *results*); and the subsequent verification and *acceptance* of it.

Peirce admits that the first process—the reasoning leading to the discovery of the result—may indeed depend on logic. In this passage, however, apparently written in 1902, he has not yet come to acknowledge that this process is genuine mathematics. Instead, he writes, "But the process of invention of the proof is not . . . mathematical. . . . It is, in fact, a piece of probable reasoning in regard to which a good logical methodeutic may be a great aid . . . " (NEM 4: 46).

However, the second type of process—the review, verification, and resultant acceptance of the proof—Peirce continues, does not depend on logic.

> The theorem which was not evident before the proof was apprehended, now becomes itself entirely evident, in view of the proof. Such reasoning . . . is not itself amenable to logic for any justification; and although logic may aid in the discovery of the proof, yet its result is tested in another way. (NEM 4: 46)

If we alter this view slightly, as Peirce later came to do, and take the invention of a theorem that depends on following logical principles of some kind to be an instance of *mathematics*, we then have a *process* version of our previous Derivability or Theorem ML Dependency Thesis, which referred simply to *results* (theorems and axioms). This version, stated broadly to include either type of process and additional features Peirce often noted as part of mathematical reasoning, is the

> *Process Derivability ML Dependency Thesis:* The process of observation, experimentation, discovery, verification, etc. of proofs of mathematics depends on following some principles of *logica docens*.

So this dependency thesis may provide a rare case in which Peirce seemed not simply to suggest, but *almost* to recognize and affirm, a dependency of mathematics on logic.

In analogy with this Process Dependency Thesis, we can generate other process dependency theses corresponding to the previous results dependency theses. These would include, for instance, a

> *Process Definition ML Dependency Theses:* The process of observation, experimentation, discovery, verification, etc. of definitions of mathematics depends on following some principles of *logica docens*,

and a

> *Process Conjecture ML Dependency Thesis:* The process of forming a mathematical conjecture before its proof is discovered (if ever) depends on following some principles of *logica docens*.

What logical principles these might be, and which Peirce had in mind, merit further investigation. Finally, for each of these process dependency theses, in turn, we can distinguish a strong and weak version of the thesis.

Note the differences between the current Process Thesis(es) and the

Table 1. Several Theses on the Dependency between
Mathematics and Logic and Peirce Passages Supporting Them

ML Dependency Thesis	*Supporting Passages*	
	Weak Thesis	*Strong* Thesis
Definition Dependency	NEM 3: 949, 4: xxiii	NEM 3: 769, 1069–70, 4: xxiii; *Century Dictionary* "Infinitesimals"
Derivation Dependency	NEM 3: 949	3.569, NEM 3: 769
Proof Technique Dependency	4.103, NEM 3: 338	
Analysis Dependency		NEM 3: 1069; *Century Dictionary* "Infinitesimals"
Process Derivability Dependency	NEM 4: 46	NEM 4: 46

previously distinguished Analysis ML Dependency Thesis. They both deal with processes as well as results, but in different ways. The Analysis Thesis has the *logical* process as the *second* component of the dependence relation, while the Process Theses have the *mathematical* process as the *first* component. Further, since a proof *result* codifies a mathematical *process* and a proof technique is a result of *logica docens*, the Proof Technique ML Dependency Thesis seems to be a consequence of the Process Derivability ML Dependency Thesis.

By way of summary, Table I lists the ML dependency theses we have distinguished and the cited passages in Peirce's work that suggest support for either the weak or the strong version of the thesis. Some passages support both the weak and strong versions because they lend themselves to both interpretations.

§3. *The Theoremic/Corollarial Distinction*

In this section, I analyze Peirce's conception of the distinction between theoremic and corollarial reasoning. We will see that the connection between that distinction and the dependencies of mathematics on logic is twofold: Some theoremic ideas in mathematics are logical, and some logical analyses engender mathematical theoremic ideas.

Peirce attached great importance to his distinction between theoremic and corollarial reasoning, at one point calling it his "first real discovery" (NEM 4: 49). (See *Ketner 1985, Hintikka 1980,* and *Zeman 1986.*) By corollarial reasoning, Peirce did not mean simply reasoning that produces the corollaries in Euclidean geometry and other branches of mathematics, although "the corollaries affixed to the propositions of Euclid are usually arguments of . . . [this] . . . kind . . . " (NEM 4: 49). No, under Peirce's conception, many *theorems* of mathematics, in fact, exemplify corollarial reasoning, for "corollarial reasoning . . . consists merely in carefully taking account of the definitions of the terms occurring in the thesis to be proved" (NEM 4: 8). This "carefully taking account" requires that you follow the principles of logic. As he put it in the *Monist* in 1908, corollaries, the results of corollarial reasoning, are "deducible from their premises by the general principles of logic" (4.613). As we will note, for the most part these principles are known and used through one's *logica utens.*

As an example of corollarial reasoning, Peirce offers a proof of the principle of associativity of addition in a paper apparently written in 1901. Using our contemporary notation for quantification, we can state this principle as

$$(x)(y)(z)((x+y)+z = x+(y+z))$$

Now, the definition of addition Peirce uses in his proof is recursive. It consists of three propositions:

$$0+0 = 0$$
$$GM+N = G(M+N)$$
$$M+GN = G(M+N)$$

where *G* means "the number next greater than *N*" (NEM 4: 2), i.e., its *successor* in contemporary terms. The chief principle of logic used in the proof is mathematical induction (Fermatian Inference): "whatever is true of *zero* and which if true of any number, *N*, is also true of *GN* the ordinal number next greater than *N*, is true of all numbers" (NEM 4: 2, Peirce's emphasis). Peirce applies this principle twice in his proof, once to show that

$$(x+0)+z = x+(0+z)$$

and again to show that, given this equation as the first part of the inductive proof, the conclusion follows. In both applications of induction, he repeatedly uses the recursive definition of addition. So although the proof is corollarial, it is not—unlike typical instances of corollaries—a simple, obvious derivation from another theorem.

There is another point of interest in Peirce's corollarial proof of the associativity of addition. The proof uses a definition of a concept (addition) already present in the original statement of the theorem. It does not introduce further definitions of concepts that arise in the definition itself. In the next section, we will see an example of a corollarial proof that, in contrast, introduces new definitions based on the definitions in preceding steps in the proof. This class of corollarial proofs requires that you even more "carefully tak[e] account of the definitions of the terms" in the theorem.

Unlike corollarial reasoning, theoremic reasoning, according to Peirce, requires the introduction of a new or "foreign idea, using it, and finally deducing a conclusion from which it is eliminated" (NEM 4: 42). By a *foreign* idea, Peirce understands "something not implied at all in the conceptions so far gained, which neither the definition of the object of research nor anything yet known about could of themselves suggest, although they give room for it" (NEM 4: 49).

Peirce offers several examples of theoremic reasoning. Perhaps the simplest examples occur in geometry, where "subsidiary lines are drawn . . . " (4.233), that is, "subsidiary lines or surfaces, not mentioned either in the proposition to be proved nor in previously proved propositions." (NEM 3: 172) There are, of course, numerous examples of this in Euclid, such as the theorem that in a triangle having two equal angles, the sides opposite them are equal (I, 6), and the familiar theorem that the sum of the angles of a triangle equals two right angles (I, 32). In the first, a subsidiary line is drawn that is equal to the length of the opposite side, and in the other, a subsidiary line is drawn parallel to one side of the triangle. In each case, the idea of the subsidiary line does not occur in the statement of the theorem.

Perhaps Peirce's favorite example of theoremic reasoning stems from Cantor's set theory. It is his proof that "every multitude is less than a multitude." In his proof, he speaks of an arbitrary set X, whose members are x's. Peirce is not always clear as to what idea in the proof is the theoremic idea, but at times it seems to be what Peirce calls the collection of all "possible collections of x's," or what mathematicians have since come to call the *power set* of X. Although Peirce does not state the point, concomitant with this new idea is the Power Set Axiom, which asserts the existence of the power set of any set. Both are new to the hypothetical state of things described in the theorem and are essential to the proof.

The proof is indirect—it assumes that there is a one-to-one correspondence between the elements of X and those of the power set of X. In this proof, still another candidate for the theoremic idea is that of the set whose introduction quickly leads to the contradiction concluding the proof. Where r is the assumed one-to-one correspondence between the elements of X and its subsets (the possible collections of X's), Peirce describes the crucial set as

a collection of X's which is not the sole collection of X's that is r to any X. Namely, this collection shall include every X (if there be any) of which

a collection of X's not containing it is the sole r, and it shall exclude every
X (if there be any) of which a collection containing it is the sole r. . . .
(NEM 4: 6)

In contemporary dress, Halmos calls r a function f, names the set A, and
describes it more clearly, thus:

Write $A = \{x \in X : x \notin f(x)\}$; in words, A consists of those elements of X
that are not contained in the corresponding set. (*Halmos 1960:* 93)

By asking of the object in X corresponding to set A whether or not it is a
member of A, we quickly generate the contradiction.

Perhaps this example illustrates the use of more than one theoremic
idea—the idea of the power set, the Power Set Axiom, the hypothesis that
a one-to-one correspondence exists between the members of a set and those
of its power set, and the set of all members and their corresponding sets.
In each case, the idea does not occur in the original statement of the theorem; nor does it appear simply by defining the notions occurring within it;
nor, finally, does it appear simply by following the principles of logic.

Still other examples of theoremic reasoning that Peirce might have
noted as such are Fermat's introduction of mathematical induction and
Lobatchevsky's and Riemann's introductions of alternatives to Euclid's fifth
postulate. Fermat's idea originally spawned the proofs of literally hundreds
of theorems in number theory; and in this century it inspired the proofs
of numerous theorems in mathematical logic and computer science. And,
of course, Lobatchevsky's and Riemann's postulates generated the proofs
of a vast wealth of theorems in the non-Euclidean geometries.

There is an important distinction to be drawn between two classes of
theoremic reasonings, a distinction that illuminates the connections and
contrasts between logic and mathematics. Peirce gives some examples that
illustrate the distinction, although he does not explicitly draw it. At one
point, his account seemed even to deny the distinction, but in fact it does
not. Thus, a few paragraphs back, we saw Peirce writing that the theoremic
idea was "not implied at all in the conceptions so far gained . . . " (NEM 4:
49). Peirce uses the term "implied" here to mean *suggested* or *mentioned explicitly*. This is evident from the rest of the sentence (cited earlier) and from
the next sentence as well. The rest of the sentence speaks of what "the definition of the object . . . [or] anything yet known . . . suggests." And the next
sentence states, "Euclid . . . will add lines to his diagram which are *not* at all
required or *suggested* by any previous proposition, and which the conclusion
that he reaches by this means *says nothing about*" (NEM 4: 49, emphasis
added).

The theoremic ideas are not suggested or mentioned, then, in the statement of the theorem under consideration or in theorems previously proved

in the theory at hand. In some cases, the theoremic idea is nonetheless logically *implied* by the axioms of the theory. For instance, in the geometric examples where a construction line or plane is the theoremic idea, the axioms or hypotheses of the theory imply its existence. In particular, the Euclidean axiom or postulate, "Two points determine a straight line,"[3] and the existence of any two unspecified pertinent points (among an uncountable infinitude of points) jointly imply that a subsidiary straight line exists. In fact, the axioms of the theory (and the existence of points) imply the existence of an uncountable infinitude of such lines, none of which are explicitly cited in the proof. (The construction line drawn in Euclid and the textbooks signifies, of course, any one of these lines satisfying the stated criteria.)[4] This is so regardless of whether or not Euclid or anyone else chooses to focus on any of these lines in the course of the proof. So although the theoremic idea is not suggested or mentioned by the current theorem or by previously proved theorems or axioms, for this class of theoremic ideas, the theoremic idea—or rather a proposition about it—is logically implied by the axioms and is derivable from them.

On the other hand, in the proof that "there is no greatest multitude," the idea of the power set of a set, or all possible collections of a given collection, is *not* one that was logically implied by the axioms of set theory as understood when Peirce and Cantor first formulated the argument. The formulation of the argument provided a threefold theoremic insight: first, to think of the idea of the power set at all; second, to formulate the axiom associated with the idea, here the Power Set Axiom itself: "For any set, its power set exists"; and third, to realize how it can aid the reasoning—how the conclusion is now implied by the newly enhanced set of axioms, whereas it was not implied by the original set of axioms (that without the axiom capturing the theoremic idea). Without this additional axiom, the theorem is not derivable. Note that this kind of tripartite theoremic idea comprises the "framing of hypotheses" or axioms that mathematics requires and that Peirce finally included in his definition of mathematics.

Fermat's invention of mathematical induction is another example of this type of theoremic reasoning. It, too, constitutes an introduction of an entirely new axiom into the subject, not a consequence derivable from previous axioms. To summarize, these examples mark a crucial distinction between theoremic ideas that are implied by and derivable from previous axioms of the subject, and those that constitute new axioms themselves.

But this is not all. There is a further distinction to be drawn, at least among those theoremic reasonings that introduce new axioms. To make it out, we must keep in mind that for the moment we are talking about the *process* or *actions* of mathematics and logic addressed at the end of Section 2 of this paper (as opposed to their products or results). In this light, some theoremic introductions of new axioms into mathematics will not simply be acts or processes of *mathematics*—they will *at the same time* be acts of *logic*. Granted, the distinction between the two subjects is difficult to draw. But we have seen illustrations of the Analysis Dependency Thesis that some advances in mathematics depend on investigations of logical analysis, where logical analysis means the analysis of unclear fundamental notions (collection or set, for example) and/or the analysis of the often hidden axioms, steps, rules, or proof techniques justifying an inference.

Now, we have acknowledged that, *a priori*, logical analysis may apply to mathematical concepts as well as to logical ones—the fact that we have an instance of logical analysis does not *ipso facto* prove that the concept under analysis is one of logic. So setting aside for the moment the question of whether the notion of set, say, is logical or mathematical, we can recognize the following: Without the logical analysis of set that Cantor, Peirce, and others provided, the Power Set Axiom would not have been introduced. So whether or not the Power Set Axiom itself is one of mathematics or logic, the act of introducing it was an effect of (the process of) logical analysis, and is thus an act of logic (*logica docens*) as well as one of mathematics. (A consequence is that any subject—law or physics, for instance—that engages in logical analysis may provide new insights that are acts of logic as well as acts of the subject in question.)

Now let us return to the question of whether the notion of set is logical or mathematical. In the spirit of Peirce's definition of logic as the science of drawing necessary conclusions, we can define logical concepts recursively as those that are used to explain directly the nature of this "drawing [of] necessary conclusions," or that, in the analysis of such concepts, define or explicate them.

By the first criterion, logical concepts include the traditional ones, such as validity, truth, implication, and derivability (drawing conclusions). As a logician, Peirce was concerned with all of these. But, by the second criterion, they also include all those that arise in elucidating these notions, such as predicate, proposition, interpretation, model, and, in particular, set or collection. (For alternative accounts of the nature of logical concepts, see *Quine 1970* and *Putnam 1971*.)

A proposition—be it an axiom, lemma, corollary, or theorem—will then be (in part) logical if it contains or deals with one or more logical concepts. (Logical definitions and logical propositions constitute some of the products or results of logic.) By these definitions, the set-theoretic axioms are logical. Further, any theorems (mathematical or logical) that derive from them and are not derivable otherwise, depend on them. We thus have a case for the Strong Derivability ML Dependency Thesis of Section 2. In any case, the results or products of acts of logic (the process of logical analysis) may or may not be logical themselves.

So Peirce's and Cantor's introduction of the Power Set Axiom provides an example of a theoremic idea that is logical on two grounds—it was the culmination of a process of logical analysis and it deals with a logical concept. Indeed, the introduction of any axiom of set theory will be an example of theoremic reasoning that is logical because it is the result of a logical analysis and it deals with the notion of set. Zermelo's discovery of the Axiom of Choice is another instance on the same grounds. And still another example is Peirce's attempts to prove the axiom of comparability (or Trichotomy)

$$(x)(y)(x \leq y \lor y \leq x)$$

which, in fact, was later shown to be equivalent to the Axiom of Choice (*Mendelson 1964 [1977]:* 198).

As for examples of theoremic reasonings that are *mathematical,* we may again cite Fermatian inference or mathematical induction. Fermat's discovery of this axiom or rule of inference sprang from his mathematical investigations into number theory. In making his discovery, he presumably used his logical instinct or *logica utens.* However, later refinements of mathematical induction—such as set-theoretic proofs of the principle itself or its extensions to transfinite induction—would probably count as logic since they stemmed from (logically) analyzing the reasoning into finer steps and they dealt with the logical concepts of set and its derivatives. Mathematical proofs based on transfinite induction would then support the Proof Technique Dependency Thesis of Section 2.

Needless to say, the boundaries between acts of logic and mathematics, and that between logical and mathematical concepts, are not sharp. The discovery of non-Euclidean geometries, for instance, may have been logic for some, mathematics for others. The denial of Euclid's fifth postulate was such a demanding insight into a "foreign idea" that it took more than two millennia before it occurred. The mathematician Saccheri developed several theorems of non-Euclidean geometry, but he did not recognize them as such because he took them simply to show the absurdity of denying Euclid's fifth postulate (*Saccheri 1733*). To the extent that his denial of the postulate was characterized by an unclear logical instinct (the *logica utens* of mathematicians) which did not perceive exactly what he was doing, it would seem to be a *mathematical* action, and his version of the fifth postulate itself was a *mathematical* theoremic idea. On the other hand, Gauss, Lobatchevsky, Bolyai, and Riemann knew more clearly that they were creating new geometries. To the extent that their work required a significant amount of logical analysis to introduce the theoremic idea of denying the fifth postulate, their action was *logical* and their associated theoremic ideas— the versions of the fifth and other postulates and the claim that theirs was a mathematically valid approach—were *logical* in origin, if not in content.

Another distinction within theoremic reasonings that Peirce emphasized was that between the abstract and nonabstract. Abstract reasonings, he argued, are superior because they promote the advancement of mathematics. Peirce distinguished two types of abstraction: the precisive and the subjectal. Precisive abstraction is "leaving something out of acount in order to attend to something else" (NEM 3: 917). Subjectal abstraction, which often follows precisive abstraction and is the sense at issue here, Peirce described in different ways. In the same passage, he described it as "making a subject out of a predicate" (*ibid.*). For a nonmathematical example, he cited a well-worn joke of the day (thought to stem from a line of Molière) and wrote, "instead of saying, Opium puts people to sleep, you say it has a dormitive virtue" (*ibid.*). At other times, he illustrated subjectal abstraction in mathematics with the notions of a line as "the place which a moving particle occupies on the whole in the course of time . . . " (NEM 4: 163), and a surface, in turn, as a moving line. In these cases, abstraction simplifies mathematical thinking and promotes its advancement. The aforementioned geometric theoremic reasoning that introduces subsidiary lines illustrate this subjectal abstraction: we speak, for instance, of the parallel line passing through a vertex instead of the point vertex moving parallel to a given side. In addition, set-theoretic examples of theoremic reasoning, such as Peirce's and Cantor's proof that there is no largest multitude, illustrate subjectal abstraction as Peirce understands it, since "all collections are of the nature of abstractions" (NEM 4: 49; see *Levy 1983*).

Besides speaking of subjectal abstraction as transforming a predicate into a subject, Peirce offered other characterizations not all of which may be equivalent. At one point, he wrote "when the mathematician regards an operation as itself a subject of operations, he is using abstraction in one of its most abstract forms" (NEM 4: 11). And in a more linguistic vein, he wrote that subjectal abstraction is a "transformation from a concrete predicate to an abstract noun" (NEM 4: 160). These accounts suggest that abstract reasonings fall into at least two pertinent classes. First, some abstract reasonings may be characterized as *metatheoretic* in that they ascribe features to relations or their corresponding predicates in another, perhaps lower-level theory. To ascribe features to a relation is to take it as a subject or transform its predicate into a noun. Thus, mathematicians and logicians speak of a relation, say *greater than*, as being transitive, or state that congruence on the set of integers is an equivalence relation.

Second, some abstract reasonings may be represented *formally within a theory* by sentences that quantify over predicates or sets. Again, the proof that there is no greatest multitude is an example because it introduced not only the notion of a power set, but also the Power Set Axiom, which requires quantification over sets:

Power Set Axiom: $(x)(\exists y)(z)(z \in y \equiv z \subseteq x)$.

So in pointing out that *some* theoremic reasonings introduce additional quantification into a branch of mathematics, Hintikka (*1980*) is correct. But to say that this characterizes *all* theoremic reasoning (or that it is essential to it) is, as Ketner (*1985*) argues, to go beyond the facts. For the theoremic reasonings that introduce subsidiary lines, we saw, are logical consequences of the axioms that do not introduce additional quantification. And even some of the mathematical theoremic ideas that introduce new axioms do not introduce additional quantification. Some non-Euclidean geometries are cases in point: they deny an axiom such as Euclid's fifth postulate and do not add another axiom that adds more quantifiers. Simply to deny an existential quantifier is to change it into a universal quantifier followed by a negation, not to introduce yet another quantifier. In any event, subjectal abstraction in one form or another seems to characterize not only the best theoremic reasoning but also most corollarial reasoning.

The nature of theoremic and corollarial reasoning is represented in the tree diagram of Figure 6.1. The diagram can be viewed on the one hand as describing the *process* of mathematical reasoning, and on the other as describing its *results*—definitions, axioms, theorems, techniques, etc. The branches coming from any node of the tree show the two classes of reasoning into which you can divide a given class. Thus, the first level of necessary or mathematical reasoning is divided into the theoremic and the corollarial. Theoremic reasoning is next divided into two classes—those where the new idea is itself a new axiom and those where the new idea is derivable from current axioms. (The bipartite division of corollarial reasoning will be explained in the next section.) The different classes of theoremic and

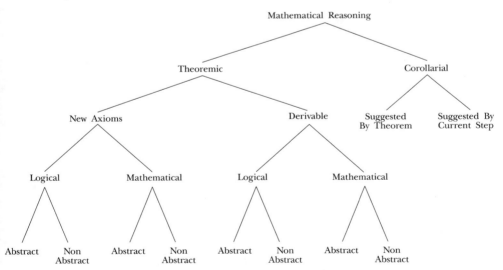

Figure 6.1

corollarial reasoning in turn each fall into two classes—the logical and the mathematical. And the last level (or leaves of the tree) derives from dividing each class into those reasonings that are abstract and those that are not.

Peirce notes further points of interest on the theoremic/corollarial distinction: that a theoremic idea often spawns corollaries and may sometimes be imitated to solve related problems. Where the theoremic idea is imitated, it may soon become unclear which problem was the originator and which the imitator. The new theoremic idea becomes a standard axiom, definition, or mode of approach in the given branch of mathematics. In this case, "a theorem may pass over into the class of corollaries, in consequence of an improvement in the system of logic" (4.613). In other words, the steps in the (corollarial) reasoning are now suggested by the newly accepted definitions of the terms in the theorem. Likewise, what was once a theoremic idea of the first kind—one that introduces one or more wholly new axioms—may spawn a theoremic idea of the second kind—one that is derivable from the newly introduced or now-accepted axioms. This evolving nature of mathematical propositions may partly explain the longstanding controversy surrounding their philosophic analysis.

In this section, we have seen how Peirce's theoremic/corollarial distinction illuminates the connections between mathematics and logic. Specifically, it shows that mathematics depends in part on logical theoremic ideas. In the next section, we will see how the distinction also illuminates the traditional distinction between analytic and synthetic truths and the question of whether or not mathematical truths are analytic.

§4. The Analytic/Synthetic Distinction

Like many philosophers before him, Peirce argued that the analytic/synthetic distinction applied to mathematical truths, but his analysis endowed the thesis with a much richer significance. In fact, in light of Peirce's theoremic/corollarial distinction, we can define *three* pertinent senses of *analytic*.

In his reply to Carus in the *Monist (1893)*, Peirce used the term *analytic* in a broad sense: "An analytical proposition is a definition or a proposition deducible from definitions. . . . Mathematical propositions . . . are in truth only analytical" (6.595). So in one sense, all mathematical truths are analytic. Later, however, Peirce refined this view and used the term *analytic* in a narrower sense, founded on his distinction between corollarial and theoremic reasonings. Thus, Peirce considered Kant's classic example of a mathematical truth and argued that it was not synthetic:

> Kant maintains that $7 + 5 = 12$ is a synthetical judgment, which he could
> not have done if he had been acquainted with the logic of relatives. For if

we write G for "next greater than," the definition of 7 is $7 = G6$ and that of 12 is $12 = G11$. Now it is part of the definition of *plus*, that $Gx + y = G(x + y)$. That is, that $G6 + 5 = G11$ is implied in $6 + 5 = 11$. But the definition of 6 is $6 = G5$, and that of 11 is $11 = G10$; so that $G5 + 5 = G10$ is implied in $5 + 5 = 10$, and so on down to $0 + 5 = 5$. But further it is a part of the definition of *plus* that $x + Gy = G(x + y)$ and the definition of 5 is $5 = G4$, so that $0 + G4 = G4$ is implied in $0 + 4 = 4$, and so on down to $0 + 0 = 0$. But this last is part of the definition of *plus*. There is, in short, no theorematic reasoning required to prove from the definitions that $7 + 5 = 12$. (NEM 4: 58–59)

This proof exemplifies corollarial reasoning in two related ways. First, repeated applications of the definitions of the numbers 0 through 12 and the principles of logic prove the theorem without appeal to any other theorems. Second, each definition introduced in the proof is immediately suggested by a term in the current step in the proof. In this respect, the example differs slightly from the associativity example discussed in the previous section, where the definitions introduced are suggested by the statement of the theorem itself instead of by the current step in the proof.

As to analyticity, the passage suggests a second definition of the notion. It begins by denying that the truth $7 + 5 = 12$ is synthetic, hence implying that it is analytic. It ends with the apparently equivalent denial that theorematic reasoning is required to prove it, hence implying that it rests on corollarial reasoning. The upshot is that the results of corollarial reasoning are analytic and those of theoremic reasoning are synthetic. This is perhaps the most straightforward interpretation of Peirce's views.

In the spirit of Peirce's account, however, there is yet a third possible definition of *analytic* that applies to mathematical truths, one narrower than the first but broader than the second. For if we restrict Peirce's definition in reply to Carus above, namely "An analytical proposition is a definition or a proposition deducible from definitions . . . " by adding the phrase *that are currently accepted within the branch of mathematics*, then the extension of this intermediate notion of *analytic* will cover mathematical truths whose derivations are either corollarial reasonings or theoremic reasonings of the second kind—those that introduce a "foreign idea" derivable (deducible) from current axioms.

So, on the basis of Peirce's corollarial/theorematic distinction, we have three senses of *analytic* in order of increasing specificity:

- deducible from axioms or definitions whether newly discovered or long-standing (theoremic or corollarial);
- deducible from long-standing axioms or definitions (theoremic of the second kind or corollarial);
- deducible from long-standing axioms or definitions where every step

introduces a definition of a term that occurred in the definition of previous terms in the proof of the proposition (corollarial).

The denials of these furnish the corresponding senses of *synthetic*. Finally, as previously noted in discussing the corollarial/theoremic distinction itself, the placement of a given truth within the analytic/synthetic dichotomy can change depending on whether the theoremic idea in its proof is newly discovered or has become a generally accepted axiom, definition, or method in the branch of mathematics at hand.

Peirce noted interesting connections between theoremic reasoning and hypothesis formation in science generally—what he called *abduction* or *retroduction* (see *Kapitan 1994*). In the Lowell Lectures of 1903 he said, in a manner reminiscent of his earlier remarks on theoremic reasoning, "abduction is the only logical operation that introduces any new idea" (5.171). Both, he argued, stem from an unconscious act of the mind. And in a later manuscript (ca. 1907), he explicitly acknowledged the connection, saying, "theoremic reasoning . . . is very plainly allied to Retroduction, from which it only differs, as far as I now see, in being indisputable" (MS 754).

Abductive or retroductive hypotheses are disputable, in that experience may prove their consequences false, and with it themselves false, too. Theoremic reasoning, on the other hand, is indisputable in that it either introduces a new axiom that (barring contradiction) engenders a new set of inevitable consequences that cannot be proven false or is itself derivable from the current set of axioms.

A related feature of abduction Peirce mentioned in this context was that "no reason can be given for abduction" (5.171). As for theoremic reasonings, this is true only of theoremic reasonings of the first kind—those that introduce a new axiom. Those of the second kind, of course, do admit of a reason, namely those axioms from which they are derivable.

The occasional dependence of mathematics on logic suggested by the analysis of theoremic reasoning has become more apparent with twentieth-century developments in the two subjects. Three examples will illustrate the point.

First, consider the notion of substitution. By instinct (*logica utens*), logicians and mathematicians have long used forms of substitution in their work. Aristotle, and others since, substituted letters for terms in analyzing the forms of the syllogism. Peirce used substitution among other ways to illustrate theoremic reasoning, if only "a little theoric step" (4.616), one of the second kind. He cites Euclid's use of letters to refer to a specific geometric figure after stating a theorem in terms of the general class of figure, such as a triangle. Logical analysis of this mode of inference by Peirce and others resulted in the statement of the familiar quantification rule of Universal Instantiation. And again, the composition or substitution of a set of functions as arguments to another function to create new functions is formalized in the definition of recursive functions. These instances show that substitution is a logical notion by the two criteria cited at the end of Section 2. First, in analyzing the detailed steps of reasoning (= doing logic), one regularly finds such acts of substitution. Second, substitution is itself used to define other logical notions, such as validity.

The notion of substitution is so fundamental, in fact, that its use is a paradigm of the mathematician's instinctive skill with logical concepts and principles (*logica utens*). Important incarnations of the logical notion of substitution occur as theoremic ideas in mathematics. Consider, for instance, integration in the calculus of Newton and Leibniz. The definition of integration relies on dividing the area under a curve into an infinite series of rectangles. In numerical analysis, substituting the finite for the infinite, one comes up with an *approximation* to the integral. Substituting, further, trapezoids for rectangles, one comes up with a typically more accurate approximation with the trapezoidal rule. And substituting still further pairs of adjacent figures for single trapezoids, and parabolas for straight line segments, Simpson arrived at the rule named after him. This is only the slightest hint at the variety of guises under which the logical notion of substitution has appeared, through *logica utens*, as a theoremic idea in mathematics.

Yet, even with these instinctive notions of substitution, logical analysis has been required to avoid errors. Rules of Existential Instantiation in the predicate calculus, for instance, note that we must exercise care in choosing the constant to be substituted for the quantified variable. It was the analysis of the details of reasoning, i.e., logical analysis, that discovered the need for such a rule. This rule then became a part of the corpus of results comprising *logica docens*. Further, to the extent that mathematics has come to depend on formulations of the predicate calculus that incorporate such a rule—for example, in deductive systems programmed on computers that have discovered new theorems and their proofs—it depends on this point of logic, in the sense of the Proof Technique Dependency Thesis of Section 2.

As a second example of the recent dependence of mathematics on logic, consider Gödel's Incompleteness Theorem. The undecidable sentence is a sign simultaneously of a mathematical result and of the logical truth that it itself is an unprovable sentence. Here we have a paramount example of the interdependence, if not blending, of the two subjects. The novel theoremic idea of Gödel numbering arose from the logical analyses of sentences and their constituent relations and functions. Further, Gödel numbering provides a logical/mathematical approach that leads to a logical result as to the inherent limits of reasoning (derivability, provability) within axiomatic systems of mathematics. The mathematical result that the undecidable mathematical truth can be demonstrated only in the meta-theoretic context, not in the (object) theory itself, depends on this logical result. This state of affairs supports the (Meta)Theorem ML Dependency Thesis of Section 2.

Note that the logical (metatheoretic) result guides the way mathematicians will seek solutions to their problems: No mathematician can prove or disprove a Gödel sentence with the axioms of the current theory alone, and—knowing this—no mathematician will even *try* to do so. Mathematicians will look elsewhere. This lends support to another process ML dependency thesis that merits formulation:

> *Process/Result ML Dependency Thesis:* The process of mathematical research is in part directed by, and to that extent depends on, some results of logic.

Finally, consider Church's Thesis, which in one form states that every computable (calculable) function is recursive (see *Mendelson 1964*). This thesis stems from a logical analysis of the notion of computability—logical because it seeks to analyze the steps in mathematical reasoning. Support for the thesis is strong, not simply because no counterexamples have been found but also because alternative attempts to analyze the notion of computability have produced equivalent forms of the thesis, such as "every computable function is Turing-computable" (*Mendelson 1964*) and "every computable function is grammatically-computable" (*Garey* and *Johnson 1979*). In other words, the proposition that every recursive function is Turing-computable implies the proposition that every Turing-computable function is grammatically-computable, which in turn implies the proposition that every grammatically-computable function is recursive.

In whatever form, Church's Thesis acquires mathematical importance because where a function can be shown not to be recursive, mathematicians conclude that it is not computable. In other words, for many classes of problems, the class of problem can be shown to be unsolvable by a single general procedure (*Mendelson 1964*: 255ff.). (Likewise, computer scientists note that a computer program that computes arbitrary values of a nonrecursive function cannot be constructed.) Consequently, we have support for a variant of the (Meta) Theorem ML Dependency Thesis in which the mathematical conclusions stemming from logical results do not have strictly complete proofs, but fall shy in that they derive from an unproven hypothesis, namely Church's Thesis. Alternatively, we can say that we do have theorems with complete proofs, one of whose hypotheses is Church's Thesis, so we have support for the Theorem ML Dependency Thesis proper, not a variant of it.

We have seen that, in spite of his oft-stated view to the contrary (for instance, 4.228, 4.242), Peirce's views implied that mathematics depends on logic in several ways. In this chapter, I have distinguished several theses of dependency of mathematics on logic, and shown how Peirce's writings recognize and support them. To summarize, the logical analysis of (logical and mathematical) concepts and reasoning promotes the discovery of mathematical concepts, theorems, and proofs that would not occur without it. In other words, logical analysis fosters the progress of mathematics. When Peirce was not considering the question of the definition of mathematics, he was able to see this point, and indeed insisted on it. So why, then, did Peirce plainly deny the point whenever the context turned to the *definition* of mathematics? And deny it thoroughly, in that in these contexts he apparently rejected virtually all the theses that have been distinguished.

A plausible explanation is that Peirce's reluctance here was a side effect of the enormous esteem and love he had for his father (see, for instance, 4.229 and NEM 4: v)—an esteem and love that any parent would long for and cherish. The resultant influence of the father's views on his son was probably felt in several ways. First, the definition of mathematics that Peirce adopted, as noted, was his father's. Second, the definition, and his father's apparent interpretation of it, seems to reject the dependence of mathematics on logic. And, third, his father devalued the importance of logic: Peirce relates that his father tried to discourage him from pursuing logic by saying that he wanted his son to be remembered and that logic was not the way to achieve that goal. This confluence of factors would probably make most individuals shy away from logic as much as possible and not give it its due. To Peirce's credit, he did not follow his father's advice on this point—logic remained one of the leading guideposts of his life. Moreover, at times he was able to criticize his father on this topic and realize the mutual dependence of mathematics and logic. One such occasion occurred a few months before he died, when he wrote:

> my father . . . had a low opinion of the utility of formal logic, and wished to draw directly upon the geometric instinct rather than upon logical deductions from the smallest number of axioms. . . . This was one of a number of allied points in which I have long disagreed with my father, mainly because I do not believe instinct is an absolutely infallible source of truths. (NEM 4: vii)

Here he acknowledges the "utility of formal logic" for geometry, if not for other branches of mathematics as well.

Shaped by twentieth-century developments in mathematics and logic, the contemporary view of most mathematicians and logicians apparently agrees with Peirce's sometimes latent view that in several areas, mathematics (during analysis, research, and discovery as well as during verification of results) depends on developments in logic and would not be what it is without them. Indeed, some would even suggest (for reasons different from Russell's) that they are the same subject. Of course, a mere vote on the matter is not decisive. It is the growth of mathematics and logic themselves, and analyses such as Peirce's theoremic/corollarial distinction, that will continue to deepen our understanding of mathematics, logic, and the relationship between them.

NOTES

I would like to thank Nathan Houser and James Van Evra for their very thoughtful comments on this paper, and express a second thanks to Nathan Houser for suggesting that I write it in the first place. In addition, I would like to thank Susan Haack for our fruitful correspondence on issues of mutual interest in her paper "Peirce and Logicism: Notes toward an Exposition" (*Haack 1993*).

1. Peirce used as synonymous *theoric, theorematic,* and *theoremic,* for the most part, the last spelling will be used here.

2. To put the matter briefly here, Peirce defined theoremic reasoning as that which introduces a "foreign idea," whereas corollarial reasoning does not.

3. In Heath's translation, "Let the following be postulated: 1. To draw a straight line from any point to any point" (Euclid, Postulates).

4. In Peirce's metaphysics, the undrawn line is an instance of firstness, while the drawn line is an instance of secondness (*Levy 1982b*).

7.

PEIRCE AND RUSSELL:

THE HISTORY OF A NEGLECTED 'CONTROVERSY'[1]

Benjamin S. Hawkins, Jr.

§1

1.1. *Introduction.* Although he was not as original as Charles S. Peirce (1839–1914), Bertrand Russell (1872–1970) was nonetheless a crystallizer, and, as Arthur Koestler points out, "the crystallizers often achieve more lasting fame and greater influence on history than the initiators of new ideas" (*Koestler 1963:* 210). Russell's work certainly seems to have fared better than Peirce's; in the short run of history, Russell's work has been successful, and though not limited to logic, his name is "inseparable from mathematical logic" (*Quine 1966a:* 657).

Now there is growing appreciation, long overdue, of the originality of Peirce's work. Given recent historical research, Peirce emerges as an extraordinary figure in logic and mathematics of the nineteenth and early twentieth centuries. As R. M. Martin observes,

> One does not approach [Peirce's] work as though it were a century old; one reads him as though he had written yesterday. His ideas are for the most part so astonishingly modern that one is shocked to encounter antiquated methods, inadequate formulations, and the like. It is a tribute to his greatness that his mode of writing solicits criticisms in accord with contemporary standards of rigor and in the light of contemporary knowledge. Peirce and Frege are unique in this respect among the logicians of the nineteenth century. (*Martin 1976b:* 244)[2]

Peirce, a trained chemist, made contributions not only in mathematical logic but also in astrophysics, spectroscopy, gravimetry, and geodesy. In addition, he extended the work of his father, Benjamin Peirce (1809–1880), on linear algebra. Indeed, many of his contemporaries attributed to him "higher mathematical attainments than his father" (*Lenzen 1969:* 6).[3]

Though Peirce liked to refer to himself as a logician,[4] his contributions to logic have been recognized only in the last thirty or so years. "Peirce's work failed," Bell remarks, "to make the immediate mark its penetrating quality should have made," so that "others retraced his steps, unaware that he had gone before" (*Bell 1945:* 556–557). Ernst Schröder (1841–1902), writing in 1896, predicted that "however ungrateful [Peirce's] countrymen and contemporaneans might prove," his fame as a logician "would shine like that of Leibniz and Aristotles into all the thousands of years to come" (L 392).[5] Christine Ladd-Franklin may have inadvertently suggested a reason for the unresponsiveness to Peirce's groundbreaking contributions to logic when she said that he "wrote his papers with the brevity and abstractness that befit a scientific journal" (*Ladd-Franklin 1892a:* 126).

1.2. *Writs on Passages.* Early this century Peirce and Russell were combatants in a *guerre de plume*, an interchange which has so far received scant attention. This is the 'controversy'—a missed opportunity in the history of logic—which I propose to examine in this chapter.

In Section 2, I outline the character of Peirce-Russell disagreement. Sections 3 and 4 contain a discussion of issues relating to Peirce's review of Russell's *Principles of Mathematics* [PM] and to work of Gottlob Frege (1848–1925), and reference is made in Section 3 to Peirce's annotations in his copy of PM now in the Houghton Library at Harvard University. Section 5 contains an account of the the two logician's cognizance of one another's work, as well as my objections to M. G. Murphey's exegesis of Peirce on Russell (*Murphey 1961:* 241–243). Finally, the appendix contains a formulation of Peirce's demur to a *falsigrafis* in PM.

§2

2.1. *Discordant Themes.* "My analyses of reasoning," Peirce caustically observes in 1903, "surpass in thoroughness all that has ever been done in print, whether in words or in symbols—all that De Morgan, Dedekind, Schröder, Peano, Russell, and others have ever done—to such a degree as to remind one of the difference between a pencil sketch of a scene and a photograph of it" (5.147). The same year Russell alleges that subsequent to

> the publication of Boole's *Laws of Thought* (1854), the subject [of Symbolic Logic] has been pursued with a certain vigour, and has attained to a very considerable technical development. Nevertheless, the subject achieved almost nothing of utility either to philosophy or to other branches of mathematics, until it was transformed by the new methods of Professor Peano. (*Russell 1903* [*1956*]: 10)

Russell, however, qualifies this comment: "Professor Frege's work, which largely anticipates my own, was for the most part unknown to me when the printing of the present work [that is, PM] began; I had seen his *Grundgesetze der Arithmetik,* but, owing to the great difficulty of his symbolism, I had failed to grasp its importance or to understand its contents" (*Russell 1903* [*1956*]: xvi).[6] He remarks elsewhere that "The enlightenment I derived from Peano came mainly from two purely technical advances [distinguishing ϕx and xRy, both] of which . . . had been made at an earlier date by Frege, but I doubt whether Peano knew this, and I did not know it until somewhat later" (*Russell 1959*: 66). Peirce also made these "technical advances," and Russell's "doubt" about Peano appears to be quite mistaken.[7] Russell recognized that while "a few hints for it [that is, the calculus of relations] are to be found in De Morgan, the subject was first developed by C. S. Peirce" (*Russell 1903* [*1956*]: 23).[8]

In 1903 Russell claimed that "the most complete account of the non-Peanesque methods will be found in the three volumes of Schröder, *Vorlesungen über die Algebra der Logik*" (*Russell 1903* [*1956*]: 10n.). Yet that work of Schröder largely codifies Peirce's logic of relatives (or relations). The Peirce-Schröder logic is actually not only closely related to the logic in Frege, Peano, PM, and Whitehead's and Russell's *Principia Mathematica* [PMa], the results of the Peirce-Schröder deductive methods and PMa are roughly the same.[9]

P. Bernays noticed that by using "a method of C. S. Peirce, Schröder introduced *identical product* and *identical sum*." Bernays also noticed that Schröder "treats the laws of general lattice theory, of which . . . [not only] the proofs of the associative law for the identical product and the identical sum are given by the method of Peirce," but also "that, in the system of *Principia mathematica*, the formula Assoc can be derived [by the method of Peirce] from the formulas Taut, Add, Perm, and Sum" (*Bernays 1975*: 610, 611).

"In working over [Peirce's] theory at large," Schröder indicated in 1898, "I have but slightly and never without intrinsic reasons modified his (or Boole's) denotations" (*Schröder 1898a*: 52 [*1899*: 320]). "The nearest approach to a logical analysis of mathematical reasoning that has ever been made," Peirce remarked in 1903, "was Schröder's statement, with improvements, in a logical algebra of my invention" (4.426).[10] He commented the same year, "I value Schröder's work highly & he was a highly sympathetic man whom it was impossible to know and not to like even more than the great merit of his work justified. But he was too mathematical, not enough of the logician in him" (MS 302). Peirce also remarked derisively in 1908 that "Such ridiculously exaggerated claims have been made for Peano's system, though not, as far as I am aware, by its author, that I shall prefer to refrain from expressing my opinion of its value" (4.617). This is apparently an allusion to Russell, and it is ironic that Peirce refrained from expressing his opinion.[11]

2.2. *Russell on Peirce's Logic of Relations.* In 1901 Russell said that Peirce's logic of relations "is difficult and complicated to so great a degree that it is possible to doubt its utility" (*Russell 1956:* 3). There seems to be a sense of the straw man in this, a sense that seems reinforced in Russell's 1903 elaboration of such objections in PM:

> Peirce and Schröder have realized the great importance of the subject [of the calculus of relations], but unfortunately their methods, being based, not on Peano, but on the older Symbolic Logic derived (with modifications) from Boole, are so cumbrous and difficult that most of the applications which ought to be made are practically not feasible. In addition to the defects of the old Symbolic Logic, their method suffers technically . . . from the fact that they regard a relation essentially as a class of couples, thus requiring elaborate formulae of summation for dealing with single relations. This view is derived, I think, probably unconsciously, from a philosophical error: it has always been customary to suppose relational propositions less ultimate than class-propositions (or subject-predicate propositions, with which class-propositions are habitually confounded), and this has led to a desire to treat relations as a kind of classes. (*Russell 1903* [*1956*]: 24)[12]

Russell particularly objected to Peirce's treatment of *relative addition*:

> Peirce and Schröder consider also what they call the relative sum of two relations R and S, which holds between x and z, when, if y be any other term whatever, either x has to y the relation R, or y has to z the relation S. This is a complicated notion, which I have found no occasion to employ, and which is introduced only in order to preserve the duality of addition and multiplication. This duality has a certain technical charm when the subject is considered as an independent branch of mathematics; but when it is considered solely in relation to the principles of mathematics, the duality in question appears devoid of all philosophical importance. (*Russell 1903* [*1956*]: 26)

Yet Russell observed, where "$dT_{ac}b$ implies $dT_{ac}e$ or $eT_{ac}b$", that

> "$dT_{ac}e$ or $eT_{ac}b$" is the relative sum of T_{ac} and $T_{ac,}$ if d, e, and b be variable. This property [of the T-relation in Projective Geometry] results formally from regarding T_{ac} as the negation of the transitive relation Q_{ac}. (*Russell 1903* [*1956*]: 387n.)

Russell considered this to be "an instance (almost the only one known to me) where Peirce's relative addition occurs outside the Algebra of Relatives" (*Russell 1903* [*1956*]: 387n.).

2.3. *Peirce on Russell or Whitehead* Peirce's reaction to Russell's criticism of his logic of relations is recorded in a 1904 letter to Lady Welby:

> As to my algebra of dyadic relations, Russell in his book which is superficial to nauseating me, has some silly remarks about my "relative addition" etc., which are mere nonsense. He says, or Whitehead says, that the need for it seldom occurs. The need for it *never* occurs if you bring in the same mode

of connection in any other way. It is part of a system which does not bring in that mode of connection in any other way. In that system it is indispensable. But let us leave Russell and Whitehead to work out their own salvation. (L 463 [12 October 1904]/SS: 30)[13]

The first and second of these quotations from PM together take the brunt of Peirce's remarks, not the first passage alone as is adduced by some editors. An omission of the second passage (from *Russell 1903* [*1956*]: 26) renders Peirce's exception to Russell nonsensical.[14]

Peirce's reference to both Russell and Whitehead seems to be an almost perverse exercise of Russell's acknowledgment that

> At every stage of my work [on PM], I have been assisted more than I can express by the suggestions, the criticisms, and the generous encouragement of Mr A. N. Whitehead [1861–1947]; he also has kindly read my proofs, and greatly improved the final expression of a very large number of passages. (*Russell 1903* [*1956*]: xviii)

In this regard, Russell's promise of a second volume of PM, "in which I have had the great good fortune to secure the collaboration of Mr A. N. Whitehead" (*Russell 1903* [*1956*]: xvi) is also worth noting,[15] as is Russell's subsequent admission in 1968 that "Whitehead provided proper mathematical proofs [in PMa] and did all the polishing" (*King-Hele 1974–1975:* 23).[16]

Peirce generally regarded neither Russell nor Whitehead in very complimentary terms. He stated in 1906, for example, that "In my opinion Russell and Whitehead are blunderers continually confusing different questions" (L 148 [8 May 1906]/NEM 3:785). It is interesting that Whitehead, without ever simply dismissing Peirce's work, cites Peirce's 1867 analysis of signs (ipso facto, relations) as "obscure" (*Whitehead 1898* [*1960*]: 3n.).[17] Russell's and Whitehead's initial collaboration is not, as is usually thought, the 1910–1913 PMa, but a 1902 paper under Whitehead's general authorship; a paper in which Whitehead commends Russell's notation for relations (*Whitehead 1902:* 367–368, 378–382).[18] Peirce's persistent linking of Russell and Whitehead appears, then, not to be a confounding of Russell and Whitehead, but a quite credible suspicion that Whitehead's hand was well into PM.

2.4. *Peirce's Ultimate Analytic of Thought.* Peirce's ironic response to Russell continued in the 1912 manuscript "Notes Preparatory to a Criticism of Bertrand Russell's *Principles of Mathematics*": "Russell has in this first volume sufficiently expounded the principles of mathematics to give those who have not at all studied the subject some notions of it that may be said to be true in the main; i.e., to say, in so far as to inform them of the important truths" (MS 12: 1^P). In 1908 Peirce observed that "If the ground [of what mathematics is] were fully covered, the book could not be a very small one," but "would have to be at least as large as Russell's Phil. of Math.,—a weak performance, by the way" (L 387 [18 September 1908]/NEM 3: 970). Peirce's reference to "Russell's Phil. of Math." cannot have been to Russell's *Introduction to Mathematical Philosophy*, which was not published until 1919, five years after Peirce's death; undoubtedly it is to PM of 1903, perhaps "Phil." being a slip of the pen for "Prin."[19]

The "weak performance" that Peirce observed in PM in 1908 seems to have been repeated in the 1912 manuscript. There Peirce claimed that in PM, Russell "betrays insufficient reflection on the fundamental conceptions of the subject," which "may be said in a general way to be due to his not having thought enough on the subject, but especially it is owing to his not having begun with a thorough examination of the elements,—the ultimate analytic of thought" (MS 12: 1[P]).

An account of the weft and warp of Peirce's and Russell's fabric of logic and mathematics is needed at this point; for in their respective schemes, logic and mathematics are of very different, indeed divergent designs. I will therefore complete the current section with a sketch both of Peirce's own "reflection on the fundamental conceptions of the subject," and of Russell on logic and mathematics (however "weak" his "performance"). (This digression will avoid inordinate parenthetical expositions of Peirce's and Russell's differences in later discussions.)

2.5. *Peirce's Use of 'Logic.'* It was Peirce's general idea that logic engages "the essential nature and fundamental varieties of possible semiosis" (5.488) and "the conditions which determine reasonings to be secure" (2.1). Peirce uses 'semiotic' and 'logic' as synonyms in the sense of a "science of the general laws of signs" (1.191).[20] He uses 'science' "in the sense of a business, that is, of a total of real acts exerting reciprocal effects one upon another, and concerned with closely analogous purposes," and he says that by "any given heuretic science, I mean the body of doings in Past and Future time not too remote from the present of the members of a certain social group" (MS 499: 13[P]). Thus, Peirce remarks that "it will be necessary for the present and for a long time to come to regard logic, not as a distinct science, but as only a department of the science of the general constitution of signs,—the physiology of signs,—cenoscopic semeiotics" (MS 499: 15[P]–16[P]). He also writes that

> A great desideratum is a general theory of all possible kinds of signs, their modes of signification, of denotation, and of information; and their whole behaviour and properties, so far as these are not accidental. The task of supplying this need should be undertaken by some group of investigators; and since pretty much all that has hitherto been accomplished in this direction has been the work of logicians, among whom may be instanced indiscriminately George Boole, Mary Everest Boole, Victoria Welby, M. Couturat, Alfred Bray Kempe, Alfred [sic] Peano, Bertrand Russell,

and since a large division of this work ought to be regarded as constituting the bulk of the logician's business, it would seem proper that in the present state [of] our knowledge logic should be regarded as coëxtensive with General Semeiotic, the *a priori* theory of signs. (MS 634: 14–15)

Yet Peirce sometimes restricts 'logic' to 'deductive methodology' and, in doing so, regards a system of logical symbols as a medium to "analyze a reasoning into its last elementary steps" (4.239).[21]

Peirce considered an argument, for example, within the province of logic in the restricted sense, to be a "a sign whose interpretant represents its object as being an ulterior sign through a law, namely, the law that the passage from all such premises to such conclusions tends to truth" (2.263). This is an anticipation of Tarski's conception of logical consequence; and indeed, various remarks by Peirce on argument, interpretation, and icon anticipate certain essentials of model theory.[22]

The point is that in the context of Peirce's semiotic, if A_n is a deductive argument, there is a relation K_n on A_n iff K_n is a relation on the illative transformation of a sign S_i into the *finis operis* S_{i+m}, so that K_n is "requisite besides the premises to determine the necessary . . . truth of the conclusion (2.465).[23] The effect of K_n is a *habit*: "The truth is that an inference is 'logical,' if, and only if, it is governed by a habit that would in the long run lead to the truth" (L 463 [23 December 1908] / *Lieb ed. 1953:* 30).[24] "Peirce," as Morris observed, "was indebted at many points in his semiotic to the scholastics' studies of sign processes" (*Morris 1970:* 20). Thus, Peirce's conception of a rule of inference or procedure (a logical leading principle K_n) is anticipated in the medieval *maximae propositiones;* and at the semantic level, the scholastic's *operari sequitur esse* is an anticipation of his pragmatism.[25]

2.6. *Peirce's Logic as Semiotic.* It is not widely appreciated that Peirce's semiotic or logic is effectively a tripartite hierarchy. Peirce's hierarchy, while not a simple division into three branches, is "similar to the medieval analysis of semiotic (then called *scientia sermocinalis*) into the fields of grammar, logic, and rhetoric" (*Morris 1970:* 20).[26]

The first division, speculative grammar—following work attributed to John Duns Scotus (1266–1308), but apparently having been written in the first half of the fourteenth century by Thomas of Erfurt—is a taxonomy of signs.[27]

The second division, critical logic or critic, is, in Peirce's *restricted sense,* logic; and though both are divisions comprising syntax and semantics, and though both divisions are behavioral, observational, and analytical, critic assumes and depends upon speculative grammar.

The third division, methodeutic or speculative rhetoric, is devoted to the appraisal of the arguments determined in critic as good or bad; it assesses arguments from the point of view of "the methods that ought to be pursued in the investigation, in the exposition, and in the application of truth" (1.191). These methods are heuristic, heuretic, or normative in the sense that truth should be the object of reasoning, and such methods are principles or canons for its attainment, so that "every principle of logic is a Regulative Principle & nothing more" (MS 469: 30).[28]

2.7. *Peirce's Use of 'Mathematics.'* Peirce's conception of logic as semiotic is apparently entirely unique in modern annals, and it is singular in its treatment of mathematics, for Peirce's view is that mathematics and deductive or necessary reasoning are simply one and the same, whereas deductive logic is the science of *drawing* necessary conclusions (4.239).

Peirce, following his father, defined 'mathematics' as "the science which [from pure hypotheses] *draws* necessary conclusions" (4.239).[29] Peirce conceived of pure mathematics as "completely abstracted from concrete reality" (4.428),[30] not as abstracted from the manipulation of concrete signs, which involves "turning what one may call adjective elements of thought into substantive objects of thought" (MS 462: 48).[31]

For Peirce, then, the propositions of mathematics are "conditional" (4.240) and "devoid of all definite meaning" (5.567).

> All features that have no bearing upon the relations of the premisses to the conclusion are effaced and obliterated. The skeletonization or diagrammatization of the problem serves more purposes than one; but its principal purpose is to strip the significant relations of all disguise. . . . Thus, the mathematician does two very different things; namely, he first frames a pure hypothesis stripped of all features which do not concern the drawing of consequences from it, and this he does without inquiring or caring whether it agrees with the actual facts or not; and, secondly, he proceeds to draw necessary consequences from that hypothesis. (3.559)[32]

Peirce equated 'mathematics' with 'deductive' or 'necessary reasoning,' and he posited functions (rhemata, terms), propositions (dicent signs), and rules of procedure as the primitive basis of an object theory, for which "the meaning of a proposition or term is all that that proposition or term could contribute to the conclusion of a demonstrative argument" (5.179).[33] Interpretation here "is a thing totally irrelevant, except that it may show by an example that no slip of logic has been committed" (4.130). Peirce construed meaning in mathematics as a definiendum and deduction as a definiens, in the sense that "necessary reasoning only explicates the meanings of the terms of the premisses, to fix our ideas as to what we shall understand by the *meaning* of a term" (5.176).[34]

Peirce's construal of meaning in mathematics is rather akin to Hilbert's investigation of "the meaning of various groups of axioms," *mutatis mutandis,* "the significance of the conclusions that can be drawn from the individual axioms" (*Hilbert 1899* [*1971*]: 2). Hilbert contended, not unlike Peirce, that formulae "in themselves mean nothing but are merely things that are governed by our rules and must be regarded as the *ideal objects* of the theory" (*Hilbert 1927:* 470).

These and other of Peirce's views on mathematics anticipate modern formalism and intuitionism. Yet Peirce neither envisaged numbers simply as variables, nor $\alpha \vee \sim\alpha$ as "downright *false*." He does "say that . . . an intermediate ground between *positive assertion* and *positive negation* which is just as Real . . . does not involve any denial of existing logic, but it involves a great addition to it" (L 224 [26 February 1909]/NEM 3: 851).[35]

2.8. *Peirce's Logic Versus Mathematics.* Peirce's distinction between mathematics and deductive logic, between "*draws*" and "*drawing* necessary conclusions," is one of differences in method and subject matter. Indeed, Peirce's distinction led him to reject logicism (see 4.239–244).[36]

Peirce noted that "mathematics has such a close intimacy with . . . logic, that no small acumen is required to find the joint between them" (1.245). The juncture so obvious to Peirce is "that very part of logic which consists merely in an application of mathematics" (1.247).[37] Thus, "formal logic is nothing but mathematics applied to logic" (4.263),[38] which, for Peirce, "is by no means the whole of logic, or even its principal part" and "is hardly to be reckoned as a part of logic proper" (4.240).[39]

However, in a 1901 contribution to *Baldwin's Dictionary*, Peirce claimed that the systemic purpose of logic differs from that of mathematics:

> The first requisite to understanding this matter is to recognize the purpose of a system of logical symbols. That purpose and end is simply and solely the investigation of the theory of logic, and not at all the construction of a calculus to aid the drawing of inferences. These two purposes are incompatible, for the reason that the system devised for the investigation of logic should be as analytical as possible, breaking up inferences into the greatest possible number of steps, and exhibiting them under the most general categories possible; while a calculus would aim, on the contrary, to reduce the number of processes as much as possible, and to specialize the symbols so as to adapt them to special kinds of inference. (4.373)[40]

Peirce judges, with some relish, "that not only have the rank and file of writers on the subject [of logic] been . . . men of arrested brain-development, and not only have they generally lacked the most essential qualification for the study, namely mathematical training, but the main reason why logic is unsettled is that thirteen different opinions are current [in 1904] as to the true aim of the science" (4.243).[41]

The impetus of Peirce's own work is revealed in his comment that "in logic, my motive for studying the algebra of the subject, has been the desire to find out with accuracy what are the essential ingredients of reasoning in general and of its principal kinds. To make a powerful calculus has not been my care" (L 237 [29 August 1891]/8.316).[42] What Peirce required of a system of logical symbols is that it be cast in a notation of optimum iconicity for analysis of deductive reasoning; a *veritas compositionis* or notation to represent transformations of necessary inferences by analogous omissions and insertions in the representation.[43]

In a 1906 compendium of the subject, Peirce collated his and others' criteria for a system of logical symbols:

> The majority of those writers who place a high value upon symbolic logic treat it as if its value consisted in its mathematical power as a calculus. In my [1901] article on the subject in Baldwin's Dictionary I have given my reasons for thinking . . . if it had to be so appraised, it could not be rated as much higher than puerile. Peano's system is no calculus; it is nothing but a pasigraphy; and while it is undoubtedly useful, if the user of it exercises a discreet [sic] freedom in introducing additional signs, few systems of any kind have been so wildly overrated, as I intend to show when the second volume of Russell and Whitehead's Principles of Mathematics appears. . . . As to the three modifications of Boole's algebra which are much in use, I invented these myself,—though I was anticipated [by De Morgan] as regards to one of them,—and my dated memoranda show . . . my aim was . . . to make the algebras as analytic of reasonings as possible and thus to make them capable of exhibiting every kind of deductive reasoning. . . . It ought, therefore, to have been obvious in advance that an algebra such as I was aiming to construct could not have any particular merit [in reducing the number of processes, and in specializing the symbols] as a calculus. (MS 499: 1P–5P)[44]

Schröder is the likely source of Peirce's remark that "Peano's system is . . . a pasigraphy": "the aim of this novel branch of Science [of pasigraphy, which] is nothing less than the ultimate establishment of a scientific Language, entirely free from national peculiarities, and through its very construction conveying the foundation of exact and true philosophy" (*Schröder 1898a:* 45).[45] It is also of note that Peirce alludes to "Peano's system" as "wildly overrated" by Russell and Whitehead—another indication of his suspicion that Russell and Whitehead are together the authors of PM.

For Peirce, then, "mathematics lays the foundation on which logic builds; and those mathematical chapters will be quite indispensable" (2.197).[46] It is "producing a method for the discovery of methods in mathematics" that is for Peirce "the resolution of one of the main problems of logic" (3.364).[47] He likewise held that "Logic can be of no avail to mathematics" (2.197), and "that the appeal will be, not of mathematics to a prior science of logic, but of mathematics to mathematics" (1.247).[48] Further, Peirce contended that

The real nature of Mathematics has only come to be understood by mathematicians during the last half century. B. Peirce in 1870 first gave the definition now substantially approved by all competent persons. Mathematics is the science which draws necessary conclusions. There is no other necessary inference than mathematical inference. Some (as Dedekind) make mathematics a branch of logic. But mathematics is synthetic of inferences, logic analytic. (NEM 1: 256)

Moreover,

Mathematics is not subject to logic. Logic depends on mathematics. The recognition of mathematical necessity is performed in a perfectly satisfactory manner antecedent to any study of logic. Mathematical reasoning derives no warrant from logic. . . . It does not relate to any matter of fact, but merely to whether one supposition excludes another. Since we ourselves create these suppositions, we are competent to answer them. But it is when we pass out of the realm of pure hypothesis into that of hard fact that logic is called for. We then find that certain modes of reasoning are sound, because they must, by mathematical necessity, be sound, in whatever universe there may be in which there is such a thing as experience. . . . But there is no more satisfactory way of assuring ourselves of anything than the mathematical way of assuring ourselves of mathematical theorems. No aid from the science of logic is called for in . . . mathematical demonstrations of mathematical propositions so long as they are not open to mathematical criticism and have been submitted to sufficient examination and revision. The only concern that logic has with this sort of reasoning is to describe it. (2.191–192)[49]

Peirce also recognized, interestingly, "that the very objects of study themselves are subject to a logic more or less identical with that which we employ" (6.189).[50] Thus Beth, for instance, puts an emphasis on "the concept of *logico-mathematical parallelism,* which was at the bottom of the results of Peirce" (*Beth 1959* [*1966*]: 172). Beth's point is important, if the parallelism be in Peirce's sense of logic and mathematics.

2.9. *Russell's Mathematics as Logic.* Peirce's work is a vista on modern formalism, intuitionism, model theory and nonstandard analysis (see *Hawkins* 1986: 68–69). The differences that Peirce found to lie between mathematics and deductive logic, together with his restricted sense of logic, also anticipate the distinction between mathematics and metamathematics, with Peirce's requirements for a system of logical symbols being those of a metalanguage.[51]

These differences are not, however, entirely congenial with some recent distinctions. Thus, second-order (or higher) quantification theory, which is incomplete, is mathematics, while first-order quantification theory with identity, being complete, is logic.[52] These are distinctions that do not readily fit with Peirce's sense of mathematics and deductive logic. Rather, they are, in Peirce's sense of mathematics, distinctions that are *ab intra* mathematics. Peirce's sense of logic and mathematics also differ in kind from Russell's sense of the two subjects. For Peirce, "Mathematical logic is formal logic," which, "however developed, is mathematics" (4.240).

Russell's agenda in PM was "first, to show that all mathematics follows from symbolic logic, and secondly to discover, as far as possible, what are the principles of symbolic logic" (*Russell 1903* [*1956*]: 9). Like Peirce, Russell construes $\alpha \supset \beta$ as the form of mathematical propositions. Unlike Peirce, he claims "that all Mathematics is Symbolic Logic" and "the remainder of the principles of mathematics [still unborn, or in their infancy] consists in the analysis of Symbolic Logic itself" (*Russell 1903* [*1956*]: 5).[53]

> Logic consists of the premisses of mathematics, together with all other propositions which are concerned exclusively with logical constants and with variables but do not fulfill the above definition of mathematics [as the class of all propositions of the form "*p* implies *q*"]. Mathematics consists of all the consequences of the above premisses which assert formal implications containing variables, together with such of the premisses themselves as have these marks. Thus some of the premisses of mathematics, *e. g.*, the principle of syllogism, "if *p* implies *q* and *q* implies *r*, then *p* implies *r*," will belong to mathematics, while others, such as "implication is a relation," will belong to logic but not to mathematics. (*Russell 1903* [*1956*]: 9)[54]

Russell held that "when Logic is extended, as it should be, so as to include the general theory of relations, there are . . . no primitive ideas in mathematics except such as belong to the domain of Logic" (*Russell 1903* [*1956*]: 429).[55] "And," he concluded, "when it is realized that all mathematical ideas, except those of Logic, can be defined, it is seen also that there are no primitive propositions in mathematics except those of Logic" (*Russell 1903* [*1956*]: 430).

Russell's program was, of course, logicism. Plato, Leibniz, Dedekind, and Frege also shared Russell's ideal of reducing "mathematics to rigorous deduction from expressly formulated logical premisses by exactly specified logical methods" (*Taylor 1926* [*1960*]: 293). It is an ideal which stands in sharp contrast to Peirce's singular distinction between logic and mathematics.

In light of modern theory, Peirce is in a stronger position than Russell in this disagreement, for the difference of levels that Russell accepted in extending logic "so as to include the general theory of relations," is a gulf between logic and mathematics. Otherwise it is strange to apply 'logic' "to a system in which the consequences of the axioms are not all accessible by inference from the axioms. Yet that, as we can now see in the light of Gödel's theorem, is what Frege [and Russell] did when [they] undertook to reduce arithmetic to logic" (*Kneale 1962:* 741). It is surprising that after his 1901 discovery of the "paradox" in Frege's "logicising" of mathematics, not only did Russell fail to recognize the "paradox" in PM as a signal failure of logicism, but he also did not realize that it is the same program he offered in PM and the one he attempted to work out later (1910–1913) with Whitehead in PMa.[56] Russell, as late as 1937, claimed that the "fundamental thesis" of PM, i.e., "that mathematics and logic are identical, is one which I have never seen any reason to modify" (*Russell 1903* [*1956*]: v).

§3

3.1. *Peirce's Review and Readings of Russell.* The lines of controversy that Russell expresses about Peirce's work are primarily drawn in PM; while Peirce's views about Russell and his work, rarely delivered without irony or sarcasm, are more extensive than Russell's, and are found in a review, correspondence, papers, lectures, and as marginalia in a copy of PM. While Russell's lines of contention are not always carefully drawn or sustained, there is the occasional impression that Peirce's initial glance through PM was aversive enough to suffice as his finished reading of the work.

Peirce's (15 October 1903) review in *The Nation* of Lady Welby's *What is Meaning?* briefly incorporates the first of his public notices of PM:

> Two really important works on logic are these; or, at any rate, they deserve to become so, if readers will only do their part towards it. Yet it is almost grotesque to name them together, so utterly disparate are their characters. This is not the place to speak of Mr. Russell's book, which can hardly be called literature. That he should continue these most severe and scholastic labors for so long, bespeaks a grit and industry, as well as a high intelligence, for which more than one of his ancestors have been famed. Whoever wishes a convenient introduction to the remarkable researches in the logic of mathematics that have been made during the last sixty years, and that have thrown an entirely new light both upon mathematics and upon logic, will do well to take up this book. But he will not find it easy reading. Indeed, the matter of the second volume will probably consist, at least nine-tenths of it, of rows of symbols. (*Peirce 1903/*8.171n.1)

Peirce's prediction that "the matter of the second volume [of PM] will probably consist, at least nine-tenths of it, of rows of symbols" was duly confirmed; the second volume of PM became the three volumes of PMa.[57] It is noteworthy too that Peirce refers to PM as "logic" and as an "introduction". With his characterization of the work as casting "new light both upon mathematics and upon logic" as an apparent reference to his distinction, Peirce goes on to refer to PM as "logic of mathematics."

Peirce's duplex review presents a view of PM in terms of gentle irony;

an irony which his sarcasm intensifies in a letter of 1 December 1903 to
Lady Welby:

> As to Bertrand Russell's book, I have as yet made but a slight examination
> of it; but it is sufficient to show me that whatever merit it may have as a
> digest of what others have done, it is pretentious & pedantic,—attributing
> to its author merit that cannot be accorded him. Your ladyship perhaps
> did not notice that I hinted at this in the *Nation,* in saying your book was
> such a contrast to his. *The* man is Dr. Georg Cantor. Besides his strictly
> mathematical presentations, there is an interesting set of letters by him to
> philosophers in the Zeitschrift für Phil. u. phil. Kritik for 1890. I am myself
> working on the doctrine of multitude & that is one reason why I have not
> read what Whitehead & Russell have written. I don't want my own train
> of thought shunted for the present. (L 463 [1 December 1903]/*Lieb ed.*
> *1953:* 2)[58]

There seems a certain sense of his having been remiss in Peirce's
description of his reading of PM; his "examination" of PM having been
"slight," he says, during the interval between his notice in *The Nation* and
his letter to Lady Welby. It is a remissness that Peirce seemed not ready to
rectify. For he subsequently (1905–1907) remarked that

> By the time Whitehead's and other works had appeared, I was so engaged
> in the struggle with my own conceptions that I have preferred to postpone
> reading those works until my own ideas were in a more satisfactory con-
> dition, so that I do not know in how much of what I have to say I may have
> been anticipated. (MS 27 [1905–1907]/NEM 3: 1069)

Peirce here refered to "Whitehead's" work and, in the letter to Lady Welby,
to "Bertrand Russell's book," if he be "its author," and to "what Whitehead
& Russell have written." These are allusions to Peirce's suspicion of the
extent of Russell's authorship of PM.

There is another passage in a draft of Peirce's Lowell Lectures (Lecture
3, 30 November 1903) that is similar to his letter to Lady Welby:

> Since I have alluded to Cantor, for whose work I have a profound admira-
> tion, I had better say that what I have to tell you about multitude is not in
> any degree borrowed from him. My studies of the subject began before
> his, and were nearly completed before I was aware of his work, and it is
> my independent development substantially agreeing in its results with his
> of which I intend to give a rough sketch. And since I have recommended
> Dedekind's work, I will say that it amounts to a very able and original
> development of ideas which I had published six years previously. Schröder
> in the third volume of his logic shows how Dedekind's development might
> be made to conform more closely to my conceptions. That is interesting;
> but Dedekind's development has its own independent value. I even incline
> to think that it follows a comparatively better way. For I am not so in love
> with my own system as the late Professor Schröder was. I may add that

quite recently Mr. Whitehead and the Hon. Bertrand Russell have treated of the subject; but they seem merely to have pre[sented] truths already known into a uselessly technical and *pedantic* form. (MS 459: 17–18/NEM 3:347)[59]

Once again the point is being made that Russell and Whitehead are co-authors of PM. Indeed, the original second clause in the last sentence is, "but I have not as yet had time to examine their work."

There is also the interesting reference to Russell in Peirce's 1906 remark that "I *think* Russell is right in what he *means* when he says that the question [of the multitudes corresponding to ω numbers] is one of logic and not of mathematics" (L 148 [8 May 1906]/NEM 3:785).[60]

It is to be noted, however, that Peirce regarded PM "as a digest of what others have done" (L 463 [1 December 1903]/*Lieb ed. 1953*: 2), as "a convenient introduction to the remarkable researches in the logic of mathematics that have been made during the last sixty years" (*Peirce 1903*/8.171 n. 1), and as giving "those who have not at all studied the subject some notions of it that may be said to be true in the main" (MS 12: 1ᴾ).

3.2. *Peirce Notices PM or Not Two Notices of Russell.* The evidence for Peirce's judgment of PM, however tenuous for his never having perused the work,[61] pales on recognition of his "examination" being "slight." There are, of course, the annotations by Peirce in a copy of PM in the Houghton Library at Harvard and there is also some evidence of his having owned two other copies (*Fisch 1974*: 100, citation 21).[62]

In a letter of 30 July 1903 (L 78),[63] J. McKeen Cattell, the editor of *Science,* asked Peirce to write a more professional review of PM. It seems that Cattell, on receiving Peirce's agreement to undertake such a review, sent a copy of PM to Peirce on 15 August 1903. The review, however, appears untried. In a 27 July 1904 draft of a letter to Ladd-Franklin that is perhaps the most sensitive, telling, and self-revealing of his references to Russell and Whitehead, Peirce reports that

> a year has passed since I agreed to notice Russell's Principles of Mathematics Vol. I and I feel its pretentiousness so strongly that I cannot well fail to express it in a notice. Yet it is a disagreeable sort of thing to say, and people may ask themselves whether it is not simply the resentment of the old man who is getting laid upon the shelf. Perhaps I might require strict searchings to be sure myself that that element was not present. The result is that I have never written the notice.
>
> I have the same opinion about Whitehead.
>
> I winked at it in the Nation by making a few commonplace remarks about the Hon. Russell's book and then turning to Lady Welby's "What is Meaning" and saying "This is a book as opposite to the other as possible. It is entirely free from pretentiousness and from pedantry," etc. I can't give

my exact words, not having kept a copy; but it must have appeared Oct. 15 last. (L 237)[64]

Peirce presumably received one copy of PM from the editor of *The Nation*, Wendell Phillips Garrison, before or soon after PM was listed in the 9 July 1903 "Books of the Week" (*Peirce 1903*). There is also a second copy of PM that Garrison mistakenly sent on 9 October 1903, since Peirce's review would have already been in his hands. There apparently were, then, three copies of PM, the two from Garrison and the one from Cattell, which Peirce received in 1903 and could have variously read. The annotated copy is in all likelihood one of the three copies; the whereabouts of Peirce's other two copies of PM are unknown.

The annotations in Peirce's extant copy of PM generally comprise marginalia on terms such as 'conditionality' and 'class,' but it is by no means certain whether these notes are from the period of his notice in *The Nation* or later.[65] These notes, nonetheless, constitute a record of Peirce's principal criticism of Russell. Recourse to all of Peirce's annotations—to be quoted subsequent to quotation of the appropriate passage in PM—is then duly in order.

3.3. *Peirce and Russell on 'Conditionality.'* Peirce's "treasure of the pure theory [of logic] itself" (2.122)[66] and his logical notation, were, unlike Russell's, an evolutionary link in the passage from the algebraic to the graphic.[67] Peirce (*1885*) employs $\alpha \prec \beta$ for philonian or conditionality *de inesse*. He also utilizes conditionals *cum modo*, but treats 'α implies β' by extending substitution of α and β in $\alpha \subset \beta$ to expressions ranging over propositions:

> Any proposition which neither requires the exclusion from nor the inclusion in the universe of any state of facts or kind of object except such as a given second proposition so excludes or requires to be included, is implied in that second proposition in the logical sense of implication. . . . Accordingly, whatever can be logically deduced from any proposition is implied in it; and conversely. . . . All that concerns logic is, whether all the facts excluded and required by the one proposition are among those so excluded or required by the other. (2.604)[68]

Russell's exposition in PM presents implication as primitive and '*a* implies *a*' as a *definitio nominis* of 'proposition':

> A definition of implication is quite impossible. If *p* implies *q*, then if *p* is true *q* is true, *i.e.* *p*'s truth implies *q*'s truth; also if *q* is false *p* is false, *i.e.* *q*'s falsehood implies *p*'s falsehood. Thus truth and falsehood give us merely new implications, not a definition of implication. (*Russell 1903* [*1956*]: 14)[69]

Peirce here comments in the margin that " 'Implication' is a class proposition about states of things."

The use-mention distinction is something of an historical morass, but it is a distinction with some precedent in the work of Peirce and Frege.[70] Russell here confuses "if p, then q" with " 'p' implies 'q'." The failure to observe the use-mention distinction is frequent in PM, and leads to a blurring, as happens later in Pma, of 'predicate' and 'statement connective,' with Russell indiscriminately explaining 'α implies β' as 'truth-functional conditionality' and as 'material implication.'

Russell's mode of exposition can be found in his ten axioms—the first statement of which being the only pertinent one—for the propositional calculus: "(1) If p implies q, then p implies q; in other words, whatever p and q may be, 'p implies q' is a proposition" (*Russell 1903 [1956]:* 16).[71] Beside it Peirce wrote: "Ridiculous modes of formulation."

Russell cites $(p)(f \supset p)$ in an exposition of Lewis Carroll's so-called "logical paradox," arguing that if $q \supset r$ and $p \supset (q \supset \sim r)$, then because $\sim ((q \supset r) \cdot (q \supset \sim r))$, then $\sim p$:

> But in virtue of our definition of negation [that not-p is equivalent to the assertion that p implies all propositions], if q be false both these implications will hold: the two together, in fact, whatever proposition r may be, are equivalent to not-q. Thus the only inference warranted by Lewis Carroll's premisses is that if p be true, q must be false, *i.e.* that p implies not-q; and this is the conclusion, oddly enough, which common sense would have drawn in the particular case which he discusses. (*Russell 1903 [1956]:* 18n.)[72]

Peirce counters with the example (Figure 7.1) that if $\sim q$, then $(q \supset r) \cdot (q \supset \sim r)$ is true whether p is true or false. "Carroll is not right for if q is absurd $q \prec r$ and $q \prec \bar{r}$ may both be true."

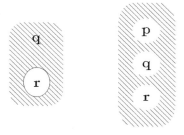

Figure 7.1

A smudging in Peirce's marginal graph makes it difficult to be sure what is and what is not shaded. Note also that Russell's $(p)(f \supset p)$ actually is equivalent to $\alpha \prec x$ in Peirce's 1885 system (3.381–385).[73]

Russell also cites in PM a dispute between Schröder (*1891*), asserting that $((p \cdot q) \supset r) \equiv ((p \supset r) \vee (q \supset r))$, and MacColl (*1878*), admitting only that $((p \supset r) \vee (q \supset r)) \supset ((p \cdot q) \supset r)$: "Schröder asserts that if *p, q, r* are propositions, '*pq* implies *r*' is equivalent to the disjunction '*p* implies *r* or *q* implies *r.*' Mr MacColl admits that the disjunction implies the other, but denies the converse implication" (*Russell 1903* [*1956*]: 22).[74] Peirce states in his note that "MacColl is wrong." Russell's reference to MacColl is interesting, for as late as 1892 Ladd-Franklin had complained, "I think it unfortunate that Mr. MacColl's valuable work in Logic should have met with almost complete neglect in England" (*Ladd-Franklin 1892b*: 527).[75] It is pertinent then that while Peirce's knowledge of Schröder's work is assured, his acquaintance with MacColl's 1877 paper can only be inferred from evidence that he was aware of MacColl's 1877–1880 papers. Peirce remarks in 1880:

> Mr. Hugh MacColl, apparently having known nothing of logical algebra except from a jejune account of Boole's work in Bain's *Logic,* published several papers on a *Calculus of Equivalent Statements,* the basis of which is nothing but the Boolian algebra with Jevons's addition and sign of inclusion. Mr. MacColl adds an exceedingly ingenious application of the algebra to the transformation of definite integrals. (3.199 n.2)[76]

Russell suggests that a "reason for the divergence is, that Schröder is thinking of propositions and material implication, while Mr. MacColl is thinking of propositional functions and formal implication" (*Russell 1903* [*1956*]: 22).[77] Peirce concurs with Schröder, who is correctly thinking of a relation ø on the set {1,0}, whereas MacColl is thinking of a relation ø' on the set {1,2,0}.[78]

3.4. *Peirce and Russell on Class.* Russell held $(p \supset (q \vee r)) \supset ((p \supset q) \vee (p \supset r))$ to be true in the calculus of propositions: "but its correlative [in the calculus of classes] is false, namely: 'If *a, b, c* are classes, and *a* is contained in *b* or *c,* then *a* is contained in *b* or *a* is contained in *c* ' " (*Russell 1903* [*1956*]: 12).

Peirce demurs, "Not so." The correlative as a class is obviously not true but it can equal a *universe X*. However, without any suggestion to the contrary, the available evidence indicates (see Appendix) that Russell is translating "If *a* is contained in *b* or *c,* then *a* is contained in *b* or *a* is contained in *c*" as $(a \subset b \vee c) \supset ((a \subset b) \vee (a \subset c))$, which is nonsensical. If the less equivocal $(a \subset b \cup c) \supset ((a \subset b) \vee (a \subset c))$ be granted as Russell's translation, it is not the correlative of $(p \supset (q \vee r)) \supset ((p \supset q) \vee (p \supset r))$ in the calculus of classes.

Russell recognized a distinction between propositions and classes. He regarded propositions as primary, and he criticized propositional functions construed as propositions, but he says that "From this criticism, Schröder is exempt: his second volume deals with genuine propositions, and points out their formal differences from classes" (*Russell 1903* [*1956*]: 13). Peirce responded that this is "utterly false."[79]

Russell regarded a propositional function ϕx as a proposition for every value of x, and "the study of propositional functions . . . to be strictly on a par with that of classes, and indeed scarcely distinguishable therefrom" (*Russell 1903* [*1956*]: 13).[80] Russell, however, confused the mention of ϕ (as a mark for an unspecified predicate) with its use as a name of an unspecified attribute in 'ϕx.'

Peirce, anticipating (anon) Russell here, observed in 1885 that "What the usages of language may be does not concern us; language has its meaning modified in technical logical formulae as in other special kinds of discourse." (3.374) And in 1902, he said that

> One needs but to turn over the leaves of a few of the first logical treatises that come to hand, especially if they are English, nor will he have to search long, in order to meet with appeals to the ordinary usages of speech as determinative of logical doctrines. Some recent books are quite crowded with this type of argument. It seems usually to be employed unreflectively; but there are works in which it is deliberately laid down as the principal basis of logical science. (2.67)[81]

Thus, Russell remarks in PM that "Verbs and adjectives occurring as such are distinguished by the fact that, if they be taken as variables, the resulting function is only a proposition for *some* values of the variable, *i.e.* for such as are verbs or adjectives respectively" (*Russell 1903* [*1956*]: 20n.).[82] About this Peirce comments with scorn, "What utter trifling!"

3.5. *Peirce on Class, Being, Collective, Identity.* Peirce described his logical realism as an example "of the most pronounced type" (6.163).[83] This requires a brief digression, for it is here that Peirce's logical and ontic differences with Russell's account of n-adic relations lie.

Peirce, unlike Russell, was not committed to extensionality. He utilized, that is, predicates *de inesse* and *cum modo*. Peirce conceded that the *extension of a property* ϕ is the class defined by 'ϕ,' and he casually noticed in 1902 that "A *class*, of course, is the total of whatever objects there may be in the universe which are of a certain [qualitative] description" (1.204).[84]

The next year Peirce advanced, in his fifth Lowell Lecture, two collective senses of ϕ:

> The one I shall call a gath which is simply the word 'gather' with the last syllable dropped. The other I call a sam which is the word 'same' with the last letter dropped. I also like this word because it is so much like the word *sum*, in the phrase *sum total*. It also recalls the German *Samlung* [*sic*]. A collection, in the sense of a *gath*, is a subject which is a pure Secondness without Firstness, and whose only mode of being is whatever existence it

may have; and this consists in the existence of certain other existents, or pure Seconds, called its *members*. . . . A *sam* is an *ens rationis* whose essence is the being of a definite quality (imputed to the *sam*) and whose existence is the existence of whatever subject there may be possessing that quality. . . . No doubt the easiest way to conceive of the *sam* is to imagine that you have a common noun, without specifying what noun it is, and to think that that noun signifies some quality which is possessed by anything to which it applies, but is not possessed by anything to which it does not apply. Now you are to imagine a single thing which is composed of parts. Nothing is done to these parts to put them into their places in the whole: their mere existence locates them in the whole. Now think of this rule as describing the whole. If any individual object can properly have that common noun predicated of it, it is a part of the single object called a *sam;* if not, it is not. That gives you the idea of the *sam*. Now to get the idea of a *gath*, you are to consider that those individual objects might change their qualities without losing their individual identity; so that limiting ourselves to any instant any individual object which at that instant forms a part of the *sam* forever forms a part of an object of which no object not at that instant a part of the sam is a part, and this individual composite whole which has nothing to do with the qualities of its members is a *gath*. . . . So for every gath there is a corresponding *sam*. But it is not true that for every *sam* there is a corresponding *gath*. . . . What has been said of qualities is equally true of relations, which may be regarded as the qualities of sets of individuals. That is to say, if any form of relation is logically possible between the members of two *gaths*, a relation of that form actually exists between them. (MS 469: 8^P–36^P passim)[85]

According to Peirce, the generality of ϕ being "peculiar to the category of quality," is "merely potential" (1.427).[86] In fact, by virtue of Peirce's professed "extreme form of realism" (NEM 4: 345),[87] a relation ø is *universale inter res*, and "it does not contain any individuals at all. It only contains general conditions which *permit* the determination of individuals" (6.185).[88] Peirce consequently neither delineates sharply between '∈' and '⊂,' nor between 'individual' and 'predicate constant' (*Martin 1969:* 147, 149).

Peirce also considered the basis of identity to be a regularity of semiosis—a symbol the instance of which is "only a replica of a symbol of the nature of an index" (4.500).[89] He considered i, the index in x_i, to be pronominal and indicatory, and held that a noun, as denotative and an interpretation of 'x', to be "an imperfect substitute for a pronoun" (2.287 n.1).[90]

Peirce applied the scholastic 'first intention' (*intentio prima*) and 'second intention' (*intentio seconda*) to ϕ:

First intentions are those concepts which are derived by comparing percepts, such as ordinary concepts of classes, relations, etc. *Second intentions* are those which are formed by observing and comparing first intentions. Thus the concept "class" is formed by observing and comparing class-concepts and other objects. The special class-concept, *ens*, or what is, in the sense of including figments as well as realities, can only have originated in that way. Of relative second intentions, four are prominent—identity, otherness, co-existence, and incompossibility. (2.548)

Thus in 1885 he adopted '1', for example, as a general relative "of second intention . . . to express identity." Using '=' for 'logical equivalence,' he defines '1' as $1_{ij} = \Pi_k (q_{ki} q_{kj} + \overline{q}_{ki} \overline{q}_{kj})$ (3.398).[91] Peirce then states that

we can dispense with the [first-order predicate or] token *q*, by using [a second-order predicate or] the index of a token and by referring to this in the Quantifier just as subjacent indices are referred to. That is to say, we may write $1_{ij} = \Pi_x (x_i x_j + \overline{x}_i \overline{x}_j)$.[92]

The properties of *q* are then "summed up in this, that taking any individuals, i_1, i_2, i_3, etc., and any individuals, j_1, j_2, j_3, etc., there is a collection, class, or predicate embracing all the *i*'s and excluding all the *j*'s except such as are identical with some one of the *i*'s" (3.399).

3.6. *Russell on Class, Being, Propositional Function.* Russell construed classes reductively, i.e. as propositional functions, and his logical and ontic account is generally at variance with Peirce's account of *n*-adic relations.

Russell counted $(\alpha)\phi$ and $(\exists \alpha)\phi$ as *proposition schemata*, and $\phi \alpha$ as a *propositional function* associated with a *class* $\hat{\alpha}\phi$, that is, the "values [of *x* which render a propositional function ϕx true] form a *class*, and in fact a class may be defined as all the terms satisfying some propositional function" (*Russell 1903 [1956]*: 20). Peirce retorts in his note that "A class can never be so defined."

Russell's stance also involves the ontic status of a class, since it is involved in the "notion" of

what is called the *existence* of a class—a word which must not be supposed to mean what existence means in philosophy. A class is said to exist when it has at least one term. A formal definition is as follows: *a* is an existent class when and only when any proposition is true provided "*x* is an *a*" always implies it whatever value we may give to *x*. It must be understood that the proposition implied must be a genuine proposition, not a propositional function of *x*. A class *a* exists when the logical sum of all propositions of the form "*x* is an *a*" is true, *i.e.* when not all such propositions are false. (*Russell 1903 [1956]*: 21)

Russell's use of 'class' engenders objects. He uses "the word *object* in a wider sense than *term*, to cover both singular and plural" (*Russell 1903 [1956]*: 55n.), and he reckons as values of *x* in '*x* is an *a*' anything that

may be an object of thought, or may occur [as a subject] in any true or false proposition, or can be counted as *one*, [which] I call a *term*. . . . I shall use as synonymous with it the words unit, individual, and entity. The first two emphasize the fact that every term is *one,* while the third is derived from the fact that every term has being, *i.e. is* in some sense. A man, a moment, a number, a class, a relation, a chimaera, or anything else that can be mentioned, is sure to be a term; and to deny that such and such a thing is a term must always be false. (*Russell 1903 [1956]:* 43)

Russell states that by 1900, he "began to believe in the reality of whatever could not be *dis*proved—e.g. points and instants and particles and Platonic universals" (*Russell 1903 [1956]:* 12).[93] Peirce pointed out in 1871 that

British philosophers have always desired to weed out of philosophy all conceptions which could not be made perfectly definite and easily intelligible, and have shown strong nominalistic tendencies since the time of Edward I., or even earlier. Berkeley is an admirable illustration of this national character, as well as of that strange union of nominalism with Platonism, which has repeatedly appeared in history, and has been such a stumbling-block to the historians of philosophy. (8.10)

Russell's use of 'class', 'object', and 'term' in PM effectively posits nonindividual and individual entities, and the system of propositions in PM commits him to *platonism* and to *nominalism.*[94]

3.7. *Russell on Extension, Intension, Class, Identity.* In PM, Russell classically assumes that $(x)(x \in \hat{x}(\phi x) \supset \phi x)$, and the axiom of extensionality, i.e., that $(x)(\phi x \equiv \psi x \supset (\hat{x}(\phi x) = \hat{x}(\psi x)))$.[95] "The extensional view of classes, in some form," Russell concludes, "is thus essential to Symbolic Logic and to mathematics, and its necessity is expressed in the above axiom" (*Russell 1903 [1956]:* 21). "Not at all," Peirce remarks, rejecting Russell's claim that "Symbolic Logic" requires an "extensional view of classes."

Russell proposed "the following [for the relation R on] classes: The class of terms which have the relation R to some term or other, which I call the class of *referents* with respect to R; and the class of terms to which some term has the relation R, which I call the class of *relata*" (*Russell 1903 [1956]:* 24). Peirce says of Russell on this point that "He considers only dyadic relations."

"The intensional view of relations here advocated," Russell states in PM, "leads to the result that two relations may have the same extension without being identical" (*Russell 1903 [1956]:* 24).[96] Peirce's note on this point recommends insertion of "existential" between "same" and "extension."

Russell also proposes an n place relation on segments and points of a straight line:

Given any three different points *a, b, c* on a line, consider the class of points *x* such that *a* and *c, b* and *x* are each harmonic conjugates with respect to some pair of points *y, y′*—in other words, *a* and *c, b* and *x* are pairs in an involution whose double points are *y, y′*. Here *y, y′* are supposed variable: that is, if any such points can be found, *x* is to belong to the class considered. This class contains the point *b* but not *a* or *c*. (*Russell 1903* [*1956*]: 386)

Peirce here prefixes "?" to the passage, apparently indicating that the last two sentences constitute an abuse of language.

"The distinction [between ∈ and the relation of whole and part between classes] is the same," Russell observes in PM, "as that between the relation of individual to species and that of species to genus, between the relation of Socrates to the class of Greeks and the relation of Greeks to men" (*Russell 1903* [*1956*]: 19). Peirce responds by remarking about relations *species infima* and *species subiicibilis* that "They are not formally distinct except for the relation of identity."

"It is to be observed that the class," Russell cautions, "must be distinguished from the class-concept or predicate [or *species praedicabilis*] by which it is to be defined: thus men are a class, while *man* is a class-concept" (*Russell 1903* [*1956*]: 19). Peirce rejects Russell's *use* of 'men', asserting that "Men are *not* a class."

3.8. *Russell on Ones, Twos, Threes or More*. In PM, Russell poses a distinction between a unit-class and the single element it contains:

> If we could identify a class with its defining predicate or class-concept, no difficulty would arise on this point. When a certain predicate attaches to one and only one term, it is plain that that term is not identical with the predicate in question. But if two predicates attach to precisely the same terms, we should say that, although the predicates are different, the classes which they define are identical, *i.e.* there is only one class which both define. (*Russell 1903* [*1956*]: 130–131)

In his extant copy of PM, Peirce has no marginal note here, although he may have annotated the passage in one of his other two copies of PM; but there seems to be a rebuttal in the fifth of his 1903 Lowell Lectures:

> Another point which I observe puzzles the Hon. Bertrand Russell in his "Principles of Mathematics" is whether a collection which has but a single individual member is identical with that individual or not. The proper answer is that if by a *collection* you mean a *sam*, the *sam* of the sun is not the sun, since it is an *ens rationis* having an essence, while the individual has no essence and is not an *ens rationis*. But if you mean the *gath*, the gath of the sun has no being at all except the existence of the sun which is all the being the individual existent sun has. Therefore, having precisely the same being they are identical and no distinction except a grammatical or linguistic one can be drawn between them. Mr. Russell's being puzzled by this is a good illustration of how impossible it is to treat of philosophy without making a special vocabulary such as all other sciences make. (MS 469: 32–33/NEM 3: 371)[97]

In a 1904 letter to Lady Welby, Peirce also objects to Russell's "pluralism," i.e., the increment of an *n* place relation to an *n* + 1 relation, in PM:

> I will only say that Mr. Russell's idea that there is a *fourthness*, etc is natural; but I prove absolutely that all systems of more than three elements are reducible to compounds of triads; and he will see that it is so on reflection. The point is that triads evidently cannot be so reduced since the very relation of a whole to two parts is a triadic relation. (L 463 [2 December 1904]/SS: 43/*Lieb ed. 1953:* 14)[98]

Indeed, if a logical view is taken of Peirce's three categories, "*Firstness* corresponds to a *monadic predicate, Secondness* to a *dyadic relation,* and *Thirdness* to a *triadic relation*" (*Wennerberg 1962:* 33).

§4

4.1. *Peirce, Russell, and Frege's Work.* There are extensive expansions of the logical and mathematical works of Frege in PM, whereas the references that Russell culls from Peirce's work are scant and cursory.[99] The intriguing questions here are whether Peirce was aware of Frege's work, or whether, given his "slight examination" of PM, Peirce noticed Russell's references to Frege. Peirce's annotated, extant copy of PM, with his marginalia never being too far from Russell's references to Peirce, seems to indicate that Peirce's 'reading' of PM was limited to the index entries under his name.

There are some circumstantial grounds, however, for speculation about Peirce's knowledge of Frege's work. There is, for example, the copy of Frege's *Begriffsschrift* [Bg] (*1879*) in the library at Johns Hopkins University.[100] There are no annotations by Peirce in this volume, but its date of acquisition is 5 April 1881, a date which falls within Peirce's ephemeral tenure at Johns Hopkins (1879–1884) as an instructor in logic.[101] There is also a copy of *Begriffsschrift* in the Rare Book Room at Princeton University which was owned by A. Marquand, a contributor to *Studies in Logic.* Of *Studies in Logic,* Peirce had remarked, "These papers, the work of my students, have been so instructive to me . . . " (*Peirce ed. 1883:* iii). There are, in addition, the citations of *Begriffsschrift* and Schröder's review of it (*Schröder 1880*) in Ladd-Franklin's bibliography of her contribution to *Studies in Logic* (*Ladd-Franklin 1883:* 70–77). There is also a copy of Schröder's review among a bound collection of offprints from Peirce's library now in the Widener Library at Harvard.[102] (The collection may have been bound after its acquisition by Harvard, with Schröder's review numbered 23 [not by Peirce] and marked [by Peirce] in green pencil "Formal Logic.")[103]

None of this, of course, proves that Peirce read Bg or even Schröder's review of it. The likelihood of either alternative should not, however, be dismissed merely as idle.

4.2. *Peirce, Schröder and His Review.* It is not clear when, or even if, Peirce received the offprint from Schröder. The earliest of Schröder's extant letters to Peirce is of 1 February 1890, and begins "Herewith I am trying to ascertain, whether your address continues to be the same that I have known some years ago" (L 392). So there was a correspondence between the two earlier than 1890. It may be, then, that Schröder sent the offprint of his review to Peirce before *Studies in Logic* went to press in 1882. It may also be, however, that Ladd-Franklin read and subsequently drew Peirce's attention to Schröder's review in the *Zeitschrift für Mathematik und Physik* of 1880, and that the Peirce-Schröder correspondence began later.

A tantalizing possibility is that, having read Schröder's review, Ladd-Franklin read either the copy of Frege's work in the Johns Hopkins Library, or perhaps Marquand's, and that she then drew Peirce's attention to it.

There is every indication that Peirce and Schröder were professionally acquainted. Peirce included Schröder's (*1877*) *Der Operationkreis des Logikkalkuls* in the list of texts for his courses of 1882–1883 at Johns Hopkins.[104] In his 1880 review of *Begriffsschrift*, Schröder cites three papers of 1867 and one of 1870, and includes references to four of Peirce's early papers. Remember, too, that in 1903 Peirce wrote that "I value Schröder's work highly & he was a highly sympathetic man whom it was impossible to know and not to like even more than the great merit of his work justified" (MS 302).[105] All of this suggests that the two were personally acquainted.

The years 1870, 1875–1876, 1877, and 1883 mark Peirce's sojourns in Germany. He also made a trip to Europe in 1880, but it was cut short before he reached Germany by his father's illness. If Peirce and Schröder met, it was probably in 1883, the year Schröder sent (or gave) Peirce an offprint of his review and, subseqently, they began their correspondence.

4.3. *Peirce, Schröder, Frege, Their Connection.* The writings of Schröder and Frege pose an extremely complicated textual connection with Peirce's work. There are various depictions of Schröder's work in the writings of Frege, while the writings of Schröder contain varying accounts of the work of Peirce and Frege.

Frege's writings contain much hostile criticism of the work of others, and especially of Schröder's work. Schröder, in turn, reciprocates Frege's criticism. If Schröder is mistaken about Frege, Frege is often as mistaken or misguided about Schröder.

In a review of 1895, Frege peremptorily dismissed Schröder's *Vorlesungen über die Algebra der Logik* (*1890–1895*). In it, he claims that Schröder's collective sense of class entails a contradiction for Λ. He also contends that Schröder's class comprising individuals, a unit-class (*eine singuläre Klasse*), coincides—which Leśniewski later showed does not follow—with an individual (*Frege 1967:* 193–210).[106] In his 1880 review of *Begriffssdchrift*, Schröder cited four of Peirce's early papers (*1867a* and *b*, and *1870*) noting that *1867a* anticipates "various results" (*verschiedene Ergebnisse*) in his *Der Operationkreis des Logikkalkuls* (*Schröder 1877*). He also cited Leibniz and, inaccurately, Boole as obviating Frege's system of notation or conceptscript (*Begriffsschrift*).[107]

Elsewhere (*1898a*), Schröder claimed that Peirce's work is "fundamental" to pasigraphy, with Boole and De Morgan being Peirce's "English precursors," but that "*Herr Frege,* who heedless of anything accomplished in the same direction by others, took immense pains to perform what had already been much better done and was therefore superseded from the outset, thus delivering a still-born child" (*Schröder 1898a:* 60–61).[108] Thus Schröder alleges that Peirce's logical work is continuous with the development of pasigraphy while that Frege's conceptscript is an isolated attempt.[109]

The interesting question is whether or not Frege saw in Schröder's writings any of the representations and developments of Peirce's work. It is almost inconceivable that Schröder's review of *Begriffsschrift*, with its references to four of Peirce's papers, would have escaped Frege's notice. Frege's more obvious secondary access to Peirce seems to have been through his review of Schröder's *Vorlesungen*. Yet Frege's writings appear quite uninfluenced by Peirce.[110]

4.4. Peirce's Awareness of Frege's Work. Peirce poignantly complains in 1893 that "for all my life my studies have been cruelly hampered by my inability to procure necessary books" (4.118).[111] Peirce's words here are particularly significant, for in 1893, Volume I of Frege's *Grundgesetze der Arithmetik* was published. Yet the circumstances seem to favor the possibility that Peirce was not aware of Frege's work.

Peirce's writings contain no mention of Frege—and the polymath Peirce was generally attentive to references.[112] Peirce, for example, seems entirely unaware of the quantification in *Begriffsschrift* when (in *1885*) he introduced his notational separation of the quantifying and indexing functions.[113]

"My analyses of reasoning," Peirce observed in 1903, "surpass in thoroughness all that has ever been done in print, whether in words or in symbols—all that De Morgan, Dedekind, Schröder, Peano, Russell, and others have ever done" (5.147).[114] The presence of "Russell" here is as significant as the absence of "Frege," for it was in 1903 that both Russell's PM and Volume II of Frege's *Grundgesetze der Arithmetik* were published or were "in print."

There are two instances where Peirce expressly gauges his "rank" as a logician, instances in which the absence of "Frege" seems to further suggest Peirce's lack of awareness of him. Thus Peirce writes that "I place myself somewhere about the real rank of Leibniz," who "had the advantage of coming to a field into which no reapers had come" (L 387 [15 November 1904]).[115] He wrote later that of "the only writers known to me who are in the same rank as I are Aristotle, Duns Scotus, and Leibniz, the three greatest logicians in my estimation, although some of the most important points escaped each" (L 482 [18 June 1909]/NEM 4: vi–vii).[116]

§5

5.1. *Peirce's and Russell's Cognizance of Each Other's Work.* The evidence that Peirce was either aware, or unaware, of Frege's work, is circumstantial. The evidence is somewhat more straightforward for the extent of Peirce's and Russell's cognizance of one another's work, and especially for Russell's knowledge of Peirce's work.

The question of the extent of Peirce's and Russell's cognizance of each other's work, of course, is not without some significance in their 'controversy' and, albeit as a footnote, the history of logic. This question, then, including M. G. Murphey's exegesis of Peirce on Russell, will be considered within this concluding section.

The scanty and cursory references to Peirce in PM would seem to suggest that in 1903 Russell was not well acquainted with Peirce's work. Russell admits this as late as 1946:

> I am—I confess to my shame—an illustration of the undue neglect from which Peirce has suffered in Europe. I heard of him first from William James when I stayed with that eminent man in Harvard in 1896. But I read nothing of him until 1900, when I had become interested in extending symbolic logic to relations, and learnt from Schröder's "Algebra der Logik" that Peirce had treated of the subject. Apart from his work on this topic, I had until recently read nothing of him except the volume entitled by its editors [*sic*] "Chance, Love and Logic." (*Russell 1946:* xv)[117]

The evidence for Russell's having read anything by Peirce on the topic of the logic of relations can be found in a set of notes (ca. 1900) which constitutes a precis of "On the Algebra of Logic" (*Peirce 1880*) and of "On the Algebra of Logic: A Contribution to the Philosophy of Notation" (*Peirce 1885*). Russell's notes comprise entries corresponding to 3.172, 184, 198–203, 215–219, 359, 402.[118] These few pages, I am convinced, represent Russell's reading of Peirce on the "topic," which may account for the inaccuracies in Russell's criticism of Peirce's relative addition.

However, the incompleteness of Russell's reading of Peirce on the "topic" seems not entirely to account for the inaccuracies. As Geach points out, "Russell, like Aristotle, so often distorts others' thought into his own mold" (*Geach 1958–1959:* 72). It is reported, for example, "that Leśniewski [1886–1939] in the last years of his life advocated a movement 'Back to

Frege' as he (Leśniewski) was discouraged by Russell's 'oversimplifications and distortions' of Frege"; in Leśniewski's view, "Russell was too vigorous and too impatient in his expansions of Frege's ideas and made a number of blunders not to be found in Frege" (*Skolimowski 1967:* 245 n. 5).[119] Thus, Russell's inaccuracies seem the inaccuracies of a superficial reading of Peirce on the "topic," and of the straw man in his criticism of Peirce's logic of relations.

It appears that Peirce's examination of PM was no more than slight during the interval between his notice in his 1903 duplex review for *The Nation* and his 1905–1907 remark on his dilatory reading (MS 27/NEM 3: 1069). Yet in his 1912 manuscript "Notes Preparatory to a Criticism of Bertrand Russell's *Principles of Mathematics*" (MS 12), Peirce—whether the marginalia in his extant copy of PM are from the period of this late manuscript or earlier—seems to have resumed his "examination."

Murphey accepts Peirce's 1903 notice of PM in *The Nation,* the annotations in Peirce's extant copy of PM, and Peirce's 1912–1913 (unfinished) manuscript "Mr. Bertrand Russell's Paradox" as "cursory" evidence that Peirce "never read" PM (*Murphey 1961:* 241–242).[120] Contrary to Murphey, however, the evidence does not seem so clear-cut.

Murphey fails to appreciate properly the significance of Peirce's manuscript on "Russell's Paradox." The passage in question is the following:

> In two recent notices of Russell and Whitehead's last volume [of PMa] I have seen the following set down as self-[contradictory]. I change only the wording.
>
> If any *x* does not include itself, it includes *w*, it does [*sic*].
>
> If any *x* includes *w* it does not include itself.
>
> But suppose that, in place of the logical relation of inclusion, we substitute an ordinary relation between individuals. We shall so get such a pair of propositions as the following:
>
> Whatever is not a lover of itself is a lover of *w*.
>
> If anything is a lover of *w*, it is not a lover of itself.
>
> These two propositions simply imply that anything of the nature of *w* is not self-identical. It is perhaps of a general nature, such as an abstraction. There are many things that have to be reasoned about that are not exactly *existent.*
>
> In mathematics such apparently self-contradictory assertions are often made, but are easily interpreted so as to remove their self-contradictory sound. (MS 818)[121]

"Peirce's rendering of the paradox," according to Murphey, "makes it equivalent to the paradox of the barber," whereby the supposition of the barber's existence leads not to a genuine paradox but to the contradiction "that there is no such barber" (*Murphey 1961:* 242).

Peirce's assumption that any "proposition besides what it explicitly asserts, tacitly implies its own truth" (3.446) escaped Murphey's notice. Peirce, that is, requires $\phi \vdash$ 'ϕ' (1) as a formal condition for assertion of any relation ϕ on the set {1, 0}.[122] Thus, Peirce—following Paul Nicollet of Venice (ob. 1429)—considers the $\alpha(\alpha \notin \alpha)$ paradox to reduce to an *insolubilia,* and "therefore, [to be] true in all other respects but in its implication of its own truth" (5.340).[123] Russell actually credits Peirce's "*Insolubilia*" (*1901*) in *Baldwin's Dictionary* with showing that an essentially correct diagnosis of the "paradox" is anticipated in logical work of the Middle Ages (*Russell 1906a:* 627–650).[124]

5.2. *Concluding a Random Harvest.* The direct value of studying a 'controversy' like that between Peirce and Russell is very considerable; it lies as much in the addition of detail to the history of logic (or mathematics) as to the disputants' biographies. The references, notices, remarks, and commentaries that are met in Peirce's and Russell's 'controversy' leave an impression that theirs was a cognizance of bits and pieces excised from each other's work. "The history of human thought," Koestler asserts, "is full of lucky hits and triumphant *eurekas;* it is rare to have on record one of the anticlimaxes, the missed opportunity which normally leaves no trace" (*Koestler 1963:* 201). Peirce's and Russell's 'controversy' is such a record of the missed opportunity—at the eventide and morn of their respective careers, as it were, a random harvest of occasional theses and antitheses.

A P P E N D I X

On a Falsigrafis in Bertrand Russell's Principles of Mathematics .

1. *Introduction.* "The propositional calculus," Russell states in PM, "is characterized by the fact that all of its propositions have as hypothesis and as consequent the assertion of material implication" (*Russell 1903* [*1956*]: 13), whereas in the calculus of classes "there are very much fewer new primitive propositions—in fact, two [operations of $\in$ and $\subset$] seem sufficient—but there are much greater difficulties in the way of non-symbolic exposition of the ideas embedded in our symbolism" (*Russell 1903* [*1956*]: 18). Thus, he designates

$$\text{S)} \quad (p \supset (q \vee r)) \supset ((p \supset q) \vee (p \supset r))$$

as true within the calculus of propositions but its correlative [in the calculus of classes] is false, namely: "If *a, b, c* are classes, and *a* is contained in *b* or *c,* then *a* is contained in *b* or *a* is contained in *c*" (*Russell 1903* [*1956*]: 12). The only known demurral to Russell here is that of C. S. Peirce. The correlative of S as a *class* obviously cannot be true, although it can be equal to a *universe X*. Russell in PM, however, takes implication as primitive and 'α implies α' as a *definitio nominis* of 'proposition' (*Russell 1903* [*1956*]: 13–18). He consequently seems to equivocate by translating 'If a is contained in b or c, then a is contained in b or a is contained in c' as

$$S') \qquad (a \subset b \vee c) \supset ((a \subset b) \vee (a \subset c)),$$

which is not the correlative of S—albeit nonsensical—in the calculus of classes.

The question, then, is whether the correlative of S is a *class* or a *statement*? Furthermore, when is 'set inclusion' or 'truth-functional conditionality' an interpretation of $\subset$, and is there a rule to decide? In what follows, the present author shall examine briefly these questions.

2. *Translation of* S. The relation denoted by '=' on S and X (mentioned above) is rendered as $Tr(S)$, where 'Tr' is an abbreviation of 'translation.' Thus,

$$1a) \qquad Tr\,(p \supset q) = 1 \text{ iff } Tr(p) \subseteq Tr(q)$$

is supported by

$$1b) \qquad Tr(p \supset q) = Tr(\sim p \vee q)$$

and

$$1c) \qquad Tr(\sim p \vee q) = \overline{Tr(p)} \cup Tr(q),$$

where $\overline{Tr(p)}$ denotes the complement of $Tr(p)$. Now because

$$2a) \qquad A \cup B = X \text{ iff } A \subseteq B$$

is true of the classes A and B (a simple Venn diagram can be employed to verify this),

$$2b) \qquad Tr(p \supset q) = \overline{Tr(p)} \cup Tr(q) = X \leftrightarrow Tr(p) \subseteq Tr(q)$$

holds and

$$3) \qquad Tr(S) = X \text{ iff } Tr(p \supset (q \vee r)) \subseteq Tr((p \supset q) \vee (p \supset r))$$

follows by applying 2b) to S. Now in as much as

$$4a) \qquad Tr(p \supset (q \vee r)) = Tr(\sim p \vee (q \vee r))$$

and

$$4b) \qquad Tr(\sim p \vee (q \vee r)) = \overline{Tr(p)} \cup (Tr(q) \cup Tr(r))$$

together with

$$5a) \qquad Tr((p \supset q) \vee (p \supset r)) = Tr((\sim p \vee q) \vee (\sim p \vee r)),$$

5b) $Tr((\sim p \lor q) \lor (\sim p \lor r)) = (\overline{Tr(p)} \cup Tr(q)) \cup (\overline{Tr(p)} \cup Tr(r))$

and

5c) $(\overline{Tr(p)} \cup Tr(q)) \cup (\overline{Tr(p)} \cup Tr(r)) = \overline{Tr(p)} \cup (Tr(q) \cup Tr(r))$

all hold. Therefore, not only does it follow that

6) $\qquad\qquad Tr(p \supset (q \lor r)) \subseteq Tr(\sim p \lor (q \lor r))$

but, *a fortiori*, it also follows that

7) $\qquad\qquad (p \supset (q \lor r)) \equiv ((p \supset q) \lor (p \supset r))$

is tautologous (*i.e.,* it passes the truth-table test). Consequently

8) $\qquad\qquad\qquad Tr(S) = X$

holds. In general,

9) $\qquad\qquad$ S is tautologous iff $Tr(S) = X$

and so there is no contradiction if 'correlative' is precisely defined.

3. *Russell's Error.* If a, b, and c are classes (following Russell above),

10a) $\qquad\qquad\qquad a \subset b \cup c$

means

10b) $\qquad\qquad (x) \; [x \in a \supset (x \in b \lor x \in c)]$

and

11a) $\qquad\qquad\qquad a \subset b \lor a \subset c$

means

11b) $\qquad (x) \; [x \in a \supset x \in b] \lor (x) \; [x \in a \supset x \in c].$

Now if one is careless and construes 11b equivalent to

11c) $\qquad (x) \; [(x \in a \supset x \in b) \lor (x \in a \supset x \in c)]$

iff

11d) $\qquad\qquad (x) \; [x \in a \supset (x \in b \lor x \in c)],$

then because 7 is tautologous, 10a and 11a may be incorrectly construed as equivalent. Russell surely means something like this move, his error being that he (most likely) assumed

12) $\qquad\qquad [(x)\phi \lor (x)\psi] \supset (x) \; [\phi \lor \psi]$

to be valid.

4. *Epilogue.* In Russell's estimation, PM (1903) corresponds to PMa

(1910–1913) as "a crude and rather immature draft of the subsequent work, from which, however, it differ[s] in containing controversy with other philosophies of mathematics" (*Russell 1959:* 74).[125] Russell considers that PM "is in the right where it disagrees with what had been previously held, but where it agrees with older theories it is apt to be wrong (*Russell 1903* [*1956*]: xiv). Yet Russell's *falsigrafis* or S' is not a matter of "older theories."

N O T E S

1. I should like to express my gratitude and thanks to the Harvard Department of Philosophy for permission to quote from Peirce's manuscripts at the Houghton Library; to Dr. Piero E. Ariotti, Professors Don D. Roberts, James Van Evra, Norman Feldman, and Nathan Houser for valuable suggestions and criticism; and to Professor Max H. Fisch for the inspiration and many of the primary sources—but for him, the present paper would not have been written.

2. See *Hawkins 1986:* 62–69.

3. See *Bell 1945:* 249–150; *Eisele 1975:* 149–158; *Fisch 1972:* 487–488; *Hawkins 1975a:* 109–115; *1975b:* 128–139; *1986:* 62–69; *Lenzen 1973:* 239–254; *1975:* 159–166, 225–226; *Murphey 1961:* 197; *Royce and Kernan 1916:* 701.

4. 2.197, 364; 3.565; 4.239; 5.181; 6.125; L 168 (13 January 1878); NEM 3: 769, 925. See also *Hawkins 1986:* 62–63.

5. Compare L 387 (15 November 1904); MS 1482 (18 June 1909), or NEM 4: vi–vii; see *Fisch 1972:* 486, 487.

6. Compare *Russell 1918* [*1949*]: 78n.1.

7. See 3.20–403M; *Peano 1895:* 122–128. Compare *Hawkins 1975a:* 109–115; *1981:* 381–389; *Russell 1959:* 39, 65, 69–70; *1919* [*1960*]: 11, 25n.2; *1967:* 91; *van Heijenoort 1967:* 3, 83–85, 124–125.

8. See 4.4. Compare *Russell 1918* [*1949*]: 76n.1; *1903* [*1956*]: 203n., 232n., 320n., 376; *1959:* 87.

9. Compare *Hawkins 1975a:* 112, 114n.53; *Ladd-Franklin 1892a:* 126–128, 132; *Lewis 1918* [*1960*]: 79–117, 222–290; *Russell 1959:* 65; *Whitehead 1898* [*1960*]: 10n., 37–44, 115–116.

10. Compare 2.333, 349–352; 3.425, 429–455, 510–525, 580, 602–603, 610, 622, 640, 643; 4.94, 346, 392, 617, 633, 660; 5.85, 178; 6.175; L 463 (12 October 1904); N 1: 91, 110–111.

11. See 4.617n.§.

12. Compare *Hawkins 1975a:* 109–115.

13. Also published in *Lieb ed. 1953:* 10–11 and *Wiener ed. 1958* [*1966*]: 388–389.

14. See *Russell 1903* [*1956*]: 24, 26. Compare *Lieb ed. 1953:* 11n.26; *Wiener ed. 1958:* 388n.3.

15. Compare *Whitehead and Russell 1910–1913* [*1959–1960*] 1: v.

16. Compare *Russell 1959:* 74.

17. Compare 1.558 and see 1.566–567. Also compare *Whitehead 1901:* 139; *1898* [*1960*]: 10n., 37, 42, 114–116.

18. Compare *Russell 1944* [*1951*]: 12–13.

19. *Russell 1919* [*1960*] and *1903* [*1956*] respectively.

20. Compare 1.444; 2.93, 227; 4.9.

21. Compare 1.191, 427–440, 444, 559; 2.93, 205, 229, 446, 461–467, 588–589; 3.154–172; 4.134, 239–250. See *Hawkins 1986:* 63–64.

22. See, for example, 2.267, 278–282, 295, 309–383; 3.559–561; 4.130, 176; 5.175–179. Compare NEM 2: 629–632. See also *Hawkins 1981:* 387–389; *Henkin 1956:* 28–32; *Tarski 1956:* 409–420.

23. Compare 2.253, 226, 461–470, 588–589; 3.154–172; L 463 (12 October 1904).

24. Compare 2.266, 461–470; 3.154–172; 4.531; HP 2: 809–811, 912; MS 12, 5^P; L 463 (12 October 1904).

25. Compare 2.446, 588–589; 3.160–172; 5.120–212; 7.500–538; 8.191. See *Bird 1962:* 175–178; *Bocheński 1961:* ix; *Boler 1963:* 102 et passim; *Patin 1957:* 243–252.

26. See also *Kneale 1962:* 230 and *Savan 1988.*

27. Compare 1.444; 2.206, 229; 3.430; 8.342. See *Kneale 1962:* 242.

28. Compare 1.539, 559, 574; 2.52–55, 93, 105–229, 599; 3.163–169, 364; 4.9, 227–249; 5.35, 40, 108–114, 129–168, 365, 446; 7.49–138, 186; HP 2: 1035; MS 499, 13^P. See *Buchler 1939:* 197–215; *Burks 1943:* 187–193; *Hawkins 1975b:* 138n.13; *Hocutt 1972:* 451.

29. Compare 1.247; 3.428, 558–559; 4.134, 229, 245–249; 5.8; 7.524–525; 8.110; NEM 4: 149–150; HP 1: 143, 147; HP 2: 720–721, 826–827, 832–833, 880–881, 892, 939–944, 1118. See *Bell 1945:* 211; *Hawkins 1975b:* 138n.13; *1986:* 63–64; *Murphey 1961:* 229–232.

30. Compare 4.233–239.

31. Compare 1.83; 2.227, 464, 778; 3.462; 4.233, 530–531, 549; L463 (12 October 1904); NEM 1: 232–240.

32. Compare 3.558.

33. Compare 1.53–54, 66, 245–249, 283; 2.250–272, 541, 544–545; 4.133, 426–429, 481; 5.161–166, 175–178; 7.524–525; L 463 (12 October 1904). See *Church 1956:* 50n.117, 76–77n.168; *Frege 1967:* 262–272, 395–422; *Hawkins 1975b:* 136n.2; *Kleene 1952 [1971]:* 53–65; *Kneale 1962:* 686–687; *van Heijenoort 1967:* 376, 380–381, 467, 475.

34. Compare 4.132; 6.474; Kant 1781–1787: A6–7, B10–11, A154–158, B193–197.

35. I am indebted to Professor Carolyn Eisele for calling my attention to this passage. Compare 1.434; 2.352; 5.447–450; NEM 1: 122; 2: 514; 3: 868. See *Bell 1945:* 211–212, *Burks 1949:* 676–677; *Kleene 1952 [1971]:* 46–65; *Murphey 1961:* 230–231, 238–240, 244, 286–288; *Patin 1957:* 243–251; *van Heijenoort 1967:* 334–345, 446–463, 480–484, 490–492.

36. See also *Hawkins 1975a:* 109; *1975b:* 134–135, 138n.13; *1986:* 63–64.

37. Compare 1.248–249; 2.191; 4.228, 239–244; HP 2: 827, 832–833; MS 469, 4–6.

38. Compare 1.247, 3.92, 4.288.

39. Compare 2.122, 195.

40. Compare 4.134, 239; 5.83–85; MS 499, 1^P–5^P.

41. Compare 4.263.

42. Compare 4.368–373, 617; MS 499, 1^P–5^P.

43. See *Hawkins 1975b:* 133; *1981:* 386–389.

44. Compare 1.562; 3.45–46, 199n.2, 359–364, 641; 4.4, 134, 239, 368, 372–373, 428–429, 481, 530–531, 561n.1, 617; HP 1: 143. See *Hawkins 1975b:* 133; *1979:* 32–61; *Martin 1976a:* 223–230.

45. Compare *Schröder 1898a:* 45–56, 60–61. See *Hawkins 1981:* 381–383.

46. Compare 1.247–249; 2.191–192; 4.134, 227–249; NEM 4: 36–37, 98–99, 225, 267–273. See *Hawkins 1986:* 63.

47. Compare 3.454; 4.368–370, 428–429; 8.316. See *Hawkins 1981:* 387; *1986:* 62–64.

48. Compare 1.248–249; 2.191–192; 4.228, 239–244; NEM 4: 57, 98–99. See *Hawkins 1986:* 63–64.

49. Compare 1.53, 247; 2.120; 4.134, 228–229, 232–250, 373; 5.8, 567; 8.110; NEM 4: 34–35, 98–99, 225–226, 267–272. See *Hawkins 1975a:* 109; *1975b:* 138n.13; *1986:* 62–67.

50. Compare 3.364. See *Beth 1959* [*1966*]: 172, 442; *Hawkins 1986:* 66–67.

51. See *Kleene 1952* [*1971*]: 63; *Tarski 1956:* 60–109.

52. See *Hawkins 1975b:* 138n.16; *Quine 1970:* 61–79.

53. See *Russell 1903* [*1956*]: 1–9. Compare 3.364, 373–374, 441–446; 4.240. See also *Hawkins 1975a:* 111, 113n.35.

54. Compare *Russell 1903* [*1956*]: 3, 106–107.

55. See *Quine 1970:* 65–68.

56. Compare *Russell 1944* [*1951*]: 13; *1903* [*1956*]: v, 101–107, 521–522; *1960:* 7, 194–195; *van Heijenoort 1967:* 124–125. See *Henkin 1962:* 788ff.

57. Compare *Russell 1903* [*1956*]: xvi; *Whitehead and Russell 1910–1913* [*1959–1960*] 1: v.

58. See *Hawkins 1986:* 64, 67–68.

59. Compare 8.168; NEM 4: xxii–xxiv, 34, 50, 55–56. See *Hawkins 1986:* 64, 67–68.

60. Compare 3.252–288, 547–549n.1, 563–570, 626–631; 4.170–226, 331–339, 639, 647–681; 5.526; 6.164–176; 7.209; 8.321–324; MS 469, 6. Compare also *Russell 1903* [*1956*]: 296–324, 361–368, 527–528, and see *Murphey 1961:* 266–288.

61. Compare *Murphey 1961:* 241.

62. See *Lenzen 1965:* 7–8 and *Murphey 1961:* 241n.6.

63. Compare *Fisch 1974:* 100, citation 21.

64. Compare 8.168; L 148 (8 May 1906); L 463 (1 December 1903).

65. For example, MS 12. Compare *Murphey 1961:* 241n.6.

66. Compare 2.195.

67. See *Hawkins 1975a:* 109–115; *1975b:* 128–139; *1981:* 386–389.

68. Compare 1.427; 2.65, 355, 433; 3.47–67, 162–172, 175, 184, 365–391, 440–446, 474, 496, 572; 4.376, 510–528, 654. See *Hawkins 1975a:* 111; *1975b:* 129–131, 135; *Lewis 1918*[*1960*]: 84–85; *Quine 1961:* 31–33; *1950* [*1972*]: 42–44.

69. Compare *Russell 1903* [*1956*]: 15.

70. Compare 3.385, 398; 4.481–482; 5.323; *Frege 1893* [*1962*] 1: §§ 1–52; 2: §§ 56–67, 139–144, 146–147. See *Carnap 1959:* 233–240, 284–292; *Hawkins 1975a:* 111; *Quine 1960:* 270–276; *1961:* 31–33; *1966c.* 163–164, 175; *1950* [*1972*]: 42–44.

71. Compare *Russell 1903* [*1956*]: 16n.

72. See *Carroll 1894:* 436–438.

73. See *Hawkins 1975a:* 111; *1975b:* 135.

74. Compare *MacColl 1878:* 16–17; *1896:* 172, 182–183; *Schröder 1890–1895* 2: 29, 68–69, 258.

75. For example, *Jevons 1880* [*1908*]: xiv–xvi. Compare *Peirce ed. 1883:* iii–v.

76. Compare 3.154n., 182n.2, 204; 4.134; 8.189; N 1: 38–40, 63–64, 86–88, 111; NEM 3: 287, 760. See *Fisch and Cope 1952:* 288. Compare *MacColl 1877a:* 9–20; *1877b:* 177–186; *1878:* 16–28; *1880:* 113–121.

77. Compare *Russell 1903* [*1956*]: 16n.

78. Compare *MacColl 1896:* 182–183; *1897a:* 555–579; *1897b:* 98–109; *1897c:* 496–510; *1900:* 75–84. See *Kneale 1962:* 549; *Lewis 1918* [*1960*]: 108.

79. Compare 2.349–352; 3.448–453, 642. See *Schröder 1890–1895* 2: 1–25, 256–276.

80. Compare *Russell 1903* [*1956*]: 13, 19.

81. See *Hawkins 1979:* 41–42.

82. Compare *Russell 1903* [*1956*]: 19–20, 42–94; *1956:* 76; *1959:* 124–125; *Whitehead and Russell 1910–1913* [*1959–1960*] 1: xxvii–xliii. See *Quine 1966b:* 20–23, 147–150; *1970:* 66–68; *Robinson 1975:* 172–182.

83. See *Hawkins 1981:* 387.

84. Compare 2.548; 3.642; 4.5, 101, 107, 532–538; 5.345; 6.361, 384.

85. Compare 1.204–220; 2.391–430; 4.234; 6.185; NEM 1: 124–126; 3: 368–371. See *Boler 1963:* 124–128; *Hawkins 1975b:* 131. Compare discussions, *Frege 1884* [*1950*]: §§ 29, 38–86, 92; *1893* [*1962*] 1: xii–xv, §§ 19–25; *1967:* 125–142, 167–210.

86. See *Hawkins 1986:* 66–67.

87. See *Hawkins 1981:* 387.

88. Compare 3.609–610; 8.191, 313; NEM 3: 97–100, 106, 463–469, 925; 4: 340–352. See *Boler 1963:* 120; *Hawkins 1986:* 66–67.

89. Compare 1.300–567, 591, 608; 2.230–264, 292–304, 440, 541, 600; 3.435, 586; 4.431, 447–448, 464, 500, 530, 551; 5.28, 119, 287, 289, 295; 6.32, 150–157, 225, 270, 373–380; 7.137, 535–586, 591–596; 8.268, 328–379; NEM 3: 1041, 1120–1121. See *Hawkins 1975a:* 109–115; *1975b:* 128–139; *1981:* 386–389.

90. Compare 1.15–34, 127, 372; 2.262, 274–282, 287, 305–307, 337, 357, 369, 393, 438; 3.360–363, 403H, 419–424, 433–445, 456–463; 4.58, 391, 456–459, 544; 5.494; 6.471. See *Hawkins 1975a:* 109–115; *1975b:* 128–139; *1981:* 386–389.

91. See *Hawkins 1975a:* 110–111.

92. 3.398. Compare 3.392–403M, 467, 636–642; 4.80–225. See *Hawkins 1975a:* 110–111; *1975b:* 138n.11; *Kneale 1962:* 432–434.

93. Compare *Russell 1944* [*1951*]: 11 and *1903* [*1956*]: xiv.

94. See *Beth 1959* [*1966*]: 362, 407, 464–477, 627; *Castonguay 1972:* 74–77; *Goodman 1966:* 33–61, 141–145, 189–192; *Henkin 1953:* 19–29; *Schilpp ed. 1944* [*1951*]: 68.

95. *Russell 1903* [*1956*]: 20–21. Compare *Quine 1966b:* 147–151; *Whitehead and Russell 1910–1913* [*1959–1960*] 1: xix–xliii.

96. Compare 1.292, 346, 446; 4.406, 416–417, 561, 580, 583.

97. Compare 2.219–226.

98. Compare 1.343–383; 3.636–642. See *Russell 1903* [*1956*]: vii–viii, 22–23, 31–32, 38, 68, 70–80, 95–100, 525–526.

99. See *Russell 1903* [*1956*]: 23, 26, 203n., 232n., 320n., 376, 387n.

100. Catalogue number BC 135.F8.

101. See *Fisch and Cope 1952:* 277–311.

102. Catalogue number Phil. 5005.4, binder's title "Formal Logic, 1866–1906."

103. See *Fisch 1972:* 488n.9.

104. Compare 3.199n.2. See *Fisch and Cope 1952:* 288.

105. Compare NEM 3: 741; 4: 150.

106. See *Sinisi 1969:* 239–246. Compare *Kneale 1962:* 443.

107. See *Schröder 1880:* 81–87, 90–94. See *Hawkins 1981:* 381–386.

108. Compare *Schröder 1898a:* 47–48, 51–53, 56, 59.

109. See *Hawkins 1981:* 381–389. Compare *Frege 1964b:* 97–114; *1969* 1: 9–59; L 237 (29 August 1891); MS 499, 1^P–5^P.

110. See *Hawkins 1975a:* 112n.2; *1981:* 387.

111. See *Hawkins 1975a:* 112n.2.

112. See *Hawkins 1975a:* 112n.2; *1981:* 386–387; *1986:* 67–68.

113. See 3.393. Compare 3.363; N 1: 111. See also *Beatty 1969:* 71–75; *Frege 1879:* §§ 9–12; *Hawkins 1975a:* 109–115; *1981:* 381–389; *Martin 1976b:* 231–245; *Mitchell 1883.*

114. See *Hawkins 1975a:* 112; *1981:* 386–387; *1986:* 64.

115. See *Fisch 1972:* 486.

116. See *Fisch 1972:* 486. Compare L 392 (*Fisch 1972:* 487).

117. Compare *Russell 1959:* 65.

118. I am indebted to The Bertrand Russell Archives (McMaster University, Hamilton, Ontario, Canada) and particularly to Mr. Kenneth Blackwell (Archivist) for providing me with a copy of Russell's notes. Regretfully, I was not granted permission to quote those passages.

119. Compare L 148 (8 May 1906)/NEM 3: 758.

120. Compare *Murphey 1961:* 243.

121. Quoted in *Murphey 1961:* 241–242. Compare NEM 4: xxii–xxiv.

122. See *Michael 1975:* 373.

123. Compare 2.352, 618; 3.446. See *Bocheński 1961:* 15, 161, 240–251; *Hawkins 1975b:* 136n.2; *Michael 1975:* 369–374; *1976:* 46–55; *Prior 1962:* 288–293; *Thompson 1949:* 513–536.

124. See 2.618. Compare *Kneale 1962:* 656.

125. Compare *Whitehead and Russell 1910–1913* [*1959–1960*] 1: v.

8.

LOGIC AND MATHEMATICS IN
CHARLES SANDERS PEIRCE'S "DESCRIPTION OF
A NOTATION FOR THE LOGIC OF RELATIVES"

James Van Evra

§1. Introduction

This is a case study of the way analogy works, and sometimes fails to work, in science. I am concerned here not with the more common use of analogy by which science is rendered intelligible to the uninitiated but with a case in which a theory itself is based analogically on another *theory*. The point, in general, is that differences in how broadly or narrowly such an analogy is conceived produce significant changes in the way the science based on it is conceived. The grounding theory in this case is algebra, the science based on it is logic. The period in question is the mid-nineteenth century.

During the nineteenth century several attempts were made to link mathematics and logic, including, most prominently, those contained in the works of George Boole, Augustus De Morgan, Charles Sanders Peirce, Ernst Schröder, and Alfred North Whitehead. They were pursuing a variety of goals. At one extreme, Whitehead sought to establish logic *in toto* as one proper part of a larger universal algebra (*Whitehead 1898:* v, 36), while Boole (*1854*), at the other, simply sought a way to use the power and generality of mathematical operations in logical inferences, while keeping logic separate from mathematics.

In one way, all such attempts to link the two, regardless of the reason, failed. The advent of the individual quantifier (due independently to Peirce and Gottlob Frege) drew logic away from algebra, and the standard theory of first-order logic now incorporates it in no essential way. In another way, however, they were just as clearly successful. The works of the nineteenth-century algebraic logicians provided the basis for a slightly different tradition, a "changing the subject" in the sense of Quine (*1970 [1986]:* 80ff.), to which interesting additions are still being made (cf. *Blok 1989; Craig 1974; Gindinkin 1985*). In what follows, I am concerned less with the distinct

branching tradition than with the earlier attempts to link logic, traditionally conceived, with mathematics.

§2

Charles Sanders Peirce described his "Description of a Notation for the
Logic of Relatives" (*1870;* hereinafter DNLR) as an extension of George
Boole's algebraic logic to include relative terms. From the way each employs
mathematics in his logic, however, it seems that Peirce's first full work on
the logic of relatives is an extension of Boole's logic in roughly the same
way that space travel is an extension of walking.

The works of the two logicians begin on shared ground, with Peirce
following Boole in initiating an analogical link between logic and mathematics at a point of strong connection between the two, i.e., between the
logical "and" and multiplication, and the logical "or" and addition. That,
together with the similarity between symbols on both sides, provided a secure contact between the two disciplines (cf. *Boole 1854:* 27; W2: 360). The
presumed benefit to be gained by establishing it was that once formed,
properties (such as associativity and commutativity) of the allied algebraic
operations could be employed in strictly logical contexts, and, in general,
logic could thereby benefit from the power and generality of mathematics.

However, where there is analogy, there is also disanalogy; and in this
case it begins at the point beyond which the logical interpretability of
mathematical operations fails. If that point could be easily established, all
would be well, for there would be an easily observed bound within which
the logically interpretable portion of mathematics could be contained.
Here, however, lies a problem. Analogy is not a simple relation, and establishing it in a complex subject area requires skillful gerrymandering of the
base science so that while meanings are preserved on both sides, the only
admissible cases will be those which carry logical meaning (in this case, for
instance, only algebra modulo two fits the analogy with bivalent logic). In
addition, analogy is a vague relation which admits of degrees. Hence there
will be some operations which bear only a partial resemblance to operations
in the other discipline. So, for instance, subtraction and division, the inverses of addition and multiplication, seem to bear *some* logical significance,
but cannot be simply linked with operations in logic.

The question which naturally arises then, is how much of mathematics
can properly be drawn into analogical connection with logic? The wider
the reach, the more mathematical power can be brought to bear in logical
contexts. But if it extends too far, the price is meaninglessness.

Boole and Peirce responded to the problem in different ways. Both
were supremely well versed in mathematics,[1] and both used a wide variety

of mathematical devices in logical contexts. The way they justified such use, however, greatly differed, and the difference indicates different conceptions of the nature and limits of logic held by the two logicians.

On the one hand, Boole attempted to solve the problem of limiting the analogy by rigidly confining it to the area of elementary algebraic operations and their inverses. While even this restricted association created some problems of interpretation (concerning, most importantly, the interpretation of the inverse operations of subtraction and division), it is at least clear that Boole was well aware that there were bounds beyond which logical interpretability failed. Just as clearly, however, Boole wished to reach beyond the strict confines of algebraic logic so conceived, and use mathematical operations in logical contexts which were not themselves logically interpretable. This he did by introducing what he called his "general method in logic," in terms of which logically uninterpretable operations could be employed in logical contexts as follows: If an inference begins with a mathematical expression which also has a definite logical interpretation, then it was possible to use *any* mathematical operation on it, whether the operation, or the results of using it, were themselves logically significant. So long as the sequence concluded with another expression which was again directly logically interpretable, the entire sequence was considered to be a legitimate logical transformation, even though it may have employed steps which, individually, were not. By invoking this device Boole attempted to have the best of both worlds: a settled logical landscape on the one hand, and on the other, the availability of the power and generality of mathematical devices (such as Taylor's Series, which Boole uses extensively in his account of development), whether they fit within the scope of the original analogy or not.[2]

DNLR, by contrast, follows a distinctly different path. Rather than beginning with psychological and logical considerations and only later introducing mathematical topics (as Boole had done), Peirce begins in the midst of mathematics itself by listing the categories of algebraic operation to be applied in logical contexts. Throughout, he makes it clear that the project at hand is one of extending the ordinary signs of algebra to cover new applications. Next, Peirce outlines his notation for relatives and demonstrates how the algebraic operations are to be applied in this context. In this section, he skillfully distinguishes mathematical operations which bear logical interpretation from those which do not. In cases in which the analogy is only partial, he identifies the logically interpretable portion with a slightly altered symbol. Once finished with his presentation of the basic notation for relatives, he moves (in a section entitled "General method of working with this notation") to an account of their logic. Here, strange things begin to appear. Soon we find him directly employing devices such as logarithms, derivatives, complex numbers, quaternions, transcendental equations, non-Euclidean geometry, and more, all within a strictly logical context and without regard to the limits of interpretability. Here are two examples:

EXAMPLE 1. When working with what he calls "infinitesmal relatives" (i.e., relatives whose correlatives are individual), Peirce first defines the difference of a function as follows:

$$\Delta\varphi x = \varphi(x + \Delta x) - \varphi x$$

where Δx is defined as an indefinite relative with no correlate in common with x (i.e., x, $(\Delta x) = 0$). Then if the second difference of x is taken, and the two increments which x successively receives are distinguished as $\Delta' x$ and $\Delta'' x$, then $(\Delta' x)$, $(\Delta'' x) = 0$. Then if Δx is "relative to so small a number of individuals that if the number were diminished by one $\Delta^n \cdot \varphi x$ would vanish, then I term these two corresponding differences *differentials*, and write them with d instead of Δ" (W2: 398).

As an example of the application of differentiation in logic, Peirce poses the following problem: "in a certain institution all the officers (x) and also their common friends (f) are privileged persons (y). How shall the class of privileged persons be reduced to a minimum?" Beginning with

$$y = x + f^x,$$
$$dy = dx + df^x = dx - f^x, (1 - f)\, dx$$

Peirce goes on to derive a general solution to the problem, using 0 and 1 now as limits rather than as logically interpretable coefficients in the solution (W2: 406ff.).

EXAMPLE 2. Given the vocabulary c: colleague; t: teacher; p: pupil; s: schoolmate, and given that a,, b,, c,, and d, are "scalars" in the following sense: a scalar is a character in "respect to the possession of which both members of any one of the pairs between which there is a certain elementary relation agree" (e.g., "supposing French teachers have only French pupils and *vice versa*, the relative f will be a scalar for the system 'colleague:teacher:pupil:schoolmate' "). If all this is the case, "Then any relative which is capable of expression in the form

$$a,c + b,t + c,p + d,s$$

I will call a *logical quaternion*." That accomplished, he goes on to meld logic with typical structures in non-commutative algebra, and renders the entire process completely abstract (W2: 409–410).

The problem is that Peirce makes no attempt to settle the independent and individual logical significance of the devices which he uses in this part of the work, and none of them now seem particularly logical. Nor did he, as Boole had done, attempt to circumvent the difficulty. Rather, in this section he seems to assume that *any* mathematical operation carries logical meaning. As a result, while he did manage to extend the mathematical side of the analogy on which logic had been grounded, the logical side did not automatically extend with it. The analogical link was thereby strained, and the result was a work full of odd analogies, which did not connect easily with the logical theory which had come before or with that which followed.

Faced with these difficulties, those who comment on DNLR at all concentrate primarily on its less problematic aspects; they pass over its less tractable parts with comments such as "It is no easy task to eke out of Peirce's logical writing just what he is trying to say" (*Martin 1980:* 45). Or, "the most valuable of Peirce's logical writings make tremendously tough reading" (*Lewis 1918:* 106). Those who do squarely face the problem of interpreting the difficult parts of DNLR, either try to explain them away by saying that Peirce was really doing a number of separate things in the work, some logical and some mathematical (e.g., *Brink 1978:* 285), or that he simply strayed from dealing with the logic of relative terms into developing a general theory of relations, all in this case mathematical (e.g., *Merrill 1984:* xlviii).

Were the troublesome parts of DNLR simply the work of a scientist unwittingly following a false lead or otherwise mistaking the point of the project, we might describe it as little more than an interesting historical curiosity. But a large part of the genius of Peirce was that whatever he did, he seemed to do wittingly, and DNLR is no exception. He freely admits, for instance, that the algebra to be used *is* more complicated and "much less manageable" than ordinary arithmetic algebra (W2: 388). In addition, within the text there is every evidence that he considered the work to be a unified whole. Thus the core of it, i.e., the series of 172 equations which constitute the body of the work, form a continuous series, beginning with the simple and obviously logical and working steadily toward the abstract and obscure. Why, then, if Peirce was aware of the difficulties involved, and if the work is a unified whole, did he think that the use of mathematical devices of higher orders of complexity was warranted in a work on the *logic of relatives?* Why, that is, did he wind up entangled in obscure analogies?

§3

The reason for Peirce's liberal acceptance of such a wide range of operations as being logical, as well as the reason for Boole's attempted conservatism in the same regard, lies in the way each author interpreted the limits of the analogy between logic and mathematics. Peirce obviously saw more in it, Boole less. The "more" that Peirce saw, I suggest below, bears a striking resemblance to views about mathematics held at the time by his father, Benjamin Peirce.

As is well known, at the same time that Charles was writing DNLR,

Benjamin was writing *Linear Associative Algebra* (hereafter LAA), later described as the "most original and able contribution" by the "leading American mathematician of his time" (*Archibald 1925:* 15; *Smith and Ginsburg 1934:* 119). While the finished work was not the product of a formal collaboration between father and son, it is clear from comments Charles made later (MS 78, 1895) that they had extensive discussions concerning the algebra and pointed discussions about foundational issues having to do with the precedence of logic or mathematics over the other.

Moreover, it is also obvious that LAA plays an important role in DNLR itself. The development of the more complex algebraic portion of the work culminates in the appearance of some of the matrices from LAA (by which Benjamin Peirce defined variant algebras). Then Charles Peirce makes the claim that

> I can assert, upon reasonable inductive evidence, that all algebras resembling LAA can be interpreted on the principles of the present notation. ... In other words, all such algebras are complications and modifications of the algebra of LAA. It is very likely that this is true of all algebras whatever. The algebra of LAA, which is of such a fundamental character in reference to pure algebra and our logical notation, has been shown by Professor Peirce to be the algebra of Hamilton's quaternions. (W2: 413)

There is also evidence that the two works are related at a deeper level. The relationship which Charles established between logic and mathematics in DNLR appears to reflect ideas which his father set forth in LAA concerning the foundations of mathematics. What Charles Peirce *does*, that is, is similar to what Benjamin Peirce *says*. While the genesis of the similarity is not known, the existence of the similarity suggests that the ground for Charles's use of mathematics in DNLR can be found in LAA.[3]

LAA occupies an important place in the lineage of a series of developments in algebra in the nineteenth century. While continental mathematics had for some time been enjoying the heights of abstraction, algebra in England was still conceived in a traditional manner in terms of which letters stood for some particular but unspecified number, and the subject was tightly tied to concrete application. As late as 1831, for instance, De Morgan was still telling students that negatively signed integers are not numbers, for the latter must designate some quantity, and there are no negative quantities.[4]

LAA by contrast is completely abstract. In an addendum published in 1875, Benjamin Peirce says

> Some definite interpretation of linear algebra would, at first sight, appear to be indispensable to its successful application. But on the contrary, it is a singular fact, and one quite consonant with the principles of sound logic,

that its first and general use is mostly to be expected from its want of significance. The interpretation is a trammel to the use. (*1870:* 216)

LAA contains not one algebra, but many, ranging from single to sextuple. Just two features are common to all of them: the associative principle is universally applicable within them, and they are linear in the sense that every expression in each is resolvable into a normal form with a single letter and single coefficient in each disjunct.

The roots of the abstract turn in algebra in the nineteenth century are now well known. Early in the century, a group of formalistically inclined Cambridge mathematicians turned their attention to algebra (as an off-shoot of other mathematical interests).[5] The main contributors were George Peacock (1791–1858), whose *Treatise on Algebra* (1830) earned him the title "Euclid of algebra." Peacock recognized that algebra was susceptible to non-numerical interpretation, and although he held to the primacy of the arithmetical interpretation, he devised one main alternate interpretation in the area of geometry. Another member of the group, Boole's Cambridge mentor, D. F. Gregory, described the classic algebraic properties (associativity and commutativity of addition and multiplication; distributive law of multiplication over addition) in 1840. Thus it became clear that algebra had an internal structure. Throughout this early period, however, the arithmetic interpretation of algebra was still primary; it was still inconceivable, that is, that an algebra could be developed which was inconsistent with arithmetic algebra. That changed in 1843, when the Irish mathematician Sir William Rowan Hamilton solved a seemingly intractable problem by devising his quaternion algebra, in which the commutative law of multiplication did not hold. In effect, algebra had been freed from its remaining tie to *any* preferred mode of interpretation.

One indication of the reason for Boole's relatively conservative approach to the relation between algebra and logic and for Peirce's more liberal stance to the same relation can be seen in the way each accommodated these developments in their respective logics. Although Boole was well aware of the recent developments in algebra, and even made some contributions of his own in the area, the algebra on which he relied as the basis for his logic remained confined to the more restricted variety established by Peacock and Gregory.[6] The analogy between logic and algebra then posed no particular problem, for algebra itself was confined within limits which approximated the limits of logical interpretability.

LAA, by contrast, is free of the traditional constraints. It is purely formal, and it is raised to a higher level of generality and abstraction than its forbears. Alfred North Whitehead later described it as an earlier attempt at a universal algebra, i.e., a "comparative anatomy" of systems of symbolic algebra which drew together in one system matrices, quaternions, and the theory of linear algebra. It is in this broad scope, I believe, that we can find the reason for the extravagance in (Benjamin) Peirce's expression of the foundations of the subject.[7] But whatever their origin, it is those same foundations which are similar to what is actually carried out in DNLR.

The idea of algebra contained in LAA is so all encompassing that any *analogical* connection between logic and algebra dissolves, and logic remains distinctively different only by tradition, not theory. Thus from the outset of LAA, (Benjamin) Peirce casts a particularly wide net by defining mathematics as the "science which draws necessary conclusions" (*1870:* 97). He then justifies such a broad definition by saying:

> This definition of mathematics is wider than that which is ordinarily given, and by which its range is limited to quantitative research. The ordinary definition . . . is objective, whereas this is subjective. Recent investigations, of which quaternions is the most noteworthy instance, make it manifest that the old definition is too restricted. (*1870:* 97)

Now, says Peirce, "the sphere of mathematics is . . . extended . . . to include all demonstrative research, so as to include all knowledge strictly capable of dogmatic teaching" (*1870:* 97). Furthermore, he defines algebra in such a way that it is virtually coextensive with mathematics itself. According to him, all that distinguishes the two is the fact that algebra uses symbols rather than common language. As he puts it, "Algebra . . . is formal mathematics" (*1870:* 98).

This conception of algebra as all-encompassing parallels Charles Peirce's lack of recognition of a strict boundary between algebra and logic. That lack is just what is evident in DNLR. At the outset of his presentation of the general method, for instance, Peirce gives an explicit indication of this view. After pointing out that the algebra of logic developed by Boole and amended by Jevons is too restricted in scope, Peirce begins his own extension by first giving a brief account of mathematical demonstration. Such demonstration, he says, is "founded on suppositions of particular cases" (W2: 389), which are then broadened to the point of perfect generality. Then he says,

> The advantage of the mathematician's procedure lies in the fact that the logical laws of individual terms are simpler than those which relate to general terms, because individuals are either identical or mutually exclusive, and cannot intersect or be subordinated to one another as classes can. Mathematical demonstration is not, therefore, more restricted to matters of intuition than any other kind of reasoning. Indeed, *logical algebra conclusively proves that mathematics extends over the whole realm of formal logic;* and any theory of cognition which cannot be adjusted to this fact must be abandoned. (W2: 389; italics added)

In another instance, once Charles shows that the algebra of LAA can be readily transformed into the proper form for Hamiltonian quaternions, and since his "logical quaternions" are representable in LAA algebra, the possibility naturally arises that there is therefore a connection between logic and geometry. Thus after generating a matrix for a pure Hamiltonian quaternion, Charles Peirce says:

> It is no part of my purpose to consider the bearing upon the philosophy of space of this occurrence, in pure logic, of the algebra which expresses all the properties of space; but it is proper to point out that one method of working with this notation would be to transform the given logical expressions into the form of Hamilton's quaternions . . . and then to make use of geometrical reasoning. (W2: 414)[8]

Here he treats abstract algebra as "pure logic"; in fact, the two are considered to be perfectly indistinguishable. In addition, shortly thereafter he talks both of *applying* geometry to the logic of relatives (W2: 415) and also claims that "for every proposition in *geometry*, there is a proposition in the pure logic of relations" (W2: 417). 'Logical algebra' for Peirce, it seems, had become a blend of the two, an amalgam rather than an analogical pairing.

There is further support for this view, again found in LAA, when (Benjamin) Peirce talks about the relationship between logic and algebra. "Mathematics," he says, "belongs to every enquiry, *moral as well as physical.* Even the laws of logic, by which it is rigidly bound, could not be deduced without its aid" (*1870:* 97). As Peirce describes it, there is a kind of reciprocity between the two: logic is primary in a sense, but it is mathematics which transmutes logical form to fit natural language. Mathematics in this sense is an arbiter with respect to the world at large. His father's unwillingness to assign either logic or algebra an absolute primacy over the other was, according to Charles's later account, the result of his dissuading his father from the view then held by Dedekind, that mathematics is a branch of logic. "I argued strenuously against it," Charles says, "and thus (Benjamin) consented to take the middle ground of his definition" (MS 78: 4), which suggests again that neither logic nor mathematics was assigned primacy.

§4

When used in a context which demands precision, analogy is difficult in the best of circumstances. When it is drawn between two formal theories which are close relatives to begin with, the problems loom even larger.

The attempts to tie logic, which was not clearly defined in the nineteenth century, to mathematics via algebra, which was also in transition at the time, produced a variety of results. In the case of Boole it is relatively easy to tell where logic ends and where his algebra begins, and it is easy to see the points of analogical contact. All of that is preserved in DNLR as well. But if algebra is construed as it is in LAA, there would be no way of systematically differentiating what is logic from what properly belongs to algebra. So it should come as no surprise that Peirce begins on relatively safe Boolean ground and then moves to ever-higher levels of abstraction as a way of animating his logic of relatives. Both logic and algebra were at points in their development at which their boundary conditions did not permit strict separation.

It is worth noting, finally, that the ambiguity found in DNLR was transitory, both for algebra and for Peirce. By 1898, when Whitehead published *Universal Algebra*, for instance, algebra had become more clearly an umbrella under which logic was subordinated to the status of just one self-contained part of algebra in toto (cf. *Whitehead 1898:* v). As for Peirce, nine years after writing DNLR his enthusiasm for analogy had waned:

> The effort to trace analogies between ordinary or other algebra and formal logic has been of the greatest service; but there has been on the part of Boole and also of myself a straining after analogies of this kind with a neglect of the differences between the two algebras, which must be corrected, not by denying any of the resemblances which have been found, but by recognizing relations of contrast between the two subjects. (MS 575; in *Houser 1985:* 77)

What I have suggested is that there is evidence that the transitory straining after analogies found in DNLR is a reflection of the state of the algebraic base of the analogy Peirce employed, as that base was formulated in LAA.

NOTES

1. At a time when little attention was being paid to them in England, Boole (beginning at age sixteen) studied, and mastered the works of the great European mathematicians. For an account of Boole's mathematical education and later work in mathematics, see *MacHale 1985,* chs. 4 and 15.
2. Boole's general method in logic formally resembles his general method in analysis. See *Boole 1844.* For a statement of the general method in logic see *Boole 1854:* 8ff. Also on this topic, see my *1977, Brink 1987:* 2ff., *Hailperin 1976* [*1986*]: 135ff., and *Putnam 1982:* 294.
3. Jacqueline Brunning (*1991:* 33–49) provides a particularly interesting and detailed account of the links between LAA and Charles's development of a theory of relations in DNLR. Her thesis is that Charles was after an independent theory of relations in DNLR rather than a further clarification of the concept of relation.
4. "[B]ut 3 − 8 is an impossibility, it requires you to take from 3 more than there is in 3, which is absurd. If such an expression as 3 − 8 should be the answer to a problem, it would denote either that there was some absurdity inherent in the problem itself, or in the manner of putting it into an equation." (*De Morgan 1831:* 104)

5. Principally the introduction of the functional notation into Britain. Cf. *Enros 1979* and *Grattan-Guinness 1988:* 74ff.

6. The lack of contact between Boole and Hamilton is referred to as "One of the great mysteries of Boole's life in Ireland" by Boole's biographer Desmond MacHale (*1985:* 182). Boole did publish a note on quaternions (*1848*) but despite this, "it must be admitted that Boole took very little interest in quaternions as an algebraic system, and also showed a surprising lack of interest in Hamilton's algebraic theory of complex numbers as ordered pairs" (*MacHale 1985:* 189).

7. Criticism which LAA received from continental sources had been "due in part . . . to the extreme generality of the point of view from which his memoir sprang, namely a 'philosophic study of the laws of algebraic operation.' " (J.B. Shaw, quoted in *Archibald 1925:* 15).

8. Note that Peirce here advocates something like Boole's general method, described above. However, no mention is made of returning the sequence to something which is independently logically interpretable.

9.

RELATIONS AND QUANTIFICATION
IN PEIRCE'S LOGIC, 1870–1885

Daniel D. Merrill

It is widely recognized that Peirce's two most important contributions to logic were the logic of relations and the theory of quantification. It is also clear that these two aspects of his logic are closely connected, since Peirce's search for an adequate logic of relations stimulated his development of quantification theory. The quantificational complexities of many relational statements cried out for quantifiers.

It is equally well known that Peirce's original logic of relations did not use quantifiers. In fact, Peirce moved from an *algebraic* logic of relations (ALR) in 1870 to a *quantificational* logic of relations (QLR) in 1885.[1] Only the latter logic of relations contained quantifiers in the modern sense. At the same time, the ALR contained ways of dealing with quantification even if it did not contain quantifiers. I would like to approach the connections between relations and quantification by asking the following question: Given the strengths of his ALR, why did Peirce think that his QLR was superior to his ALR?

Working up to this question requires that I first describe some crucial features of Peirce's ALR and exhibit its strengths. These will include its systematic treatment of relations and its quantificational resources. Against this background, I will examine the grounds on which Peirce preferred his QLR to his ALR.

§1. The Algebraic Logic of Relations

The QLR, which Peirce developed in 1885, is roughly equivalent to modern first-order logic, and thus it is quite familiar. His ALR, though, requires some explanation. As developed in Peirce's 1870 memoir, "Description of a Notation for the Logic of Relatives" (DNLR), it is a brilliant combination of Boole's logical algebra and De Morgan's logic of relations. Despite its many idiosyncratic features, often motivated by unhappy algebraic analogies, it is a system of great power and originality.

DNLR contains an algebraic treatment of the logic of one-place, two-place, and three-place predicates. Peirce called these absolute terms, simple relative terms, and conjugative terms respectively, and I will follow his usage. Peirce and others developed the logic of simple relative terms most fully, and it is that part which will be emphasized in this section. Its key ideas can be understood most easily by thinking of relations extensionally— that is, as classes of ordered pairs. These classes of ordered pairs are combined by using the usual Boolean operations on classes, yielding the complement of a relation and the logical product and sum of two relations. In addition, "1" and "0" stand for the universal and null relations respectively.

All of this is Boolean, aside from the fact that the classes dealt with are sets of ordered pairs rather than sets of individuals. The specifically relational part can be introduced by adding other operations on relations. In DNLR, Peirce, following De Morgan,[2] introduced four new relational operations. Using some of Peirce's examples, they can be stated as: *Converse:* the converse of servant is *master or mistress*; *Relative product:* the relative product of lover and servant (*ls*) is *lover of a servant*; *Involution:* the involution of lover and servant (*l^s*) is *lover of every servant*; and, *Backwards involution:* the backwards involution of lover and servant (*l_s*) is *lover of only servants*.

These operations are not independent of each other, for the two involutions can be defined using complementation and relative product. Also definable in these terms is an additional operation, which Peirce introduced in the late 1880s and which lacks the simple intuitive meaning of the two involutions (*Peirce 1883c/* 3.332): *Relative addition:* the relative sum of lover and servant (*l † s*) is *lover of every non-servant*. Symbols for the identity and diversity relations are also included. The terms of the logic of simple relatives are generated recursively from its primitive relative terms by means of these operations.

The sentences of this language use Peirce's "illation" sign to state inclusions between simple relatives, as in $R \prec S$. The identity of two relations is just inclusion in both directions.

Although the title of DNLR suggests that Peirce was providing only a notation for this logic, DNLR also states a great many logical laws in this notation. Peirce disavowed any claim to having axiomatized this logic, but the resulting system (if that is not too strong a word) has a great deal of deductive power.

The crucial difference between this ALR and the later QLR is that the former contains no individual variables and no quantifiers; aside from its logical and relational constants, it contains only relational variables, and they are never quantified. With these limitations it is surprising that the logic of DNLR is as expressively and deductively powerful as it is. It can express a great many of the statements which the QLR can express. Furthermore, its purely algebraic form gives it a certain mathematical elegance and power that have made it the object of mathematical study even to this day. In this context, our question is a natural one: Given all the good features of this ALR, why did Peirce move on to the now-standard QLR?

Two other contextual points remain to be made. The first is that many of the notations and laws of the logic of simple relatives apply equally well to absolute terms. In fact, DNLR contains few examples of logical laws which involve only simple relatives. The product of a simple relative term and an absolute term, such as "lover of a woman," is much more common than the relative product of two simple relatives, as in "lover of a servant." Even relative products usually occur as part of a relationally defined class term, such as "lover of a servant of a woman."

The other point to remember is that Peirce also dealt with three-place predicates, or "conjugative terms," in DNLR. His treatment of these terms is puzzling and less explicit than his treatment of simple relatives, and only now is it becoming well understood.[3] Conjugative terms greatly expand the expressive possibilities of the language of DNLR; and even though Peirce did not fully present the algebraic treatment of conjugative terms, it is important to remember that they lurk in the background. Their presence may significantly affect the issues about the expressive power of the language of DNLR as a whole. Nevertheless, we shall tend to follow not only Peirce but his commentators and emphasize his theory of simple relatives.

§2. Relatives and Relations

My exposition of Peirce's logic of relatives has assumed that relative terms stand for relations. This assumption has not gone unchallenged, so that it will first be necessary to make sure that DNLR does, in fact, deal with the logic of relations. Peirce's use of the term "relatives" rather than "relations," which he later regretted, along with several other features of DNLR, have understandably led to confusion. Perceptive interpreters of Peirce have reached quite different conclusions about the real subject matter of DNLR. Several writers have claimed that DNLR deals with the logic of *relatives* rather than the logic of *relations* (Cf. *Lewis 1918, Brink 1978,* and *Martin 1978*). On these interpretations, relative terms stand not for classes of ordered pairs but for special kinds of classes of individuals—those which involve relations in their definition. Thus, the relative term "servant" stands for the class of servants rather than the relation of being a servant; alternatively, a relative term would be a term such as "servant of a man," which stands for a relationally defined class of individuals. I believe that these interpretations are incorrect. Despite certain inconsistencies of usage and formulation, Peirce's relative terms stand for relations; DNLR actually presents a logic of relations. While Peirce's later QLR, with its subscripts and quantifiers, allowed relational statements to be expressed more clearly, relative terms stand for relations even in DNLR.

What, then, is a relative? To begin with, Peirce's basic use of the word "relative" is as an adjective: a relative is *a relative term*.[4] The lack of a clear distinction between use and mention, which Peirce shared with a great many logical writers, muddies the situation somewhat. But in most instances, "relative" just abbreviates "relative term."

It is then natural to ask what relative terms stand for; and, we might even call these objects "relatives" as well. At least three candidates for the referents of relative terms have been explored in the literature: a relative term is a term which stands for the domain of a relation, such as the class of lovers; or the result of compounding a relation and a class, such as the class of lovers of men; or a relation, as expressed by "x loves y" or "x is a lover of y," which is a class of ordered pairs. As already assumed in Section 1, I believe that the third definition is the correct answer: relative terms stand for relations. The case for this depends in part on examining the other alternatives.

The first alternative is supported by the fact that Peirce adhered to subject-predicate logic, so that he used nominalized forms of relation terms rather than verb forms: he used "lover" rather than "loves." His initial list of simple relative terms contained only nominalized forms, and some of his examples were stated in this way as well. For instance, he explained backwards involution by saying that "l_s" denotes "everything which is a lover, *in whatever way it is a lover at all*, of a servant." This explanation seems to take both "l_s" and "s" as referring to classes of individuals. This suggests that Peirce was dealing with the classes of lovers and servants rather than the relation of loving. It appears to have led C. I. Lewis to interpret Peirce in this way (*Lewis 1918*: 85–90).

These usages would also support Chris Brink's detailed interpretation of Peirce's notation, which includes additional evidence for this approach (*Brink 1978*). One of the most dramatic facts is that it can be proved in DNLR that every relative is included in the universal class. DNLR carries no suggestion that this must be the universal relation rather than the universal class of individuals. This would require that simple relatives refer to classes of individuals.

Despite the textual basis for this position, I believe that it cannot be sustained, for it requires that too many obvious features of DNLR be explained away as mere confusions between relatives and relations. This can be found in a close reading of Lewis's account. Lewis saw that if "servant" stands for the class of servants, then one must be able to form, say, the union of that class and the class of men; but this is something Peirce never did. He never took the Boolean sum or product of a simple relative term and an absolute term. Furthermore, basic laws fail on this interpretation. Lewis translated Peirce's law (93),

$$\text{If } l \prec s, \text{ then } l^w \prec s^w,$$

as "If all lovers are servants, then a lover of every woman is a servant of every woman" (*Lewis 1918*: 88). Yet this principle is not even true. Every lover could be a servant, yet some of those who love every woman might well not serve women at all.

In fact, Peirce would not have translated the antecedent of this conditional by "every lover of a servant"; rather, it would be, "every lover of anything is a servant of the same thing,"[5] On this relational interpretation, the conditional is clearly true.

More basically, *l* and *s* cannot stand solely for classes of individuals, since then the results of using the relative operations would not make sense. For instance, *ls* would have to stand for the class of lovers of servants; yet this is not a function of the class of lovers and the class of servants. It must, instead, be explained in terms of the *relations* expressed by *l* and *s*. For Lewis to make sense of the above law, *l* and *s* would have to stand for the domain of a relation in their first occurrence and for a relation in their second occurrence.

Peirce's initial explanation of "simple relative term" is consistent only with the relational account of relatives. This class of terms "embraces terms whose logical form involves the conception of relation, and which require the addition of another term to complete the denotation" (3.63). This means that the proper form of a relative term is, for instance, "lover of _____ ," rather than "lover." A term must be placed in the blank to "complete its denotation"—that is, to make it a term which stands for a class of individuals, such as "lover of a man."[6]

This may seem an odd way to express a dyadic relation, for is contains only one blank rather than two. Today we would use either "_____ is a lover of _____" or "_____ loves _____," rather than "lover of _____."[7] This usage by Peirce reflects the influence of traditional logic. His adherence to subject-predicate logic made him prefer "lover" to "loves"; and given that preference he dropped the first correlate to obtain "lover of _____." This latter step is on a par with the traditional use of "animal" rather than "_____ is an animal." Despite this unusual form for relative terms, the basically relational nature of simple relative terms is shown throughout DNLR, and I conclude that simple relative terms stand for dyadic relations and not for the domains of relations.[8]

The late R. M. Martin suggested that relative terms stand for another kind of class of individuals—not for domains of relations but for classes resulting from compounding relations and classes (*Martin 1978*: 27). A relative term would stand, say, for the class of lovers of women rather than the relation *lover of*. It would be *lw*, rather than *l*. This is the second interpretation of "relative term" and it also fails to fit Peirce's definition of "relative term" given above: the definition says that something must be "added to" a relative term to "complete its denotation," while on Martin's interpretation this added "something" is already included in the relative term. At the same time, Martin does have an important point. In DNLR, it is very rare for relative terms to "stand on their own feet" (his phrase); rather, they almost always occur in compound class terms such as "lover of a woman."

It is hard to know what to make of this fact. Peirce's Logic Notebook shows that relation-class composition was what first interested him in the logic of relatives, and this did carry over to DNLR (*Merrill 1978*: 266–269). At the same time, operations on relations are the main feature of DNLR, even though these usually occur in the context of relation-class compounds. The confusion is made worse by the fact that, as Martin admitted, purely relational formulas occasionally occur in DNLR (*Martin 1978*: 32). Furthermore, Peirce's generalized formulas contain variables with no specification of whether the variables stand for absolute or relative, or even conjugative, terms. A great many of the formulas remain correct when interpreted in either way.

My tentative conclusion is that Peirce was quite clear that his relative terms stood for relations, but that Boole's legacy led him almost always to embed their logic within compound class terms. In his papers of the 1880s, this restriction was dropped.

§3. Quantification in the ALR of DNLR

DNLR expresses *quantification* in several ways, even though there are only hints of what we would call *quantifiers*. I will explore these ways in order to compare them later with the quantificational forms of the mid-1880s. My discussion of quantification will be divided into two parts: monadic quantification and multiple quantification.

First, then, monadic quantification: DNLR has a Boolean substructure, so it contains some Boolean forms of quantification. Specifically, Peirce retained Boole's way of expressing universal propositions, even though he added another way of his own. Let us take Peirce's own example (3.141–143). The proposition that all horses are animals is represented in a Boolean way by $h,(1 - a) = 0$, where the comma represents Boolean multiplication, and 1 and 0 stand for the universal and null classes respectively; alternatively, Peirce used his newly introduced inclusion (illation) sign to represent it by $h \prec a$. These two forms for universal propositions are interderivable in DNLR.

Particular propositions are another matter, for Boole's insistence on using equational forms meant that he had no way to negate propositions. This forced him to represent particular propositions with his special elective symbol v, so that "some horses are black" would be expressed by $v,h = v,b$. Peirce saw the problems with this method, including the fact that it allows the deduction of "Some X's are not Y's" from "Some Y's are not X's." Peirce's innovations in DNLR gave him two ways to get around this problem.

The simplest way for Peirce to express particular propositions was to drop Boole's restriction and allow propositional negation (3.143). If one lets $A < B$ mean the same as $A \prec B$, and it is not true that $B \prec A$, then "some h are b" can be expressed by $0 < h,b$ (equivalently, $h,b > 0$); and this is interderivable in DNLR with the denial of $h,b = 0$—that is, with the denial of "No h are b." The familiar syllogistic forms can be shown valid in this notation. In using this simple method, Peirce would have had to give up the fundamental principle that all propositional forms must be expressed in terms of equations or inclusions, for the definition of $<$ contains the denial of an inclusion. But this new sign does provide a way to express particular propositions in an intuitively appealing way.

At the same time, this method makes no use of the logic of relatives. For this, we need a second method. We are surprised to read Peirce's claim that Boole's problems with hypothetical and particular propositions were what "first led me to seek for the present extension of Boole's logical notation" (3.138). Without considering Peirce's treatment of hypothetical propositions, let me note that this is a striking claim about the connection between the logic of relatives and issues of quantification. It is natural to suggest that problems in expressing relational propositions were the main source for modern quantification theory; but Peirce's claim reverses the order of influence. Within the constraints imposed by the Boolean program, issues even of *monadic* quantification motivated his logic of relatives.

Peirce's claim is plausible if we remember Boole's determination to express all propositions as equations. The logic of relatives allowed Peirce to retain that program by expressing particular propositions as equations— and not, as in the first method, by *negating* an equation. This is an amazing result. If we use Peirce's involution and let 0 stand not only for the null class of individuals, but also for the null relation (depending on context), "some h are b" can be expressed by $0^{h,b} = 0$. In this formula, the first 0 stands for the null relation; the second, for the null class. To see why this equation works, we first note, with Peirce, that the class of x has members if and only if $0^x = 0$ is true (3.82). 0^x is the class of all things bearing the null relation to every member of x. If x has members, this will be the null class; and, if x has no members, it will be the universal class. Since "some h are b" just says that the class h , b has members, it will be true if and only if the above equation holds. This result allowed Peirce to deny an equation h , $b = 0$ by using another equation $0^{h,b} = 0$. Such equations can then be combined with either equational or inclusional formulations of universal propositions to yield the traditional syllogistic forms.

We can thus see that maintaining Boole's equational program provided a quantificational reason for Peirce's logic of relatives. But all of this is within the realm of monadic quantification. To deal with *multiple* quantification, especially mixed quantification, we must turn to other features of Peirce's logic of simple relatives.

The algebra of simple relatives which Peirce developed in DNLR contains very rich techniques for expressing multiple quantification. These come from at least three sources:

1. Inclusions between relations are implicitly quantified. To say "lover $\prec$ servant" is to say that for all x and y, if x is a lover of y, then x is a servant of y.
2. All three of Peirce's two-place relative operations contain implicit quantifiers. With the relative product, x is a lover of a servant of y if and only if there is some z such that x is a lover of z and z is a servant of y. Each of the two involution operations uses universal quantification: x is a lover of *every* servant of y, and x is a lover of *only* servants of y.
3. By using the four relational constants which stand for the universal, null, identity, and distinctness relations, many common quantified propositions can be formulated. For instance, to say that the identity relation is included in the product of R and R-converse means the same thing as "for all x, there is a y, such that xRy."

In 1883, of course, Peirce replaced the two involutions by relative addition; but the expressive possibilities remained essentially the same. It is very important to emphasize that these are not limited to doubly quantified propositions. Löwenheim, for instance, gave a quintuply mixed quantified proposition which can be written in this algebraic form (*Löwenheim 1915*: 233).

We may conclude, then, that DNLR contained the quantificational resources for dealing significantly with statements containing both absolute terms and simple relative terms with large numbers of quantifiers. At the same time, we should note that the application of these methods to conjugative terms is unclear and goes largely unexplored by Peirce.

§4. From the ALR to the QLR

By 1885, Peirce had moved from ALR to the QLR. I will not attempt to trace in detail the steps by which quantifiers evolved in Peirce's thinking.[9]

Let me merely state that Peirce's use of sum and product signs to represent "some" and "all" went through three stages:

1. To begin with, these were term-forming operators. The summation sign was a short way of treating a class as the logical sum of its members or of its subclasses. Similarly, the product sign represented a class as the logical product of other classes.

2. In 1882 and 1883, Peirce came to think of a class or relation as the logical sum of terms, each with a numerical coefficient attached—a 1 if it was a member of the class or relation and a 0 if it was not. Quantified expressions in English are then translated into statements about these coefficients. For instance,

$$\Pi_i \Sigma_j \, (l)_{ij} > 0$$

means that everything is a lover of something. But Peirce quickly noted that > 0 can be omitted from these, leaving an expression that clearly resembles a standard quantificational proposition.

3. Finally, in 1885, he reached something very close to modern quantification theory. Letting $(l)_{ij}$ mean that i is a lover of j,

$$\Pi_i \Sigma_j \, (l)_{ij}$$

just says that everything loves something.

It is important to note that the relative operations of 1870, and even of 1883, disappeared entirely by 1885. While these operations can be easily defined in the language of 1885, Peirce apparently saw no need to do so.

Peirce clearly believed that the QLR was superior to the ALR. But why? This is an intriguing question, given the power of the ALR. Unfortunately, Peirce's own comments on this are very sketchy. But I would like to outline some of the relevant considerations in a systematic way.[10]

The progression from the ALR to the QLR is so natural that it may even seem odd to ask for a further explanation of Peirce's preference for the QLR. When we examine the evolution of his treatment of sum and product signs, we are inclined to believe in logical predestination. In "twenty-twenty hindsight," Peirce's treatment of numerical coefficients in 1883 cries out to be transmuted into our modern quantifiers. But to note the naturalness of a certain progression of thought does not really get at the reasons for preferring its later stages. This caution is reinforced when we remember the dangers of imputing the naturalness which this notation has for us today to Peirce over a hundred years ago.

Let us examine the advantages of the QLR over the ALR from two different standpoints: that of the *deductive* usefulness of the notations, their capacity for being used in carrying out deduction; and, their *expressive* usefulness, their ability to express clearly a wide variety of propositions. Within each of these two categories, I will look, in turn, at three different aspects of the notation: its *power*, its *convenience*, and its *analytical depth*.

§5. Deductive Advantages of the QLR

Peirce made some strong claims for the power of his QLR. In 1885, for instance, he said that he would "purpose to develop an algebra adequate to the treatment of all problems of deductive logic" (3.364), thus extending "the power of logical algebra over the whole of its proper domain." Yet, as far as I know, Peirce did not claim that, strictly speaking, QLR had greater *deductive power* than ALR. He did not say, for instance, that there are valid arguments which can be formulated in both the QLR and the ALR, but whose validity can be shown only in the QLR. Such a claim would have amounted to saying that the ALR is deductively incomplete and, perhaps, suggesting that QLR is complete; and such a claim would have been ahead of its time even for Peirce.[11] Some years later, in 1897, he damned the ALR with faint praise, saying that it "has a moderate amount of power in skillful hands"; but this was not followed by the claim that the QLR had greater power (in whatever sense of power he had in mind). Rather, he said that the "great defect" of the ALR was the "vast multitude of purely formal propositions which it brings along"(3.497).[12]

In fact, Peirce's statements about the advantages of the QLR tend to emphasize its deductive *convenience* rather than its power. This was evident in his paper of 1883, where he complained:

> The [algebraic] logic of relatives is highly multiform; it is characterized by innumerable immediate inferences, and by various distinct conclusions from the same sets of premisses. . . . The effect of these peculiarities is that this algebra cannot be subjected to hard and fast rules like those of the Boolian calculus; and all that can be done in this place is to give a general idea of the way of working with it. (3.342)

In the same article, he also said,

> When the relative and non-relative operations occur together the rules of the calculus [the ALR] become pretty complicated. In these cases, as well as in such as involve *plural* relations (subsisting between three or more objects), it is often advantageous to recur to the numerical coefficients. . . . (3.351)

In 1897, he would say that QLR is "the most convenient apparatus for the study of difficult logical problems," though he would still use the ALR in "the simpler cases in which it is easily handled" (3.502).

The ALR thus appeared to be deductively complex and chaotic to Peirce. This contrasts with the fact that in both 1883 and 1885, his introduction of the QLR was followed by a *method* of working with the symbolism. In 1883, he even said that this consists in using certain "simple rules." The full 1885 method is, of course, very powerful and systematic. It uses a normal form to place all quantifiers at the beginning of a sentence, followed by a "Boolian" part, which is a purely Boolean combination of open sentences. This Boolean part is then operated upon with well-understood Boolean methods. There are no relative operations to gum up the works. As Peirce commented in 1897, the QLR contains only two operations, neither of which is a relative operation and both of which are "easily manageable" (3.502). While this is not a mechanical procedure for constructing derivations, its highly methodical character contrasts vividly with Peirce's perception, at least, of the ALR.

Peirce was interested in working out a philosophically satisfying system of logic, one which gives the most revealing account of basic inferential steps. I will call this the search for *analytical depth*. His conception of logic as a science often stressed this objective over mere manipulative ease. He thought that the QLR was superior on this score as well. In 1885, he said that one object of the QLR was,

> the enumeration of the essentially different kinds of necessary inference; for when the notation which suffices for exhibiting one inference is found inadequate for explaining another, it is clear that the latter involves an inferential element not present to the former. Accordingly, the procedure contemplated should result in a list of categories of reasoning, the interest of which is not dependent upon the algebraic way of considering the subject. (3.364)

In the same paragraph, he added that he did this without giving any "facile methods of reaching logical conclusions." Another aspect of analytical depth may be referred to when, at the end of this article, Peirce commented on the "minutely analytical" character of the system presented therein (3.403).

The concept of deductive analytical depth is intuitively appealing, yet it is very difficult to make it precise or to specify criteria for it. Applied to the contrast between the ALR and the QLR, the issue would be whether representing an inference in the language of the QLR would give a better idea of "what's really going on" than one which uses the ALR.

The problem is that such a claim may well be relative to the language in question. If one is thinking in terms of the ALR, then its rules may seem basic, so that the QLR is no deeper than the ALR. Who is to say that QLR-deductions are "really" taking place when the ALR is used? But the partisan of the QLR can say just the opposite. This would seem to drive the issue of the deductive analytical depth of the QLR back to prior questions of its expressive virtues. Perhaps it is because of its expressive virtues that the

QLR gives a better analysis of deduction. But this presupposes that the expressive virtues of the QLR and the ALR can be determined independently of their deductive virtues. This claim goes too far, since, as Peirce pointed out above, an important feature of a notation is its use in evaluating inferences. One notation may well be better than another if it is the foundation for a system of logic which generates more valid inferences.

One is tempted, then, to ask a typical chicken-and-egg question: which comes first, deductive analytical depth or expressive analytical depth? It seems more promising to suggest that they are so intertwined that they cannot be determined independently of each other. Nevertheless, it will be possible to discuss some of the expressive advantages of the QLR in comparative abstraction from deductive issues.

§6. Expressive Advantages of QLR

The first issue is one of expressive *power:* are there propositions which can be expressed in the QLR but not in the ALR? Here one must distinguish between the three types of terms in the ALR.

Absolute terms. This case is genuinely puzzling. In 1885, Peirce made the statement that "All attempts to introduce this distinction [between "some" and "all"] into the Boolian algebra were more or less complete failures until Mr. Mitchell showed how it was to be effected" (3.393). This presumably included Peirce's own attempt in DNLR to use the ALR for this purpose; but he did not say what was wrong with it. We have seen that this ALR does give a way to express particular propositions while retaining Boole's equationalism. It seems to be formally satisfactory, even though it may be unsatisfactory on some deeper level.

Simple relative terms. Peirce made no claims about the greater power of the QLR to express propositions using only simple relatives. It seems not to have been a factor in his thinking. In fact, we know that this is a problem for ALR only from Korselt's result, which Löwenheim reported thirty years later, in 1915 (*Löwenheim 1915*: 233ff.). This result states that the proposition that there are at least four individuals cannot be expressed in the ALR for simple relatives, even though it can be expressed in the QLR.[13]

Plural relatives. Here it is not even clear that questions of expressive power are properly asked, since Peirce never fully developed an ALR for plural relatives. He believed, of course, that all predicates with four or more argument places can be reduced to those with three or fewer places; but even an ALR for conjugative terms remains undeveloped. In 1870 and 1882 he commented on the possibility of such an ALR, but he did not actually construct it (3.69–72, 3.317). And well he shouldn't, for it would have been extraordinarily complicated. Peirce himself noted in 1882 that it would contain five forms of conversion alone. This may have been the source of the need for QLR. Peirce's reference to plural relatives in 1883 (3.351) may well show his recognition of the need for a better way to handle plural relatives. The mere fact that he never fully developed an ALR language for plural relatives suggests that these problems may have been an important stimulus for the QLR. At the same time, this could have been more a matter of expressive convenience than of expressive power.[14]

Another issue of expressive power concerns the ability of a notation to combine these several types of predicates in one sentence. DNLR already contained some ways of doing this, since many of its expressions combine two-place and one-place predicates, as in "lover of a woman." But it is not clear how far this expressivity goes; and when three-place predicates are thrown in, the situation becomes even more murky. It is important to note that Peirce's examples of the QLR in both 1883 and 1885 (3.351–357, 3.394–397) contain just such combinations of predicates which might have been hard to handle in the ALR. Once again, though, this could be more a matter of convenience than of expressive power.

The convenience feature of expressive adequacy might just be called *readability*. To what extent does a notation express propositions in a natural and readable way? It often appears that the ALR uses very odd ways to express propositions that can be expressed more simply and intuitively in the QLR.

Here is one example from 1883. The proposition that everyone serves someone can be expressed in many ways in the ALR. Among these are, $1 \prec (0 \dagger (s1))$ and $I \prec (s\check{s})$. (I depart here from Peirce's notation of 1883 by continuing to use 1 for the universal relation, and by using I for the identity relation. $\check{s}$ stands for the converse of s.) Contrast this with the version in the QLR,

$$\Pi_i \Sigma_j (s)_{ij}.$$

Peirce has systematic reasons for liking the first ALR form, but its meaning is hardly evident. Even the second form, which he could easily have used, seems to express the proposition less clearly than does the QLR form, with its explicit quantifiers to express "everybody."

I am not sure how to deal with the apparent subjectivity of appeals to the readability of a notation. Yet, I cannot help but think that the ALR versions of "everyone serves someone" are less perspicuous than the versions in the QLR. I know of no place where Peirce used this as a criterion explicitly, but he may have done so implicitly. We have seen that in both 1883 and 1885, he followed the introduction of quantifiers with increasingly complex examples of their use. In doing so, he may have been daring the reader to write these complex propositions in a readable ALR form.

At the same time, one must acknowledge that Peirce would be wary of simple appeals to readability. Such appeals might appear to take ordinary language as the norm for logic, and Peirce was often suspicious of ordinary ways of thought. He would probably be more interested in the third expressive virtue of the QLR, its *analytical depth*. It is likely that Peirce saw in the QLR a better form of logical analysis than the ALR. At the same time, my discussion of the relation between deductive analytical depth and expressive analytical depth should alert us to possible obscurities in this claim.

Some have suggested that this deeper analysis lies in Peirce's recognition that the logic of relatives must deal with genuine relations rather than with relationally defined classes (*Martin 1978*: 27). For the reasons already suggested, I do not find this convincing. In DNLR Peirce already knew that relative terms stand for relations. Whatever expository problems there were in DNLR are absent from the ALR portion of 1883, in which relational terms are fully extracted from relation-class compounds. Still, Peirce could well claim that his QLR provided a better notation for relations and that the use of indices and quantifiers gets at a deeper level of logical analysis than does his ALR.

Here are two signs of Peirce's respect for the analytical depth of QLR. First, he noted that relative operations can be easily defined in the QLR. In 1883, relative product and relative addition are defined by,

$$(ls)_{ij} = \Sigma_x[(l)_{ix}(s)_{xj}]$$
$$(l \dagger s)_{ij} = \Pi_x[(l)_{ix} + (s)_{xj}].$$

By 1883, then, Peirce saw that quantificational notation provides a uniform and intuitively meaningful way to capture these operations.[15] We may either retain them, as in 1883, or just dispense with them, as in 1885. But the quantificational formulation is more fundamental.

The comparison is actually more dramatic than this, for from this standpoint one of the greatest weaknesses of the ALR is the sheer variety of quantificational notations which it contains (see Section 3). The ALR contains too many ways of doing the same thing, while the QLR has one unified method.

This suggests that a second type of analytical depth lies in the uniformity and generality of quantification theory. In addition to providing an elegant theory of quantification, it deals in a uniform way with predicates of any number of places, and it allows one to mix them at will. Peirce's attempts to get a uniform notation by applying the ALR to one-place predicates in 1870 and 1883 were both very artificial; and the extension of this to three-place predicates is merely hinted at. It is noteworthy that Peirce's examples of using the QLR to formulate complex propositions contain mixtures of one-, two-, and three-place predicates (3.356, 3.394–397). All of these examples are handled in the same way by the new notation. This uniformity and generality of QLR is much more than a matter of convenience;

as in the empirical sciences, it is one sign of the theoretical power of a theory.

Peirce's algebraic logic of relations is a logic with many virtues, and his reasons for preferring the QLR to the ALR in 1885 seem heterogeneous and not clearly articulated. It appears, though, that issues of convenience and analytical depth figured more prominently in his thinking than did matters of sheer deductive and expressive power.[16]

N O T E S

1. Peirce later called these the Algebra of Dyadic Relatives and the General Algebra of Logic, respectively (*Peirce 1897/* 3.492–502).

2. For a discussion of the relation between De Morgan and Peirce, see *Merrill 1978.*

3. See, especially, *Brunning 1981, Burch 1991,* and *Herzberger 1981.*

4. Peirce persists in this usage. See *Peirce 1897/*3.466.

5. This mimics Peirce's rendering of $m \prec l$ at 3.66.

6. The use of blanks is explicit only in Peirce's treatment of conjugative terms in 3.63–64, but it seems clearly intended for simple relatives as well.

7. For a later account of the relation between noun forms and verb forms, see *Peirce 1897/*3.458–459.

8. It is interesting to note that Peirce still used this form in *1883c/*3.353 in the very same paragraph where he also used the notation of the later QLR.

9. See *Beatty 1969.*

10. Some of these issues are also dealt with in *Dipert 1984.*

11. For a full account of such issues, see *Tarski and Givant 1987.*

12. In his long reviews of Schröder, Peirce criticized Schröder several times for his undue fondness for the ALR. These criticisms, though, concern not the power of the ALR but its overly algebraic character. See *Peirce 1896/*3.451 and *Peirce 1897/*3.497–499, 512–519.

13. Burch's important claims about the expressive power of the full notation of the DNLR should be consulted on this point (*Burch 1991*). Since his account requires Peirce's three-place "teridentity" relation, these claims would seem not to affect our assessment of the algebraic logic of *simple* relatives.

14. On the expressive power of the full DNLR notation, see *Burch 1991.*

15. The Kneales suggest that Peirce might first have been attracted to the QLR because of its "clear and simple definitions" of the relative operations, only seeing later that "it could be employed to deal with all sorts of problems which could not be expressed in the language of relative multiplication and relative addition" (*Kneale 1962:* 430–431). If the argument of this paper is correct, this claim is not true. It should be remembered that Peirce gives complex sentences as examples of the QLR notation from the very beginning (*Peirce* 1883c/3.351ff.).

16. This chapter has benefitted from discussions with other participates at the Peirce Congress, as well as with my Oberlin colleagues Robert H. Grimm, Barbara Horan, Peter McInerney, and Phyllis Morris.

10.

FROM THE ALGEBRA OF RELATIONS
TO THE LOGIC OF QUANTIFIERS

Geraldine Brady

Peirce introduced his system of quantificational logic in his paper of 1885. Within this system he distinguished a first-order part, which he called "the first-intentional logic of relatives" (3.392), from second-order, i.e., "second-intentional," logic. Remarkably, the first-order fragment he presents is pure, general first-order logic, in prenex normal form, free of the operations of relative product and its dual, relative sum, which are essential to his algebra of relations. The restriction to first-order logic has been enormously fruitful in modern mathematical logic, and although modern foundation literature ascribes first-order logic primarily to Frege's fundamental paper (*1879*), we support the case that it was Peirce's work, as systematized and extended by Schröder (*1895*), that was a primary influence on Löwenheim (*1915*) and Skolem (*1920, 1923*), as reflected by their notations and methods.[1]

How, then, was first-order logic extracted from the calculus of relatives? Why did Peirce, when he focused on the first-order fragment of the calculus of relatives in 1885, choose the particular fragment that he chose? To answer these questions we need first to consider two of Peirce's previous papers in logic, *1870* and *1880*. His investigations in these two papers were driven in part by a problem left dangling in Boole's work. Boole had devised a poor system for representing existential assertions, upon which Peirce sought to improve. By 1870 Peirce had hit upon the idea that existentials could be represented by the composition of binary relations in his nascent system of the algebra of relations. In 1880 he returned to the theory of the syllogism and made repeated attempts to give a satisfactory treatment of particularity. From 1870 onward he continued to emphasize the role of the relative product, which is a natural generalization of the composition of functions to binary relations, and it was not until 1883, when his student O. H. Mitchell presented a system which successfully treated the notions of "for some" and "for all" using quantifier notation, that Peirce began to realize that quantifiers might be as basic as binary operations for mathematical reasoning. Peirce at first simply adopted Mitchell's quantifiers as an addition to his calculus of relatives, but in 1885 he focused on quantifiers as a direct object of investigation, thereby making the step from an algebra of binary relations to the idea of a first-order language with variables, predication, quantifiers, and stand-alone propositions.

§1. Background of the Quantification Problem

1.1. *The Problem of "Some" in Boole's Algebra of Logic.* Although Leibniz was the first to envision logic as a calculus of reasoning, it was Boole (*1847, 1854*) who made the first convincing implementation of this idea. Boole wanted to develop logic as a subbranch of mathematics, expressing logical concepts in mathematical terms and then analyzing them with the available mathematical technologies. Aristotle was Boole's source, and Boole tried to make the theory of the syllogism algebraic, employing like methods to cover all of its cases. In a sense he succeeded for universal formulas, but the special devices he introduced did not apply as well in the existential cases.[2]

Boole represented the premisses of the syllogism as equations between class terms. He symbolized the universal proposition "all y's are x's" as $y = vx$, where v is an indefinite (nonempty) class term expressing the notion "some" (*Boole 1854*: 61). (Note that Boole understands the universal proposition to be "all y's are some x's," reflecting the tendency of the times toward the quantification of the predicate.)

Writing $y = vx$ as $y - vx = 0$ and eliminating v by an algebraic rule, he obtains $y(y - x) = 0$; $y(1 - x) = 0$, or $y = xy$ for the universal proposition.

Boole's treatment of the particular proposition is less effective. He expresses "some x's are y's" by the equation $vx = vy$. Here he cannot eliminate v without arriving at an identity $0 = 0$; instead he represents the particular proposition by $v(x(1 - y) + y(1 - x)) = 0$, with the stipulation that v cannot be eliminated.

It is worth noting that although Boole's special symbol v engendered much subsequent criticism (*Peirce 1870, 1880; Schröder 1880; Venn 1881 [1894]*), his theory is not inconsistent. As Hailperin (*1976 [1986]*) shows, Boole's extra variable v is, in fact, an indeterminate in a polynomial ring over a Boolean algebra, a device no different than an indeterminate over the field of real numbers. Boole relies on polynomial equations similar to the one given above to express particular propositions in his algebra. The notions of modern algebra, as Hailpern observes, explain Boole's original formalism fairly well, but this was left to a later and more algebraically sophisticated generation to understand.

1.2. *Peirce's Initial Attempts to Express Particularity.* Peirce shared with

Boole the view that logical reasoning could be done by algebraic rules but found fault with Boole's treatment of particularity. Several times in his writings he made clear his view that Boole's representation of particular propositions was a failure and proposed solutions of his own.

In *1870*, he criticized Boole's treatment of particular propositions (in 3.138) in detail. He noted that Boole expressed "some y's are not x's" as $vy = v(1 - x)$, and he observed that from $vy = v(1 - x)$, $vy = v - vx$ and $vx = v - vy = v(1 - y)$ follow; however, since $vx = v(1 - y)$ means "some x's are not y's" in Boole's system, he notes that the result is the invalid inference of "some x's *are not* y's" from "some y's are not x's." He seems to say that Boole went wrong because his symbol v does not denote any particular subset and, therefore, as in the above argument, can be used to denote two different things, resulting in a contradiction if they are equated.

In fact, it is the algebraic interpretation of $vy = v (1 - x)$ which is incorrect. Indeed, $vy = v(1 - x)$ is $vy = v - vx$, and so $vy + vx = v$, or $v(y + x) = v$. But '+' is symmetric difference (exclusive 'or') in Boole's algebra. Thus, $v(y + x) = v$ basically says that v is a subset of the symmetric difference $y + x$, which is nonempty: that is, it either has elements in y but not in x or it has elements in x but not in y, but we do not know which. Thus, Boole's expression is meaningful, but it only represents the symmetric case; it does not represent the asymmetric case in which some x are not y.

In an earlier paper (*1867*), Peirce had simply dismissed Boole's treatment of particular propositions and proposed his own solution. He introduced "logical addition" (i.e., inclusive disjunction or union) to supplement Boole's operation of symmetric difference and defined its inverse, "logical subtraction," by $x = a - b$, if $b +, x = a$, where '=' denotes identity and '+,' denotes union. After so modifying Boole's system, he declared that these operations were added to facilitate expressing particular propositions (3.18). He represented "some a" by $a - (i,a)$, where the comma indicates intersection and i denotes an indeterminate singleton subclass of a. The expression $a - (i,a)$ is hard to make precise: it seems to be an operator choosing the singleton of a class a such that $a - (i,a)$ holds if at least one such singleton exists, although it is indeterminate; if no such singleton exists, the expression is nondenoting. Peirce never again employed this representation in his work. Within a year, he was exploring a different means of expressing the notion of "some," as we shall next see.

In an unpublished note written in November-December 1968 (Note 4; W2: 88–92), anticipating material in *1870*, Peirce proposed another solution, one using the exponential to express particularity. He begins Note 4 with the observation: "The mode proposed for the expression of particular propositions is weak. What is really wanted is something much more fundamental" (W2: 88). This remark obviously reflects dissatisfaction with his treatment of particular propositions in *1867*, and he here experiments with a new idea. Taking w to denote the relation "wiser than," n to represent the relation "not" or "other than," and m to stand for "man," he seeks to determine which of the two definitions for w^m, namely "wiser than every man" or "wiser than some man," will give the best theory. He decides that if the law for addition of exponents, $w^{a+b} \leftrightharpoons w^a, w^b$, from ordinary algebra is to hold, then w^m must denote "wiser than every man." Then n^x will mean the complement of x, and "wiser than some man" can be expressed by $n^{((n^w)^m)}$. Thus, using ordinary algebra as a guide, he finds a simple way to express the notions of "some" and "all."

Note that exponentiation, as Peirce uses it, relates a binary relation (the base) to a class (the power); this operation is not directly in Boole's algebra of classes but is, in modern terms, in an extension of it. Peirce was experimenting here in mixing Boole's algebra of classes with binary relations. These explorations are a precursor to his investigations of *1870*, in which he uses the exponential as a primitive. In later refinements of his system of relational algebra, Peirce introduces an operation assigning to each class its identity relation, thus eliminating the necessity for taking the exponential, a mixed operation, as primitive.

He concludes Note 4 with an algebraic analysis of a syllogism having a particular proposition as a premiss. He introduces the symbol Ψ and lets Ψ^a denote "the case in which a does not exist or in which $a \leftrightharpoons 0$" (W2: 92). The proposition "some a is b" can thus be given by the equation $\Psi^{a,b} \leftrightharpoons 0$. Representing the premisses "some a is b" and "no b is c" by $\Psi^{a,b} \leftrightharpoons 0$ and $b, c \leftrightharpoons 0$, he obtains $b \leftrightharpoons b, 0^c$ from $b, c \leftrightharpoons 0$, and concludes:

$$\Psi^{a, 0^c, b} \leftrightharpoons 0$$
$$\therefore \Psi^{a, 0^c} \leftrightharpoons 0;$$

that is, "some a is not c." (In this derivation, 0 should be replaced by n wherever 0 is raised to an exponent.) His derivation is correct if Ψ is understood to be 0, as we will see presently.

§2. Peirce's Vision: The Algebra of Relations (1870)

Peirce thought that binary relations and the operations that combined them could be used to represent all reasoning. In *1870*, he initiated a grand experiment to test his hypothesis. Following Boole and using ordinary algebra as a guide, he undertook to make an algebra for binary relations. Starting with Boole's operations for unions and intersections of classes, he extended these operations to relations. He added relative product, exponential, and converse as the basic binary operations in his algebra. Relative product, which he emphasized, was a generalization of the composition of functions, commonly used in mathematics without formalization. Converse (inverse) also was widely used without formalization at the time. He also looked for analogs of common arithmetical operations, such as exponentiation and logarithm, and explored their properties. Some of the operations, notably relative product and exponentiation, are mixed, relating a binary relation to a class. Much as in the case of linear algebra, his algebra contains two types of objects: "absolute terms" or classes, which can be added by union and multiplied by intersection; and "relative terms" or relations, which can be applied to absolute terms. He experimented throughout the paper to find out what notations and definitions would give the best theory, and at its conclusion arrived at a provisional version of his algebra.

Peirce at this point used an intensional interpretation of natural language expressions and algebraic expressions for binary relations, and, generalizing Boole, introduced an extensional interpretation of such expressions as denoting, in effect, propositional functions of two variables. An n-variable propositional function on a domain is simply a function of n arguments from the domain with values $\{0,1\}$ or false, true. In the case where the domain is itself $\{0,1\}$ (and Boole used these functions extensively), this is the algebra of truth functions. In the case where the functions are functions of one variable, it is arguable that this is a good interpretation of Boole's algebra of subclasses of a domain. Peirce generalized Boole and used propositional functions of two variables on a domain to represent intensional relations, arguably as the denotation of the expression for the binary relation. It is not clear, however, how fully Peirce understood the scope of this idea. He also wrote down the list of all ordered pairs constituting a relation in the sense that we now understand it, although it is unclear whether or not he regarded this to be the denotation of the expression. Whatever the case, Peirce continually returned to the propositional function interpretation, which he borrowed from Boole.

Relative product is the cornerstone of Peirce's algebra, and his first tool for representing the quantifiers, although for first-order logic (see Korselt [in *Löwenheim 1915*]; *Schröder 1895, Löwenheim 1915*), this turns out not to be as general a representation of quantifiers as he perhaps first thought.

Multiplication, as Peirce defines it (3.68), is the relative product. In the example he gives, lw is the application of a relation l to an "absolute term," that is, a unary relation (i.e., a class or property), w. Representing l and w as atomic formulas, we can see that if $l(i, j)$ means "i is a lover of j" and w(j) means "j is a woman," then the relative product lw(i) means "there is a j such that i is lover of j and j is a woman." Thus the relation l operates on the property "is a woman" to produce a property "is a lover of a woman"; in other words, relative product acts as an operator on unary properties or sets. For example, if f were a function and s a set, then fs is the inverse image of s under f in modern terms. Thus, the modern equivalent of Peirce's "application" is the inverse image of a function.

Peirce intended to use relative product as a way of handling many quantification problems because, contained within it, there is an implicit existential quantifier. Its formula, in modern terminology, reveals this quite clearly:

$$l\text{w}(i) \text{ if and only if } \exists j \ (l(i,j) \wedge \text{w}(j)).$$

If the relative product is sl, "servant of a lover of," we have

$$sl(i,j) \text{ if and only if } \exists k \ (s(i,k) \wedge l(k,j)).$$

At this point (3.68) his notation is somewhat ambiguous. Having permitted relative product to operate on relations and classes, he decides (3.73) to restrict relative product to relations only. To do this he introduces a notation using commas which increases the number of variables by one. By writing a comma after an absolute term, he indicates that it is to be taken as a relative: that is, "m," means "man that is —." In other words, if $m(i)$ is a property, write m,(i, i). He observes that in this way any absolute term can be regarded as a relative and any relative can be regarded as a ternary (potentially n-ary) relation. This allows him, with no formal use of variables, to bring relations of any arities up to the same arity. Thus, if the relative product lw is written lw,:

$$l\text{w},(i,j) \text{ if and only if } \exists k \ (l(i,k) \wedge \text{w},(k,k) \wedge k = j).$$

It is likely that, in his discussion of the operations of his algebra, Peirce identifies the classes w with the binary relations w, but suppresses the comma notation because he thinks the examples can be read in English without coding the properties as binary relations.

With this in mind we can look at his relative involution, that is, exponentiation. In the example he gives, l^w is, again, the application of a relation to an absolute term. The absolute term, or class, can, in fact, be replaced by the binary identity function on the class, as we just noted.

Again representing l and w by the atomic formulas $l(i, j)$ and $\text{w}(j)$, we can see that $l^w(i)$ means "for every j such that j is a woman, i is lover of j." Thus $l^w(i)$ means "i is a lover of every woman" and contains within it an implicit universal quantifier, as can be seen by its modern formula:

$$l^w(\text{i}) \text{ if and only if } \forall j \ (\text{w}(j) \rightarrow l(i,j)).$$

These experimentations are all part of an early attempt at quantification by Peirce, for which he uses the exponential at times. De Morgan (*1860a*) had investigated relative product and involution before him, and to some extent he was simply following De Morgan's lead in exploring the algebraic properties of these operations. Although all of these operations are correct, he later found relative product to be by far the most useful.

Drawing upon De Morgan's findings, Peirce observes that relative product and involution are connected via complementation: $\overline{ls} = \overline{l}^s$. De Morgan had also used a second method, *backward involution,* to take the complement of the relative product: $\overline{ls} = {}^l\overline{s}$, where backward involution can be defined by

$${}^ls(i,j) \text{ if and only if } \forall k(l(i,k) \rightarrow s(k,j)).$$

It was the exponential, along with these results concerning complementation, that Peirce used to tackle Aristotle's syllogism and the unsolved case of the problem of "some."

As we have said, Peirce viewed existential assertions to be the stumbling block preventing a mathematical treatment of Aristotle's syllogism. He saw the key to solving the problem to lie in expressing the nonemptiness of a class in equational form. The exponential provided him with the device to accomplish this. He observed that for an arbitrary class x, if x is nonempty, $0^x = 0$; if x is empty, $0^x = 0^0 = 1$ (3.82). Then, by taking complements, he proved that x is nonempty if and only if $1x = 1$, where the 1 on the left is the universe for relations, and the 1 on the right is the universe for classes (3.85–86).

He then employs this representation of nonemptiness to express particular propositions and illustrates his method by analyzing a syllogism with a particular affirmative conclusion (3.142). The argument he considers— "Every horse is black," "Every horse is an animal," "There are some horses"; therefore "Some animals are black"—he represents in symbolic terms as: $h \prec b$; $h \prec a$; $1h = 1$; $\therefore 1(a,b) = 1$, where '$\prec$' denotes inclusion and comma is intersection. His proof runs as follows. From the premises, $h \prec b$ and $h \prec a$, he obtains $h \prec a,b$. Hence, by monotonicity of relative product (see [90], 3.91), $lh \prec 1(a, b)$. Therefore if h is nonempty, $0^h = 0$ or $1h = 1$, and $1 \prec 1(a, b)$; $\therefore 1(a, b) = 1$. He notes that this syllogism could not be treated in Boole's algebra without modification (3.142). In fact, Peirce did not figure out how to justify Boole's treatment of the existential quantifier (some) in Aristotle's syllogisms, and thus he proposed his own solution, based on the operation of composition of relations, which has an implicit existential quantifier, and also based on exponentiation, which operates on a monadic predicate (class) and a binary predicate (relation).

§3. Peirce Returns to the Syllogism (1880)

In *1880,* Peirce returned to his study of the syllogism. We can see evidence that he was still searching for a systematic way to treat the notions of "for all" and "for some," but he seems to have made little progress in this paper.

The problem of treating the notion of "some" in Aristotle's syllogisms can be regarded as the problem of negating class inclusion. Thus the negation of "All A's are B's" is "Some A's are not B's." In *1880* Peirce made three attempts to develop a theory in which class inclusion is negated.

In his first attempt (3.162–172), Peirce examined Diodoran, that is, "formal" implication. He considered propositions P_i and C_i depending on a free variable i, which ranges over a universe of individuals. He wrote $P_i \prec C_i$ to mean "$P_i \prec C_i$ for all i," where '$\prec$' here denotes material implication. To negate the expression "$P_i \prec C_i$ for all i," Peirce wrote $P_i \overline{\prec} C_i$. The notation $P_i \overline{\prec} C_i$ is defective since it can suggest that the expression be read "for all i, P_i does not imply C_i"; in other words, the overbar negates material implication, which was not Peirce's intention. In his review of *Schröder 1895*, Peirce indicates that he anticipated the quantifiers at this point in his investigations:

> Properly to express an ordinary conditional proposition, the quantifier Π is required. In 1880, three years before I developed that general algebra, I published a paper containing a chapter on the algebra of the copula. . . . I there noticed the necessity of such quantifiers properly to express conditional propositions. (3.448)

The "conditional proposition" in modern terms is $\forall i(P_i \prec Q_i)$. There is no explicit indication in *1880* that he had the quantifiers in mind, since he does not use them or anything resembling them.

In his second attempt (3.173–193), Peirce negated the class inclusion $A \prec B$ simply by writing $A \overline{\prec} B$. Here he worked on the theory of the syllogism, using propositions of the form $A \prec B$, $A \overline{\prec} B$, and $(A \prec B) \prec (C \prec D)$, but his results were fragmentary. (Note that in the immediately preceding formula the middle occurrence of '$\prec$' stands for material implication.) However, his investigations are of interest since they represent the initial stages of his system of implicative propositional logic (cf. 3.365–391).

His third attempt (3.196) was based on the idea of "quantification of the predicate." This was a method of treating the syllogism which, as we have seen, influenced Boole.[3] In it, the terms "all" and "some" are applied to both the subject and the predicate. Here Peirce, to quantify the subject A, rewrote $A \overline{\prec} B$ (i.e., "some A are not B") as $\check{A} \prec \bar{B}$. Then, in order to quantify the predicate, he translated $\check{S} \prec \check{P}$ as $(S \prec \bar{x}) \prec (P \prec \bar{x})$, for all x. This formula is obviously equivalent to $P \prec S$. The theory he developed is rather cryptic, and Peirce himself said that he left the subject in an unfinished state (3.196).

§4. Mitchell's Advance: A New Algebra of Logic

The first real advance since *Peirce 1870* came with the publication of *Mitchell 1883*. Mitchell presented a system of logic which introduced the quantifiers. Although he was one of Peirce's students, the spirit of Mitchell's paper is different from Peirce. There is no role for the relative product. Relations are primitive, and Boolean operations with the quantifiers are the subject of the paper.

Mitchell's theory is divided into two parts. The first is the algebra of classes. To every term F in the Boolean algebra of classes Mitchell associates two propositional formulas F_1 and F_u. For example, F might denote $\overline{ab} + \overline{a}b$; then the statement F_1 is true if and only if the class denoted by F is the universe U, and the statement F_u is true if and only if the class denoted by F is nonempty. As he states (*1883*: 74):

> The following are respectively the forms of the universal and particular propositions:
>
> > All U is F, here denoted by F_1,
> > Some U is F, here denoted by F_u.

He relates these forms by the principle $\overline{(F_1)} = (\overline{F})_u$, where the equality sign means logical equivalence. This observation is a basic tool of quantification theory. Its modern equivalent is $\neg \forall x \phi(x) \leftrightarrow \exists x \neg \phi(x)$.

Accordingly, these forms can be combined:

$$F_1 + (\overline{F})_u = \infty, \quad F_1(\overline{F})_u = 0$$

where '+' is disjunction, juxtaposition is conjunction, and '∞' and '0' are the truth values true and false respectively.

Mitchell gives two laws governing the combinations of subscripts:

$$F_\in G_{\in'} \prec (FG)_{\in\in'} \text{ and } F_\in + G_{\in'} \prec (F+G)_{\in+\in'}$$

where $\in$ and $\in'$ can take on the values 1 or u, with

$$1 + 1 = 1, \quad 1 \cdot 1 = 1,$$
$$1 + u = u + 1 = u, \quad 1 \cdot u = u \cdot 1 = u,$$
$$u + u = u, \quad u \cdot u = \text{undefined},$$

and '$\prec$' is material implication. Note that multiplication denotes conjunction on the left side of the first law and intersection on the right; and, in the second law, addition denotes disjunction on the left and union on the right. He then lists the cases:

(1) $F_1 G_1 = (FG)_1$ $F_u + G_u = (F+G)_u$ (1')
(2) $F_1 G_u \prec (FG)_u$ $F_u + G_1 \prec (F+G)_u$ (2')
(3) $F_u G_u \prec \infty$ $F_1 + G_1 \prec (F+G)_1$ (3')

He points out that the product $u \cdot u$ must be left undefined; otherwise the law $F_\in G_{\in'} \prec (FG)_{\in\in'}$ would yield $F_u G_u \prec (FG)_{u \cdot u}$, but if F and G are nonempty classes with empty intersection, then $u \cdot u$ can be neither u nor 1. Note also that $(FG)_u \prec F_u G_u$ holds, but is not derived from this arithmetic of indices.

Mitchell (*1883*: 82–86) gives several applications of his theory. He represents the Aristotelian propositions as:

$$(\bar{a} + \bar{b})_1 = \text{All of U is } \bar{a} + \bar{b} = \text{No } a \text{ is } b,$$
$$(ab)_u = \text{Some of U is } ab = \text{Some } a \text{ is } b,$$
$$(\bar{a} + b)_1 = \text{All of U is } \bar{a} + b = \text{All } a \text{ is } b,$$
$$(a\bar{b})_u = \text{Some of U is } a\bar{b} = \text{Some } a \text{ is not } b,$$

addition and multiplication representing union and intersection. The canonical form of Aristotle's syllogism (i.e., the figure *Barbara*) becomes

$$(\bar{a} + b)_1(\bar{b} + c)_1 = (\bar{a}\bar{b} + \bar{a}c + bc)_1 \prec (\bar{a} + c)_1.$$

Mitchell also shows that his theory treats with ease the syllogism Darapti that Peirce labored over in *1870*. Writing the premises in conjunctive normal form, applying the distributive law, and simplifying using the rule $(\bar{m} + F)_\epsilon = (\bar{m} + mF)_\epsilon$ derived earlier, he obtains:

$$(\bar{m} + p)_1(\bar{m} + s)_1 = (\bar{m} + sp)_1.$$

(Note that the formulas on the left have the same form as the premises in Peirce's example: "Every horse is black"; "Every horse is an animal.") Mitchell observes that it is necessary to add a third premiss, $(m)_u$, that is, "some of U is m," or "there is some m," corresponding to Peirce's premiss "There are some horses," in order to infer the conclusion, since a particular conclusion cannot be inferred from universal premises alone. Again conjoining the premises, distributing, and simplifying, he concludes:

$$(\bar{m} + sp)_1 \, (m)_u \prec (spm)_u \prec (sp)_u;$$

namely, "some s is p," corresponding to Peirce's "Some animals are black."

Mitchell treats many other examples from the class calculus. His system incorporates Aristotle's theory of the syllogism and in this respect exceeds Boole's calculus of classes and Peirce's treatment (*1870*; 3.142).

The second part of Mitchell's theory (*1883*: 87ff.) includes binary relations. He calls them "propositions of two dimensions," in contrast to the one-dimensional propositions of the Boolean algebra of classes, just examined. His two-dimensional logic is based on two universes, U and V, representing the universe of class terms and V the universe of time respectively. He treats quantification in the two-dimensional cases in two slightly different ways. In the first, he represents the six two-dimensional cases of quantification as propositions $F_{\alpha\beta}$:

F_{uv},   meaning 'some part of U, during some part of V, is F,'

F_{u1},   meaning 'some part of U, during every part of V, is F,'

F_{1v},   meaning 'every part of U, during some part of V, is F,'

$F_{u'1}$,   meaning 'the same part of U, during every part of V, is F,'

$F_{1v'}$, meaning 'every part of U, during the same part of V, is F,'

F_{11}, meaning 'every part of U, during every part of V, is F,'

where F is a polynomial function of class terms.

In these two-quantifier forms, u and v function as existential quantifiers, and '1' is the universal quantifier. The prime is used only on u and v, and it serves to invert the order of quantification. The convention is that if a '1' appears as a subscript, universal quantification is performed first, with respect to the coordinate indicated by the position of the '1', unless a lowercase primed letter appears as a subscript, in which case existential quantification is performed first. The modern equivalents of $F_{u'1}$ and $F_{1v'}$ are $\exists x \forall y$ and $\exists y \forall x$.

It should be noted that Mitchell does not write the subscripts in the order in which the corresponding operations are applied. In fact, there is no explicit mention of separate operations. Instead, it is as if the combination of subscripts were taken as indicating simultaneous quantification in two variables.

If these two-quantifier forms are interpreted as operations on propositional functions, Mitchell's notation becomes less mysterious. The coordinates are the variables, and the existential quantifier is a projection on a coordinate. Algebraically, the existential quantifier is the supremum, or least upper bound, of propositional functions over a domain. In other words, when $F(i, j)$ is assigned a propositional function $f(i, j)$ on the domain D, $F(i, j)$ being true or false as $f(i, j) = 1$ or 0, then the existential quantifier on i corresponds to holding j fixed and taking the least upper bound of $f(i, j)$ as i ranges over the domain D. This affords a clear explanation of Mitchell's treatment of two quantifiers.

In Mitchell's second, modified version of the two-quantifier forms (*1883*: 92), the quantifiers can be taken in either order. Parentheses delimit the scope of the operators, permitting quantifier nesting. The formula

$$\{(ab)_u + (\bar{c} + \bar{d})_U\}_V,$$

equivalent to the predicate logic formula

$$\forall j\{\exists i[a(i, j) \land b(i, j)] \lor \forall i[\overline{c(i, j)} \lor \overline{d(i, j)}]\},$$

illustrates this improvement. Here u and v act as projection operators, with dual operators U and V respectively. In fact, this formula can be expressed in cylindric algebras[4] as

$$\forall_2\{\exists_1(ab) + \forall_1(\bar{c} + \bar{d})\} = 1 .$$

Thus, Mitchell introduced a symbolic notation for propositional functions and for quantifiers as operations on propositional functions. It is less clear, however, that there is a formal system of syntactical expressions underlying this notation in Mitchell.

§5. Transition from the Algebra of Relations to the Logic of Quantifiers

Mitchell (*1883*) appeared in *Studies in Logic*, of which Peirce was editor and a contributor. Peirce's first contribution (Note A) concentrates on matters relating to the syllogism and does not touch on quantifier logic; his second contribution (Note B; hereafter *1883c*), however, responds to Mitchell's results and tries to comprehend the operations Mitchell had discovered by adding them to his own system without making them primitives in it.

In the first part of *1883c*, Peirce attempts to adopt Mitchell's quantifiers as an addition to his algebra of relations. To each of Mitchell's two-quantifier forms he associates a term $T(f)$, with free variable f, such that, in any interpretation, the two-quantifier form obtained by assigning a subscript to F is true if and only if $T(f)$ represents the universe 1 and false if $T(f)$ is 0. (In these formulas '1' has been substituted for Peirce's symbol '∞'.) In this interpretation, Mitchell's term F stands for some relation. Peirce assumes that this relation is now assigned to the variable f. For example, Peirce associates to the form F_{1v} the term $0 \dagger (f1)$, where the dagger denotes the relative operation "relative sum." (Relative sum, $x \dagger y$, is defined as the dual to relative product: $a \dagger b = \overline{\overline{a}\overline{b}}$.)

To illustrate this, we consider the relative sum $0 \dagger (f1)$. It has the property of being 1 or 0, depending on whether F_{1v} is true or false. To verify this property we first note that $h(0 \dagger (f1))k$ if and only if, for all i, the pair (h, i) is a member of 0 or the pair (i, k) is a member of $f1$. This is true if and only if for all i there exists j such that the pair (i, j) is a member of f and the pair (j, k) is a member of 1. This, in turn, is true if and only if for all i there exists j such that the pair (i, j) is a member of f, but that is exactly what F_{1v} says. Thus, $h(0 \dagger (f1))k$ is satisfied by all pairs (h, k) or no pairs (h, k), depending on whether F_{1v} is true or false.

Peirce recognizes that his representation is unwieldy (3.351). He drops the discussion of the relative calculus in (3.351–358) and tries to find expressions for quantifiers in a different way, without reference to his relative calculus.

He considers the numerical coefficients $(l)_{ij}$ given earlier (3.329) in his formula for a general relative l: $l = \Sigma_i \Sigma_j (l)_{ij} (I:J)$. He now (3.351) interprets $\Sigma_i \Sigma_j (l)_{ij}$ and $\Pi_i \Sigma_j (l)_{ij}$ as propositions concerning a pair of individuals (i, j); that is, l_{ij} is a propositional function in which the variables i, j are entered as subscripts, and Σ_i and Π_i are existential and universal quantifiers. Reference to a single domain is intended so that Σ_i, the existential quantifier, is interpreted as a supremum (least upper bound) of propositional functions evaluated over that domain and Π_i, the universal quantifier, as an infimum (greatest lower bound) over the domain. In the examples he presents from 3.352 onward, the relative product has been dropped and the operations performed on the propositions are Boolean (i.e., first-order).

For instance, in the example he discusses in 3.352, we can suppose the domain is people. Then $\Sigma_i \Sigma_j l_{ij} > 0$, in modern notation $\exists i \exists j \, l(i,j)$, means the proposition is true in the domain of people. If we consider this expression in algebraic terms we can arrive at an interpretation close to Peirce's intent. Thus, if we look at l_{ij} as a propositional function $l(i, j)$, defined on all i, j in the domain of people, with $l(i, j)$ being 1 if i is a lover of j, we can then use the operation $(\sup_j)$ on the function l of two variables to produce a function of one variable: namely, $f(i) = (\sup_j) l(i, j)$. That is, as i is fixed and j ranges over all people, we can take the sup of all the values of $l(i, j)$ (each of which is 0 or 1) to obtain $f(i)$. Then the second quantifier $(\exists i)$ is applied by taking $(\sup_i)$ applied to $(\sup_j) l$, that is, to f; that is, we obtain $(\sup_i) f$. This expression says to evaluate $f(i)$ at all arguments i which are people, getting always a 0 or a 1, and take the supremum. Then the assertion $\Sigma_i \Sigma_j l_{ij} > 0$ simply says this series of two sups, yielding first a function of one variable f and then a constant, yields the constant 1, not 0. Similarly, the universal quantifier is interpreted as an infimum (greatest lower bound) of propositional functions, obtained by suppressing the quantifier by instantiating its variable in all possible ways in the fixed domain.

This interpretation via propositional functions is consistent with Peirce's statement that $\Sigma_i \Sigma_j l_{ij} > 0$ is the proper statement and $\Sigma_i \Sigma_j l_{ij}$ is an abbreviation for it. Of course, $\Sigma_i \Sigma_j l_{ij} > 0$ could be interpreted in an algebra of statements instead, but there is no evidence that he introduced one. Rather, the inequality $\Sigma_i \Sigma_j l_{ij} > 0$ is his basic assertion of truth.

In any case, Peirce experimented with algebraic operations on propositional functions or relations corresponding to quantification, as he experimented with many other operations for their algebraic properties. He was seeking a calculus of quantifiers similar to the calculus of relations.

§6. The Logic of Quantifiers

In one brief passage (3.392–397) on "First Intentional Logic," in *1885*, Peirce presents a system of first-order logic that includes individual variables, quantifiers, predication, and stand-alone propositions. This passage, along with the evidence we have assembled thus far, suggests that Peirce looked over his calculus of relatives in light of what Mitchell had done and isolated the fragment that we recognize today as first-order logic.

There are two stages to Peirce's discussion in this passage: he credits Mitchell for quantifying the notions of "some" and "all" in both Boolean algebra and the logic of relatives; and he presents a new system of his own, which highlights individual variables and quantifiers and omits relative product and relative sum.

According to Peirce's own analysis (3.392), Boole's formulation provided the language for describing compounded unary properties; thus, a Boolean expression can only make an assertion about an individual class, but there was no means to talk about what one might call parameterized classes. On the other hand, what the logic of relatives enabled Peirce to do is to describe classes as having a functional relationship with individuals or other relatives. Stated somewhat differently, Boole's original presentation was not expressive enough to encompass mathematics, and the calculus of relatives looks to Peirce to be an analog that was sufficiently expressive.

It seems that in (3.393–394) Peirce is doing something that is strikingly different from anything that he had done before. The key feature that leaps off the page is the quantification structure. Notice the quantifiers, which he attributes to Mitchell, in 3.393:

$$\text{Any } (\bar{k} + h) \quad \text{Some } (\bar{k} + h)$$
$$\text{Some } (kh) \quad \text{Any } (kh)$$

They are implicitly over an entire universe of discourse. As a result, it is effectively impossible to describe anything very complicated because quantifier alternation would be required to do so; moreover, Peirce says, we can perform alternations once Mitchell's quantifiers are extended to the logic of relatives. Thus, it appears that he viewed the logic of relatives as the formal mode of what might be called parameterized Boolean algebra.

So far everything up to 3.394 seems perfectly clear. And it also seems clear what has happened: in working with Boolean algebra and the calculus of relations, Mitchell added quantification to Boolean algebra and, at least as Peirce explains it here, to the logic of relations, actually to a fragment of it. This fragment, which he called two-dimensional logic, is in fact first-order logic. First-order logic is what results when one adds variables and quantifiers to the algebra of relations restricted to Boolean operations.

Boolean algebras can be regarded as the theory of propositional functions of any number of variables. These are now called truth functions; their arguments and values are 0 and 1. Their theory was due to Boole and is formalized in his theory of Boolean algebra. But one cannot really generalize to quantifier logic until one has a clear concept of a propositional function of several arguments on a domain D, with values 0, 1. With this notion, the universal and existential quantifiers on any one of the arguments of a propositional function are operators on propositional functions. That is, when applied to an argument of a propositional function, one obtains a propositional function of one less variable. This does not require a formal language or the formal notion of variable, merely the traditional notion of a function of several arguments. This seems to be what Mitchell and then Peirce did.

Mitchell appears to have had a clear understanding of quantifying a propositional function of one argument and getting the propositional value 0 or 1. Mitchell's quantifiers "some" and "any" (i.e., "all") over the expression $(\bar{k} + h)$ are in fact quantifiers as long as there is only one variable. This is acceptable as long as we are referring to properties of individuals, the statements "everyone has this primitive Boolean factor," "no one has this primitive Boolean factor," "someone has this primitive Boolean factor" being either true or false. However, when we want to say that there are some sorts of interactions between individuals—that every element has an inverse, for instance—then we need not only properties of individuals, but also binary relations between individuals and some way of handling more than one variable at a time. We have already seen that Peirce (*1870*) invented a device that acted on relations to increase their number of variables and bring the number always to be the same. Here we will see that he concentrated on a logic of many variables and provided a semantics for first-order logic in doing so. Peirce clearly was quite comfortable with the general case of quantifying an argument of a propositional function of many variables. After him, Schröder worked out the algebra of quantifiers more extensively in this context. But neither Peirce and Mitchell nor Schröder introduced a formal system based on an inductive definition of formula, as did Frege.

In *1885* Peirce was mainly interested in investigating the algebraic properties of quantifiers. Thus, he begins the exposition of his own system by explaining the notation he is adopting for them and what the quantifier operations mean (3.393):

> Here, in order to render the notation as iconical as possible, we may use Σ for some, suggesting a sum, and Π for all, suggesting a product. Thus Σ_i means that x is true of some one of the individuals denoted by i or
> $$\Sigma_i x_i = x_i + x_j + x_k + \text{etc.}$$
> In the same way, $\Pi_i x_i$ means that x is true of all these individuals, or
> $$\Pi_i x_i = x_i x_j x_k, \text{etc.}$$
> If x is a simple relation, $\Pi_i \Pi_j x_{ij}$ means that every i is in this relation to every j, $\Sigma_i \Pi_j x_{ij}$ that some one i is in this relation to every j, $\Pi_j \Sigma_i x_{ij}$ that to every j some i or other is in this relation, $\Sigma_i \Sigma_j x_{ij}$ that some i is in this relation to some j.

He specifies the word *quantifier* as the official term in his system (3.396): "If the quantifying part, or Quantifier, contains $\Sigma_x \ldots$," and he emphasizes that a crucial aspect of the idea of the quantifiers consists in using individual variable, which he calls *indices*. He stresses their linguistic significance; that is, they are pronouns (3.361):

> The index asserts nothing; it only says "There!" It takes hold of our eyes, as it were, and forcibly directs them to a particular object, and there it stops. Demonstrative and relative pronouns are nearly purely indices, because they denote things without describing them; so are the letters on a geometrical diagram, and the subscript numbers which in algebra distinguish one value from another without saying what those values are.

He realizes that he is working not merely with an algebra but with a language.

Rather than providing formal definitions of variables, predicate symbols, and quantifiers and specific rules for proposition formation, he cites examples to explain his notation and how it is to be interpreted. His examples are not mathematical but instead are translations into his system of informal expressions for relations in English; namely,

> Let l_{ij} mean that i is a lover of j, and b_{ij} that i is a benefactor of j. Then
>
> $$\Pi_i \Sigma_j l_{ij} b_{ij}$$
>
> means that everything is at once a lover and a benefactor of something; and
>
> $$\Pi_i \Sigma_j l_{ij} b_{ji}$$
>
> that everything is a lover of a benefactor of itself.

And

> Let g_i mean that i is a griffin, and c_i that i is a chimera, then
>
> $$\Sigma_i \Pi_j (g_i l_{ij} + \overline{c_j})$$
>
> means that if there be any chimeras there is some griffin that loves them all.

The propositions he considers are all written in prenex form, with the interior quantifier-free part in clausal form, that is, a disjunction of conjunctions of elementary formulas of relational expressions.

In 3.396–397 he provides quantifier transformation rules, reduction and transformation rules for relatives, and rules for the elimination of variables.

Peirce extends this system to "second-intentional logic" (3.398), which includes a second family of variables, ranging over relations. The variables representing relations occur as subscripts, as they did in the first-order case. It is in second-intentional logic that mathematical notions appear. For example, he defines one-to-one correspondence using second-order quantifiers (3.401):

> However, the best way to express such a proposition is to make use of the

letter c as a token of a one-to-one correspondence. That is to say, c will be defined by three formulae,

$$\Pi_\alpha \Pi_u \Pi_v \Pi_w (\bar{c}_\alpha + \bar{r}_{u\alpha v} + \bar{r}_{u\alpha\omega} + 1_{vw})$$
$$\Pi_\alpha \Pi_u \Pi_v \Pi_w (\bar{c}_\alpha + \bar{r}_{u\alpha\underline{\omega}} + \bar{r}_{v\alpha\omega} + 1_{uv)}$$
$$\Pi_\alpha \Sigma_u \Sigma_v \Sigma_w (c_\alpha + r_{u\alpha v} r_{u\alpha\omega} 1_{vw} + r_{u\alpha w} \ r_{v\alpha w} 1_{uw}).$$

These defining formulas are second-order, since they quantify over all relations α; note also that r is a relation of higher order, with $r_{u\alpha v}$ meaning that v is in the relation α to u. His definition of the identity 1 (3.398), $1_{ij} = \Pi_x(x_i x_j + \bar{x}_i \bar{x}_j)$, is also second-order, x being a general relative. In general, we find definitions and computations freely passing into second-order in his work and higher-order quantifiers to be frequently used as a tool for expressing mathematical reasoning. It is likely that he singled out first-order quantifiers for study, recognizing that the rules he found for them also apply to second-order quantifiers.

Finally, in summarizing Peirce's contributions to first-order logic, we do not wish to give the misleading impression that his system is a well-formed system like Frege's. Peirce did create a language for first-order logic and specified rules for its use, but his is not a formal language since he does not introduce the notion of a formula by an inductive definition. Moreover, although he did clearly distinguish between first- and second-order logic, in practice he moved freely between the two (e.g., in his analysis of De Morgan and the syllogism of transposed quantity; 3.402). The reason was simply that delineating first-order logic as the well-behaved part of mathematical logic he did not see clearly. This was left to later generations to understand.

§7. The Relation between the Two Systems

We now consider some connections between the algebra of logic and the logic of quantifiers.

Since subclasses of a domain and their characteristic functions are indistinguishable, Boole's algebra of classes is a theory of operations on propositional functions of one variable. However, when Boole tried to utilize the variables present to treat quantifiers in Aristotle's syllogisms, he did not do a very clear or complete job.

Since binary relation operations on a domain and (0, 1) matrices on the domain are essentially identical, and (0, 1) matrices and propositional functions of two variables are also essentially identical, Peirce's algebra of relations is an algebra of propositional functions of two variables.

The question of how to construct an algebra of ternary and higher-order relations was the next level for Peirce. He handled this in two ways: (1) By his argument-duplicating operation $r(i, j)$ to $r_i(i, i, j)$ if and only if

$r(i, j)$, so that all relations can be made to denote propositional functions of the same number of variables, to be defined in the same Cartesian product; this allows Peirce to apply Boole's operations to all objects in his algebra. And (2) by holding all but two variables fixed, and then taking the relative product on binary relations.

The introduction of Mitchell's quantifiers provided a different route for Peirce: first, he experimented with their algebraic properties as operations on propositional functions, only slightly extending Mitchell's results; then he added them as a new operation to his algebra of relations.

The essential second-order feature in the definition of the integers is the induction axiom, and the essential second-order feature of the real numbers is the least upper bound axiom. To formulate these properties using the algebra of relations requires that we quantify over all relations on a domain. This is the way both Peirce and Schröder handled this part of the foundation of mathematics within their theory, in their attempt to show that the algebra of relations was an adequate basis in which to develop mathematics. Thus, they both went well beyond first-order logic.

Schröder followed up both systems: the algebra of relations and the logic of quantifiers.

§8. Brief Remarks on the Influence of Peirce's Quantificational Logic

Peirce had hoped to develop logic and mathematics within his algebra of relatives, where existential quantification is contained in the relative product, and he did not follow up on his work in quantifier logic after 1885. He did not include a formal notion of formula or of proof in his quantifier logic of 1885, but he did develop the semantics of this system.

Peirce's calculus of relatives was developed systematically by Schröder (*1895*), who emphasized binary relations and their combinations, as Peirce did. He viewed his relational logic as a foundation for mathematics. Schröder also developed quantifier logic, following Peirce, and distinguished first- and higher-order quantificational logic (*1895*: 500–550). He explicitly tried to code first-order statements into Peirce's relational connectives, which was the origin of Korselt's counterexample showing that this elimination of quantifiers in favor of relative product and the other operations of the algebra of relations does not always work. Korselt produced a first-order statement not expressible in the condensed form of the calculus of relatives (i.e., the fragment restricted to relative product, converse, and the Boolean operations): the statement that the domain has at most four elements. Löwenheim (*1915*) started with Korselt's counterexample. Löwenheim compared the expressiveness of condensed algebra of relatives, first-order quantificational logic, and higher order logic and proved his celebrated theorem about first-order logic, namely, if a first-order statement has an infinite model, then it has a countable model. Skolem concentrated on Löwenheim's now-famous theorem, using the same notation at first (*1920, 1923*). He extended Löwenheim's result, proving that every countable set of statements in first-order logic has a model, and observed in *1923* that set theory, if it is expressed in first-order logic, therefore has a countable model, even though it proves there are uncountable sets. Skolem (*1923, 1928*) explained the tree of propositional valuations used by Löwenheim to prove his theorem. Löwenheim's tree proved to be very important, being used by both Herbrand (*1930*) and Gödel (*1930*) in their theses.

Thus, Peirce's algebra of relatives led, through the work of his student Mitchell, to the notion of statements of first-order logic true in a domain. This treatment was expanded to great length by Schröder and was the primary influence on Löwenheim's introduction of model-theoretic ideas into logic.[5]

N O T E S

1. Schröder acknowledges the influence of Peirce in many places, in particular *1895*: 4: "Peirce has given us one such foundation [i.e., an independent foundation for the algebra of relatives], which takes its departure from the consideration of 'elements' or individuals; the comparison of the thereby created, totally peculiar basis of the entire logic with other foundations can only be instructive. We therefore follow this course." Löwenheim (*1940:* 1), in turn, states that he worked entirely within Schröder's system: "When endeavoring to analyze mathematics logically, the paradoxes discovered by Russell and others have always appeared as the most formidable obstacle. . . . I myself have never encountered such difficulties, because to analyze logically has always meant to adapt to Schröder's relative calculus." Finally, Skolem (*1920:* 254) acknowledges that he is improving on Löwenheim's results: "In volume 76 of *Mathematische Annalen* Löwenheim proved an interesting and very remarkable theorem on what are called 'first-order expressions.' The theorem states that every first-order expression is either contradictory or already satisfiable in a denumerably infinite domain. Löwenheim proves his theorem by means of Schröder's 'development' of products and sums, a procedure that takes a Π sign across and to the left of a Σ sign, or vice versa. But this procedure is somewhat involved and makes it necessary to introduce for individuals symbols that are subsubscripts on the relative coefficients. In what follows I want to give a simpler proof, in which such subsubscripts are avoided, and, . . . I shall also establish some generalizations of Löwenheim's theorem."

2. See Hailperin (*1976* [*1986*]) for a thorough study of Boole's theories from the vantage point of contemporary mathematics; see Styazhkin (*1959:* 26–33) for an analysis and explanation of Boole's algebraic methods.

3. For an account of the history of quantification of the predicate, see Styazhkin (*1969:* 148–160).

4. See Henkin, Monk, and Tarski (*1971*) for the abstract theory of cylindric algebras. See Henkin and Monk (*1974*: 105–107) for a brief summary of the contributions of Peirce and Schröder to the subject.

5. Special thanks are due to Anil Nerode for extensive discussions that greatly strengthened this chapter. I would also like to thank Saunders Mac Lane for his help and encouragement and Else M. Barth for her interest and support. A preliminary but different version of this chapter was earlier prepared with the help of William A. Howard.

11.

THE ROLE OF THE MATRIX REPRESENTATION
IN PEIRCE'S DEVELOPMENT OF THE QUANTIFIERS

Alan J. Iliff

§1. From Mathematics to Logic, Part I: Mathematics Makes a Contribution to Logic

In the Preface to the second edition (1787) of the *Critique of Pure Reason*, Immanuel Kant wrote these oft-quoted words: "since Aristotle [logic] has not [been] required to retrace a single step. . . . It is remarkable also that to the present day this logic has not been able to advance a single step, and is thus to all appearance a closed and completed body of doctrine" (*Kant 1781* [*1958*]: 17). But the appearance of being closed and completed did not indicate perfection; rather, the traditional logic that preceded the development of mathematical logic was trapped in a limited system of ideas, notwithstanding attempts by Sir William Hamilton and others to extend or to renew it. Within a century of when Kant wrote the words above, an immensely more powerful logic had been developed.

Peirce was only twelve years old when he read Richard Whately's *Elements of Logic* in the fall of 1851 and decided to devote his life to logic (2.663; SS: 85, W1: xviii); Whately's 1826 textbook on traditional logic was still used during Peirce's junior year at Harvard (W1: xix). Francis Bowen, Peirce's philosophy teacher at Harvard, was a student of Hamilton and a promoter of his ideas (*Flower and Murphey 1977*: 382). Peirce described how his father, Benjamin, criticized philosophical reasoning according to mathematical standards of rigor: "Before I came to man's estate, being greatly impressed with Kant's Critic of the Pure Reason, my father, who was an eminent mathematician, pointed out to me lacunæ in Kant's reasoning which I should probably not otherwise have discovered" (1.560). In trying to reconcile the traditional logic of philosophers with mathematics instruction from his father, Peirce had to recapitulate the intellectual revolution of the new mathematical logic in his own mental development. Although Peirce became critical of Hamilton, Whately, and Bowen (for example, W4:

173; W1: 359–360; W3: 4), he never completely renounced non-mathematical traditional logic, and it always remained a part of his outlook.

Leibniz, however, had already introduced a new mathematical approach to logic a century before Kant wrote the *Critique of Pure Reason* (*Scholz 1961*: 50–57). This approach had only limited success until 1847, when George Boole published *A Mathematical Analysis of Logic: Being an Essay Towards a Calculus of Deductive Reasoning*. Boole's system allowed logical inferences to be made by means of an equational calculus; it was both a contribution toward settling the controversy between Hamilton and Augustus De Morgan over the "qualified predicate" (*Boole 1847*: 1) and a reaction to Hamilton's claim that the study of mathematics is harmful to the intellect (*Hamilton 1852* [*1853*]: 273n.). Hamilton's definition of the subject matter of logic as "the laws of thought"—for example, see his review of Whately's book (*Hamilton 1852,* [*1853*]: 135–136)—was highly influential, providing the title for numerous traditional logic books and for Boole's treatise, *An Investigation of the Laws of Thought* (*Boole 1854*).

De Morgan gave a symbolic treatment of binary relations in his 1860 paper, "On the Syllogism: IV; and on the Logic of Relations." He observed that syllogism was inadequate for the relational reasoning that occurs often in mathematics:

> it is *not* the truth that all inference can be obtained by ordinary syllogism, in which the terms of the conclusion must be terms of the premises. If any one will by such syllogism prove that because every man is an animal, therefore every head of a man is a head of an animal, I shall be ready to—set him another question. (*De Morgan 1966*: 216)

Like Peirce, De Morgan never renounced traditional logic, and he even preferred to give examples of binary relations taken from common experience instead of mathematical ones (*De Morgan 1966*: 220).

Peirce discussed Boole's work for the first time in his 1865 Lectures on British Logicians (W1: 189–199, 223–239). He wrote about Boole's algebra of classes in an 1867 paper, "On an Improvement in Boole's Calculus of Logic," wherein he expressed his concern that Boole had not succeeded in symbolizing particular propositions (3.18/W2: 21). Nevertheless, since most of Peirce's other lectures and papers at that time dealt with the traditional logic, it seems that Boole's work had not yet affected Peirce very deeply. The turning point appears to have resulted from his study of De Morgan's 1860 paper. Peirce recalled, around 1905, the effect that the paper had on him:

> I at once fell to upon it; and before many weeks had come to see in it, as De Morgan had already seen, a brilliant and astonishing illumination of every corner and every vista of logic. . . . his was the work of an exploring expedition, which every day comes new forms for the study of which lei-

sure is, at the moment, lacking, because additional novelties are coming in and requiring note. He stood indeed like Aladdin (or whoever it was) gazing upon the overwhelming riches of Ali Baba's cave, scarce capable of making a rough inventory of them. (1.562)

It was probably not this simple (*Merrill 1978; Michael 1974*). Peirce recollected first seeing De Morgan's paper in 1866, but based on his first published reference to it (5.322n.1; W2: xxxi, 245n.2) and on calculations and remarks in his Logic Notebook (W2: xliii–xlv, 88–92)—especially his solution to the above problem posed by De Morgan (W2: 90)—it is now accepted that he first saw the paper in late 1868.

Peirce's own 1870 paper, "Description of a Notation for the Logic of Relatives . . . ," or DNLR, repeatedly takes up new forms and additional novelties, and it too suggests the metaphor of a report from an expedition into a new world of logic based on mathematics. In DNLR, Peirce combined the class operations of Boole with the relational operations of De Morgan to get a two-sorted algebra of relations and classes. An important operation in this algebra is the combining of relations by means of their relative product. For example, given that s is "servant of" and l is "lover of," then the relative product sl is "servant of lover of." That is to say, i bears the relation sl to j if and only if there is some k such that i bears the relation s to k and k bears the relation l to j. Peirce observed that this product is associative. Another important operation is the product of a relation and a class. For example, given that w is "women" then i belongs to lw if and only if i loves j for some j in w (3.68/W2: 369–370).

§2. From Logic to Mathematics: Logic Makes a Contribution to Mathematics

Near the end of DNLR, Peirce considered the effect of the relative product on elementary relatives, which he later called individual relatives (3.121–130/W2: 408–414). By a "relative," he meant what we mean today by a relation on a universe U; by "individual relatives," he meant the singleton sets of ordered pairs denoted $(u_i{:}u_j)$, where u_i and u_j are elements of the universe or domain of individuals. Although our conception is more purely extensional than Peirce's, it is close enough to what Peirce meant, and it provides a coherent interpretation of his work.

Since the relative product is associative, the system of individual relatives taken over a universe U is an associative system under the following law of multiplication:

$$(u_i{:}u_j)*(u_k{:}u_l) = \begin{cases} (u_i{:}u_l) & \text{if } j = k \\ 0 & \text{if } j \neq k \end{cases}$$

where 0 stands for the empty relation. There are n^2 ordered pairs for a finite universe of n individuals. Hence, including 0, our associative system has $n^2 + 1$ elements, and the relative multiplication table has $n^2 + 1$ rows and columns (the 0 row and column containing only zeros). Peirce's simplest example was based on a two-element universe $U = \{u, v\}$ wherein u is a teacher and v is a pupil (3.126/W2: 410–411). The individual relatives are: $(u : u) = c$, "colleague of"; $(u : v) = t$, "teacher of"; $(v : u) = p$, "pupil of"; and $(v : v) = s$, "schoolmate of." There are 2^2 or 4 individual relatives:

$$
n^2 = 4 \quad
\begin{array}{c|cccc}
 & c & t & p & s \\
\hline
c & c & t & 0 & 0 \\
t & 0 & 0 & c & t \\
p & p & s & 0 & 0 \\
s & 0 & 0 & p & s \\
\end{array}
$$

Including the 0 relative, there are $2^2 + 1$ or 5 rows and columns in the complete table:

$$
n^2 + 1 = 5 \quad
\begin{array}{c|ccccc}
 & 0 & c & t & p & s \\
\hline
0 & 0 & 0 & 0 & 0 & 0 \\
c & 0 & c & t & 0 & 0 \\
t & 0 & 0 & 0 & c & t \\
p & 0 & p & s & 0 & 0 \\
s & 0 & 0 & 0 & p & s \\
\end{array}
$$

Thus, each universe of n individuals leads to an associative multiplication table for the $n^2 + 1$ individual relatives. Peirce then took formal linear combinations of these $n^2 + 1$ individual relatives and thereby obtained an associative algebra P_n of dimension n^2. This completed the transition from logic to mathematics. (The technical mathematical details appear in Section 3, below.)

In subsequent years, P_n was for Peirce not only an algebra, but indeed an algebra of linear transformations. Already in *MS* 230, an unpublished manuscript of 1873 (W3: 93–95), he regarded the $(u_i : u_j)$ as mappings of U into U:

$$
(u_i : u_j)u_k = \begin{cases} u_i & \text{if } u_j = u_k \\ \text{absurd} & \text{otherwise} \end{cases}
$$

He should have said, as in the example from the 1870 paper above,

$$
(u_i : u_j)u_k = \begin{cases} u_i & \text{if } u_j = u_k \\ 0 & \text{otherwise} \end{cases}
$$

Peirce had difficulties here because he lacked a clear distinction between element and singleton set; besides, it appears that he was uneasy about the empty class. He was now in a position to take formal linear combinations of elements of U and to extend the $(u_i{:}u_j)$ to linear transformations on these formal linear combinations. Thus P_n is not only an algebra, but an algebra of linear transformations.

In a short note of 1875 (3.150/W3: 177) and in Peirce's more thorough 1881 appendix to his father's "Linear Associative Algebras" (3.294/W4: 320–321), this work culminated in the following important result: *Theorem 1. Every associative algebra of dimension n is isomorphic to a subalgebra of P_{n+1}.*

In the following technical discussion (which may be skipped without loss of continuity) we will discuss some of the details of this work from the standpoint of present-day mathematics.

The $n^2 + 1$ individual relatives, including the 0 relative, form an associative system or semigroup S_n. By taking formal linear combinations of elements of this semigroup, with coefficients from a given field (for Peirce the real or complex numbers), we get the semigroup algebra A_n, with its three operations of scalar multiplication, vector addition, and vector multiplication. This is the technique of the "group algebra," which actually only requires that the underlying system be a semigroup. Regarding the origins of this idea, B. L. van der Waerden wrote,

> In the same paper of 1854, in which Cayley introduced the notion of an abstract group (Phil. Mag. 7, p. 40–47), he also introduced what we today call the "Group Algebra" of a finite group G. The basis elements of this algebra are just the group elements $g_1, \ldots, g_n$. In the multiplication rule
>
> $$g_j\, g_k = \Sigma\ g_i\, a_{ijk}$$
>
> the coefficients are
>
> $$a_{ijk} = \begin{Bmatrix} 1 & \text{if } g_j g_k = g_i \\ 0 & \text{otherwise} \end{Bmatrix}$$
>
> Every representation of the group G by linear transformations can be extended to a representation of the group algebra. Conversely, every representative of the group algebra yields a representation of the group. Therefore the study of the structure of the group algebra is of primary importance in the theory of group representations. (*van der Waerden 1985*: 190)

Peirce was a pioneer in this technique, which is important today in representation theory (*Naimark and Stern 1982*: 87–92).

But A_n has dimension $n^2 + 1$, so we form a quotient algebra that lowers the dimension to n^2, thereby obtaining the desired P_n. Namely, let B consist of all scalar multiples of the 0 element of S_n. Then B is an ideal, and $P_n = A_n/B$.

As stated above, Peirce saw P_n as an algebra of linear transformations. Given $U = \{u_1, \ldots, u_n\}$ and $U^* = U \cup \{0\}$, then U^* has $n + 1$ elements and each $(u_i : u_j)$ is a mapping $U^* \to U^*$. Let V_{n+1} be the vector space consisting of all formal linear combinations of elements of U^*. Then each $(u_i : u_j)$ can be extended in a unique way to a linear transformation on V_{n+1}.

Now let H be the subspace of V_{n+1} consisting of all scalar multiples of the special element $\{0\} \in U^*$. Let E_n denote the quotient space V_{n+1}/H considered as a vector space. As transformations, all the $(u_i : u_j)$ leave H invariant, and thus they act as linear transformations on E_n. Hence, P_n is the set of all linear combinations of these transformations $(u_i : u_j)$.

Proof of Theorem 1. Let L be an associative algebra of dimension n. We proceed in two steps:

Step 1: Observe that each element of L acts on L as a linear transformation by left multiplication. This represents L as an algebra of linear transformations. But this representation may not be faithful, which is to say that some element of L may be represented by the zero transformation. The representation is faithful, however, if L has a multiplicative identity. Whether or not L already has a multiplicative identity, we can adjoin a multiplicative identity to L by a technique that is standard today, thereby getting an algebra L^*. Now L acts faithfully on L^* as an algebra of linear transformations.

Step 2. First, observe that every algebra of linear transformations of E_n is a subalgebra of P_n, because a linear transformation $f: E_n \to E_n$ is determined by its behavior on the basis elements $f(u_i) = \Sigma_j \lambda_j u_j$; hence, $f = \Sigma_{jk} \lambda_{jk}(u_j : u_k)$. Now let the algebra L have the basis $\{a_1, \ldots, a_n\}$. In the 1881 paper, Peirce associates a linear transformation of the form

$$(a_i : a_0) + \Sigma_{jk} \lambda_{jk}(a_j : a_k)$$

to each basis element a_i (3.293/W4: 320). By hindsight, the addition of the term $(a_i : a_0)$ can be understood as corresponding to the extension of L to L^* as in Step 1.

Matrices. Peirce realized by the spring of 1882 that his algebra P_n is mathematically identical to the algebra of all $n \times n$ matrices over the given field. Indeed, writing $(i : j)$ for $(u_i : u_j)$, direct calculation yields

$$(\Sigma_{ij} \alpha_{ij}(i : j))(\Sigma_{kl} \beta_{kl}(k : l)) = \Sigma_{il} \gamma_{il}(i : l),$$

where

$$\gamma_{il} = \Sigma_x \alpha_{ix}\beta_{xl}, \tag{1}$$

which is the formula of matrix multiplication. Since Peirce used the word "matrix" in his 1881 appendix (3.294/W4: 321), he had recognized the connection between his algebra P_n and matrices before the appendix was published in early 1882 (*Wiener and Young, eds. 1952*: 299/*Fisch 1986*: 58–59).

Peirce's hasty private printing in early 1882 of his *Brief Description of the Algebra of Relatives* and his two controversies with Sylvester, one over the addition of a credit to Peirce in a paper of Sylvester's and the other over the priority for the discovery of nonions, indicate that he was very concerned about how his work would be received in light of Cayley's much earlier work on matrices. Peirce recounted the story of showing the *Brief Description* to Sylvester:

> When it was done and I was correcting the last proof, it suddenly occurred to me that it was after all nothing but Cayley's theory of matrices which appeared when I was a boy. However, I took a copy of it to the great algebraist Sylvester. He read it, and said very disdainfully—Why, it is nothing but my umbral notation. I felt squelched and never sent out the copies. But I was a little comforted later by finding that what Sylvester called "my umbral notation" had first been published in 1693 by another man of some talent named Godfrey William Leibniz. . . . (*Wiener and Young, eds. 1952*: 299/*Fisch 1986*: 58)

Thus, although the *Brief Description* was printed and used in Peirce's course at Johns Hopkins, it was not effectively published (*Weiner and Young, eds. 1952*: 288, 357n.36, 299, 358n.65/*Fisch 1986*: 47, 58, 76n.56). Evidence of Peirce's resentment over this affair can be seen in an unpublished paper of 1902 (4.305). For thorough and interesting accounts of these events, see Fisch (*Wiener and Young, eds., 1952*: 294–302/*Fisch 1986:* 54–62) and Houser (W4: xxxvi–lxx, especially lii–lix).

§3. From Mathematics to Logic, Part II: Mathematics Makes Another Contribution to Logic

The transition from mathematics back to logic began when Peirce and two of his students, Christine Ladd-Franklin and Oscar Mitchell, began to work on the problem of systematically incorporating the notions of "for some" and "for all" into a mathematical theory of logic, publishing their work in *Studies in Logic by Members of the Johns Hopkins University* (*Peirce, ed., 1883*). Although Mitchell used subscripts to express these notions in his paper "On a New Algebra of Logic," these subscripts do not stand for individual variables; furthermore, Mitchell's notation is extremely cumbersome for relations with several arguments. Peirce was probably stimulated by the ideas in Mitchell's paper to develop his own solution to the problem in an appendix to *Studies in Logic* called Note B.

In Note B Peirce used truth values, instead of field elements, in the matrices—giving what is called today the graph of a relation. Peirce had already mentioned this idea in *MS* 230 of 1873 (W3: 93–95). These truth values, represented by 0 and 1, are added and multiplied in the usual way, except that $1 + 1 = 1$ (3.331/W4: 454–455). In formal linear combinations, the operation of addition is actually taken as the addition in the scalar field. Since any relative is the union of individual relatives, these new linear combinations (with truth value coefficients 0 and 1) give precisely the unions of individual relatives, and thus the operation of addition in this context is actually set-theoretic union. Furthermore, the matrix corresponding to the relative product of two relations l and b is precisely the matrix product of the individual matrices corresponding to l and to b.

The connection between the matrix representation and the ideas of quantification is expressed in Note B (3.332–333/W4: 455): the relative product, lb for lover of a benefactor, is "defined" by means of the equation

$$(lb)_{ij} = \Sigma_x (l)_{ix} (b)_{xj}, \tag{2}$$

which is the same as equation (1) for matrix multiplication, arrived at in section 2. Peirce stated that the relative product lb "is called a particular combination, because it implies the *existence* of something *loved by* its relate and a *benefactor* of its correlate" (3.332/W4: 455). Similarly, the relative sum involves a product and the idea of universality. Later in Note B he returned to these sums and products of numerical coefficients:

> Any proposition whatever is equivalent to saying that some complexus of aggregates and products of such numerical coefficients is greater than zero. Thus,
>
> $$\Sigma_i \Sigma_j l_{ij} > 0$$
>
> means that something is a lover of something; and
>
> $$\Pi_i \Sigma_j l_{ij} > 0$$
>
> means that everything is a lover of something. (3.351/W4: 464)

Then he dropped the equational form of the expression:

> We shall, however, naturally omit, in writing the inequalities, the >0 which terminates them all; and the above two propositions will appear as $\Sigma_i \Sigma_j l_{ij}$ and $\Pi_i \Sigma_j l_{ij}$. (3.351/W4: 464)

Further examples of the notation follow (3.352–356/W4: 464–466). Thus Peirce expressed the notion of existence or "for some" as equivalent to the non-vanishing of a sum, and he used the subscripts as variables for individuals of the universe. Similarly, the notion of universality or "for all" is equivalent to the non-vanishing of a product. Presumably, the universe is allowed to be infinite in this paper. In the case of a countable universe there is no problem with the meaning of equation (2) because an infinite series of truth values always converges. What this formula might mean in the case of an uncountable universe is another question.

In "On the Algebra of Logic," his great paper of 1885, Peirce reinterpreted the letters given as subscripts as propositions instead of as truth values (3.392). In today's terms these subscripts can be regarded as atomic formulas. He also reinterpreted Σ and Π as denoting the logical notions of "for some" and "for all":

> Here, in order to render the notation as iconical as possible we may use Σ for *some*, suggesting a sum, and Π for *all*, suggesting a product. Thus $\Sigma_i x_i$ means that x is true of some one of the individuals denoted by i or
>
> $$\Sigma_i x_i = x_i + x_j + x_k + \text{etc.}$$
>
> In the same way, $\Pi_i x_i$ means that x is true of all of these individuals, or
>
> $$\Pi_i x_i = x_i \, x_j \, x_k, \text{etc.}$$
>
> . . . It is to be remarked that $\Sigma_i x_i$ and $\Pi_i x_i$ are only similar to a sum and a product; they are not strictly of that nature, because the individuals of the universe may be innumerable. (3.393)

By "innumerable" Peirce probably meant infinite here, since that is the way it is used in the *Brief Description* (3.306/W4: 328), although it does mean uncountable in an unpublished manuscript of 1893 (4.113). Peirce overloaded the terms "sum" and "product" insofar as they may mean arithmetic sum and product, logical sum and product, or set-theoretic union and intersection. We avoid this confusion today by using "disjunction" for logical sum and "conjunction" for logical product. Since Peirce used "addition" and "multiplication" to mean disjunction and conjunction in the 1885 paper (3.390–391), perhaps we should read the above passage this way: "Σ for *some*, suggesting a [disjunction], and Π for *all*, suggesting a [conjunction]. . . . $\Sigma_i x_i$ and $\Pi_i x_i$ are only similar to a [disjunction] and a [conjunction]; they are not strictly of that nature, because the individuals of the universe may be innumerable." It is doubtful that Peirce meant literally infinite disjunction and conjunction here.

To an extent heretofore unrecognized, Peirce's discovery of the quantifiers was based on his expertise in sophisticated techniques of abstract algebra; it was not merely a simple generalization of the Boolean sum and product. Thus an application of *mathematics* led to an advance in *logic*. His discovery of the quantifiers provided a vast extension of the program Peirce had inherited from Boole for developing a calculus of deductive reasoning.

§4. What Is the Relation between Logic and Mathematics?

Before Boole's 1847 *Mathematical Analysis of Logic* was published, the disciplines of logic and mathematics had little to do with each other. Certainly, mathematicians, including De Morgan, had long since realized that the syllogism is inadequate for mathematical demonstrations. Boole and De Morgan tried to give mathematical accounts of the traditional logic: Boole's work included probability, and De Morgan's incorporated binary relations into traditional logic. Gottlob Frege gave a clear mathematical account of the logic used in mathematics itself. Peirce tried to arrive at a powerful logic for general or even traditional subject matters by applying the ideas and methods of mathematical reasoning.

In the logicist program, exemplified in Frege's writings and in *Principia Mathematica* by Russell and Whitehead, mathematics is to be based on logic. For Frege, mathematical objects are to be manufactured out of functions and concepts, and logic supplies the means for developing the theory of concepts. Moreover, this logic is developed within an artificial language, a "Begriffsschrift" (*Frege 1879*). For Russell, mathematics was claimed to be identical with logic, although his logic seems to have included set theory. These works can be characterized as deduction-theoretic: inference is conceived syntactically as formal deduction.

In the pragmaticist program of Peirce, logic is to be based on mathematics. "Exact" logic is an instrument arising from mathematical thought, and it is to be applied to science and philosophy. Moreover, these applications are to be carried out in a symbolic calculus that represents the instrument of "exact" logic. Peirce's work can be characterized as model-theoretic: there is an underlying set, there are relations on that set, and inference is conceived semantically—notwithstanding that the semantics is naive or informal—as proceeding from true statements to other true statements (for example, 3.168/W4: 168).

Of course, a rigorous model theory must assume enough set theory to express the necessary semantics, including the concept of logical consequence. Skolem, in his 1923 paper, "Some remarks on axiomatized set theory," criticized the apparent circularity involved in basing a foundation of set theory on logic (that is, model-theoretic logic), because some set theory has already been assumed by that logic. Specifically, he mentions, "The peculiar fact that, in order to treat of 'sets', we must begin with 'domains' that are constituted in a certain way" (*van Heijenoort 1967*: 291). The mathematical definition of logical consequence—the model-theoretic analogue of deduction—first appeared in Tarski's 1936 paper, "On the Concept of Logical Consequence" (*Tarski 1956* [*1983*]: 417).

Following De Morgan and Boole, and in sharp contrast to Frege, Peirce began with a set of individuals—a universe. This is a basic idea of model theory, and it can be traced back to an 1846 paper of De Morgan.

> Writers on logic, it is true, do not find elbow-room enough in anything less than the whole universe of possible conceptions; but the universe of a particular assertion or argument may be limited in any matter expressed or understood. And this without limitation or alteration of any one rule of logic. . . .
>
> By not dwelling on this power of making what we may properly (inventing a new technical name) call the universe of a proposition, or of a name,

matter of express definition, all rules remaining the same, writers on logic deprive themselves of much useful illustration. (*De Morgan 1966*: 2)

Frege did try to embrace a universe of all conceptions; herein lies a fundamental difference between his approach and Peirce's. The well-known term "universe of discourse" appears in Boole: "Now, whatever may be the extent of the field within which all the objects of our discourse are found, that field may properly be termed the universe of discourse" (*Boole 1854*: 42). Peirce affirmed this basic idea of De Morgan and Boole in DNLR, his 1870 paper: "I propose to use the term 'universe' to denote that class of individuals *about* which alone the whole discourse is understood to run. The universe, therefore, in this sense, as in Mr. De Morgan's, is different on different occasions" (3.65/W2: 366). Peirce's approach was model-theoretic, both in the sense that the reasoning was about a model and in the sense that the validity of inference was to be measured by truth with reference to that model.

Peirce, at the beginning of his 1896 review of Ernst Schröder's *Exact Logic*, gave a sketch of his program for mathematics and logic (3.425–430). He said that mathematics is the most abstract science because it deals exclusively in hypotheses. The mathematician observes examples in thought and makes inductions from them. Even though the mathematician's thought is fallible, it is so easy to repeat these thought experiments and to make new ones, that mathematics is,

> the only science in which there has never been a prolonged dispute concerning the proper objects of that science. (3.425)

> [W]e homely thinkers believe that, considering the immense amount of disputation there has always been concerning the doctrines of logic, and especially concerning those which would otherwise be applicable to settle disputes concerning the accuracy of reasonings in metaphysics, the safest way is to appeal for our logical principles to the science of mathematics, where error can only long go unexploded on condition of its not being suspected. (3.427)

> Logic may be defined as the science of the laws of the stable establishment of beliefs. Then, *exact* logic will be that doctrine of the conditions of establishment of stable belief which rests upon perfectly undoubted observations and upon mathematical, that is, upon *diagrammatical*, or, *iconic*, thought. (3.429)

Thus, for Peirce, "Logic depends upon mathematics" (4.240); or at least, logic is to be based on mathematical thought and mathematical reasoning (4.227–249).

For various reasons, the views of the logicist school have dominated accounts of the history of mathematical logic: If logicism is not promoted and extolled outright, then at the very least, the logicist interpretation of crucial terms is assumed for the sake of historical inquiry. For example, Peirce's logic has all too often been investigated by first attempting to translate his notations into the notation of *Principia Mathematica*, and then raising questions as to how many or how few anticipations of *Principia Mathematica* can be found in his writings. Given such a context, it is understandable that the work of Peirce and Schröder is often dismissed as an early "algebra of logic"—a dead end. In fact, Peirce's philosophy of mathematics and logic was so totally antithetical to logicism that today it is difficult, if not impossible, to understand him until we interpret the crucial terms, such as "logic" and "mathematics," in a sympathetic way. We can summarize the opposed answers given by Frege and Peirce to the question, "Is logic based on mathematics, or is mathematics based on logic?" as follows:

In Frege's logicist program mathematical objects are manufactured out of concepts; logic is the means to develop the theory of concepts; and logic itself is developed in an artificial language. In Peirce's pragmaticist program logical principles are derived from mathematical thought; logic is the means to develop philosophy and the sciences; and logical work is carried out in a symbolic calculus. If a reader is not aware of this vast difference in orientation, or if a reader is unsympathetic to Peirce's basic approach, Peirce may seem to be a confused, mediocre, or even incompetent writer on logic.

At the end of his 1896 review of Schröder, Peirce made a prophetic prediction for his program:

> Finally, the calculus of the new logic, which is applicable to everything, will certainly be applied to settle certain logical questions of extreme difficulty relating to the foundations of mathematics. Whether or not it can lead to any method of discovering methods in mathematics it is difficult to say. Such a thing is conceivable. (3.454)

§5. Peirce's Influence on Contemporary Logic

There are two lines of development in mathematical logic: The tradition of Frege, Peano, Whitehead, and Russell is essentially deduction-theoretic; by contrast, the tradition of Boole, De Morgan, Peirce, Schröder, Löwenheim, and Skolem is essentially model-theoretic. In five important technical papers, published between 1870 and 1885, Charles Sanders Peirce made a fundamental and enduring contribution to mathematical logic: He laid down the main elements of the framework of quantificational logic within which Schröder, Löwenheim, and Skolem worked—namely, the model theoretic framework. To refer to the tradition of Peirce as "the algebra of logic" is to ignore its culmination in the theorems of Löwenheim (*1915*) and Skolem (*1920; 1923*), which were a natural outcome of Peirce's program. Skolem, in his 1923 critique of axiomatic set theory, applied the Löwenheim-Skolem Theorem to get the "Skolem paradox," and thus logic finally made a significant contribution to mathematics.[1]

NOTE

1. I wish to express my appreciation to Professor William A. Howard for suggesting the general structure of this chapter. The research and writing were partially supported by a North Park College sabbatical.

12.

PEIRCE ON THE APPLICATION OF
RELATIONS TO RELATIONS

Robert W. Burch

Early in his great work of 1870, entitled "Description of a Notation for the Logic of Relatives, resulting from an Amplification of the Conceptions of Boole's Calculus of Logic" (DNLR), Peirce introduces an operation on logical terms that he calls *the application of a relation* and that he initially presents as a kind of "multiplication" of two terms (W2: 369). Although the nature of this operation has been extensively discussed, discussions of it have raised a number of questions about Peirce's views. It has been alleged, for example, that Peirce's account of his multiplication confuses relations with relatives, relations with classes, or individuals with universals.[1] The goal of the present chapter is to shed light on the meaning of Peirce's "multiplication" operation in DNLR. My main thesis is that although Peirce's notion of the application of a relation is not without its difficulties for the interpreter, when the notion is properly illuminated, the clouds of confusion associated with it can be seen not to be in Peirce's head at all. I shall also indicate briefly that in Peirce's concept of the application of a relation the embryo of his reduction thesis is contained, and furthermore that DNLR contains a logic of relations at least as powerful in expressive capability as first-order predicate logic with identity.[2]

§1. Relations, Relatives, and Classes

In Sections 2 and 3 it will be argued that Peirce's notion of the application of a relation involves a systematic ambiguity that first emerges in the context of his discussion of whether or not his operation is *associative*. Peirce explicates two rather different, albeit very closely affiliated, notions of the application of a relation; the first is the notion of a binary operation analogous to algebraic multiplication; the second is not a binary operation at all but is a unary operation. Despite the ambiguity in Peirce's notion of the application of a relation, there still seems to be no reason to consider that Peirce is confusing any concepts with any others. Rather, he is just including two operations (more exactly: two *types* comprising collections of operations) under one heading. Section 1 of this chapter seeks to dispel several specific worries concerning Peirce's views. It also seeks to indicate that it makes little difference whether we talk of relations or of relatives. In its overall contentions lies the point of the title of this chapter, which could just as well have been "Peirce on the Application of Relatives to Relatives," or even "Peirce on the Application of Relations (*to* Nothing At All)."

A detailed critical discussion of the worries that have been raised about DNLR must be reserved for another occasion. At the moment, by way of considering these worries, let us merely take note that the concepts in terms of which they are formulated are themselves not very clearly delineated. For example, relations, when they are understood in the usual fashion as relations *in extension*, simply *are* (kinds of) classes, so that, in this sense of "relation," it makes no sense whatsoever to accuse Peirce of *confusing* relations with classes. Of course, if relations are understood non-extensionally, relations might be distinct from classes, but it is not obvious from the various expositions of DNLR what relations should be taken to be. Moreover, these expositions do not contain a clear line of *argument* that Peirce mixed up relations *in any sense* with classes; what they rather seem to contain for the most part is iteration of the assertion that he does do so, as if repeating a point many times in succession will make it true. Now, what has just been said of the alleged distinction between relations and classes can also be said of the alleged distinction, to which Peirce's commentators often appeal, between relations and relatives: for the commentators *do* seem to have a clear account of relatives, namely, an account according to which relatives are just certain sorts of *classes*.

These very brief remarks should at least hint that something has gone wrong in the accounts of DNLR that find serious potential problems in it. Now, part of what *has* gone wrong, is that the accounts have not fully appreciated the basic structure and fundamental goal of Peirce's 1870 work. In particular they have undervalued the fact that at its outset Peirce erects a very clear distinction between (what logicians currently call) *syntax* and (what logicians currently call) *semantics*. Once one appreciates the presence of this distinction, the fog begins to lift.

The very title of DNLR indicates unambiguously that the work is an endeavor to create a *notation*, that is to say, a *syntactical* structure. In particular the syntax is to be a syntax for the logic of relatives. Peirce is not only acutely aware that his task is the creation of a syntax but is also utterly clear about the general character of the syntax he is going to create: it must be *algebraic* in form. What this means is that the logical language Peirce intends to create is a language that duplicates in specific ways—which ways Peirce takes great pains to explicate—the syntax of *algebra*, of algebra as the subject was understood in 1870. The language will contain *terms* analogous to the terms of algebra and *operations* on these terms analogous to the operations of algebra. But the language is not to be the language *of* algebra, even though it is to be patterned syntactically after the language of algebra: the language is to be a language of or for the logic of relatives. Peirce describes his project as "extending the use of old symbols to new subjects" (W2: 360).

Now this procedure of setting up a language of or for a portion of logic is so familiar in twentieth-century logic that it is difficult to understand how one might miss the fact that the procedure Peirce follows in DNLR is the same one that is typically followed whenever any logic is set forth. In elaborating a first-order predicate logic, for example, the initial definitory activity is to say what is and what is not a well-formed formula. To this end, a vocabulary of symbols is introduced, containing predicate symbols, variables, logical connectives, and quantifiers, along with parentheses and perhaps also function symbols and constants; then rules of formation are given that specify how well-formed formulae are to be constructed recursively out of symbols in the vocabulary. At this syntactical level, no question exists of whether or not relations, classes, etc., are involved: the only question is how the linguistic structures are to be defined, how, for example, the various symbolic strings of the syntax are properly combined with one another. The further question about what they *mean* is a question not at the level of syntax but at the level of *semantics*.

Of course, logicians usually provide a good deal of informal discussion to accompany and "motivate" their syntactical definitions, and in such discussion a considerable amount of informal semantical exposition is quite typical. Thus, for example, in presenting first-order predicate logic, it will typically be explained that the predicates are to be informally understood to denote or stand for relations defined on some non-empty set. Now Peirce engages in this typical sort of informal semantic exposition in presenting his algebraic syntax, and perhaps it is this "motivational" discussion that has bothered commentators. Peirce often speaks, for example, of relative terms as *denoting classes*. But Peirce should not be taken to be exhibiting any confusion about the significance of relative terms: he is merely trying to indicate perspicuously to the reader *why* he constructs the sort of notation for the logic of relatives that he does construct. And of course relative terms *do* denote classes, when the semantics is extensional in structure and when the sense of the word "denote" is understood to be given by some interpretation function which connects syntax with this semantics.

Peirce in fact exerts considerable effort, in his discussion of what he calls "the universe," to indicate to readers both that he is aware of semantical considerations, along with their difference from syntactical considerations, and that his main goal in the paper is to erect the *syntax* for the logic of relatives, while more or less ignoring the semantics.[3] Thus, he says:

> I propose to use the term "universe" to denote that class of individuals *about* which alone the whole discourse is understood to run. The universe, therefore, in this sense, as in Mr. De Morgan's, is different on different occasions. In this sense, moreover, discourse may run upon something which is not a subjective part of the universe; for instance, upon the qualities or collections of the individuals it contains. (W2: 366)

It is difficult to imagine how any logician could possibly be clearer or more forthcoming about his task.

When we take into account the distinction between syntax and semantics in DNLR, the worry as to whether Peirce may confuse particulars with universals can similarly be seen to evaporate. Peirce uses capital English letters to "denote individuals," as he says, whereas he uses lower-case letters to denote "generals" (W2: 365). What this means is that Peirce introduces a distinction in his syntax between terms that are understood to be inherently *singular terms* (his capital letters) and terms that are understood not to be inherently singular terms (his small letters). One example Peirce gives of such a singular term is "The second Philip of Macedon" (W2: 390). This convention is exactly the convention of standard English, in which, for example, "White House" is inherently a singular term, whereas "white house" is not. Of course, in a particular "universe" there might be just one white house; but this point does not obviate the distinction Peirce makes; nor does it indicate any confusion on Peirce's part. For the fact that in a given "universe" there is just one white house does not preclude there being many white houses in some other "universe." What Peirce is clearly *not* saying is that the individuals of the "universe" are in the syntax of the logical language he is erecting.

Of course, an important question about DNLR is what Peirce means by the word "relative," and what he understands its affiliation to be with the contemporary word "relation." The word "relatives," that occurs in the title of DNLR, is not further elucidated, except obliquely and by means of examples, which might perhaps have been expected, given that Peirce's main task in the work is to elaborate a *notation* for the logic of relatives. We may, however, begin to understand the word "relative" in its context by noticing that Peirce's examples clearly indicate his understanding of the phrase "relative term." On pp. 364–366 of DNLR Peirce indicates that there are (more exactly: that he only needs to discuss) three sorts of what he calls "logical signs" or—equivalently—"logical terms." There are "absolute terms," "simple relative terms," and "conjugative terms." Peirce most often uses the phrase "relative term" to indicate only terms of the last two varieties, and occasionally he explicitly contrasts relative terms with absolute terms (W2: 372–373). Nevertheless, there is no harm in referring to all three sorts of terms as being relative terms in the broad, inclusive sense (just as predicates may be called relations of adicity 1); and this inclusive usage would coincide better with the title of DNLR than the contrastive usage.

What, then, does Peirce mean by the word "relative"? Occasionally, he uses it simply as shorthand for "relative term." When he writes, for example, "*A term multiplied by two relatives . . . ,*" he can hardly mean by "relative" anything but "relative term," since terms cannot be multiplied by anything other than terms (W2: 371).[4] As distinct from this usage as shorthand, however, the word "relative" must be understood to mean whatever is the *signification* of relative terms, that is to say, as *whatever* the logic whose syntax contains *relative terms* is the logic *of*. What exactly this signification is will appear only when relative terms are correlated, by means of an interpretation of the syntax, with structures in the semantics for this syntax. Relatives might consistently be interpreted as classes given one semantics, as functions given another semantics, or perhaps as some other sort of objects. Since DNLR does not provide a systematic semantics, the answer to the question "What is a relative?" is not explicitly given in the work. But this omission does not imply that Peirce is confused about any fundamental issues.

Even though Peirce does not develop an explicit semantics in DNLR, it is still possible to provide a rough-and-ready informal account of the notion of a *relative* in this work by discussing *relative* terms as such terms can be elicited from and expressed by ordinary discourse. In such a manner Richard Beatty has accurately characterized relatives (*Beatty 1969:* esp. 66, 77). Relatives are simply the significations of a certain sort of linguistic items (viz., relative terms), which items are obtained from nominalizations of verbs and verb phrases by extracting from these nominalizations various nouns and noun phrases and replacing them with *blanks*. The resultant linguistic items are then understood to occupy noun positions (at least when their blanks are filled with nouns and noun phrases) in propositional structures that are basically subject-copula-predicate (-nominative) in form. Thus, for example, let us consider the verb phrase "gives a horse to a count." This phrase would be substantivalized in the noun phrase "giver of a horse to a count." Then the corresponding relative term would be obtained by extracting the noun phrases "a horse" and "a count" from this nominalization and replacing them with blanks: "giver of _____ to _____." The relative in question would be the signification (whatever it is understood to be) of this linguistic item.

That Peirce's logical analysis in his early works is accomplished by attending to relative terms rather than to verbs and verb phrases (as, of course, first-order predicate logic does) might perhaps be explained by the fact that in his early logical works, Peirce is still tied to the idea that all propositions are basically subject-predicate in form, no matter how complicated the subject or predicate. Also this approach might be argued to be an attempt to pour new wine into old bottles. Nevertheless, it is clear that Peirce really is, even though concentrating his analysis upon relative terms and relatives, still talking about something that is recognizably the logic of relations.

To see that this is so, one only need bear in mind that Peirce's procedure of obtaining relative terms from substantivalizations of verb phrases by replacing certain nouns within them by blanks is simply a minor variation of the procedure whereby in quantificational (i.e., first-order predicate) logic, the linguistic structure "_____ gives _____ to _____," or equivalently, the primitive well-formed formula $Gxyz$ is obtained from the proposition "John gives a horse to Sally" by replacing its nouns and noun phrases by blanks (or by variables). Since in Peirce's logic relative terms are the basic foci of analysis, what would appear in quantificational logic as a well-formed formula with n free variables will appear in Peirce's logic as a relative term with $n-1$ blanks. Thus, instead of "_____ gives _____ to _____" or $Gxyz$, we find "giver of _____ to _____." Instead of "_____ loves _____" or Lxy, we find "lover of _____." And instead of "_____ is a horse" or Hx, we find "horse." Whereas first-order predicate logic focuses on verbs, Peirce's logic of 1870 focuses on nouns: that is all there is to the matter. It should be evident from this that there is no more reason to deny that Peirce is constructing a logic of relations than there would be to deny that first-order predicate logic itself is a logic of relations. Peirce is discussing relations by concentrating attention upon their bearers.

The tiny difference between Peirce's manner of understanding relative terms and the manner in which first-order predicate logic represents relations suggests a symbolization of relative terms akin to the familiar syntax of quantificational logic. All we need do is employ a formal device that captures the substantival character of Peirce's relative terms, in particular that feature of it according to which what would appear in quantificational logic as a well-formed formula with n free variables will appear in the symbolism as a relative term with $n-1$ blanks (for which we might just as well use variable symbols). Let us, therefore, use a notation similar to a common notation for classes, omitting the external braces used in the notation for classes. Thus, just as the relation of giving would be symbolized in quantificational logic as the triadic relation symbol G flanked by three variables: $Gxyz$, so also the relative term whose meaning is the same as that of "giver" will be symbolized in this notation as $x|Gxyz$, which is read "x such that x gives y to z." In this symbolization, we have two parts, a "prefix" $x|$, and a "suffix" $Gxyz$, which *is* a well-formed formula of first-order predicate logic.

It should need no special argument to show that all relative terms can be symbolized in this fashion.

This symbolization encodes the fact that the relevant relative term is understood to be a substantival term with, as Peirce writes it, two blanks. In the symbolization the variables that appear *only in* the suffix correspond to blanks; in the above example the variables y and z correspond to blanks. In this notation, we have in effect *distinguished* or *favored*, by repeating it in the prefix, one of the variables in $Gxyz$ in order to indicate that it is inherently the *main* referential variable of the term, the variable indicative of the bearer of the relation associated with the term. In this notation, the difference between, for example, the relative term "lover of _____" and the relative term "loved by _____" could be written as the difference between $x|Lxy$ and $y|Lxy$.

Of course, we might also just employ a different predicate symbol for the same purpose, so that, for example, the relative term "loved by _____" might be symbolized as $x|Bxy$. Similarly, "Given by _____ to _____" could be written as $y|Gxyz$, and "Recipient from _____ of _____ " could be written as $z|Gxyz$. Or, we might introduce different predicate symbols to do the same job. Peirce himself seems to prefer introducing different predicate symbols:

> We must be able to distinguish, in our notation, the giver of A to B from the giver to A of B, and, therefore, I suppose the signification of the letter equivalent to such a relative to distinguish the correlates as first, second, third, etc., so that "giver of _____ to _____" and "giver to _____ of _____" will be expressed by different letters. (W2: 370.)

This usage, as Peirce says, enables the "correlates," i.e., the variables that occur in the symbolism here used, to be numbered in order from left to right. For example, in $x|Gxyz$, x indicates the first correlate, y indicates the second correlate, and z indicates the third correlate.[5]

In the symbolism employed in Peirce's paper, ordinary (monadic) predicates may also be symbolized. For example, the term whose meaning is that of "horse" may be symbolized as $x|Hx$.

Now to generalize. Peirce understands by a term of his logic both what we shall call "primitive (relative) terms" (viz. absolute terms, simple relative terms, and conjugative terms for which there is a single letter in the logic) and what we shall call "general (relative) terms" (viz., any terms constructible from primitive terms by means of applying the various operations of his logic, in a recursive fashion, to primitive terms). Peirce's understanding of his notation is that it should be capable of representing any substantival expression of a natural language. We may then present Peirce's thinking, in the notation just introduced, in the following manner.

There will be a stock of relation symbols of all adicities. (By 1870, Peirce has become assured of his reduction thesis, so that he thinks that relation symbols of the first three adicities only are necessary; but in generalizing the ideas, we can ignore for the time being this feature of Peirce's thinking.[6]) Then, we may, in the notation just introduced, symbolize a primitive relative term of adicity n with the symbolization $x_1|Rx_1x_2 \ldots x_n$, where all the variables are distinct and where each variable in the suffix occurs only once in the suffix. Then, a general relative term of adicity n may be symbolized by $x_1|\Psi(x_1x_2 \ldots x_n)$, where $\Psi(x_1x_2 \ldots x_n)$ is a well-formed formula of first-order predicate logic with identity, containing the n distinct free variables $x_1, x_2, \ldots, x_n$ (each perhaps occurring more than once in the suffix), as well as perhaps various bound variables. The n variables that are free in $\Psi(x_1x_2 \ldots x_n)$ will be called the *correlates* of the term, with x_1 being the first correlate (or: favored variable) of the term, x_2 being the second correlate of the term, and so on. The variables $x_2, x_3, \ldots, x_n$ will also be called the *blanks* of the term. The use of the notion of "adicity n" is simply one way of capturing what Peirce means by associating absolute terms with *first*, simple relative terms with *second*, and conjugative terms with *third* (*W2: 365*).

It is also useful to symbolize Peircean relative terms by employing a *graphical syntax* that is similar to the graphical syntax of Peirce's existential graphs.[7] For this purpose, a relative term of adicity n is depicted as a spot with n lines or rays radiating outward from it, with the line corresponding to the first correlate of the term radiating horizontally toward the left and the $n-1$ lines corresponding to any blanks of the term radiating toward the right at various angles from the horizontal. Thus, for example, the primitive term $x_1|Rx_1x_2 \ldots x_n$ would be given in graphical syntax by Figure 12.1.

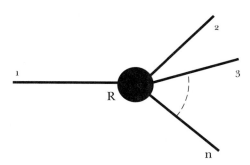

Figure 12.1

§2. The Application of a Relation

By employing the notation for relative terms introduced in Section 1, we may now begin to explicate the operation Peirce calls "the application of a relation." We shall eventually see in Section 3 that he seems to understand this notion in such a way that it includes what would ordinarily be considered two operations (more exactly: *types* of operations), one binary and one unary.

Peirce initially introduces application of a relation as a kind of algebraic "multiplication" of one logical term by another. This binary operation is indicated by combining two relative terms, given in a certain order, in such a way that one of the blanks in the first relative term is "filled in" by the second relative term. Thus, for example, the blank in the relative term "Lover of _____" might be filled in by the relative term "horse" to get the term "Lover of (a) horse"; and this would signify the operation of *applying* the relative signified by the term "Lover of _____" to the relative signified by "horse". In symbols we could indicate the filling in of a blank in a first relative term by a second relative term as follows. We would first list, in the appropriate order, the symbolizations of the two relative terms involved in the operation, taking care that each favored variable (i.e., first correlate) in the list occurs only twice in the list (once in a prefix and once in that prefix's corresponding suffix) and that each unfavored variable occurs only once in the list. (The idea is to assure that all variable-occurrences are, with the limited exception of favored variables, occurrences of *distinct* variables.) With regard to the present example, the list is $x|Lxy, z|Hz$. Now, the variable y in the symbolization $x|Lxy$ (i.e., the second correlate, or first blank, of the first of the two terms) is "filled in with" (i.e., substituted by) the term $z|Hz$. Substituting in this way yields the term $x|Lx (z|Hz)$.

This manner of indicating how relative terms are combined to signify the operation of "the application of a relation" is not particularly helpful, however, in perspicuously representing Peirce's thinking about the meaning of this operation. A much better method for indicating Peirce's intent is to symbolize "Lover of (a) horse" as $x|(\exists t)(Lxt \& Ht)$. One may see from this symbolization that Peirce's multiplication, as originally introduced, has the effect—in terms of the notation being used—of identifying two free variables and then quantifying existentially with respect to them in a conjunction. The variables so identified correspond respectively to the second correlate (i.e., first blank) of the first term in the list and the first correlate (i.e., the favored variable) of the second term in the list.

In graphical syntax Peirce's multiplication, as originally introduced, is depicted by *joining* the ray or line corresponding to the second correlate of the first term to the ray or line corresponding to the first correlate of the

L   H

Figure 12.2

second term. Thus, in graphical syntax, the term $x|(\exists t)(Lxt \ \& \ Ht)$ is depicted in Figure 12.2.

The same pattern—viz., identifying two free variables (the second correlate of the first term and the first correlate of the second term) and then existentially quantifying with respect to them in a conjunction—may also be seen in several other examples Peirce discusses in "the application of a relation." Suppose, to take another Peircean example, that the symbolization $z|Szw$ indicates the relative signified by "Servant of _____." Then, the symbolization that indicates the relative signified by "Lover of a servant of _____" is symbolized by $x|(\exists t)(Lxt \ \& \ Stw)$. We may note in this case that the resultant relative term has two correlates, corresponding to the two *free* variables x and w of its suffix. The variable x is its first correlate or favored variable, and the variable w is its second correlate. In graphical syntax the term $x|(\exists t)(Lxt \ \& \ Stw)$ is depicted in Figure 12.3.

Figure 12.3

Obviously this term could be applied to yet another term. So, let the absolute term $u|Wu$ indicate the relative signified by "Woman." Then, the relative signified by "(Lover of a servant of) a woman" is indicated by the symbolization "$x|(\exists v)[(\exists t)(Lxt \ \& \ Stv) \ \& \ Wv]$." In graphical syntax this term is depicted in Figure 12.4. If we regard the application of a relation as a kind of "multiplication"—as Peirce says that he does—then we have in effect multiplied three terms together by applying a binary multiplication operation twice; and thus the question naturally arises whether this operation of "multiplication" is an associative one.

Figure 12.4

In order to answer this question, we must "multiply" the three terms in accord with the only other possible binary association of them. The relative signified by "Servant of a woman" is indicated by the symbolization $z|(\exists v)(Szv \ \& \ Wv)$, which in graphical syntax is depicted in Figure 12.5. Thus, the relative signified by "Lover of (a servant of a woman)" is indicated by the symbolization $x|(\exists t)[Lxt \ \& \ (\exists v)(Stv \ \& \ Wv)]$, which appears graphically in Figure 12.6.

From first-order predicate logic it is clear that the well-formed formula

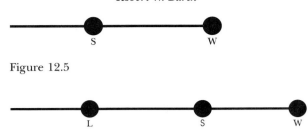

Figure 12.5

Figure 12.6

$(\exists v)[(\exists t)(Lxt\,\&\,Stv)\,\&\,Wv]$ and the well-formed formula $(\exists t)[Lxt\,\&\,(\exists v)(Stv$ $\&\,Wv)]$ are logically equivalent. This equivalence shows that the relative signified by "(Lover of a servant of) a woman" and that signified by "Lover of (a servant of a woman)" are, as Peirce puts it, "identical," and that in this sense, as Peirce also puts it, his multiplication operation, as originally introduced, is associative (W2: 361, 366, and esp. 371). The identical graphs of the two terms is another way of indicating the "identity" (in Peirce's sense) of the two terms.

We may pause to note here that there is no problem in applying a relative to itself as long as the general scheme for symbolizing "the application of a relation" is followed. Thus, the relative signified by the term "Lover of _____" may be applied to itself to yield the relative signified by the term "Lover of (a) lover of _____." In symbols this would appear as $x|Lx(z|Lzw)$, or, more perspicuously, as $x|(\exists t)(Lxt\,\&\,Ltw)$, or as the graph in Figure 12.7.

Figure 12.7

We may also note here that, given the account of the application of a relation so far, relatives with no blanks (which are, of course, signified by absolute terms), such as $x|Hx$, may not be applied to other relatives at all, even though they may have other relatives applied to them. To be applied to anything, a relative must have at least one blank, that is to say, it must correspond to what in quantificational logic would be a relation of adicity ≥ 2. This limitation on the generality of the multiplication is, obviously, a by-product of focusing on terms that are inherently substantival in character. Peirce overcomes this limitation by introducing an ingenious construction, his Comma Operator, which will be discussed in Section 3.

We have already seen that Peirce's "multiplication" operation, as originally introduced, is associative in special cases: namely, when simple relative terms (those with two correlates) and absolute terms (those with only one correlate) are the *only* sorts of terms involved. But now we need to take note, as Peirce does, that when conjugative terms (those with three correlates) enter the picture, serious complications involving associativity arise. Consider the very example over which Peirce worries: Let the term *xlGxtu* indicate the relative signified by "Giver to _____ of _____". (Note carefully that we now have a *different* meaning of the letter *G* than in Section 1, above, when this letter was correlated with "Giver of _____ to _____.") And let the term *vlOvw* indicate the relative signified by "Owner of _____." Also let the term *slHs* indicate the relative signified by "Horse." Then the question arises how we are to multiply the three terms in the order of the following list: *xlGxtu, vlOvw, slHs*. Peirce, of course, would like to have his "multiplication" operation to remain associative in such "multiplications" of three terms (for this is demanded by the desideratum that his syntax should be algebraic in structure). But, with regard to the "multiplication" originally introduced, associativity in contexts involving conjugative terms must be given up. To see why this is so, let us compare the two possible different ways of multiplying the three terms in the given order, when the idea of multiplication is that of the application of a relation as originally explicated.

The relative term "giver to (an) owner" is symbolized as *xl(∃y)(Gxyu &* *Oyw)*, or in graphical syntax as shown in Figure 12.8, so that the relative term "(giver to (an) owner) of (a) horse" is symbolized as *xl(∃z)[(∃y)(Gxyz* *& Oyw)& Hz]*, which is logically equivalent to the term *xl(∃y)(∃z)(Gxyz &* *Oyw & Hz)*, which appears in graphical syntax as in Figure 12.9. We may notice that this relative term does not specify what the owner owns, but it does specify what the giver gives (namely, a horse).

But the relative term "owner of (a) horse" is symbolized as *vl(∃z)(Ovz* *& Hz)*, which appears in graphical syntax as Figure 12.10, so that the relative

Figure 12.8

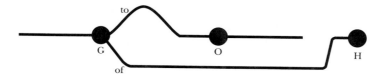

Figure 12.9

term "giver to (an) (owner of (a) horse)" is symbolized as $x|(\exists y)[Gxyu$ & $(\exists z)(Oyz$ & $Hz)]$, which is logically equivalent to the term $x|(\exists y)(\exists z)(Gxyu$ & Oyz & $Hz)$, which appears in graphical syntax as Figure 12.11. We may notice that this relative term does specify what the owner owns (namely, a horse), but it does not specify what the giver gives.

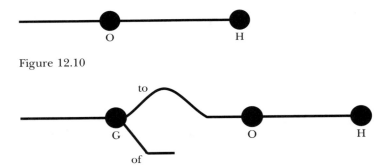

Figure 12.10

Figure 12.11

Since the two ways of multiplying the three terms in the given order do not yield the same result, Peirce's originally introduced operation of multiplication does not associate in this case. It is therefore not associative in general.

It is in the context of discussing the associativity of his multiplication operation that Peirce begins to say things that indicate he has in mind a different notion of the application of a relation than the notion originally introduced. For example, he remarks that his notation for the "multiplication" of the three terms $x|Gxtu$, $v|Ovw$, $s|Hs$ must be understood in such a manner that, when the three terms are multiplied together in the given order, the result indicates the relative signified by the relative term "Giver of a horse to an owner of *that* horse" (W2: 370). That is to say, Peirce wants the preposition "of" in "Giver to an owner of a horse" to do double duty, going simultaneously with both "Giver" and "Owner." Now, in the symbolism of the present chapter, this relative is indicated by $x|(\exists y)(\exists z)(Gxyz$ & Oyz & $Hz)$, which in graphical syntax appears as Figure 12.12.

Figure 12.12

But, this relative term is not either of the results of the two different ways of multiplying the three terms that were just discussed; also this result cannot even be obtained by any form of the multiplication (application of a relation) that was originally introduced. (To see that this is so, reflect that the multiplication originally introduced identifies variables *two at a time* before existentially quantifying over them in a conjunction, whereas the result Peirce now wants is obtainable only by identifying *three* variables and then existentially quantifying.) What we have to rely on to obtain the result Peirce wants seems to be a new operation. If we are to consider this new operation as being also a version of "the application of a relation," then this notion has taken on a greatly expanded meaning since it was originally introduced.

Peirce is in fact aware of this difficulty: "A little reflection will show that the associative principle must in some form or other be abandoned at this point" (W2: 371). Rather than finding this fact disturbing, Peirce presses forward with his discussion. It seems apparent that he intends to extend, expand, or generalize his originally introduced multiplication operation so as to make it capable of unambiguously expressing all possible combinations of a finite number of relatives, which combinations can loosely and informally be described as relatives signified by combinations of a finite number of relative terms in which one or more relative terms fill blanks in other relative terms. What is not so apparent is exactly how Peirce intends for his notion to be expanded. Let us, then, attempt to clarify his intentions.

In dealing with the issue of the associativity of his operation, Peirce seems to move in two quite different directions, one which seems on the strength of Peirce's examples to be more closely linked to his originally introduced multiplication operation, while the other seems to be more closely linked to some new, expanded notion of "the application of a relation." This dual tendency introduces an ambiguity into the overall notion of the application of a relation in DNLR, of which it is important to be aware. (This ambiguity, by the way, is perhaps the one feature *internal* to Peirce's 1870 work that may have generated certain commentators' remarks to the effect that Peirce is confused.) The ambiguity is roughly correlated with two different ways Peirce attempts to provide a notation for the generalization of his multiplication: by *subjacent numbers* and *marks of reference*.

In order to appreciate this ambiguity, let us review the three chief features of Peirce's multiplication as it was originally introduced. First, it was a binary operation. Second, it was an operation in which two variables were identified and then quantified over existentially in a conjunction. Third, the two variables identified were the second correlate of the first term and the first correlate of the second term. Now both devices, subjacent numbers and marks of reference, depart from this narrowly defined scheme of combination. Both amount to generalizations of the originally introduced operation. But now, it seems that the second device, marks of reference, is the more closely tied of the two, at least in Peirce's examples, to the original conception of application. For the device of subjacent numbers departs so widely from this original conception that it is impossible not to consider subjacent numbers as providing for an entirely new dimension of the concept of the application of a relation. Let us see why.

The device of subjacent numbers allows integer subscripts to be attached to terms of Peirce's algebra in such a way that, given the rule that Peirce describes for interpreting them, they unambiguously indicate the correlates (i.e., the variables) that are to be identified in a combination of two, three, or even a greater number of relative terms. The details of Peirce's scheme we may ignore here; but the upshot of the scheme is that the original restriction that variables must be identified *two at a time* is no longer operative: nothing now prevents variables from being identified any number (two *or more*) at a time. In effect, this means that operations using subjacent numbers need not even be binary operations. Furthermore, Peirce even allows that a *single* subjacent number may be attached to a *single* relative term in at least two ways, so that the device of subjacent numbers commits us to *unary* operations. Thus, a subjacent number 0 attached to, for example, the relative term "Lover of _____" yields the absolute term "lover of itself." In our symbolization, this operation takes us from $x|Lxy$ to $x|Lxx$. In graphical syntax this term would be depicted as in Figure 12.13.

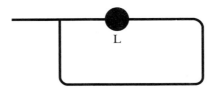

Figure 12.13

In this operation, not only has the idea of a binary operation been abrogated altogether by the use of the device of subjacent numbers, but also, as the example shows, the idea of the essential involvement of an existential quantification in the application of a relation may seem to have been abrogated as well. When the subjacent "number" ∞ is attached to, for example, the relative term "Lover of _____," this operation is again unary, taking us from the symbolization $x|Lxy$ to $x|(\exists y)Lxy$, which would appear in graphical syntax as Figure 12.14. There can be little doubt, then, that the device of subjacent numbers symbolizes an operation (more exactly: a *type* of operations) that is sufficiently far removed from Peirce's original conception of "the application of a relation" that it must be considered as a wholesale change. Peirce's intentions in connection with subjacent numbers will be clarified in Section 3 below.

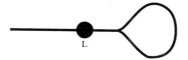

Figure 12.14

By contrast with the use of subjacent numbers, the use of marks of reference potentially provides only a modest generalization of the original "multiplication" of Peirce's algebra. For, at least if we are to judge the device by Peirce's examples, the device of marks of reference involves the use of *two inscriptions* of a given symbol in such a way as to indicate which *two* correlates of *two* relative terms are to be identified and quantified over existentially in a conjunction. Consider, for example, the relative term "Lover of (a) horse," which results from applying the relative signified by the relative term "Lover of _____" to the relative signified by the term "Horse." In the list $x|Lxy, z|Hz$, y corresponds to the second correlate of the first term, whereas z is the first correlate of the second term. It is these two variables which are in effect identified and quantified over existentially in a conjunction to obtain the relative term $x|(\exists t)(Lxt \& Ht)$, which indicates the relative signified by the relative term "Lover of (a) horse." With the device of marks of reference, this operation of the application of a relation could be written, in the style Peirce actually uses in DNLR as $L_y{}^y H$.

As the example shows, the left-most mark of reference of a pair of such marks is placed as a right-hand subscript on the first of two terms that are to be "multiplied" (and it is also placed in a list of marks in such a way as to indicate which blank it concerns), and the remaining mark of the pair is placed as a left-hand superscript on the term whose first correlate is to be identified with the variable indicated by the first mark before conjoining and existentially quantifying. The device of marks of reference thus allows an unambiguous specification of the variables, always two at a time, that are to be identified and then quantified over existentially in a conjunction, when a series of binary operations of the application of a relation is performed. Peirce shows how versatile this device can be with an example (W2: 372). Let "Giver to _____ of _____" have "$x|Gxyz$" as its symbolization; let "Lover of _____" have "$u|Luv$" as its symbolization; let "Woman" have "$w|Ww$" as its symbolization; and let "Horse" have "$t|Ht$" as its symbolization. Then the relative term "Giver of a horse to a lover of a woman" will have "$x|(\exists y)(\exists w)(\exists t)(Gxyt \& Lyw \& Ww \& Ht)$" as its symbolization. Obviously, this can be written in the style Peirce actually uses in DNLR using marks of reference: $G_{yt}{}^y L_w{}^w W^t H$.

It is important to note that this result can be obtained by three "multiplications," that is to say, operations of the application of a relation as originally introduced, as follows. We multiply "$x|Gxyz$" and "$u|Luv$" to obtain

$$x|(\exists y)(Gxyz \,\&\, Lyv).$$

This we multiply with "$s|Hs$" to obtain

$$x|(\exists t)\,[(\exists y)(Gxyt \,\&\, Lyv) \,\&\, Ht].$$

This we multiply with "$r|Wr$" to obtain

$$x|(\exists w)\{(\exists t)\,[(\exists y)(Gxyt \,\&\, Lyw)\,\&\, Ht] \,\&\, Ww\},$$

which is the same as

$$x|(\exists y)(\exists w)(\exists t)(Gxyt \,\&\, Lyw \,\&\, Ww \,\&\, Ht).$$

In graphical syntax this term appears as Figure 12.15. We thus see that the device of marks of reference—understood so that marks of reference always occur in pairs—gives an authentic and extremely modest generalization of Peirce's originally introduced notion of the application of a relation.

Figure 12.15

Let us, then, define in general terms Peirce's originally introduced multiplication operation, as generalized by the device of marks of reference (understood as always occurring in pairs), as follows. Let a symbolization of a general relative term of adicity n be defined, as in Section 1 above, by a string of symbols of the type "$x_1|\Psi_1(x_1x_2x_3 \ldots x_n)$." And, consider any two such terms "$x_1|\Psi_1(x_1x_2x_3 \ldots x_n)$," and "$y_1|\Psi_2(y_1y_2y_3 \ldots y_m)$." Now, let any binary operation APP2 on relative terms be given such that APP2 behaves according to the form:

$$\text{APP2}[x_1|\Psi_1(x_1x_2x\,_3 \ldots x_n),\, y_1|\Psi_2(y_1y_2y_3 \ldots y_m)] =$$
$$x_1|(\exists z)\,[(\Psi_1(x_1x_2 \ldots x_{i-1}zx_{i+1} \ldots x_n) \,\&\, \Psi_2(zy_2y_3 \ldots y_m)].$$

In graphical syntax APP2 operations may be represented as Figure 12.16.

Informally, the idea here is that *some* correlate *from* the second *to* the nth (not necessarily just the *second* correlate) of the first term is identified with the first correlate of the second term; then the conjunction of the two suffixes of the terms with the variables thus identified is taken, after which

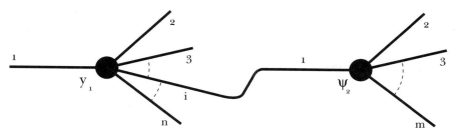

Figure 12.16

an existential quantification is accomplished, yielding a resultant term. Then, we may say that any such binary operation is an operation of the application of a relation to a relation (or: of a relative to a relative) in the sense of the notion as originally introduced by Peirce and then modestly generalized in accord with the device of marks of reference, understood as always occurring in pairs. We see that the application of a relation in this sense (APP2) is really a *type* or *kind* of binary operation, comprising a collection of binary operations.

In connection with Peirce's reduction thesis, as we shall see in Section 3, below, it is important here to take note of two matters. First, the result of any operation of application of a relation to a relation in this sense (APP2) is a single general relative term. Second, this term is a general term of adicity $n + m - 2$, with its correlates being all the variables that are *free* in its suffix.

With this understanding of the application of a relation as APP2, we can understand Peirce's meaning when he says: "Now, considering the order of multiplication to be: —a term, a correlate of it, a correlate of that correlate, etc.,—there is no violation of the associative principle" (W2: 372). And we can also see that Peirce is, given his meaning, correct. Peirce is not saying that any particular member of the collection of APP2 operations is associative in the ordinary sense. (Indeed, we have already seen that this is not the case.) Rather, what Peirce means is that any finite string of multiplications of relatives by means of operations of the type APP2 can be unambiguously expressed *without using parentheses* to group the string into a particular binary association. This is correct, since the device of marks of reference (understood as always occurring in pairs: term to correlate, correlate of that correlate, etc.) will unambiguously indicate exactly how the multiplication could be carried out two terms at a time.

It is, of course, perfectly possible to generalize the device of marks of reference in such a way as to allow for the identification of and existential quantification over more than two variables at a time. And doing this would make the device of marks of reference equivalent to the device of subjacent numbers. But Peirce's examples suggest that Peirce seems to lack the intention to use marks of reference in this way, and that he intends to use them so that they always occur in pairs. Now, whether this thesis about Peirce's intentions with regard to the device of marks of reference is correct or not, we must still note that, as long as marks of reference do always occur *in pairs*, then the term "Giver of a horse to an owner of *that* horse" cannot be constructed from "Giver to _____ of _____," "Lover of _____," and "Horse." For that relative could only be given by the expression

$$G_{yz}{}^{y}O_{z}{}^{z}H,$$

which involves *three*, not two, inscriptions of the mark of reference z.

This fact shows that the account given so far of Peirce's concept of the application of a relation, even though it is apparently adequate as far as it goes, is still only a partial account of Peirce's full notion. For the account given so far manifestly does not explain the examples Peirce provides in his discussion of subjacent numbers. Obviously, something is still missing. What *is* missing will be discussed in Section 3, below.

§3. Teridentity, the Comma Operator, and the Reduction Thesis

In Section 2 the fact that absolute terms cannot be applied to any terms at all was discussed. Peirce, however, in driving to make his notation as general as possible, attempts to remedy this limitation of his syntax by introducing an operator on terms (and by implication on the relatives that these terms signify) that he writes with a comma placed immediately after the letter used to symbolize the term upon which the operator operates (W2: 372–375). Let us call this operator the comma operator. And let us write it in operator notation, as COMMA(T), where "T" is a term of Peirce's logic. When applied to a term of adicity n, the comma operator produces a term of adicity $n + 1$, which term has essentially the same meaning as the original term. Thus, as Peirce describes it in English, the term "man" becomes, when the comma operator is applied to it, the term "man that is _____." "Lover of _____" becomes "Lover of _____ that is _____," and so forth.

We may easily represent the comma operator in the symbolism introduced in this paper for relative terms. Thus, if $x_1|Rx_1x_2 \ldots x_n$ is a (primitive or general) relative term of adicity n, then COMMA$(x_1|Rx_1x_2 \ldots x_n)$ is the $(n + 1)$-adic relative term $x_1|(Rx_1x_2 \ldots x_n \ \& \ x_1 = x_{n+1})$. In graphical syntax, this term would appear as Figure 12.17.

The comma operator, as Peirce shows by example, provides among other things the means for forming Boolean products, for in general the

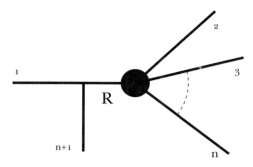

Figure 12.17

term $COMMA(x_1|Rx_1x_2 \ldots x_n)$ applied to the term $y_1|Sy_1y_2 \ldots y_m$ yields the term

$$x_1|(\exists t)(Rx_1x_2 \ldots x_n \ \& \ x_1 = t \ \& \ Sty_2 \ldots y_m),$$

which is "identical" in Peirce's sense to the term

$$x_1|(Rx_1x_2 \ldots x_n \ \& \ Sx_1y_2 \ldots y_m).$$

("Identical in Peirce's sense" means, we recall, that the suffixes of the two terms are logically equivalent.) Peirce's example is the construction of the term "man that is black." Let $x|Mx$ be the symbolization of the absolute term "man." Then $COMMA(x|Mx)$ is the term $x|Mx \ \& \ x = y$. Now, when this term is applied to the term $z|Bz$, which symbolizes the absolute term "black," the result is the term $x|(\exists t)(Mx \ \& \ x = t \ \& \ Bt)$, which is "identical" in Peirce's sense to the term $x|(Mx \ \& \ Bx)$, which *is* the term "man that is black." In graphical syntax this term is depicted in Figure 12.18a.

Figure 12.18a

The interconnections between Peirce's comma operator and the simple relative term "identical with _____," for which Peirce uses the symbol "1" are of very great interest, as we shall now see. In our symbolism this relative term appears as $y_1|y_1 = y_2$. In graphical syntax this term appears as Figure 12.18b. That this term is an identity for the notion of the application of a relation, as it has so far been explicated (APP2), is easily verified. For when the term is applied to the (primitive or general) term $x_1|Rx_1x_2 \ldots x_n$, the

Figure 12.18b

result is the term $y_1|(\exists t)(y_1 = t \,\&\, Rtx_2 \ldots x_n)$, which is "identical" in Peirce's sense to the term $y_1|Ry_1x_2 \ldots x_n$, which in turn is "identical" to the term $x_1|Rx_1x_2 \ldots x_n$. In graphical syntax this term is shown in Figure 12.19.

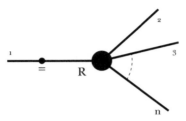

Figure 12.19

Moreover, when the term $x_1|Rx_1x_2 \ldots x_n$ is applied to the term $y_1|y_1 = y_2$, the result (taking x_2 to be the relevant correlate, with of course no loss of generality being involved) is the term $x_1|(\exists t)(Rx_1tx_3 \ldots x_n \,\&\, t = y_2)$. This in turn is "identical" to the term $x_1|Rx_1y_2x_3 \ldots x_n$, which in turn is "identical" to the term $x_1|Rx_1x_2x_3 \ldots x_n$. In graphical syntax this term depicted in Figure 12.20. And, with this, the status of the term "identical with ____" as an identity element for Peirce's multiplication, as so far explicated, is verified.

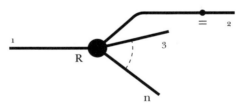

Figure 12.20

Now, clearly, when the comma operator is applied to the term "identical with ____," the result is the conjugative term "1,": $y_1|y_1 = y_2 \,\&\, y_1 = y_3$, and it would be given in ordinary language as "identical with ____ and with ____." Let us call this conjugative term the "teridentity term."[8] In graphical syntax this term appears as Figure 12.21. We have just seen that this term is definable or constructible from the simple relative term of identity (the dyadic identity term) and the comma operator. We may now see that, conversely, the comma operator may be defined in terms of the teridentity term and the notion of the application of a relation as it has so far been explicated.

Figure 12.21

For, when the term $y_1|y_1 = y_2 \;\&\; y_1 = y_3$ is applied to the term

$$x_1|Rx_1x_2 \ldots x_n,$$

the result (taking y_2 to be the relevant correlate without loss of generality) is the term

$$y_1|(\exists t)(y_1 = t \;\&\; y_1 = y_3 \;\&\; Rtx_2 \ldots x_n).$$

This term is "identical" to the term

$$x_1|(\exists t)(x_1 = t \;\&\; x_1 = y_3 \;\&\; Rtx_2 \ldots x_n),$$

which is in turn "identical" to the term

$$x_1|(\exists t)(x_1 = t \;\&\; x_1 = y_3 \;\&\; Rx_1x_2 \ldots x_n),$$

which is "identical" to the term

$$x_1|(\exists t)(Rx_1x_2 \ldots x_n \;\&\; x_1 = y_3 \;\&\; x_1 = t),$$

which is "identical" to the term

$$x_1|(Rx_1x_2 \ldots x_n \;\&\; x_1 = y_3),$$

which is "identical" to the term

$$x_1|Rx_1x_2 \ldots x_n \;\&\; x_1 = x_{n+1}),$$

which, of course, just *is* the term

$$\mathrm{COMMA}(x_1|Rx_1x_2 \ldots x_n).$$

In graphical syntax this term appears as Figure 12.22. We have now seen that the application of a relation in the sense so far defined (APP2), plus the comma operator, will enable us to define or construct the teridentity term; and that the application of a relation in this sense (APP2), plus the teridentity term, will enable us to define the comma operator. The obvious question at this point is whether the application of a relation in the

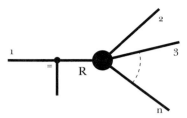

Figure 12.22

sense so far defined, *plus* either the teridentity term or the comma operator, will provide sufficient resources for constructing all the terms that Peirce mentions in his discussion of the device of subjacent numbers, namely: "$xl(\exists y)(\exists z)(Gxyz$ & Oyz & $Hz)$," "$xlLxx$," and "$xl(\exists y)Lxy$." The answer to this question, which may be proved by methods that lie outside the scope of this chapter, is that these terms *cannot* be constructed from the primitive terms that compose them simply by using the operation of the application of a relation in the sense so far defined (APP2), even if we do add teridentity or the comma operator to our constructive resources. Very informally stated, the proof depends on the fact that no APP2 operation can introduce a new loop or *cycle* in graphical syntax, whereas each of the three terms in question contains in graphical syntax a cycle that was not in the terms operated upon.[9]

What this fact means is that, even when we add teridentity or the comma operator to the constructive resources given by the application of a relation as so far defined (APP2), we still do not have sufficient resources for capturing in full Peirce's notion of the application of a relation. Something is still missing. We shall now see, however, that, if we extend the notion of application of a relation in one very minor and completely natural way, then application of a relation, as so extended, plus either the teridentity term or the comma operator, suffices for the construction of all of the examples Peirce discusses. It is difficult to imagine how the notion of the application of a relation, as so extended, could possibly fail to be the full notion of the application of a relation that Peirce has in mind.

Let us then extend the notion of the application of a relation, as so far explicated, by considering the concept of the application of a relation to include not only the *binary* (type of) operation of application, which we have already designated by the operator symbol APP2, but also a *unary* (type of) operation, which we may designate by the operator symbol APP1. APP1 signifies any operation on single terms of Peirce's logic that have adicity 3 or higher, which operation behaves as follows. Given any such term

$x_1|\Psi(x_1x_2\ldots x_n)$, an operation of application APP1 applied to this term is any operation whereby the term

$$x_1|\ (\exists t)\Psi(x_1\ldots x_{i-1}\ tx_{i+1}\ldots x_{j-1}tx_{j+1}\ldots x_n)$$

is produced. The understanding here is that two correlates of a single term, neither of which is the first correlate, are identified and then quantified over existentially. Graphically, APP1 operations may be represented as Figure 12.23.

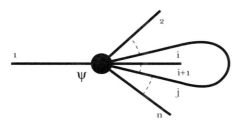

Figure 12.23

The thesis of this chapter is that Peirce understands the notion of the application of a relation to include both APP2 and APP1. As evidence, it will now be shown that the application of a relation in this inclusive sense, plus either the comma operator or the teridentity term, will enable us to construct all the examples that Peirce gives in his discussion of subjacent numbers, namely: "$x|(\exists y)(\exists z)(Gxyz\ \&\ Oyz\ \&\ Hz)$," "$x|Lxx$," and "$x|(\exists y)Lxy$."

Let us begin with the first and least-obviously constructible term. It will be constructed from the three primitive terms $x|Gxtu$, $v|Ovw$, $s|Hs$. "COMMA($s|Hs$)" is the term $s|Hs\ \&\ s=r$, shown graphically as Figure 12.24.

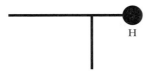

Figure 12.24

Applying (APP2) to this term, the term $v|Ovw$ results in the term $v|(\exists z)(Ovz\ \&\ Hz\ \&\ z=r)$, shown graphically as Figure 12.25.

Applying (APP2) to this term the term $x|Gxtu$ yields the term

$$x|(\exists y)[Gxyu\ \&\ (\exists z)(Oyz\ \&\ Hz\ \&\ z=r)],$$

shown graphically as Figure 12.26.

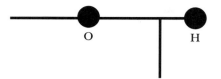

Figure 12.25

Figure 12.26

Now, operating on this term by APP1, with the free variables u and r being concerned, we get the term

$$x|(\exists q)(\exists y)[Gxyq \ \& \ (\exists z)(Oyz \ \& \ Hz \ \& \ z = q)].$$

Graphically this term appears as Figure 12.27. This term, however, is "identical" in Peirce's sense to the term

$$x|(\exists y)(\exists z)(\exists q)(Gxyq \ \& \ Oyz \ \& \ Hz \ \& \ z = q),$$

and thus to the term

$$x|(\exists y)(\exists z)[(\exists q)(Gxyq \ \& \ z = q) \ \& \ Oyz \ \& \ Hz],$$

which is in turn "identical" in Peirce's sense to the term

$$x|(\exists y)(\exists z)[Gxyz \ \& \ Oyz \ \& \ Hz],$$

which *is* the term we were to construct.

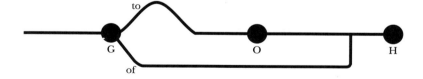

Figure 12.27

Peirce's remaining two examples are easier. The term $x|Lxx$ may be constructed from $x|Lxy$ as follows. We first apply the comma operator to this term to get $x|Lxy \ \& \ x = z$, shown graphically as Figure 12.28. Now, operating on this term by APP1, with the free variables y and z being concerned, we obtain the term $x|(\exists t)(Lxt \ \& \ x = t)$, which is "identical" in Peirce's sense to the term $x|Lxx$, shown graphically as Figure 12.29.

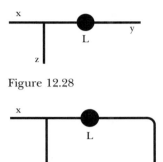

Figure 12.28

Figure 12.29

The term $x|(\exists y)Lxy$ may be constructed from the term $x|Lxy$ and the teridentity term, written as $z|z = u \ \& \ z = v$, as follows. We first apply (APP2) the term $x|Lxy$ to the teridentity term to obtain the term $x|(\exists t)(Lxt \ \& \ t = u \ \& \ t = v)$. Graphically, this term appears as Figure 12.30. Now, operating on this term by APP1, with the free variables u and v being concerned, we obtain the term

$$x|(\exists y)(\exists t)(Lxt \ \& \ t = y \ \& \ t = y).$$

This term is "identical" in Peirce's sense to the term

$$x|(\exists y)(\exists t)(Lxt \ \& \ t = y),$$

and thus to the term

$$x|(\exists y)Lxy.$$

Graphically, this term appears as Figure 12.31.

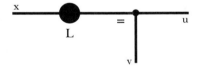

Figure 12.30

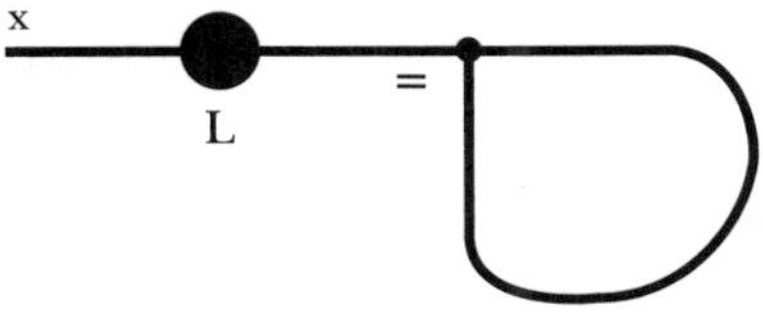

Figure 12.31

The understanding of Peirce's notion of the application of a relation that is being defended here thus has the virtue that it allows all of Peirce's examples to be explained in a fashion that is smoothly coherent with his own wording. It represents, moreover, a very minimal departure from the originally introduced notion of a binary operation. But these are not the only virtues of this understanding. For this understanding also provides a basis for comprehending the reasons behind Peirce's reduction thesis, which has both a negative and a positive aspect. Negatively, the thesis means that from *firsts* and *seconds* alone, neither *thirds* nor relatives of higher adicities may be constructed.[10] Positively, the thesis means that from relatives of the first three adicities, all relatives may be constructed.

The rationale for the negative part of the reduction thesis resides in the fact that when a relative term of adicity n is combined by application with a relative term of adicity m, the adicity of the resultant term is $n + m - 2$. This we may call "the valency formula." Now, by the valency formula, a *second* combined by application with a *first* will yield a *first*, while a *second* combined by application with a *second* will again yield a *second*. Thus, there is no possibility of reaching a *third* or a relative of higher adicity starting merely with *firsts* and *seconds*.

The full rationale for the positive part of Peirce's reduction thesis involves the notion of "hypostatic abstraction," and is thus outside the scope of this chapter.[11] But part of this rationale is the following fact, whose proof is also outside our scope.[12] Given the concept of the application of a relation, that is to say, given APP2 and APP1, and given the single triadic relative signified by the teridentity term, Peirce's logical algebra is adequate to contain a translation of every well-formed formula of first-order predicate logic with identity.[13]

Even aside from the bearing of this fact on the reduction thesis, it is significant in itself. For it shows that by 1870 Peirce had invented and elaborated a logic of relations at least as powerful in expressiveness as the standard contemporary logic of today.

N O T E S

1. See *Martin 1978*, esp. p. 27; also, *Martin 1976a, 1976b;* also *Merrill 1984*, esp. p. xlviii. In *Merrill 1984*, Peirce is merely said to be "confusing," rather than "confused," but one gathers that the confusing character of Peirce's work is still being attributed to Peirce.

2. In this thesis, the work of Emily Michael is both endorsed and extended. See, e.g., *Michael 1974, 1979.*

3. Peirce does not entirely ignore semantics. Thus, for example, his remarks

at W2: 418ff. have mainly semantical import. But, aside from sporadic remarks, Peirce leaves semantics aside in his work of 1870.

4. Again, for example, at W2: 374, Peirce writes "we must regard every term as being a relative requiring. . . . "

5. It is a trivial matter of mere terminology whether we speak of first, second, and third correlates, or alternatively of relate, first correlate, and second correlate. Peirce himself alternates between the two ways of speaking.

6. See, for example, Peirce's remarks concerning the adequacy of the first three adicities at W2: 365.

7. The main difference between the graphs of this chapter and Peirce's existential graphs is in the interpretation of *unconnected* lines of identity, that is to say, lines of identity with at least one free end. These do not indicate existential quantification in the graphs of this chapter, but merely identity. For existential quantification to be indicated by the graphs of this chapter, a line of identity must be understood to have no free ends. In the existential graphs, unconnected lines of identity also indicate existential quantification.

8. "Teridentity" is a word Peirce commonly uses in connection with the existential graphs. It is of considerable interest to notice that the prototype of the graphs' "line of identity" is already present in DNLR, namely as "1,1,1,1,1,1,etc." See W2: 374.

9. The method of proof involves using the concept of the "cyclosis" of a term, a concept that was originally introduced in a topological setting by Johann Benedict Listing in *Listing 1861*. The basic idea is this: the cyclosis of any primitive term is 0. Combining terms by using merely APP2, together with teridentity or the comma operator, preserves cyclosis, so that any term constructible with these resources from primitive terms will have cyclosis 0. The three terms Peirce explicitly discusses in connection with the device of subjacent numbers, however, all have cyclosis 1. Hence they cannot be constructed from primitive terms using merely the resources mentioned.

10. Actually the thesis is that no *non-degenerate* relations of adicity 3 or greater may be constructed from relations of the first two adicities alone. But in the interest of simplicity this qualification may be ignored in the present context.

11. This topic is discussed in the next chapter and in *Burch 1991*.

12. This fact has been proved by the author in *Burch 1991*.

13. We may note here that all conjunctions can be represented, and that Peirce represents negation by means of the simple relative term "Not ____," whose symbolization is $x \mid x \neq y$. With conjunction, negation, and existential quantification at hand, Peirce's logic may be straightforwardly used to translate every well-formed formula of first-order predicate logic with identity.

13.

PEIRCE'S REDUCTION THESIS

Robert W. Burch

By "Peirce's Reduction Thesis" I understand a doctrine that has both a positive and a negative component. The positive component of the Thesis says that from relations of adicities (or: arities) 1, 2, and 3 exclusively, *all* relations—of all non-negative (and, of course, integer) adicities—may be constructed. Equivalently, it says that all relations of adicity greater than 3 may be reduced to relations of adicities 1, 2, and/or 3. The negative component of the Thesis says, first, that relations of adicity 2 may not *in general* be constructed from (reduced to) relations exclusively of adicity 1; and, second, that relations of adicity 3 and greater may not *in general* be constructed from (equivalently: reduced to) relations exclusively of adicities 1 and/or 2. The negative component of Peirce's Reduction Thesis may be stated in a more finely-grained manner by making use of Peirce's concept of a *degenerate* relation. In exact terms, what the Thesis says is, first, that a relation of adicity 2 may be constructed from relations exclusively of adicity 1 if and only if the relation of adicity 2 is *degenerate*; second, that a relation of adicity 3 or greater may be constructed exclusively from relations of adicities 1 and/or 2 if and only if the relation of adicity 3 or greater is *degenerate*; and, third, that there do exist *non-degenerate* relations of all adicities ≥ 2. Although textual analysis is not the burden of the present chapter, Peirce's Reduction Thesis in the sense of the doctrine just specified is, I contend, a thesis that careful textual exegesis will confirm to be actually maintained by Peirce from at least as early as 1885.

The sense of Peirce's Reduction Thesis obviously depends on the exact meanings of the fundamental concepts in terms of which it is framed: the concept of a relation, the concepts of construction and reduction as these apply to relations, and the concepts of degeneracy and non-degeneracy as these apply to relations. The explication of these concepts that is provided in the present chapter, I contend, is an account of Peirce's understanding of them that careful textual exegesis will confirm.

In *Burch 1991*, I have given a detailed mathematical exposition of

Peirce's Reduction Thesis and proved it to be correct.[1] The task of the present chapter is to provide the outlines of my exposition-and-proof in a fairly non-technical manner in order to make its basic ideas—as well as those of Peirce—accessible to the general philosophical community.

It is natural to begin with the concept of a relation. Peirce's full conception of a relation involves the ideas of a mind, a sign, and an ontological structure. It is a conception that might be argued to be too rich to be *completely* captured in the sterile formalism of mathematical logic. Nevertheless, I contend, mathematical logic is capable of defining structures that do represent without undue distortion minds, signs, and ontological structures. By signs, the logician can understand a logical language or syntax. By ontological structures the logician can understand structures in the *semantics* for this syntax. And by minds, the logician can understand interpretation functions that connect logical syntax with logical semantics. In Peirce's own systems of logic, dating at least from 1870, the ideas of syntax, semantics, and interpretation are unambiguously at hand. For understanding Peirce's Reduction Thesis, we shall therefore require a logical syntax, a logical semantics, and an idea of interpretation functions whereby syntax is mapped to semantics.

I submit that it turns out to make very little difference which syntax, of the various systems of syntax that Peirce historically introduced, is used as the syntax in which the Reduction Thesis is presented. In the preceding chapter in this volume, for example, I indicate that Peirce's system of 1870 is already adequate for the purpose. In *Burch 1991*, I have chosen to present the Reduction Thesis in connection with one of Peirce's late systems of syntax: the system of Existential Graphs of 1896. I have done so in order to affiliate Peirce's Reduction Thesis with Peirce's interest in topology and especially with the Census Theorem of Johann Benedict Listing (a theorem that is a version of what nowadays is known as the "Euler-Poincaré Formula"). To this end, I have constructed an algebraic equivalent of Peirce's Existential Graphs, which I call Peircean Algebraic Logic, or: PAL, for short. PAL contains primitive terms of all adicities ≥ 1, which are intuitively understood to denote relations; and it contains operations on terms whereby further terms result, including terms of adicity 0 (Peirce's "medads"). These operations are the bases on which the crucial notions of construction and reduction are defined. The terms of PAL are to be understood as corresponding in a precise manner with the well-formed formulae of first-order predicate logic with identity, which I shall refer to as "quantificational logic."

The semantics for PAL is elaborated with the idea in mind of explicating a fully *intensional* concept of relations and not merely an *extensional* concept. Such a concept is not only demanded by Peirce's thinking; it is also requisite for being able to explicate the structure of Peirce's modal logic. To this end, I introduce a Kripke-style model structure, similar to part of the semantics for Montague grammar.[2] A model structure is a pair (W,D), consisting of a set of indices W and a set D of sets D_w, each indexed by a member of W. Intuitively, W is understood to be the set of all possible worlds, and D_w is understood to be the domain of the possible world $w \in W$. An n-adic (or: n-ary) *relation* in the intensional sense, then, is understood to be a function with arguments in W and taking each $w \in W$ to what I call a "class of n-tuples over D_w." A "class of n-tuples" is similar to the idea of a subset of the Cartesian Product $(D_w)^n$, with the difference being that I incorporate into the notion of a "class of n-tuples" the idea that such a class can consist of an "empty n-tuple" alone; I also introduce two "classes of 0-tuples," which are defined to be two "truth-values" $\top$ and $\bot$. The two truth-values are simply primitives of the semantics, and they make possible a semantic interpretation of 0-adic terms, which correspond to the closed sentences of quantificational logic. Empty n-tuples and the two truth values are necessary for a smooth definition of the semantical implications of the Peircean operations of construction. The notion of an empty n-tuple is similar to a structure used by Paul Bernays (for roughly the same purpose I use it) in his own algebraic logic (*Bernays 1959*). I define an n-tuple over D, for $n \geq 1$, to be a function f from the set $S^n(D)$ of all sequences of length n of members of D to the set $\{\top, \bot\}$ such that for *at most one* $s \in S^n(D)$, $f(s) = \top$. The unique n-tuple f over D such that for all $s \in S^n(D)$, $f(s) = \bot$ is the empty (or: null) n-tuple over D.

In the semantics for PAL great use is made of a notion of the "Cartesian Product" (or, as logicians often call it, "Concatenation") of classes of n-tuples over a domain D. The notion of Cartesian Product employed in the semantics for PAL is slightly more complicated than the ordinary, familiar notion of Cartesian Product; nevertheless, it differs sufficiently little from the ordinary notion that readers of the subsequent accounts of semantics in this paper may follow the main drift of these accounts by assuming that the ordinary notion of Cartesian Product rather than the more complicated notion is in question.

Let us now see, in sketch at least, how the logic works. In order to avoid complexities in explicating the full intensional semantics, let us focus on one index $w \in W$. The result will be in effect an extensional semantics for PAL. Rather than talking about D_w, I shall now simply talk about a Domain D of "enterpretation" (that is, an extensional interpretation). In this way I can speak of an n-adic relation as (if it were simply) a "class of n-tuples" over D. Given the syntax for PAL, we have a denumerably infinite collection of primitive terms R_1^n, R_2^n, R_3^n, etc. for each adicity $n \geq 1$. An "enterpretation function" is a function $*$ taking the primitive terms of PAL as arguments and mapping each primitive term R_i^n of adicity n to a class of n-tuples over D. In quantificational logic the term R_i^n would be given by $R_i^n(x_1, x_2, \ldots, x_n)$, where R_i^n is a predicate (or: relation) symbol and where all variable occurrences are occurrences of *distinct variables*. The term is also, and most perspicuously, given in a graphical syntax that closely matches the syntax of Peirce's Existential Graphs, except that it disambiguates an ambiguity concerning lines of identity that Peirce seems to have hurried over:

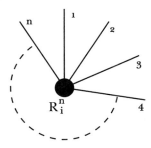

Figure 13.1

In graphical syntax the distinct free variables that would appear in the quantificational equivalent of the term R_i^n of PAL are depicted as line-segments or "rays" that radiate from a vertex or "spot" that is labeled by the term of PAL in question. The end of each such line-segment that is not attached to a vertex may be called a "loose end" or a "free hook." The "enterpretation" $*(R_i^n)$ of the term R_i^n as a class, represented as a matrix, of n-tuples over D may be written as

$$\left\{ \begin{array}{c} (d_{11}, d_{12}, \ldots d_{1n}) \\ (d_{21}, d_{22}, \ldots d_{2n}) \\ \ldots \\ (d_{\alpha 1}, d_{\alpha 2}, \ldots d_{\alpha n}) \end{array} \right\}$$

Occupying a special position among the terms of PAL is the triple identity or "teridentity" term, which is written in PAL as 1^3 and which is absolutely essential in the proof of Peirce's Reduction Thesis. The term 1^3 of PAL is written in quantificational logic as, for example, $x_1 = x_2$ & $x_2 = x_3$. In graphical syntax it appears as

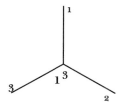

Figure 13.2

The "enterpretation" $*(1^3)$ of the term 1^3 as a class, represented as a matrix, of 3-tuples over D is

$$\begin{Bmatrix} (d_1,d_1,d_1) \\ (d_2,d_2,d_2) \\ \cdots \\ (d_\alpha,d_\alpha,d_\alpha) \end{Bmatrix}$$

The Peircean operations of construction, whereby at the level of syntax terms are operated upon, and whereby at the level of semantics relations are operated upon, are as follows. There is first the operation of *negation*, which I symbolize at the level of syntax by the operator NEG. NEG should need no special introduction, except to say that in PAL NEG operates only upon terms of "Chorisis 1," that is to say, terms that appear in graphical syntax as graphs that consist of 1 connected component (1-piece graphs). NEG may be extended to a negation operator capable of negating terms of Chorisis ≥ 2, but in order for this to be accomplished special technical constructions are involved. We may see from the following example that NEG is represented in graphical syntax by drawing a simple closed curve—which may be called a "cut" or a "sep"—around the relation to be negated and allowing the free hooks to extend to outside the curve. The following example is the graphical depiction of the term $\text{NEG}(R^5)$.

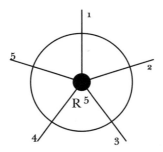

Figure 13.3

The second operation of construction of PAL is really a class of operations, the *permutation* operations. These could be defined in terms of the remaining operations of PAL together with teridentity, as *Burch 1991* proves. There is, however, no loss of generality in leaving the permutations as primitive operations. Most important among the permutation operations are the various cycles, and one may note that every permutation is a product (that is, a composition) of disjoint cycles. Like negation, the permutations operate on terms of Chorisis 1, that is to say on terms that appear in graphical syntax as 1-piece graphs. Otherwise, like negation, they should need no special introduction. The following example illustrates that the permutation operations are handled in graphical syntax simply by an orderly renumbering of free hooks in a graph. Let the operator PERM_i^5 correspond to the permutation

$$\begin{bmatrix} 1,2,3,4,5 \\ 3,2,1,5,4 \end{bmatrix}$$

Then the term $\text{PERM}_i^5(R^5)$ would be represented in graphical syntax as

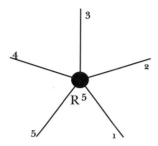

Figure 13.4

In addition to NEG and the permutation operations, PAL has three remaining operations of construction. These are called Join₁, Join₂, and Arraying.

Join₁ is an operation on a single term of PAL of Chorisis 1 and adicity ≥ 2; in graphical syntax it is an operation on a 1-piece graph that has at least two free hooks. The operation has two superscripts, whose purpose is to indicate the two free hooks in the term that are singled out for a sort of "selective deletion" that amounts to what in quantificational logic would be represented by existentially quantifying over two identified variables. The following example shows how Join₁ works. (In this and in later examples the usual subscripts on terms of PAL will be omitted.) Let the term of PAL R^n be given in quantificational logic by the well-formed formula $R(x_1, x_2, \ldots, x_n)$. Then the term $\text{Join}_1^{ij}(R^n)$ will be given in quantificational logic by the formula

$$(\exists t)\, R(x_1, \ldots x_{i-1},\, t, x_{i+1}, \ldots, x_{j-1},\, t, x_{j+1}, \ldots, x_n).$$

In graphical syntax the term is represented by drawing a line between the ith free hook and the jth free hook of the graphical representation of the term R^n, and then renumbering the remaining free hooks in an obvious fashion. The procedure can be seen from the graph of the term $\text{Join}_1^{ij}(R^n)$.

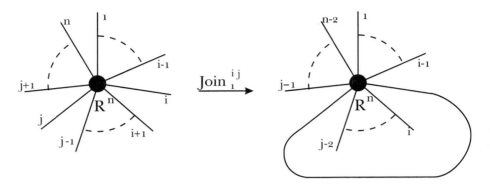

Figure 13.5

At the level of extensional semantics, the enterpretation of $\text{Join}_1{}^{ij}(R^n)$, that is to say $*[\text{Join}_1{}^{ij}(R^n)]$, is obtained by selecting out all the n-tuples of $*(R^n)$ that contain the *same* element of D in both the ith and j^{th} places; then *deleting* both these places from the thus-selected n-tuples; then closing them up to form $(n-2)$-tuples. If $n = 2$, then the result will be one of the truth-values; if there is no n-tuple in $*(R^n)$ that contains the same item in both the ith and j^{th} places, then the result is the class of $(n-2)$-tuples containing *only* the empty $(n-2)$-tuple.

Join_2 is an operation on a pair of terms (in a given order) of PAL, each of Chorisis 1 and each of adicity ≥ 1; in graphical syntax it is an operation on two 1-piece graphs (in a given order) each of which has at least one free hook. Like Join_1, Join_2 has two superscripts, whose purpose is to indicate two free hooks, one in the first term and one in the second term of the pair, that are singled out for "selective deletion." As with Join_1, "selective deletion" refers here to what would be represented in quantificational logic by identifying two variables and then existentially quantifying over them. The following example shows how Join_2 works. Let a term of PAL R^n be given in quantificational logic by $R(x_1, x_2, \ldots, x_n)$ and let S^m be given in quantificational logic by $S(y_1, y_2, \ldots, y_m)$. Then $\text{Join}_2{}^{ij}(R^n, S^m)$ will be given in quantificational logic by

$$(\exists t)\, [R(x_1, \ldots, x_{i-1}, t, x_{i+1}, \ldots, x_n)\ \&\ S(y_1, \ldots, y_{j-n-1}, t, y_{j-n+1}, \ldots, y_m)].$$

In graphical syntax the term is represented by drawing a line between the ith free hook of the graphical representation of the term R^n and the $(j-n)^{\text{th}}$ free hook of the graphical representation of the term S^m, and then renumbering the remaining free hooks in an obvious fashion. The procedure can be seen from the graph of the term $\text{Join}_2{}^{ij}(R^n, S^m)$. At the level of extensional semantics, the interpretation of $\text{Join}_2{}^{ij}(R^n, S^m)$, that is $*[\text{Join}_2{}^{ij}(R^n, S^m)]$, is obtained by selecting out all the $(n+m)$-tuples of the

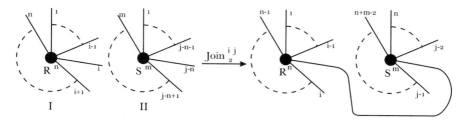

Figure 13.6

Cartesian Product (i.e. the Concatenation) $*(R^n) \times *(S^m)$ that contain the *same* element of D in both the ith and j^{th} places; then *deleting* both these places from the thus-selected n-tuples; closing them up to form $(n+m-2)$-tuples. If $n = 1$ and $m = 1$, then the result will be one of the truth-values; if there is no $(n+m)$-tuple in $*(R^n) \times *(S^m)$ that contains the same element of D in both the ith and j^{th} places, then the result is the class of $(n+m-2)$-tuples containing *only* the empty $(n+m-2)$-tuple.

The operation of *arraying* is an operation whereby two or more terms of PAL of Chorisis 1 are placed in a linear order (an array) and combined by means of the symbol ∘. The following example shows how arraying works. We might array the two terms R^n and S^m, in that order, by writing $R^n \circ S^m$. Similarly the three terms R^n, S^m, and T^p, in that order, might be arrayed by writing $R^n \circ S^m \circ T^p$. Graphically, this last array, for example, would be depicted by drawing the graphs for all three terms, along with some indication of the order in which they are understood to be arranged. Thus, the following graphical picture would depict the array:

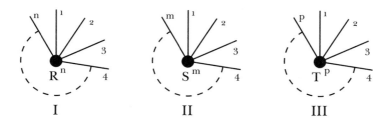

Figure 13.7

At the level of extensional semantics, the interpretation of an array is given by the Cartesian Product (Concatenation) of the interpretations of the terms arrayed. So $*(R^n \circ S^m \circ T^p) = *(R^n) \times *(S^m) \times *(T^p)$, for example.

This completes the introduction of the operations of construction of PAL. With the idea of them at hand, we may quickly sketch the meaning of *construction* and the meaning of *reduction* that are germane to Peirce's Reduction Thesis. A relation is constructible from a given set of relations

if and only if that relation may be built up from relations in the given set
by iteratively applying the operations of construction. The relation is reducible to the relations in the given set if and only if it is constructible from
the set *and* each of the relations in the set is of adicity strictly less than it.

We may now introduce the idea of a *degenerate* relation. A relation of
adicity $n \geq 2$ is degenerate if and only if there is an integer $k \geq 2$ and there
are k integers $j_1, j_2, \ldots, j_k$; with $1 \leq j_i \leq 2$ for each j_i; and with $j_1 + j_2 + \ldots$
$+ j_k = n$; such that for every $w \in W$ the value of the relation is a class of
n-tuples $X_w{}^n$ such that $X_w{}^n$ is a Cartesian Product (Concatenation) of k
classes of tuples $X_w{}^{j_1}, X_w{}^{j_2}, \ldots, X_w{}^{j_k}$, with each $X_w{}^{j_i}$ being a class of j_i-tuples
over D_w. The basic idea is that a degenerate relation of adicity 2 is one that
is Cartesian factorable into two relations of adicity 1; and a degenerate relation of adicity ≥ 3 is one that is Cartesian factorable into relations exclusively of adicity 1 and/or adicity 2. It is easily seen that there are no degenerate relations of adicity 1 but there are indeed degenerate relations of all
adicities greater than 1. It also emerges immediately from the definition of
degeneracy that there are two types of degenerate triadic relations: those
that are a Cartesian Product of three monadic relations and those that are
a Cartesian Product of one monadic relation and one dyadic relation.

In *Burch 1991* I prove that a relation of adicity 2 is degenerate if and
only if it is constructible by means of the operations of construction previously discussed from relations exclusively of adicity 1. I also prove that a
relation of adicity ≥ 3 is degenerate if and only if it is constructible by means
of these operations from relations exclusively of adicities 1 and 2.

Although the proofs are technical and involve mathematical induction,
their fundamental ideas are easily grasped informally by recurring to the
graphical syntax for PAL. In graphical syntax the adicity of a relation appears as the number of free hooks in the graph that represents the relation.
Moreover, distinct Cartesian factors of a relation may be made to appear
graphically as distinct connected pieces of the graph that represents the
relation. If, therefore, it were possible to construct a non-degenerate dyadic
relation from a set of exclusively monadic relations, then it would be possible to begin with some graphical array composed exclusively of pieces each
of which contains one free hook, for example the array

Figure 13.8

and, simply by successively joining together or "bonding" various free
hooks, to produce eventually a graph in which the number of free hooks

in some one connected piece of the graph is 2. (We may safely ignore the operations of negation and permutation here.) By trial-and-error, however,
anyone may quickly convince himself or herself that the task is an impossible one.

Similarly, if it were possible to construct a non-degenerate triadic relation from a set of exclusively monadic and dyadic relations, then it would
be possible to begin with some graphical array composed exclusively of
pieces each of which contains either one free hook or two free hooks, for
example the array

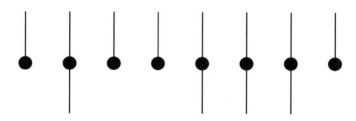

Figure 13.9

and, simply by successively joining together or "bonding" various free
hooks, to produce eventually a graph in which the number of free hooks
in some one connected piece of the graph is 3. (We may again safely ignore the
operations of negation and permutation here.) Once again, however, anyone may quickly convince himself or herself that the task is an impossible
one.

Now, to complete the proof of the negative component of Peirce's Reduction Thesis, all we need to show is that there do exist non-degenerate
relations of all adicities $n \geq 2$. But now, as it turns out, the n-adic *identity
relations* for all $n \geq 2$ are non-degenerate. At the level of extensional semantics, the interpretation $*(1^n)$ of the n-adic identity relation is the class of all
n-tuples over D, excluding the null n-tuple over D, of the form $(d,d,d,
\ldots,d)$. Such relations are not factorable into Cartesian Products at all un-
less the underlying domain has but a single member, and we can exclude
this possibility by *fiat* with only the tiniest sacrifice in generality. It also turns
out to be the case that all the n-adic diversity relations (the negations of
the identity relations) for $n \geq 2$ are non-degenerate. Starting, then, from
any relations exclusively of adicity 1, there are relations of adicity 2 that
cannot be constructed. And, starting from any relations exclusively of adicities 1 and 2, there are relations of all adicities ≥ 3 that cannot be constructed. The negative component of the Reduction Thesis, then, is established.

Now let us proceed to consider the positive component of the Thesis.
What this component says is that, starting from relations of adicities 1, 2,

and 3, *all* relations are constructible. In fact, as I have proved, all we need to add to relations of adicity 1 and adicity 2 in order to make all relations constructible is a *single* triadic relation: *the triadic identity relation.* This relation Peirce calls the "teridentity" relation, and—as I said earlier in this paper—it is of absolutely crucial importance in understanding his Reduction Thesis.

From the teridentity relation we may first construct the identity relations of all adicities $n \geq 2$. And by using the teridentity relation, we may construct, in a series of steps several very important *derived operations* of Peirce's logic. Among these derived operations, four are crucial for the Reduction Thesis. They are the COMMA operator, the QUANT operator, the HOOKID operator, and the PRODUCT operator. The COMMA operator is an operator Peirce introduced in 1870, and it has the effect of "doubling up" one of the adicity places of a relation. How it works is shown in the following example. Let R^n be given in quantificational logic by $R(x_1,x_2,\ldots,x_n)$. Then $\mathrm{COMMA}^i(R^n)$ is given in quantificational logic by $S(x_1,\ldots,x_i,x_{i+1},\ldots,x_n,x_{n+1})$, defined to hold if and only if $R(x_1,\ldots,x_i,x_{i+2},\ldots,x_n,x_{n+1})$ & $x_{i+1} = x_i$. Graphically, The COMMA operator is applied by attaching a spot of teridentity (by means of Join_2) to the ith hook of the spot for R^n, thus:

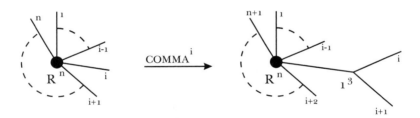

Figure 13.10

At the level of extensional semantics, the COMMA operator has the effect of doubling up the ith entry of each n-tuple in the interpretation of R^n. In the preceding chapter, I discuss in some detail the historical origins in Peirce's thought of the COMMA operator, as well as its interdefinability with the teridentity relation.

The QUANT operator is a device for accomplishing in PAL what is accomplished in quantificational logic by existential quantification over a variable. It works as follows. Let R^n be given in quantificational logic by $R(x_1,x_2,\ldots,x_n)$. Then $\mathrm{QUANT}^i(R^n)$ is given in quantificational logic by $(\exists x_i)R(x_1,x_2,\ldots,x_n)$, or in other words by $S(x_1,x_2,\ldots,x_{n-1})$ which is defined to hold if and only if $(\exists t)R(x_1,\ldots,x_{i-1},t,x_{i+1},\ldots,x_{n-1})$. Graphically, the QUANT operator is depicted by attaching a spot of teridentity (by means of Join_2) to the ith hook of the spot for R^n, thus:

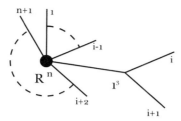

Figure 13.11

and then applying Join$_1$ to the two open hooks remaining on the teridentity spot, thus:

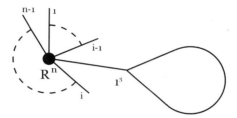

Figure 13.12

At the level of extensional semantics, the QUANT^i operator has the effect of deleting the ith entry of each n-tuple in the interpretation of R^n.

The HOOKID operator is a device for accomplishing in PAL what is accomplished in quantificational logic by the identification of free variables. How it works is shown in the following example. The example shows that the HOOKID operators are simply a certain sort of multiple attachment to a graph of some n-adic identity relation. In the example, the variables x_1, x_3, and x_5 of R^6 are identified by applying HOOKID^{135} to R^6. Graphically this identification is depicted thusly:

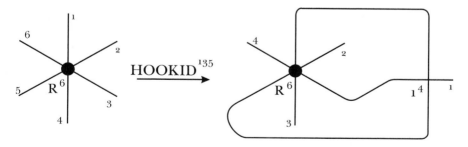

Figure 13.13

The semantics of the HOOKID operator is complicated even when under-stood merely extensionally, and for this reason I shall omit discussing it here.

The PRODUCT operator is a device for constructing, for any array, of whatever Chorisis, a term of PAL *of Chorisis 1* that is equivalent to it. Thus, the PRODUCT operator allows the NEG operator to be applied, in effect, to any array of PAL and not merely to terms of Chorisis 1. Because of its complexity, I shall avoid discussing the PRODUCT operator here any fur-ther.

With COMMA, QUANT, HOOKID, and PRODUCT at hand, the posi-tive component of Peirce's Reduction Thesis is only a few steps away. In order to prove the positive component correct, I prove two theorems. The first I call the "Representation Theorem for PAL." And the second I call the "Reduction Theorem" for PAL.

The Representation Theorem says that any well-formed formula of first-order predicate logic with identity, with the exclusion of formulae that contain function letters and/or constants, may be *translated* in a natural and obvious manner into a term of PAL. The exclusion of formulae that contain function letters and/or constants is, of course, an utterly insig-nificant exclusion because—as a standard metatheorem of logic attests—function letters and constants may always be eliminated from a first-order predicate logic with identity. The details of the Representation Theorem are too complicated to discuss here, but the fundamental idea is simple. In order to translate any well-formed formula of predicate logic with identity into a term of PAL we first correlate each primitive wff $R(x_1, x_2, \ldots, x_n)$ of predicate logic with a primitive term R^n of PAL of the same adicity n, taking care that identity wffs like $x_1 = x_2$ are correlated with the term 1^2 of PAL for the dyadic identity relation. Now, given a wff f of predicate logic, we begin by expressing f in prenex normal form, and furthermore in such a way that the matrix of f's prenex normal form expression contains only the propositional operators $\neg$ (negation) and & (conjunction). We then alter this matrix by making each occurrence of a variable in it an occur-rence of a *distinct* variable. Now, the translation of f in PAL is obtained in the following way. We first, in an obvious way, write in algebraic syntax or graphically draw, the translation of this last wff of predicate logic; then, by using the HOOKID operators, we in effect re-identify the variables that were previously made distinct. Because it may be shown that the HOOKID and NEG operators commute with each other, no special problems arise in this re-identification. We now have a translation in PAL of the matrix of f as expressed in prenex normal form using only the connectives $\neg$ and &. The only thing remaining to be done is to express in PAL the quantifica-tions involved in the prefix of this wff. But this is no problem at all. Exis-tential quantifications may be expressed with the QUANT() operator, and universal quantifications may be expressed by the combined operator NEG{QUANT[NEG()]}. In *Burch 1991* I prove that the procedure of translation just outlined is a completely general one.

As an example of this general process of translating a wff of first order predicate logic with identity into a term of PAL, let us translate the wff $(x)(\exists y)(Pxy \supset Qxy)$. First, since it is already in prenex normal form, all we have to do is to express its matrix using only the propositional operators $\neg$ and &, thus:

$$(x)(\exists y)[\neg(Pxy\ \&\ \neg Qxy)].$$

The matrix of this wff is $\neg(Pxy\ \&\ \neg Qxy)$. Second, we systematically alter this matrix by making all the occurrences of variables in it occurrences of distinct variables; by doing so, we obtain $\neg(Pxy\ \&\ \neg Qzw)$. Now, this matrix may be directly translated in an obvious way into a term of PAL. Using graphical rather than algebraic syntax and ignoring, as we may do, the use of the PRODUCT operator, this matrix may be translated as:

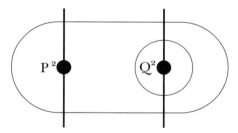

Figure 13.14

Third, by using the HOOKID operators, we in effect re-identify, as many times as is necessary, the variables that were made distinct in the second step, as follows:

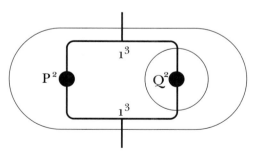

Figure 13.15

Since it may be shown that the NEG operator and the HOOKID operator(s) commute, the previous graph may also be expressed with the identity spots involved in the HOOKID operators placed outside rather than inside the sep of the graph. Thus the previous graph is equivalent to the following one:

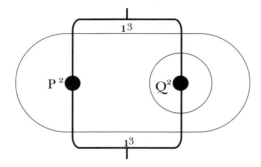

Figure 13.16

We have now translated the matrix of the original wff into a term of PAL. All that remains to be done, in order to express in PAL the original wff, is to express the quantifiers in its prefix. In doing this, we take advantage of the fact that universal quantification may be defined in terms of existential quantification and negation. Note that the original wff is logically equivalent to

$$\neg(\exists x)\neg(\exists y)\,[\neg(Pxy\ \&\ \neg Qxy)]\,.$$

We first express the innermost (the second) quantifier $(\exists y)$ as follows:

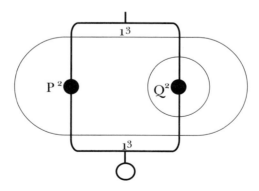

Figure 13.17

Then we express the second-innermost (the first) quantifier (x) as follows:

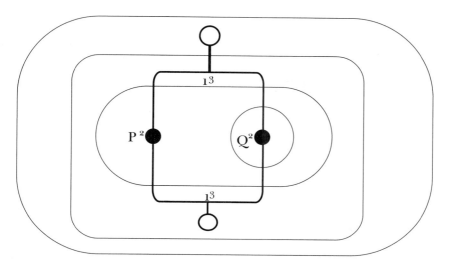

Figure 13.18

And the original wff is translated into a term of PAL (graphical syntax).

The Reduction Theorem for PAL is based on the idea of Hypostatic Abstraction. And it is therefore crucial to understand this idea. As it turns out, there is not one form of hypostatic abstraction, but rather many forms. Moreover, there are many forms that may be used in proving a reduction theorem. Here I shall present the *simplest* form of hypostatic abstraction and indicate how this simplest form may be used to prove that, by means of hypostatic abstraction, all relations may be reduced to relations exclusively of adicities 1, 2, and 3.

Let us begin with an example from Peirce. Suppose we assert: "*a* gives *b* to *c*." We may also express this assertion by saying that *There is a giving* (that is to say, an *act* of giving, an *instance* of giving, an *occurrence* of giving) with respect to which *a* is donor, *b* is given, and *c* is recipient. When we use this second form of expression, we are hypostatically abstracting; and it should be clear that doing so involves asserting the existence of something. In effect, then, once we begin talking about *givings*, we are by abstraction bringing *new* entities onto the scene. And we are replacing assertions of the form $Gabc$ by assertions of the form $(\exists x)(Gx \ \& \ I_1xa \ \& \ I_2xb \ \& \ I_3xc)$, where Gx means "x is a giving," I_1xa means "a is donor with respect to x," I_2xb means "b is given with respect to x," and I_3xc means "c is recipient with respect to x." Now in general hypostatic abstraction of the simplest sort that is here being discussed works in this same fashion. We replace a relation of adicity n with another, equivalent one, constructed out of one monadic relation, n dyadic relations, and $n+1$ occurrences of the triadic identity relation (teridentity) that serve to identify variables and to quantify existentially. In doing so, moreover, at the level of extensional semantics we *augment* the domain of extensional enterpretation by adding to the original domain the n-tuples of the enterpretation of the n-adic relation. (Each such n-tuple is considered informally as an "instance" of the original relation.) What the Reduction Theorem shows is that this procedure may *always* be accomplished.

The following example shows a bit more about how hypostatic abstraction works. It should be noted that, contrary to first appearances, and contrary to Arthur Skidmore's criticisms of Peirce (*Skidmore 1971*), the reduction is *not* to monadic and dyadic relations *alone* but rather ineliminably to monads, dyads, *and* triads. For we may not either identify variables or quantify existentially without appealing to the triadic relation of teridentity. The graph in the example displays the presence of this crucial and remarkable triad that inevitably will be involved in every case of reduction by hypostatic abstraction.

Let the term R^n of PAL correspond to the wff $R(x_1, x_2, \ldots, x_n)$ of quantificational logic. Let an enterpretation $(D, *)$ be given; and let D^+ be $D \cup {}*(R^n)$. Then one can define the monadic term R^1, and n dyadic terms I_1^2, $I_2^2, \ldots, I_n^2$, so that, when $*$ is extended to D^+ in the obvious way, the term R^n and a certain term t, which is constructed from R^1 and the n dyadic terms $I_1^2, I_2^2, \ldots, I_n^2$, will have the same enterpretation. The term t corresponds to what would be written in quantificational logic as

$$(\exists y)\,[R^1(y) \;\&\; I_1^2(y,x_1) \;\&\; I_2^2(y,x_2) \;\&\; \ldots \;\&\; I_n^2(y,x_n)].$$

In graphical syntax, this amounts to replacing the graph

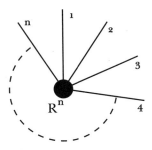

Figure 13.19

with the graph for t, which is

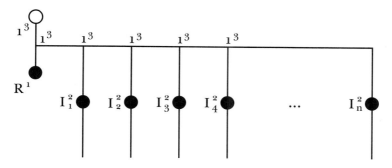

Figure 13.20

The presence of $n+1$ occurrences of the teridentity term in this graph is obvious. It is also obvious that the graph is composed of terms of PAL exclusively of adicities 1, 2, and 3.

Several years ago in a well-known paper, Hans G. Herzberger proved that, by relying for an account of the notions of construction and reduction upon certain operations of construction closely related to those that have been employed in *Burch 1991*, a relation may be reduced to triadic relations provided that the domain of the relation is "sufficiently large" (*Hertzberger 1981*). The domain's being "sufficiently large" means that the domain of the relation must contain at least as many members as the relation to be reduced. Some philosophers may perhaps have worried that the restriction in Herzberger's result made it less general than it should be in order to be genuinely relevant to Peirce's Reduction Thesis. The account I have given of hypostatic abstraction and its role in Peirce's Reduction Thesis, however, shows that such worry is out of place. For in order to be faithful to Peirce's Reduction Thesis, a reduction result like Herzberger's *must* contain a restriction to "domains sufficiently large." The reason is that hypostatic abstraction is crucial to Peirce's Reduction Thesis, and hypostatic abstraction necessarily involves augmentation of the relevant domain. But, when the domain is properly augmented by hypostatic abstraction, the domain *must* contain at least as many members as the relation to be reduced.

NOTES

1. The proof has been checked by several mathematicians, and so far no errors have been discovered in it.

2. Cf. for example *Kripke 1959* and *Montague 1973*.

14.

GENUINE TRIADS AND TERIDENTITY

Jacqueline Brunning

§1

Peirce's doctrine of categories is a logico-metaphysical claim that all thought and all reality is partitioned into three mutually exclusive and jointly exhaustive classes: monads, dyads, and triads. This doctrine is a conjunction of two claims. The first is a definitional claim that all polyads of degree greater than three are constructed from triads. Peirce often cited this as demonstration that no further categories are needed. The following is a typical citation.

> A thorough study of the logic of relations confirms the conclusions which I had reached before going far in that study. It shows that logical terms are either monads, dyads or polyads and that these last do not introduce any radically different elements from that found in triads. (1.293)

This claim remains untouched by developments in logic. The second is an irreducibility claim for dyads and triads. The irreducibility of triads Peirce took as justification for the third category. An example of such a claim is: "I prove absolutely that all systems of more than three elements are reducible to compounds of triads. . . . The point is that triads evidently cannot be so reduced." (SS:43)

The joint assertion of these claims gives the theory its particularly Peircean character. To make this character more vivid one simply needs to note that the reduction of polyads to dyads is commonplace in present logical theory. This has no real force as an objection to Peirce's third category if we realize that, in spite of apparent similarities, Peirce's logics are subtly different from ours.

If we assume that Peirce had no pretheoretic prejudice about what kinds of things there are in the world, the simplest explanation for his theory of categories may be found in his claim that logic ought to determine metaphysics, not vice versa. A careful study of all Peirce's logical machinery, starting with the algebra of relations through quantification theory to the existential graphs, can render the categories less metaphysically mysterious. If Peirce's theory of categories is driven by his formal logic, as I believe, then definitional peculiarities of his logic, intriguing in their own right, can serve to disarm many objections to the theory of categories.

This is far too comprehensive a project to demonstrate in this chapter. I propose instead, to focus on one small aspect dealing with teridentity.

The categories suggested the existential graphs. Peirce stated: "It was in fact considerations about the categories which taught me how to construct the graphs" (MS 439). The algebras, especially the algebra of relations, suggested the categories. Peirce often claimed that a thorough study of the algebras would convince one of the categories. So the algebras are related to the graphs by way of the categories. The graph are two dimensional arrays while the algebras are one dimensional. What does this difference in dimensionality reflect? One dimension gives us a line with two end points. Two dimensions gives two lines with three end points: $\curlyvee$ This is the graph of teridentity.

I want to show that while Peirce's algebras failed to provide a compelling explicit demonstration that the third category was necessary, the existential graphs via teridentity make explicit the necessity for the third category. Many of Peirce's claims for triads are then corollaries of this demonstration, which will be given in the following three sections. Peirce had two distinct notions of teridentity. In Section 2, teridentity of the algebras will be briefly discussed. Section 3 is a similarly brief discussion of genuine triads. In Section 4, teridentity of the existential graphs is treated.

§2

Teridentity of the algebras, which anticipates teridentity of the graphs and functions in noticeably similar ways, was introduced in 1870, in Peirce's first paper on the algebra of relations. It is related to the comma notation, also introduced in the same paper. This is not the appropriate place for a systematic study of the comma notation, but it is one of the keys to the link between Peirce's logic and theory of categories. The comma is a degree increasing operation which Peirce applied to both predicates and terms. Teridentity and the comma both figure prominently in Peirce's efforts to eliminate non-relative operations from his algebras. Peirce most actively pursued this project between 1870 and 1885.

One of the reasons motivating this project was Peirce's desire to establish the primacy of relative operations. But the project took on added significance as Peirce began to realize the richness of the definitional resources of his algebras. In 1882, in a letter to Mitchell, Peirce stated:

> We can write ⟨⟩ to express that something is lover, benefactor & servant of something. A triple relative of identity here enters, but the symbol (1,) is omitted. . . . We thus do away with the distinction of relative & non-relative operations, by discarding the latter altogether. It is true we might do this well enough with any notation that was adequate for triple relatives, but it is precisely the advantage of the spread formulae that they make the treatment of triple relatives easy. (W4:395–397)

There are three points in the above quotation that are relevant. First, Roberts takes this to be Peirce's first use of graphical displays (*Roberts 1973*: 18). This letter is considered the ancestor of the existential graphs. The "spread formulae" is certainly two dimensional. Peirce's claim that this added dimension makes treatment of triple relatives easy is frequently repeated in his discussion of the graphs. The difference is that in discussing the graphs, as we will see, Peirce claims that the two dimensional aspect makes representation of triads *possible*.

The second point is that teridentity is required for the expression of Boolean product as a case of relative product. The graphs are full of similar examples in which teridentity is the device employed to construe Boolean product as relative product.

The third point is that this teridentity is definable in the algebras. It is equivalent to relative product, comma, and permutation. This has dire consequences. Complete reduction of triads to dyads is possible when teridentity is definable in the logic. This overthrows the irreducibility claim for triads and eliminates the need for the third category.

Teridentity, the comma notation, and relative product underwent significant modification. The teridentity of the algebras seldom appeared after 1882. The comma notation evolved into a co-existence claim that in the graphs becomes simply conjunction. I will indicate the significance of this change in the next section. Relative product, Peirce's central definitional process, was expanded so that flexibility in bonding was maximized. By 1886, after studying Kempe's memoir, Peirce was aware, or suspected that the definitional resources of his algebras were too strong to support the irreducibility claim for triads. Peirce also became convinced that the true nature of triadic relations was often masked by the algebras. It is the existential graphs that make the "triadic" character explicit.

§3

To sharpen the problem of Peirce's third category, one should note that his algebras reflect the historical complication of having two types of operation, non-relative and relative. Peirce's particular insight into the nature of triads was governed by the relative operations. The non-relative operations do gradually become less prominent in the algebras; however, Peirce was never able to completely eliminate them, so they affect the definability results.

To account for triads that were definable in the algebras but did not fit his third category, Peirce introduced the genuine/degenerate distinction. This distinction applies equally to dyads and triads that are not primitive and irreducible, but my remarks will here be restricted to triads.

Peirce had two distinct notions of relation. One was the standard notion of a relation as a set of n-tuples. Peirce often referred to these as merely formal relations. The other was a relation as a relative concept. Relative concepts function somewhat like equivalence classes. There are three relative concepts and they correspond to Peirce's three primitive, irreducible categoric: monads, dyads, and triads.

The set of formal triads divides into degenerate and genuine triads. A triad is genuine if it is the same degree as the relational concept it expresses. Genuine triads are irreducible and the basis of all higher n-ads. A degenerate triad is of higher degree than the relational concept it expresses. While formally a 3-tuple, it can be shown by analysis (reduction) to be either a compound of dyads and hence, dyadically degenerate, or a compound of monads and monadically degenerate.

A genuine triad for Peirce was clearly more than a 3-tuple. He warned us that "three things do not necessarily make a triad" (MS 942). He repeatedly referred to degenerate triads as "mere combinations" (1.363), "mere juxtaposition" (1.371), and "congeries" (MS 717). All of Peirce's examples of degenerate triads are constructed from non-relative operations, particularly, Cartesian product and Boolean product, some of the operations that he had tried to eliminate from his algebras.

Degeneracy is a function of the mode of combination. I have argued elsewhere (*Brunning 1993*) that relative product was the "privileged mode" of combination. Relative product has an algebraic type of $(m + n - 2)$. All of the degenerate triads are formed by operations of different algebraic type.

Peirce repeatedly used "A gives B to C" as an example of a genuine triad, even after study of Kempe's memoir showed him that this could be expressed as a combination of three dyads thus (1.363):

> In a certain act D, something is given by A
> In act D something is given to C
> In the act D, to somebody is given B

Graphically this would have the following diagram:

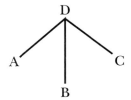

This compound of three dyads has an algebraic type $(n + m + r) - 3$. This algebraic type is different from the algebraic type of relative product. Peirce remarked that "to include this . . . as one of the regular operations of logical algebra is to make an intrinsic transmutation of that algebra" (3.424). Peirce is not completely justified in making this claim, however, after he maximized flexibility of bonding for relative product. The above operation can be seen as a variant of relative product in which the triple bonding is legitimate. This is of course, sufficient to overthrow the irreducibility claim for triads. This caused Peirce to remark that he was led into error about the capability of his algebras but not to give up his theory of categories. Peirce attempted to explain his intuitions governing triadic relations by claiming that in the pseudo-reductions, the three dyads were *not* welded into one fact. Other such claims were: a genuine triadic relation involves meaning, which (given Peirce's theory of signs) is itself a triadic relation; or significantly, every triadic relation requires a triad for genuine combination; likewise, reduction of a triadic relation cannot take place without using a triadic relation. None of these claims are readily intelligible in the algebraic framework but they all indicate something about triads that the algebras did not reveal.

I have suggested elsewhere (*Brunning 1993*) that Peirce, following some medieval logicians, considered conjunction to be a pseudo-connective, more properly used as a listing device since conjunction fails to produce any true combination of elements. It is often used this way with no suggestion of any connection other than juxtaposition between the combined elements. Cartesian product, which is ordered conjunction, and Boolean product, which involves conjunction, failed to supply the kind of connectedness that Peirce required for genuine triads. I take Peirce to have been looking for a kind of "connectedness" in the relations in his three categories that we require when we speak of relations in everyday contexts. The formal difficulties in formulating this notion are immense. Peirce's existential graphs are a way around these difficulties. The challenge for the graphs is to demonstrate the need for the third category and to establish its independence. It really doesn't matter whether we think that this counts as a compelling reason to accept the theory of categories. It is enough that Peirce found a vehicle to demonstrate the irreducibility of triads to dyads.

§4

Teridentity (sometimes called co-identity or comidentity) is a structure of the Beta part of the existential graphs. The Beta part of the graphs has the same theorems as first order predicate calculus with identity. However, theorem isomorphism is a weak condition and the structures embedded in the graphs make this a very different system.

Peirce said: "A point upon which three lines of identity abut is a graph expressing relation of Teridentity" (MS 478). Instances of this relation are represented two dimensionally by a three way branching line, $\curlyvee$. Teridentity is, however, more properly the point from which the tree lines branch. Although this graph-instance is read as "_____ is identical with _____ and _____", Peirce warned us that: "The concept of teridentity is not mere identity. It is identity and identity, but this 'and' is a distinct concept, and is precisely that of teridentity" (MS 296). Peirce explicitly noted that "teridentity cannot be formed out of binidentity (SS: 199). In spite of this warning, teridentity has sometimes been misconstrued as merely identity and identity. The reason is that Peirce's examples, as we will see, are misleading. He uses teridentity in the graphs, as in the algebras, to construe Boolean product as relative product and this, of course, is read as conjunction.

Study of the graphs indicates that teridentity has special status. It is not definable in the graphs and it must be considered a primitive constant relation. Peirce stated:

> there must be an elementary triad. For were every element of the phaneron a monad or dyad, without the relative of teridentity (which is, of course, a triad), it is evident that no triad could ever be built up. (1.292)

Peirce noted, however, a fact he claimed would please logicians in love with his algebra of dual relatives: "no other triad than that of teridentity seems to be needed" (MS 490).

We need only a few graphical terms and no actual experience with the graphs to understand why teridentity is not definable. In the graphs individuals are represented by dots which can be extended as lines of identity. Predicate terms (rhemes) are called spots, and are usually represented by letters or words. On the periphery of every spot are hooks or pegs (invisible) to which are attached lines of identity. A monad would be represented as —W. A dyad —L—. Writing (scribing) a graph on the sheet of assertion (the graphical surface, which in fact, is a kind of postulate set whose content depends upon what one chooses to reason about with the graphs) is to assert it or to diagram it. The blank sheet is also a graph which, as Zeman suggests, can be interpreted in Fregean fashion, as The True (*Zeman 1964*: 7). To assert the identity of the individuals denoted by different dots we simply connect them by a line of identity. It is important to note that distinct lines of identity may never intersect and two lines of identity may not be attached to the same hook. Given these restrictions and the important fact that the graphs have the elegance of but one mode of logical composition, it is easy to demonstrate that teridentity is not definable. Peirce stated that this single mode of composition is relative multiplication: "that composition by which 'lover of a woman' is composed of 'lover of something' and that 'something is a woman' " (MS 292). This is the same example Peirce gave in 1870, in his first paper on the logic of relations. There he claimed relative multiplication was "*the application of a relation*, in such a way that, for example, *lw* shall denote whatever is lover of a woman" (3.68/W2: 369) The bonding in this operation can be represented as follows: letting the letter "*L*" stand for "is a lover of" and "*W*"stand for "is a woman," gives, — *L*—*W*. This representation is not an existential graph but simply a pictorial device for displaying the bonding for relative product. Peirce stated that this operation "consists in indefinitely identifying a subject of the one with a subject of the other, every correlate being regarded as a subject" (1.294).

The algebraic analogue of "indefinitely identifying" is existential quantification. Algebraically Peirce's above example would be the open expression $(\exists z)(Lxz \ \& \ Wz)$. A generalization of relative product, letting "*R*" and "*S*" represent polyads, associates a unique polyad $p[R, S]$ with any R and S as follows:

$$p[R, S](x_1, \ldots x_n, y_1, \ldots, y_m) \equiv (\exists w)[R(x_1, \ldots x_n, w) \ \& \ S(w, y_1 \ldots, y_m)]$$

Peirce also gave a clear statement of the graphical analogue of relative product, his single mode of combination, as: "a hook of a graph is joined to a single other hook of a graph" (MS 296). He diagrammed the same example as follows:

—lover of —a woman

This joining of spots by lines of identity will never give a triad.

Peirce describes this bonding as two explicit indefinites mutually, at least partially, defining each other. The same diagram without the bonding reads, "something loves something" and "something is a woman." Once scribed on the sheet of assertion, Peirce said, it actually says more. It says: "something loves something coexistent with something that is a woman" and is diagrammed as:

— lover of — — a woman

Graphs on the same sheet of assertion are read as conjunctions and conjunction here is *merely* a statement of coexistence with no suggestion of combining. The graphs, therefore, rule out the formation of relations by simple conjunction.

In the graphs, "a graph with three tails cannot be made out of graphs each with two or one tail, yet combinations of graphs of three tails each will suffice to build graphs with every higher number of tails" (1.347). This is both an irreducibility claim for teridentity and a definitional claim for polyads greater than three in terms of teridentity. Peirce gave a very clear statement of the latter: "it is permitted to scribe an unattached line of identity on the sheet of assertion and to join such unattached lines in any number of spots of teridentity" (MS 478). Teridentity cannot be defined in the graphs, given the single mode of logical composition. Hence, it must be a primitive relation.

Peirce expressed Boolean product as relative product using teridentity. The "assertion that *M* is the loving servant of *N*" is diagrammed as follows (MS 300):

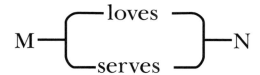

This is the same type of diagram used in his 1882 letter to Mitchell, but this time teridentity is a primitive relation. With this teridentity, Peirce can get all the constructions of his algebras without the reduction of triads to dyads.

We need to demonstrate why teridentity is a constant relation. Peirce made this claim when he stated:

> it is impossible to represent that *A, B, C* are all identical without the triadic rhema "_____ is identical with _____ and with _____" which is a simple rhema. But the assertion that *A, B, C, D* are all identical can be made by two triads. (MS 492B: 202)

The graphs are rich with examples of diagrams of genuine triads that, according to Peirce, do not constitute reductions but do demonstrate the role of teridentity. Consider again the triad "*A* gives *B* to *C*." Peirce drew the following two diagrams (MS 292):

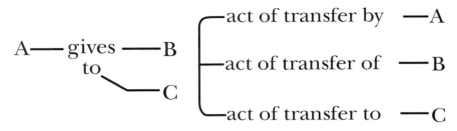

Figure 14.1   Figure 14.2

He stated that both diagrams say the same thing except that in Figure 14.2. teridentity is represented as the only triadic element. The function of teridentity is to assert the identity of the three transfers, thus, insuring the triad-ness by making explicit the three-way identity of the act of transfer. According to Peirce, this did not count as reduction of the triad to dyads. Without the graph-instance of teridentity, Figure 14.2. would merely be a statement of the coexistence of three dyads and would strikingly resemble the algebraic analysis. This gives us a feeling for what Peirce meant when he said that in the algebraic reduction the three dyads merely co-exist and are not welded into one fact. It is now easy to understand what Peirce meant when he claimed that a triad was required for the reduction of a genuine triad. This can be formulated as a rule: A triad is genuine, if teridentity is the only triadic relation that remains after analysis.

Consider one more example, this time of two triads. Figure 14.3 and Figure 14.4 below both assert: "Somebody steals something from a person and gives it back" (MS 292).

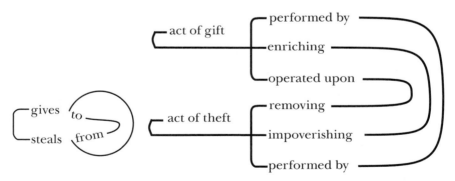

Figure14.3   Figure14.4

According to Peirce, the only difference is that in Figure 14.4 the only triadic graph instances are teridentity. Peirce added: "there is here no reduction of triadic relations of giving and stealing into dyadic relations with the notion of teridentity." He then added: "I leave the precise formulation of the reason as a highly instructive exercise" (MS 292). Several pages later he did provide a hint.

Peirce often repeated a criticism of the algebra of dual relatives, noting that: "the very triadic relations which is does not recognize, it does itself employ" (8.331). He made this complaint more explicit when he said: "A triadic relation cannot be reduced without the use of a triadic relation" (1.346). In the algebras, it is very difficult to make sense of this claim. In the existential graphs a genuine triad when diagrammed makes explicit the "triad-ness" by the presence of *only* triads of teridentity. This feature of the graphs so pleased Peirce that he stated: "indeed we may once and for all lay down the principle that any supposed analysis that can be represented in the existential graphs . . . must be true analysis provided the representation . . . involves no iteration" (MS 292).

This Peirce claimed was the clue to the analysis of "gives" and "steals." Iteration is one of the rules of transformation (rules of inference) of the graphs. These rules enable one to produce proofs of theorems. The rule of beta iteration, given its clearest statement by Zeman, has three clauses (*Zeman 1964*: 17). A study of this rule would take us too deeply into the graphs. All we need to do now is figure out what Peirce meant when he claimed that analysis of a graph is true analysis, if the graph involves no iteration.

One form of beta iteration Peirce describes as follows: "To iterate a graph means to scribe it again, while joining by Ligatures every Peg of the new Instance to a corresponding Peg of the Original instance" (4.566). Zeman gives the following schematic graph as an example of iteration across cuts (*Zeman 1964*: 18):

Peirce also provided an example as follows (4.566):

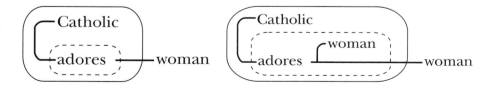

Each of these examples of iteration share a feature that might lead to false analysis. The iteration is legitimate within Peirce's system, but Peirce tells us that:

> Indeed we may once for all lay down the principle that any supposed analysis that can be represented in existential graphs as being an analyses must be a true analyses, provided the representation in existential graphs involves no iteration. (MS 292)

Peirce's following comment explains the restriction on iteration. He says "The reason for this exception is that iteration is a device whereby two graphs may be combined without their combination being diagrammatized as such " (MS 292). Iteration is very suggestive of the comma notation of the algebras. The comma notation, a degree increasing operation, gave rise to degenerate dyads and triads by merely annexing the additional correlate with expressions like "coexists with" or "and." To annex a term is not necessarily to logically combine that term. As mentioned earlier, Peirce viewed "and" and "coexists with" as listing devices not logical connectives.

In the above two graph examples the two graphs with iteration have structures different from those of the original graphs. The differences however, are a result of merely annexing another correlate and not of diagammatization (logical combination). Peirce's restriction on iteration does not affect the resources of the existential graphs but seems to prevent the formation of degenerate dyads and triads.

One closing remark about analysis in the graphs. Peirce considered the analysis to be correct analysis, if as mentioned, the graph involved no iteration. In fact, he claimed false analysis cannot be carried out in graphs. He clarified this even further using the following diagrams (MS 296):

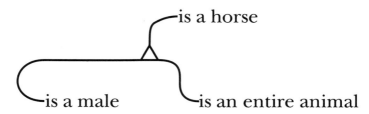

He stated that Δ represents teridentity and that the above must represent true analysis. But the analysis in the algebra of dyadic relations, which is (MS 266):

[Horse • Male • Entire]

where the dots represent simple identity and not teridentity, is wrong. He then turned to the universal algebra of logic. Letting "*h*" signify "is a horse,"

"m" signify "is male," and "e" signify "is entire," the assertion "Some horse is both male and entire" is written (MS 296):

$$\Sigma_i \, h_i m_i e_i$$

This, Peirce noted, is equivalent to the following graphical use of selectives (MS 296):

S—is a horse
S—is male
S—is entire

These last three diagrams, according to Peirce, are all imperfect representations of teridentity (MS 296). He concluded these remarks saying that the use of selectives in the graphs is analogous to the algebraic formulation and is not just superfluous, but misleading and can result in false analysis. Teridentity, however, Peirce claimed never produces a false analysis. Perhaps this, in addition to the reasons suggested by Zeman and Roberts, is why Peirce referred to his graphs as "My Chef D'Oeuvre."

15.

PEIRCE'S INFLUENCE ON LOGIC IN POLAND

Henry Hiż

§1. C-calculus

Peirce had a fundamental insight when he used implication as the only primitive connective of propositional logic. In his system of 1885, 3.375–384, Peirce accepts as axioms four formulas which he calls icons:

(icon 1)	*Cpp*	(Id)
(icon 2)	*CCpCqrCqCpr*	(Comm)
(icon 3)	*CCpqCCqrCpr*	(Syl)
(icon 5)	*CCCpqpp*	(Peir)

The names "Id", "Comm," etc. are from Wajsberg. In Warsaw this system was studied in depth, improved, and applied to other systems, mainly by Łukasiewicz, Leśniewski, Tarski, and Wajsberg. I include in the Warsaw school logicians like Arthur Prior who, though not residing in Warsaw, were influenced by those four.

What was the verdict of the Warsaw research about the original axiomatization by Peirce? It was rather positive. The system is a complete system of implicational logic. The implicational logic was studied on its own and, when augmented by quantifiers, led to many interesting systems with negation as defined. All this was inspired by Peirce. It was assumed, with some understandable tolerance, that Peirce employed the rules of substitution and of detachment (the meta-logical rule of modus ponens). Peirce seemed aware of the distinction between a procedure and a logical statement. From the beginning he was using the rule of detachment, without stating it, but he had proved what he called *modus ponens, CpCCpqq.* (Following Wajsberg, I will call this formula "Pon.") He wrote that the formula of transitiveness, *CCpqCCqrCpr*, justifies the inference *Cpq → CNqNp*. Even though he did not make much use of it, his distinction between a formula and an inference foreshadowed the later distinction between a rule of inference and a logical theorem.

If Peirce understood the difference between a formula and an inference, he did not master the ways to go from one to the other. This is evident in the derivations from his axioms where early on he commited a blunder. He wrote and I quote changing only the symbolism: "To say that *Cpp* is generally true is to say that it is so in every state of things, say in that in which *q* is true; so that we can write *CqCpp*, and then, by transposition of antecedents, *CpCqp*, or from *p* we can infer *Cqp*" (3.378). Many commentators, most clearly Prior (*1958*), have written that *CpCqp* (called by Russell, Simplification, or Simpl) can follow neither from {Id, Comm}, nor from {Id, Comm, Syl}, because Id, Comm, and Syl accept the interpretation of *C* as equivalence while Simpl does not. To some it seemed that Peirce should have added Simpl as a new axiom. But the system based on all four icons is complete and Simpl is provable in it. This is evident from the Wajsberg-Łukasiewicz theorem. Here it its short history. A formula *CpCCpqq* was called by Peirce the *modus ponens* of hypothetical inference. Wajsberg named it Pon. Peirce showed that {Id, Comm} → Pon, in other words, that Pon is deducible from Id and Comm jointly. But Wajsberg (*1937*) has shown that in Peirce, axiomatization {Id, Comm} can be replaced by Pon. Thus {Syl, Peir, Pon} constitutes a complete axiomatization of *C*-calculus. Wajsberg presented several variants of this axiomatization. These results were generalized by Łukasiewicz into what I call the Wajsberg-Łukasiewicz theorem: {Syll, Peir, *CpCab*} is a complete axiomatization of *C*-calculus where *p* is a variable and *CpCab* any valid formula.

From the Wajsberg-Łukasiewicz theorem it follows that Peirce's original axiomatization is complete, for Comm self applied gives *CqCCpCqrCpr*, which is in the required form. But it also follows that Peirce axiom Id is superfluous. The remaining axioms happened to be independent (*Prior 1958*). Peirce's axiomatization shares the faute de beauté of a superfluous axiom with that of Frege's (*1879*), where one of the six axioms for *CN*-calculus was dependent, namely Comm (see *Łukasiewicz 1934*). Note that in Frege implication is not treated on its own. There are valid purely implicational formulas which cannot be proved without the help of axioms that contain negation. As for the completeness of *C*-calculus, the published proofs can be similarly divided into those which reduce the problem of the completeness of *C*-calculus to the completeness of *CN*-calculus, for instance, Schütte's (*1933*), and those which deal only with *C*-calculus, such as Wajsberg's (*1937*), or mine (*1973*).

§2. Negation

It is implicit in material implication that any proposition follows from a false proposition. Therefore, 'not *p*' is for Peirce an interpretation of "*Cpa* whatever *a* may be," as Peirce put it. This is Peirce's fourth icon. Negation is thus introduced as an interpretation of some expressions containing only implications. The content of any formula built of implications and negations is given by a purely implicational formula. The question remains: What does it mean "whatever *a* may be?" Although in the same paper Peirce writes extensively about quantifiers, he uses only verbal expressions for propositional quantification. This fails to show where exactly that quantifier appears in the formula and what its scope is. Logicians filled this gap later. Russell (*1903* and *1906b*) suggested two definitions: "$Np = \Pi q Cpq$" and "$Np = Cp\Pi qq$," but he did not use either in his subsequent publications. Leśniewski (*1929* and *1931*) and Tarski (*1923*) used both definitions in their work on protothetics and in place of Peirce's "interpretations" they developed suitable rules of definition, following Frege. The progress in logic often consists in supplying the details for what was sketched intuitively by an initiator. Peirce suggested that by using the fourth icon one can obtain $CCpqCNqNp$ from Syl. Here is his derivation with some more details:

1. $\Pi rCCpqCCqrCpr$ from Syl, making explicit a tacit quantifier
2. $CCpq\Pi rCCqrCpr$ from 1, distributing the quantifier
3. $CCpqC\Pi rCqr\Pi rCpr$ from 2, distributing the quantifier and Syl
4. $CCpqCNqNp$ from 3, taking $\Pi rCqr$ as Nq and $\Pi rCpr$ as Np

This way of interpreting Peirce is fruitful for systems which have quantifiers, in particular, sentential or propositional quantifiers. Leśniewski (*1929*) introduced rules for operating with quantifiers which govern substitution and distribution of quantifiers over implication. This led to protothetics. Other commentators take Peirce's icon 4 to be a definition of negation by means of implication and 0, so that $Np = Cp0$. This is how Prior (*1958*) reads Peirce. The *C*0-system with implication and false as primitive terms was investigated by Łukasiewicz and Wajsberg, and it influenced many other works where 0 appears as a constant value. Which way of understanding icon 4 is closer to the intention of Peirce is hard to decide. His use of logical constants favors Prior's reading. The fact that for him $CCpqCCqrCpr$ is saying the same as $CCpqCNqNp$ speaks against it. In this case the distribution of a tacit quantifier binding *r* is more natural.

§3. Peirce's Law

It seems that Peirce's law is the first published purely implicational thesis which is valid classically and not valid intuitionistically. And it is Peirce's law that constitutes the exact difference between the intuitionistic implicational logic, often called the positive logic, and the classical logic. The positive logic can be axiomatized by {Simpl, Comm, Syl}. The addition of Peir to the positive logic gives the classical logic. But the difference between the two logics can be analyzed in further detail. For instance, the addition of

(A₁) *CCCqrpCCCpqpp* to the positive logic does not yield Peir and therefore does not result in the classical logic. Similarly, (A₂) *CCCrspA₁* does not give (A₁). Generally, by preceding (Aₙ) with *CCCxyp* where *x* is the second variable of (Aₙ) and *y* a new variable, an infinite sequence of non-equivalent formulas is formed between the positive and the intuitionistic logics. These are, as it were, weaker and weaker Peirce's laws (*Hiż 1946*). There are other classically valid propositions which are not valid in the positive logic and are not in this sequence, like *CCCpqrCCprr*. That Peirce's law was unusual and that it extended the traditional ways of reasoning was understood by Peirce. He spoke about it as dilemmatic or non-syllogistic. He knew that Peir, together with "syllogistic" logic leads to the law of double negation and excluded middle.

Peirce suggests a derivation of the law of excluded middle from his fifth icon. Here is a more formal derivation:

4.	*CCCpqpp*	Peir
5.	*CΠrCCpqrCCpqp*	substitution rule *r/p*, deduction theorem
6.	*CΠrCCpqrp*	Syl, 5, 4
7.	*CNCpqp*	6 taking Π*rCCpqr* as *NCpq*
8.	*CNCpΠqqp*	7, *q*/Π*qq*
9.	*CNΠqCpqp*	8, *Cp*Π*qq* = Π*qCpq*
10.	*CNNpp*	9 taking Π*qCpq* as *Np*

Because *CNab* = *Aab*, 10 = *ANpp*, the law of excluded middle.

§4. Logical Values

Peirce introduced the values that may be taken by propositions. For him these values are constants. They can be used as propositional terms in logical formulas. They are values in the same sense in which the value of cos 0 is 1. In the binary system of logic with only two values, **v** and **f**, they are treated by him as quantities, as elements of a rudimentary counting system. The letters **v** and **f** stand for verum and falsum, and this terminology was adopted by Leśniewski. Peirce's characterization of the conditional (in his terminology, hypothetical) proposition in terms of logical values is now generally accepted: *Cpq* = **v** except when *p* = **v** and *q* = **f**. His explanation of logical necessity clarifies the issue which had been obscure for a long time. "Though we limit ourselves to the actual state of things, yet when we find that a formula of this sort [a conditional] is true by logical necessity, it becomes applicable to any single state of things throughout the range of logical possibility." For example, '*CNCpqCrq*' means that "in whatever state of things we find *p* true and *q* false, in that state of things *r* is false or *p* is true. In that sense, it is not limited to the actual state of things, but extends to any single state of things." There are true and false propositions. No matter whether we know which of '*p*,' '*q*,' or '*r*' are true and which false, '*CNCpqCrp*' will be true. Peirce treats '*p*' as a specific proposition the logical value of which we neither know nor need to know. The set {**v**,**f**} is treated by Peirce quite formally without entering into the philosophy of truth and falsehood. He was the first to use logical matrices. This led to the Łuka-siewicz-Tarski (*1930*) theory of logical matrices and to the concept of veri-fiability of a formula by a matrix. It is natural to extend the number of logical constants. This was suggested by Peirce when he wrote: "According to ordinary logic, a proposition is either true or false, and no further dis-tinction is recognized. This is the descriptive conception, as the geometers say; the metric conception would be that every proposition is more or less false, and that the question is one of amount." This passage was often quoted by Łukasiewicz in his lectures to show that Peirce was a precursor of many-valued logics.

The Warsaw logicians used the term "logical values." Peirce used just "values." In the English literature the term "truth value" is popular. It is Russell's translation of Frege's "Wahrheitswert." It seems to me a misnomer (why should falsity be a truth value?), particularly if one accepts many val-ues and without intuitive interpretation. Anyway, Peirce's concept of value is substantially different from that of Frege's Wahrheitswert. Wahrheitswert is neither a proposition, nor a formula, nor any other linguistic entity. It is an abstract object which is supposed to be the denotation of sentences; the Truth the denotation of any true sentence, the Falsehood the denotation of any false sentence. This is a fantastic metaphysical conception without a real use in logic. No one among the most creative logicians in Poland treated Frege's conception of truth values seriously, in spite of the great respect which they had for Frege as a logician. It was Peirce's logical values which were generally used. Peirce's value is a proposition or a sentence, a symbol or a formula; it can be plugged into another logical formula, which Frege cannot do. By that property Peirce's conception of value proved to be technically useful in logic. Leśniewski often took 0 to be the sentence Π*pp* and 1 to be the sentence *C*00. Any other suitable choice of a false and a true formula would be acceptable. It seems to me advisable to abandon the use of the term "truth value" for a propositional constant and to retain "logical value."

§5. Quantifiers

There are two originators of quantifiers: Frege and Peirce. Formally their theories are equivalent but the commentaries in the two cases differ. The term "quantifier" and the symbols "Π" and "Σ" are from Peirce. Those symbols have been preserved by many logicians in Poland, presumably be-cause they suggest that quantification is an extension of the infinite (per-

haps non-denumerable) product and of the infinite sum in other branches
of mathematics. In that spirit, following a suggestion of Schröder, a geometrical interpretation of quantification was developed, mainly by Tarski
and Mostowski, and this idea was applied to topology. Mostowski's (*1957*)
generalization of quantifiers was a high point in this development.

Peirce acquired the skill in operating with quantifiers only gradually.
In his 1885 paper he already had the distinction between the "first intention" and "second intention" logic which we would call today first and second order quantification. The difference is shown in his treatment of identity. He first stated

   11. $\Pi a \Pi b \mathrm{Cl}\, abCXaXb$

where "1" stands for the identity relation and "*X*" for a one-place property.
In addition he introduced quantifiers binding variables that range over
properties and he asserted

   12. $El\, ab \Pi X A K X a X b K N X a N X b$

In Peirce's notation:

   12'. $1_{ab} = \Pi_x(x_a x_b + \bar{x}_a \bar{x}_b)$

The formula 11 states that no matter what *a* and *b* are, if they are identical
and *a* has the property *X*, *b* has it too. The formula 12 states that *a* is identical with *b* if and only if any property either is a property of both of them
or is a property of neither. Higher order quantification was developed by
Tarski and Leśniewski. In the Leśniewski systems there are no free variables;
all occur bound. And of whatever semantic category there is a constant in
the system, one can introduce a variable of that category. Peirce introduced
two levels of quantification; Leśniewski extended this to any number of levels. More precisely, there are no levels of quantification but a multiplicity
of semantic categories; quantifiers are the same for all levels. The fundamental rules of operating with quantifiers established by Peirce apply to all
semantic categories. The fruitfulness of this freedom is shown in Tarski's
(*1923*) definition of conjunction by means of equivalence alone:

   13. $\Pi p \Pi q\, EKpq \Pi f EpEfpfq$

Or, in a more familiar notation:

   13'. $\Pi p\, \Pi q[(p\ \&\ q) \equiv \Pi f\, [p \equiv (fp \equiv fq)]]$

§6. Theory of Relations

The last but not least field in which Peirce profoundly influenced logicians in Poland was the logical theory of relations. In 1940 Tarski delivered

a lecture to a meeting of the American Philosophical Association in Philadelphia on the calculus of relations and I shall quote from his talk. Although De Morgan realized the necessity of a theory of relations

> the title of creator of the theory of relations was reserved for C. S. Peirce. In several papers published between 1870 and 1882, he introduced and made precise all the fundamental concepts of the theory of relations and formulated and established its fundamental laws. Thus Peirce laid the foundation for the theory of relations as a deductive discipline; moreover he initiated the discussion of more profound problems in this domain. In particular, his investigations made it clear that a large part of the theory of relations can be presented as a calculus much like the calculus of classes developed by G. Boole and W. S. Jevons, but which greatly exceeds it in richness of expression and is therefore incomparably more interesting from the deductive point of view. (*Tarski 1941*)

After Peirce and Schröder, there was not much study of the calculus of relations on its own. Rather, what Peirce and Schröder wrote about it was adapted to this or another comprehensive system of logic without prusuing the intrinsic problems of the calculus of relations. This was the way of *Principia Mathemtica*, of Leśniewski's *Ontology* and *Mereology*, and Quine's *Mathematical Logic*. Tarski distinguished the theory of binary relations as submerged in the restricted functional calculus from the proper calculus of relations. The first has individual variables bound by quantifiers and free relational variables. The second is a fragment of this theory, the fragment which contains only those formulas in which individual variables do not occur. Each system was axiomatized by Tarski. He developed the second calculus very much in the algebraic spirit with which Peirce treated it. Tarski proved that every sentence of the calculus can be transformed into an equivalent sentence of the form "$R = S$," in other words, into an equation between relations. This amounts to the elimination of the entire apparatus of propositional logic. Tarski also raised several methodological questions about the two systems. One was: How do they compare in their ability to express the properties of different relations? For instance, we may think of the relations that in an infinite domain satifsy the formula

14. $\Pi x \, \Pi y \, \Pi z \, \Sigma u [xRu \ \& \ yRu \ \& \ zRu]$

Surprisingly, this property, as well as many others, cannot be expressed in terms of the calculus of relations.

16.

TARSKI'S DEVELOPMENT OF PEIRCE'S LOGIC OF RELATIONS

Irving H. Anellis

Tarski and Givant's formalism $\mathcal{L}_3$ (*Tarski and Givant 1987*) is a three-variable fragment of first-order logic. It develops an axiomatic system which presents set theory and number theory as sets of equations between predicates constructed from two binary atomic predicates denoting the identity and the set-theoretic elementhood relations. It is also a partial positive reply to the question posed by Schröder (*1895:* 551), of whether the algebra of relatives can express all statements about relations as equations of the calculus of relations. Indeed, in the preface to their work (*1987:* xv), the authors specifically indicate that the historical source of the inspiration for their work lies in the work of Peirce and Schröder. Although their historical survey of the early development of their subject is far from complete, it is worth noting as the starting point for our historical investigation. Thus, they begin by remarking that the mathematics of their work "is rooted in the calculus of relations (or the calculus of relatives, as it is sometimes called) that originated in the work of A. De Morgan, C. S. Peirce, and E. Schröder during the second half of the nineteenth century," and go on to declare that

> both Peirce and, later, Schröder, who extended Peirce's work in a very thorough and systematic way in [*Schröder 1895*], were interested in the expressive powers of the calculus of relations and the great diversity of the laws that could be proved. They were aware that many elementary statements about (binary) relations can be expressed as equations in this calculus. . . . Schröder seems to have been the first to consider the question whether all elementary statements about relations are expressible as equations in the calculus of relations, and in [*Schröder 1895*], p. 551, he proposed a positive solution.

The remainder of their historical discussion outlines the contemporary history behind *Tarski and Givant 1987*, and in particular the work of Tarski and his colleagues in the algebra of relations. Givant (*1991*) gives a detailed sketch of this aspect Tarski's work couched in the language of $\mathfrak{L}^\times$, a simple formal language that is the analog of the algebraic theory of relations, and notes the historical antecedents and influences on the problems with which Tarski dealt. Givant (*1991:* 190) also refers specifically to Tarski's early acquaintance with Schröder's work on algebraic logic and Löwenheim's 1915 paper on calculus of relations.

The starting point for the work undertaken by Tarski and Givant (*1987*) is Peirce's paper "The Logic of Relatives" (*1883c:* 187–203). In that paper (as described in *Tarski and Givant 1987:* xv), we are presented with a fixed arbitrary set, together with binary relations. The calculus of relations is the collection of all binary relations on U, together with four binary operations and two unary operations on these relations, and four distinguished relations. Thus, the universe of discourse for this calculus is the set of all subsets of $U \times U$, that is, of ^2U.

For any two binary relations R and S in this system, their absolute sum is the union $R \cup S$, and their absolute product is the intersection $R \cap S$. The relative sum (Peirce sum) of the relations R, S is $R \uparrow S$, comprised of all pairs $\langle x, y \rangle$ such that for every z, xRz or zSy. Similarly, their relative product (Peirce product) $R|S$ is comprised of all pairs $\langle x, y \rangle$ such that for every z, there exists both xRz and xSy. The unary relations on R are complementation, $\sim R$, with respect to $U \times U$, and the converse R^{-1} comprised of all pairs $\langle x, y \rangle$ such that yRz.

The four distinguished relations are absolute zero, presented by the empty relation $\emptyset$, the absolute unit, that is, the universal relation $U \times U$, designated by $\mathbf{1}$, relative zero, given by the diversity relation D_i on U composed of all pairs such that $x + y$, and the relative unit, given by the identity relation Id on U composed of all pairs such that $x = y$.

Many elementary statements about (binary) relations can be readily expressed in this calculus of relations. For example, the elementary statement $\forall x \forall y \forall z\ (xRy \wedge yRz \rightarrow xRz)$, that the relation R is transitive, can be expressed by the equation $(R|R) \cup R = R$.

As is well known, an *algebra* is defined as any structure $\mathfrak{A} = \langle A, O \rangle$ determined by a universe $\mathfrak{A}$, a nonempty set A, and a system O of operations on A, where each O_i ($i \in \mathbb{I}$, O_i having any cardinality) has a definite finite rank, and distinguished elements are identified with operations having rank 0. The algebras of this type are developed within a formalism $\mathcal{P}$ of predicate logic, and all nonlogical constants are operations symbols. In the formalism $\mathcal{P}$, we consider in particular equations containing universal quantifiers and identity. The class K of similar algebras which is capable of being characterized as the class of models of a set of universally quantified equations is said to be an *equational class* or *variety*; and the set of all universally quantified equations which are true for a given algebra $\mathfrak{A}$, or of some class K of such algebras is called the *equational theory* of $\mathfrak{A}$, or of K. Examples of two varieties of algebras include the variety of the class **BA** of all Boolean algebras and the variety of the class **RA** of all relation algebras. A *proper relation algebra* is defined by Tarski and Givant (*1987*: 239) as an algebra of the form $\mathfrak{A} = \langle A, +, \bar{\ }, \otimes, \cup, i \rangle$, where A is a nonempty family of relations contained in a largest relation $\mathfrak{R}$ with field U, $i \in A$, and A is closed under $+$, $\bar{\ }$, with respect to $\mathfrak{R}$, $\otimes$, and $\cup$. A relation algebra $\mathfrak{A}$ is *representable* if it is isomorphic to a proper relation algebra. Thus, representability is the algebraic counterpart of completeness of the axiom system of **RA**. **RRA** is the class of representable relation algebras. *Tarski and Givant 1987* is particularly concerned with the **RA**s, and much of it (in particular chapter 8) is concerned with the application of the theory $\mathfrak{L}^\times$ to relation algebras and to varieties of algebras, including in particular representable relation algebras and Q-relation algebras, i.e. algebras whose universe includes some conjugated quasiprojections. (For a general discussion of the basic concepts presented here, see *Maddux 1991*). A *relation set algebra* $\mathfrak{R}$ is an algebraic structure, $\mathfrak{R} = \langle \mathfrak{R}, \cup, \cap, \bar{\ }, 0, {}^2U, |, {}^{-1}, U|\text{Id}\rangle$, where $\mathfrak{R}$ is the set of relations over a nonempty domain U, and $U|\text{Id}$ is the identity relation consisting of all pairs $\langle u, u\rangle$ for every u in U. Much of the work of Peirce and Schröder was devoted to the determination of the equations holding in every relation set algebra, while Schröder also worked on determining the solvability of the equations of relation set algebras.

Tarski began his own work by first presenting a formalization of the calculus of relations in 1941 (*Tarski 1941*); by 1943, he had done much of the work of developing a formalization of set theory without variables; publication of many of these early results occurred in the early 1950s, beginning the discussion with the abstracts *Tarski 1953a*, *Tarski 1953b*, *Tarski 1954a*, *Tarski 1954b*, *Tarski 1954c*, and in particular in *Tarski 1953a*. He completed it in his posthumously published work with Givant in 1987. In *Tarski and Givant 1987*, $\mathfrak{L}^\times$ is presented as an equational theory closely related to abstract relational algebra as presented by Chin and Tarski (*1951*). $\mathfrak{L}^\times$ consists of two binary relations, identity i, and set membership **E**, and the binary operations of relative product $\otimes$, conversion $\cup$, Boolean addition $+$, complementation $\bar{\ }$, and equality $=$. If, for example, $\overset{\circ}{i} + \overset{\circ}{i}^- = 1$, we have $[\mathbf{E}^{\cup -} \otimes \mathbf{E} \otimes \mathbf{E} + \mathbf{E}^\cup \otimes (\mathbf{E} \otimes \mathbf{E})^-] \mathbf{E} \otimes i = 1$. In addition to the usual logical axioms for first-order logic with identity and nonlogical binary relation **E** (FOL, $\mathfrak{L}$), the following logical axioms obtain for $\mathfrak{L}^+$:

$$xR \otimes Sy \leftrightarrow \exists z(xRz \wedge zSy)$$
$$xR + Sy \leftrightarrow xRy \vee xSy$$
$$xR^\cup y \leftrightarrow yRx$$
$$xR^- y \leftrightarrow \neg xRy$$
$$R = S \leftrightarrow \forall y(xRy \leftrightarrow xSy)$$

with the axiom of union, for example, given as

$$\forall s \exists u \forall x (x\mathbf{E}s \leftrightarrow \exists y (x\mathbf{E}y \wedge y\mathbf{E}s)).$$

Although it follows from *Quine 1969* that $\mathfrak{L}^+$ is an equipollent extension of $\mathfrak{L}$, it was known by Tarski as early as 1943.

$\mathfrak{L}^\times$ is the fragment of $\mathfrak{L}^+$ which contains only variable-free equations, and including only equations of $\mathfrak{L}^+$ of the form $R = S$, with the inference rule of replacement of equals for equals and nonlogical relation $\mathbf{E}$ and the nonlogical axiom $\mathbf{E} \otimes \mathbf{E} = \mathbf{E}$. $\mathfrak{L}^\times$ is weaker than $\mathfrak{L}^+$, since $\mathfrak{L}^+$ is equivalent to FOL, whereas $\mathfrak{L}^\times$ does not correspond to FOL. Translation from $\mathfrak{L}^\times$ to FOL is therefore easy, whereas translation from FOL to $\mathfrak{L}^\times$ is impossible. Thus, for example, Löwenheim (*1915*: 448), reports the example, given by Korselt, that there exist sentences expressible in $\mathfrak{L}^\times$ which are not, to use Schröder's terminology (*1895*: 550), "condensable," i.e. not translatable from $\mathfrak{L}^+$ into the calculus of relations; specifically, it is claimed that

$$\forall x \forall y \forall \exists u [\neg(x \overset{\circ}{\mathbf{1}} u) \wedge \neg(y \overset{\circ}{\mathbf{1}} u) \wedge \neg(z \overset{\circ}{\mathbf{1}} u)]$$

according to which there exist four elements, is not expressible in the calculus of relations, contrary to Schröder's claim (*1895*: 551). Proponents of Peirce's reduction thesis, who hold that all tetradic relations and higher polyadic relations are reducible to combinations of monadic, dyadic, and triadic relations, would argue that this result is equivalent to Herzberger's theorem T2, according to which there exist some tetrads (tetradic relations) which are not reducible, by the operation of relative product alone, to triads, in a domain of three elements (*Herzberger 1981*: 45). It is not clear, from the standpoint of modern algebraic logic, however, that Korselt's example is equivalent to T2.

Korselt's result measures expressive power. A stronger result was given by Tarski (*1941*: 89), Lyndon (*1950*), and McKenzie (*1970*); they supplemented this result of Löwenheim's, by proving that the postulates given for abstract relation algebras presented by Chin and Tarski (*1951*) are not sufficient to derive all of the equations that are true in every concrete relation algebra. This measures not just expressive power, but proof power. Tarski (*Tarski and Givant 1987*: 54) was able to show that by adding new operators one could translate Korselt's example into $\mathfrak{L}^\times$; he also generalized Korselt's example to show that ($\mathfrak{L}^\times$ plus finitely many more logical operations) is still insufficient to express in $\mathfrak{L}^\times$ every formula of FOL. In particular, he showed (*1987*: 54) that a sentence of the form

$$\forall x \forall y \forall z \exists u (x\mathbf{E}u \wedge y\mathbf{E}u \wedge z\mathbf{E}u)$$

is not infinitely equivalent to any equation of $\mathfrak{L}^\times$. Henkin, for example, (*1973*; see *Maddux 1982*: 501) proved that, although the associativity of relative product can be expressed by an FOL-formula with binary relation symbols and although that FOL-formula

$$\forall x \forall y [\exists z (\exists y (Rxy \wedge Syz) \wedge Tzy) \leftrightarrow \exists z (Rxz \wedge \exists z (Rxz \wedge \exists x (Szx \wedge Txy)))]$$

has only three variables, it cannot be expressed in the usual FOL-axioms without using four variables.

On the other hand, Marcin Schroeder (1991) has announced his result that Aristotelian syllogistics (as a fragment of the monadic FOL) may be defined quite readily in relation algebras. Letting $\mathfrak{R}(S)$ be a concrete (binary) relation algebra on a set S, each relation T and U left residual in $\mathfrak{R}(S)$ is defined by $T : . U = {}^-({}^-TU^{-1})$. Then for every relation in $\mathfrak{R}(S)$ we define relations the $(\leq_T) = : . T$ (which is a quasi-order on S) and $(\perp_T) = T : . \sim T$ (which is an abstract orthogonality relation on S; i.e. is symmetric and commutative). In fact, $\perp_T$ is weakly orthagonal with respect to the ordering $\leq_T$. The structure $\langle S, \leq_T, \perp_T \rangle$ defined on S is a *syllogistic* generated by T; meaning that when T is defined with the set-theoretic predicate $\in$, $\langle S, \leq_T, \perp_T \rangle$ is a set-theoretic interpretation of Aristotelian syllogistic. It follows that syllogistics bear the same relationship to Boolean algebras as Aristotelian syllogistic bears to the propositional calculus.

In general, there are many logically valid equations which are not provable in $\mathfrak{L}^\times$. An example of the shortest such equation known is

$$\mathbf{1} = \mathbf{1} \otimes ((\mathbf{E}^- \otimes \mathbf{E}^-) + [\mathbf{E} \otimes \mathbf{E} + \overset{\circ}{\mathbf{1}} + (\mathbf{E} + \mathbf{E})^- \otimes (\mathbf{E} \otimes \mathbf{E}^\smile) \bullet \mathbf{E}^{\smile-}] \bullet$$
$$\mathbf{E}^- + (\mathbf{E} \otimes \mathbf{E}^\smile)) \otimes \mathbf{1}$$

So we see, as already noted, that Tarski showed (*Tarski and Givant 1987*: 54) that, even adding finitely many more logical operations to $\mathfrak{L}^\times$, it is still not possible to express in $\mathfrak{L}^\times$ every FOL-formula. The failure to express every FOL-formula in $\mathfrak{L}^\times$ by the addition of finitely new operations and axioms to $\mathfrak{L}^\times$, however, does not obviously contradict Peirce's conjecture, or claim (*Peirce 1897*: 169; 183) and Herzberger's theorem T3 (*1981*: 47), purporting to prove Peirce's claim, which states that every tetradic relation is reducible to the relative products of triads and dyads in a sufficiently large domain, that is, in a domain of k–many elements for a k–adic relation. Indeed, Herzberger's theorem T5 (*1981*: 48) might appear to the uncritical reader to be a restatement of Tarski's result that every locally finite cylindric algebra of infinite dimension is representable, although from the stand-point of modern algebraic logic, it most probably is not. Tarski's general theorem, however, clearly contradicts Herzberger's theorems T6 and T8 (*1981*: 50; 51), according to which all polyadic relations are absolutely reducible in bonding algebras with sufficiently large universes, that is, in algebras containing the linking element w, the element X belonging to the relation T

defined on a set of relations K closed under relative product, permutation, relative complement, and the bonding operation $g(T)$, defined by

$$g(T) = \{X : \exists w(X^\wedge w^\wedge w \in T)\}$$

(Whether Tarski's general theorem in fact disproves Herzberger's theorems T6 and T8 will of course depend upon whether Peirce's bonding algebra, as presented by Herzberger, is equipollent to $\mathfrak{L}^\times$, and in particular to $\mathfrak{L}^+_3$. It would appear, however, that Herzberger's theorem T2 shows either that his theorems T6 and T8 are not negations of Tarski's general theorem, but are false, or that the bonding algebra is not equipollent to $\mathfrak{L}^+$. Further study will therefore be required.)

Despite these results, one can still get a complete formalization of set theory and number theory in $\mathfrak{L}^\times$. In particular, we begin by showing that $\mathfrak{L}^\times$ is equivalent to the weakenings $\mathfrak{L}_3$ and $\mathfrak{L}^+_3$ of $\mathfrak{L}$ and $\mathfrak{L}^+$ obtained by using only the first three variables in $\mathfrak{L}$ and $\mathfrak{L}^+$ respectively; that is, the equational calculus $\mathfrak{L}^\times$ is equivalent to the three variable fragments $\mathfrak{L}_3$ and $\mathfrak{L}^+_3$ of FOL.

To determine that set theory and number theory are expressible in $\mathfrak{L}_3$, we begin by showing that pairing is expressible in $\mathfrak{L}^\times$, by the following translation rule P (*Tarski and Givant 1987*: 129):

$$P = \forall x \forall y \exists p \forall z (z\mathbf{E}p \leftrightarrow z \,\mathring{\mathbf{1}}\, x \vee z \,\mathring{\mathbf{1}}\, y)$$

Taking Π as the set of predicates of $\mathfrak{L}^\times$ and Q_{AB} as the equation correlated with arbitrary pairs of predicates A, B of Π according to the equation

$$Q_{AB} = [((A^{\cup -}\otimes A + B^\cup \otimes B)^- + \mathring{\mathbf{1}}) \otimes (A^\cup \otimes B) = 1],$$

we obtain the result that there are A, B in Π such that $P \equiv Q_{AB} \equiv (A^\cup \otimes B = 1)$. The proof that $P \vdash (A^\cup \otimes B = 1)$ is based on Kuratowski's construction of an ordered pair (see *Tarski and Givant 1987*: 129).

Such equations are obtained by setting $D = \mathbf{E}^\cup \otimes \mathbf{E}^\cup \bullet (\mathbf{E}^{\cup -} \otimes \mathring{\mathbf{1}})]$, $\mathfrak{F} = \mathbf{E}^\cup \otimes \mathbf{E}^\cup$, with $A = D \bullet (D^- \otimes \mathring{\mathbf{1}})$ and $B = F \bullet (F^- + A\otimes\mathring{\mathbf{1}})$, where A, B are conjugated quasiprojections or "pairing functions." In this case, D and F contain at most three distinct variables, x, y, z, and only two free variables, and with constants and predicates the same as in $\mathfrak{L}$. The two binary relations defined by D and F form a pair of conjugated quasiprojections; that is, D and F are functions having the additional property that, for every pair of elements x, y in the universe of our model, there is an element z which is mapped to x by the first function and to y by the second function, so that z is a representation of the ordered pair $\langle x, y \rangle$.

The following transformations serve as examples (*Tarski and Givant 1987*: 64):

$$\forall x \exists y \forall z (z\mathbf{E}y \leftrightarrow z \,\mathring{\mathbf{1}}\, x) \equiv \mathbf{E} \bullet (\mathring{\mathbf{1}} \oplus \mathbf{E}^-) \otimes 1 = 1$$
$$\exists z \forall x \forall y (x\mathbf{E}y \wedge y\mathbf{E}z \rightarrow x\mathbf{E}z) \equiv 1 \otimes [(\mathbf{E}^- \oplus \mathbf{E}^{-\cup} \oplus \mathbf{E}) \bullet \mathring{\mathbf{1}}] \otimes 1 = 1$$

Next, we obtain Tarski's main theorem. Defining a set Ψ as a set of axioms of $\mathfrak{L}^\times$, $\mathfrak{F}$ as the set of formulae of FOL, X as the equations of $\mathfrak{L}^\times$, and K as a translation function, we have

Theorem. $\exists K\colon \mathfrak{F} \to X$ such that if $\Psi \vdash$ "pairing," then $\Psi \vdash X$ if and only if $K(\Psi) \vdash_x KX$.

Letting Ψ be the axioms of set theory, e.g. **ZF**, the "Main Theorem" means that set theory can be formalized in $\mathfrak{L}^\times$, and in fact that any theory having "pairing" can be formalized in $\mathfrak{L}^\times$. In particular, defining *Q-relation algebras* as algebras whose universe includes some conjugated quasiprojections, and noting that a relation algebra is a *Q*-relation algebra if it contains two pairing elements A and B satisfying the equation Q_{AB}, we see that a set of equations Γ of $\mathfrak{L}^+$ is a *Q*-system if $\Gamma \vdash Q_{AB}$. A system $\mathcal{S}$ of $\mathfrak{L}$ is a *Q-system* if there are formulae D and E of $\mathfrak{L}$ such that D and E contain at most three variables and just two free variables, and where, in every model of $\mathcal{S}$ the two binary relations defined by D and E form a pair of conjugated quasiprojections. Number theory, Peano arithmetic, and real arithmetic are all *Q*-systems, and thus have models in $\mathfrak{L}^\times$.

One interesting application of the Main Theorem is the theorem that relation algebras with quasiprojections are representable. The proof of this result follows directly from the Main Theorem (and in fact is a corollary of the Main Theorem). An algebraic proof of this, slightly different therefore from the metamathematical proof given by Tarski, is due to Maddux (*1978*). Let the Main Theorem be designated **M** and the theorem on representability of relation algebras with quasi-projections be designated **QRRA**. One result in the folklore, that **M** $\equiv$ **QRRA** is occasionally, but probably erroneously, ascribed to McNulty, who did not publish a proof (see *Anellis 1987*: 4), was proved by Tarski (*Tarski and Givant 1987*: 242–244).

Now we define a *cylindric algebra* as a Boolean algebra together with three commuting closure operations. Thus, $\langle \mathfrak{B}, C_0, C_1, C_2 \rangle$ is a cylindric algebra if and only if $(\forall_{ij}C_3)(C_i$ is a closure operation and $C_{ij} = C_j\,C_i)$ and $\mathfrak{B}$ is a Boolean algebra. More generally, the theory of cylindric algebras is just the algebraic theory of FOL. (Similarly, a *polyadic algebra* is a Boolean algebra having α-many variables ($\alpha > 3 \geq \omega$), with α-many commuting closure operations, and with universe of discourse $^\alpha U$.) In this case, Tarski's Main Theorem says that set theory, and thus mathematics, can be constructed in cylindric algebras. It also means that Tarski's theorem (*Tarski and Givant 1987*: 54), which says that the addition of finitely many more logical operations to $\mathfrak{L}^\times$ is still not enough to permit the expression of every FOL-equation, is equivalent to the theorem that the addition of finitely many non-logical constants to cylindric algebra gives only finitely many (new) axioms. This result was strengthened by Németi (*1985*; see also *Anellis 1987b*: 3), to get the result that it is enough to add one constant, the discrete element d, to cylindric algebra in order to obtain finitely many axioms. A proof of Maddux, however, eliminates the need for d. Additionally, Sáin (*1988*) has presented cylindric algebraic versions of the downward Löwenheim-Skolem theorem; this helps to confirm the result presented by Tarski and Givant (*1987*: 54), that by the addition of new logical operations to $\mathfrak{L}^\times$, it is possible to obtain the Löwenheim-Skolem theorem in $\mathfrak{L}^\times$.

The proof of the Main Theorem, that is, the proof of the equipollence of $\mathfrak{L}^+{}_3$ and $\mathfrak{L}^\times$, was merely sketched in *Tarski and Givant 1987*, on the historically interesting ground that it is too "cumbersome" to attempt in full (*1987*: 87). Here is an echo of Peirce's professed ground for failure to provide a proof of the controversial full law of distributivity in his 1880 axiomatization of algebraic logic on the ground that it is "too tedious" (*1880*: 33). The proof of the distributivity law was finally given by Huntington (*1904*: 300–302), who reproduced it from Peirce's 24 December 1903 letter. The presentation of the proof in the letter to Huntington was in turn based upon Peirce's work of 31 January 1902 in his *Logic Notebook* (MS 339: 437; see *Houser 1989b:* xlvii). The proof for the generalization of the Main Theorem (of the equipollence of $\mathfrak{L}^+{}_3$ and $\mathfrak{L}^\times$) as the Main Mapping Theorem of $\mathfrak{L}^+$ to $\mathfrak{L}^\times$ required fourteen pages (*Tarski and Givant 1987*: 110–124). A history of the proof was outlined by Maddux (see *Anellis 1986–1987*: 3; also *Tarski and Givant 1987*: 109), and runs as follows. Tarski had a written "short", somewhat different, mapping theorem sometime around 1976 or 1978, but apparently lost it. The short version, however, was discovered around 1960 by J. D. Monk. D. Pigozzi, who learned of it from Monk, presented it at a universal algebra seminar at University of California at Berkeley; Tarski was either absent at the time, or, less likely, was present but failed to recognized it. In either case, it is not included in *Tarski and Givant 1987*. Tarski's attempts to reconstruct the original theorem were unsuccessful, although several others, including a shorter mapping rediscovered by Maddux, slightly different from the one developed by Monk, were developed. The construction given in *Tarski and Givant 1987*: 110–124 is essentially one discovered in 1974 by Maddux, similar to one discovered, but left unpublished, by Monk. As has been noted elsewhere (*Anellis and Houser 1991*: 6), Tarski considered his work to be directly descended from the work of Peirce; this is clearly evidenced from the brief historical remarks made by Tarski (*1941*: 74). It was also shown (*Anellis and Houser 1991*: 6) that Tarski's historical perception was rather limited. I want now to show that the historical connection between the results presented in *Tarski and Givant 1987* are much deeper than has previously been thought. I will therefore present a detailed discussion of the historical and mathematical connections between the work of C. S. Peirce and A. B. Kempe in the 1880–1890s on relative triples or linear triads as products of two binary relations between logical atoms, R. C. Lyndon's work in the 1950–1960s on cycles in relation algebras, and Tarski and Givant's formalism $\mathfrak{L}_3$ as a three-variable fragment of first-order logic. Indeed, it would appear that Tarski's mathematical progress in developing the system $\mathfrak{L}_3$ closely replicates Peirce's work on relative triples. In fact, the study of proper relation algebras was well under way prior to Tarski's work.

It has already been pointed out that Tarski and Givant (*1987*: xv) held that the mathematics of their work "is rooted in the calculus of relations (or the calculus of relatives, as it is sometimes called) that originated in the work of A. De Morgan, C. S. Peirce, and E. Schröder during the second half of the nineteenth century." This work was initiated in the period 1867–1873. Peirce began investigating laws on operations of relation algebras at least as early as 1870 in "Description of a Notation for the Logic of Relatives, Resulting from an Amplification of the Conceptions of Boole's Calculus of Logic" (W2: 359–429), and in this limited sense may be said to have undertaken an abstract algebraic characterization of relation algebras, although abstract relation algebras have their official roots in Tarski's *1941*. Peirce also considered concepts which led to cylindric algebras in his papers "On the Algebra of Logic" (*1880*), "The Logic of Relatives" (*1883c*), which introduced quantifiers into the algebra of relations, and "On the Algebra of Logic: A Contribution to the Philosophy of Notation" (*1885*). This work reached a particularly crucial point in 1880 with Peirce's formalization of algebraic logic and presentation of Boolean algebra in terms of the algebra of relations based upon the dyadic relation or binary operator ($\prec$), which could be variously interpreted as material implication, a partial ordering, class inclusion, and set elementhood. Thus, as C. Ladd-Franklin (*1892a*: 127–129) points out, while *Schröder 1895* can be understood as a detailed exposition of the subject of algebraic logic and an account of the work of Peirce through the 1880s, it was nevertheless incorrect in the assertion (*Schröder 1895*: 290) that *Peirce 1880* treated terms only as statements, but not as classes. It is equally true, however, that *Peirce 1885* presents, along with much more, a fully axiomatized propositional calculus equivalent to the Tarski-Bernays implicational calculus presented by Łukasiewicz and Tarski (*1930*). By 1885, and indeed possibly as early as 1880, Peirce had been able to provide, through his work of formalization of Boolean algebra in terms of his algebra of relations, interpretations of Boolean algebra as a propositional calculus, as a theory of quantification, and as a class calculus. The claim by Henkin and Monk (*1974*: 105) that the first abstract algebraic theory of a substantial portion of predicate logic is the theory of *relation algebras* initiated in *Tarski 1941,* must be read cautiously, then, noting that Henkin and Monk (*1974*: 106) specifically modify their claim, restricting its scope to relation algebras with a small finite number of equations holding for all relation set algebras, thus to relation algebras as abstractions of the systems presented by Peirce and Schröder.

In "A Brief Description of the Algebra of Relations" (*1882a*), Peirce studied dyadic and triadic relations. He provided a geometric interpretation, based upon matrices, according to which dyads are ordered pairs arrayed in squares or blocks, and triplets are ordered triples arrayed in cubes. It is argued there that dual relations can therefore be shown to be equivalent to absolute terms, and that a dual relation can be regarded as being equivalent to a triple relation, so that relations of tetradic order or greater are expressible in terms of relative products triple relations. In this case, we write the dyadic and triadic relations as sums of all of the atoms of the blocks and cubes respectively. Thus, for example, a dual relative in an n–ary universe is a system of n^2 ordered pairs arranged in an $n \times n$ matrix. Letting A be an atom of the system and letting $A{:}B$ be an individual dual relative, Peirce obtained

$$A = A{:}A + A{:}B + A{:}C + \ldots + A{:}N$$

and

$$A{:}B = A{:}B{:}A + A{:}B{:}B + A{:}B{:}C + \ldots + A{:}B{:}N$$

to show how dual relatives may be rewritten as triple relatives. The arithmetic of such systems is well known, and left to the reader. A modern presentation of the details of the arithmetic of this system, particularly as developed in *Peirce 1882a*, is presented by Copi (*Copilowish 1948*). Indeed, Copi (in a lengthy historical discussion in *Copilowish 1948*: 193–196) stresses the roots of his work in the work of Cayley, but more particularly in *Peirce 1870*, and sets himself the task of developing a modern matrix logic for the algebra of relations using the notation of *Principia* rather than the Peirce-Schröder notation. He notes (*Copilowish 1948*: 194) that Peirce's object appeared to have been to introduce matrices "partly as an aid in his classification of relations, and partly for the sake of illustrations or examples," citing in particular *Peirce 1882a* and MS 539 to support his conjecture concerning Peirce's purposes. The 2-dimensional case, that is, the modern matrix-theoretic presentation of Boolean algebras, was given by Parker (*1964*). (Stern [*1988*], who claims incorrectly, albeit unknowingly, to be absolutely original, in fact merely presents the first textbook exposition of matrix logic.) So Peirce (*B. Peirce 1870* [*1881*]; see notes and appendices) had developed a matrix logic in terms of which all of the algebras presented by Benjamin Peirce (as well as those presented by Cayley and Sylvester) can readily be expressed as special cases. Thus, for example, as Houser (*1989b*: lii) notes, Peirce (*1882a*) also argued that Sylvester's universal multiple algebra is just a special case or interpretation of his own logic of relatives (see *Sylvester 1884*, the published version of Sylvester's Johns Hopkins University lectures to which Peirce was responding in *Peirce 1882*). The work trying to show that universal algebras are interpretations of matrix logic presented within the formalization of the calculus of relations was carried out in a series of papers, including "On the Application of Logical Analysis to Multiple Algebra" (W3: 177–179), "Notes on Associative Multiple Algebra" (MS 75), "Linear Associative Algebra[:] Improvement in the Classification of Vids" (W3: 161–163), "Notes on the Fundamentals of Algebra" (W3: 186–187), "Notes on the Fundamentals of Algebra" (W3: 187–188), "Sketch of the Theory of Non-Associative Multiplication" (W3: 198–201), "Note on Grassmann's Calculus of Extension" (W3: 238–239), "Nilpotent Algebras" (MS 97), and "Nilpotent Algebras" (MS 80), as well as in the notes and appendices to *B. Peirce 1870 [1881]*. Indeed, much of the work of both Benjamin and Charles Peirce on linear algebras may well have been inspired and initiated specifically in order to show that the algebra of relations has mathematical applications outside of logic. Associative algebra is included in the theory of matrices (although there is some disagreement about this point; compare, for example, *Brunning 1981*: 96–97, 137–138; and *Lenzen 1973*; also see *Taber 1890* and *Hawkes 1902* for evaluations by Peirce's contemporaries). Peirce's colleague Henry Taber (*1890*: 353) wrote that "Charles Peirce has shown that the whole theory of Linear Associative Algebra is included in the theory of matrices. He has also shown that every linear associative algebra has a relational form. . . . " Some work in this direction was also undertaken by Whitehead (*1901*). Moreover, Copilowish (*1948*) has shown that, just as representable relation algebras are matrix algebras, so every proper relation algebra can be thought of as a Boolean matrix algebra.

A forerunner of *Peirce 1882a* is the ca. 1880 paper, "A Boolian Algebra with One Constant" (W4: 218–221), which presents the binary connective Peirce's arrow, as a precursor of the Sheffer stroke, in a system containing a single logical connective in terms of which all other connectives may be defined.

In the 1890s, C. S. Peirce and A. B. Kempe began a process of axiomatizing geometry on the basis of the algebra of relations, based upon Peirce's work, representing dyadic relations as blocks and triadic relations as cubes. This was followed by Schröder's extensive treatment in *Vorlesungen über die Algebra der Logik* (*1905 [1966]*: 56–592) of Kempe's work in axiomatizing geometry on the basis of the algebra of relations, and especially of Kempe's work on the connection of the calculus of identity (or identity calculus) with the geometry of position. Their work led to one of the first modern presentations, by Mautner (*1946*), of the geometry of symmetric groups in terms of Boolean tensor algebras, according to which logic, and particularly Boolean algebra, is considered to be a theory of invariants. Although there are no direct references to Mautner's work by his successors and only one direct reference by Mautner to Peirce, *Mautner 1946* can be seen as the link between the early work of Peirce, Kempe, Schröder, and Huntington on the logic of relations and invariant theory and the work on Jónsson and Lyndon on representations of relation algebras and projective geometry. Copi (*Copilowish 1948*: 203) helps establish the link between this work of Peirce and Kempe on the one hand and on the other his own and that of Tarski (and thus also Jónsson and Lyndon). The representation of dyadic relation D on atoms A, B in projective geometry has only one possible algebraic expression, as $D = (A{:}B)$; the representation of triadic relation T_2 on atoms A, B, C can be expressed algebraically as either $(A{:}B){:}C$ or as $(A{:}C){:}B$ or as $A{:}(B{:}C)$, and the logical equivalences between these equations can be described using the calculus of relations; similarly for the representation of the tetradic relation T_3 on A, B, C, D, where it is easy to show, however, that some of the resulting equations in the calculus of relations are reducible to other equations in the calculus, although some of the linear combinations presented in T_3 are not translatable into equations of the three-variable calculus of relations (their proofs are left to the reader).

Peirce (MS 589 [*1892*]) argued that all relations could be defined in terms of dyadic and triadic relations, and in particular that all polyadic relations of tetradic order or greater are expressible in terms of relative products of triadic relations. Thus, in particular, letting A be a triple relative with matrix values (i, j, n) and letting B be a triple relative with matrix values (n, k, l), their relative product C is defined as the tetradic relation $(C)_{ijkl} = \Sigma_n \{(A)_{ijn} \times (B)_{nkl}\}$. The geometric representation of $(C)_{ijkl}$ is given as

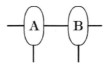

Indeed, Peirce, in "Notes on Kempe's Paper on Mathematical Forms" (MS 714 [1889]/NEM 3: 440n.) took Kempe strongly to task for attempting to "get rid of" triads, effectively arguing that triple relatives are absolutely essential to, as well as sufficient for, the full expressive power of the logic of relations. Much of MS 708: 5–6, 11, 15–19, and NEM 3: 436 [1897], is devoted to showing that despite the differences in detail between Kempe's graphs and Peirce's, Kempe's diagrams, contrary to Kempe's claims, in fact show that tetrads can be rewritten as triads (but not as monads or dyads), and that "accordingly, when we follow out the idea of Mr. Kempe's system, we arrive at the result that each graph consists of *monads*, *dyads*, and *triads*." These remain claims, however, as the "proofs" are graphical, rather than algebraic (see MS 708: 6, 11, and NEM 3: 436, where versions of the following diagrams are presented):

(graph of tetrad)

and

(graph of tetrad rewritten as two connected triads)

In the paper "The Logic of Relatives" (*1897*: 168–183), Peirce shows that logical combinations of connecting "tails" of classes or dyadic or triadic relations yield propositional equations. The "tails" are the "hooks" or "joints" by which logical connectives build the terms of the relations into propositional or class equations. The tails graphically or geometrically describe or represent the logical connections of terms or classes, in a manner analogous to the geometric representations of combinations of dyadic and triadic relations. This technique permitted Peirce to borrow the diagrammatic tools of chemistry to graph logical relations not only between relational terms and classes but also between equations. Peirce's logical graphs, therefore, became a new means to geometrically represent logical implications among propositions, in a more sophisticated and complex manner than could be done by Euler diagrams or Venn diagrams. Moreover, the apparatus is sufficiently sophisticated to permit an analysis or logical parsing of the propositions themselves (although, I suggest, in a much more untidy manner even than that presented by the apparatus of Frege's *Begriffsschrift*). There is justification, however, in Quine's assertion (*1935*: 292–293) that Peirce's attempts to reduce all tetradic relations to triadic relations will fail to yield the desired result, as Mertz (*1979*: 168–174) also showed, correctly anticipating Tarski's proof that the equation expressing the existence of four elements is not expressible in the calculus of relations (*Tarski 1941*: 89) and thereby implicitly if unknowingly alluding to the Löwenheim-Korselt example. The difficulty in the use, whether by Peirce (for example in *1897*) or by present-day Peircean scholars (especially Herzberger [*1981*] and Burch [*1991*]), of such so-called "valency proofs" for the reducibility of tetradic and polyadic relations to products of monadic, dyadic, and triadic relations, which are diagrammatic, is that they have not yet been shown to be algebraically justified.

In a large number of papers, published and unpublished, written during the period from the 1880s through the first half of the first decade of the twentieth century, Peirce worked out the details of the three-variable calculus. Much of the work on triadic logic was apparently intended to provide an algebraic system capable of supporting a trivalent logic with a modal interpretation (see, e.g. *Fisch and Turquette 1966*). Thus, although most writers who concern themselves with Peirce's triadic logic do so from the perspective of a trivalent propositional logic, our concern is strictly and specifically with the algebraic aspect, on the basis of which we are able to note the contributions which Peirce made towards Łukasiewicz and Post algebras as algebras corresponding to 3-valued and *m*–valued logics respectively.

Kempe, to a large extent in interaction with Peirce, worked out his theory of linear triads in a series of papers (*Kempe 1886, 1887, 1889–1890, 1890, 1897*). The most important of these papers was the first, "A Memoir on the Theory of Mathematical Form" (*Kempe 1886*), which Peirce, in "Consequences on Common-Sensism" of *Pragmatism and Pragmaticism,* called a "great memoir . . . the most solid piece of work upon any branch of the stecheology of relations that has ever been done" (5.505). Peirce's copy is heavily annotated, and much of his concern, both in the marginal annotations and in separate notes, regards questions primarily of terminology. The "Memoir" is a lengthy and rather informal discussion of the various ways in which collections of various arities of mathematical atoms (units) can be arrayed and of the graphical or geometrical, as well as the tabular representations of these collections. In particular, it introduces and serves as the basis for Kempe's more formal, axiomatic, presentation of these systems of atoms. The fundamental relation among the collections of undistinguished atoms is the linear triad. A crucial step in defining systems in terms of the triads is the definition of the apparatus of "aspect" by which collections of units, a, b, c, . . . come to be represented as single units $(abc . . .)$, although, if a, b, c, . . . are distinguished from l, m, n, . . . , then $(abc . . .)$ are $(lmn . . .)$ are distinct units. Thus, the pairs which $(abc . . .)$ forms with a, b, c, . . . is different from the pairs which $(abc . . .)$ forms with l, m, n, . . . (*Kempe 1886*: 27). On this basis, Kempe (*1886*: 27) defined the "unified aspect" $(abc . . .)$ as "a discrete pair associate of the collection a, b, c, ... A *projection* is a complete correspondence between two systems" (*Kempe 1886*: 29), and in modern terminology may therefore be understood as the surjective homomorphism of the cartesian product $(S_1 \times S_2 \times . . . \times S_n)$ of the systems S_1, S_2, . . . , S_n onto some arbitrary coordinate S_i $(1 \leq i \leq n)$ of that product. In general, Kempe can be said in the "Memoir" to be searching for and describing the combinatorial laws of algebra common to geometry, graph theory, group theory, and linear and multilinear algebras. Algebras possessing one or more of these features are called *primitive algebras* (*Kempe 1886*: 53). These laws in particular are commutation, association, distribution, idempotence, and the rule that $(ab) \neq a$ and $(ab) \neq b$. ("Ordinary logic," meaning Boolean algebra, is singled out as the primitive algebra with idempotence (*Kempe 1886*: 53).) Triads are then introduced as composed of three units which give rise to primitive equations, that is, as equations expressing the laws characterizing the primitive algebras. The triad (abc) can be interpreted as the pair (ab) of undistinguished atoms with their product c (*Kempe 1886*: 54). A *quadrate algebra* is an algebra in which, for any two systems S_1 and S_2, all combinations yield cartesian products and each product is itself an ordered pair taken as a unit, rather than a simple unit. The remainder of the "Memoir" is devoted to the mathematics of triads applied to geometry and to algebraic logic. The next paper, "Note to a 'Memoir on the Theory of Mathematical Form' " (*Kempe 1887*), introduces some technical corrections (largely concerning the differences between n–ads which are distinguished, where, for example, $(n_1 n_2 \ldots n_n) \neq (n_n \ldots n_2 n_1)$ and those which are undistinguished) to the "Memoir," in response to criticisms in a letter of 17 January 1887 from Peirce on the matter. *Kempe 1897* is likewise an informal response in part to this same letter of Peirce's, as well as to similar comments in "The Logic of Relatives" (*Peirce 1897*: 168ff). MS 708/NEM 3: 431–440, which is undated but probably also written around 1897, is a reply to Kempe's reply in his 1897 note in *The Monist* to Peirce's *1897*. It deals primarily with the nature of the graphs developed by Kempe to represent the n–ads. The detailed and formal work of clarifying the characteristics of distinguished and undistinguished n–ads is carried out in a rigorous and axiomatic manner, along with matrix-theoretic algebraic proofs, in the paper "On the Relation between the Logical Theory of Classes and the Geometrical Theory of Points" (*Kempe 1889–1890*). The apparatus hinges upon the fully developed concept of a *base system*, that is a system of (distinguished and undistinguished) linear triads whose form is fully determined by the logico-algebraic laws (*Kempe 1889–1890*: 149):

(K1) If $ap \cdot b$ and $cp \cdot d$ exist, then there is a q such that $ad \cdot q$ and $bc \cdot q$

(K2) If $ab \cdot p$ and $cp \cdot d$ exist, then there is a q such that $aq \cdot d$ and $bc \cdot q$

(K3) If $ab \cdot c$ and $a = b$, then $c = a = b$

(K4) If $a = b$, then $ac \cdot b$ and $bc \cdot a$ for any c

(K5) (Continuity): No entity is absent from the system which can consistently be present in the system

It follows that there should be entities in the system capable of forming linear triads with any two given entities of the system, that is, that, given any three entities *a*, *b*, *c* of the system, there exists an entity *x* in the system, such that $ab \cdot x$, $bc \cdot x$, $ca \cdot x$, . . . all exist.

The remainder of *Kempe 1889–1890* is devoted to showing that the algebraic theory of form has an interpretation in the class calculus and in geometry, and presenting that interpretation. G. J. Stokes (*1900*: 7), no doubt having *Kempe 1889–1890* specifically in mind, summarized Kempe's work as showing "that between the mathematical theory of points and the logical theory of statements, a striking correspondence exists. Between the laws defining the form of a system of points, and those defining the form of a system of statements, perfect sameness exists with one exception," namely that the system of points includes the parallel postulate. Peirce (MS 708: 15) argued, however, that the fundamental property of plane projective geometry presented by Kempe in his "Memoir" (*1886*) was simply too indefinite to be of any use. R. M. Robinson (*1959*: 68) noted that Pieri showed that a ternary relation—that of a point being equidistant from two other points—can serve as the only primitive notion of Euclidian geometries of dimension two or greater. Robinson went on to ask whether one or more binary relations could serve as primitives for some geometries, and gave several examples from elliptic geometries (e.g. the binary relation $AB = \pi/2$). He showed that there are no such binary primitives for Euclidian or hyperbolic geometry. Robinson's work in this area has much in common with the geometric aspect of Kempe's work, as well as with the work of Lyndon (*1961*).

For his part, Peirce, in his notes on *Kempe 1889–1890* (MS 1584), is concerned primarily to summarize Kempe's work and examine some of the possible permutations on linear triads. Many of Peirce's attempts to understand Kempe's work are found in a series of undated manuscripts, including "Note on Kempe's Paper in Vol. XXI of the *Proceedings of the London Mathematical Society*" (MS 709), "Notes on Kempe's Paper" (MS 710: 2–8), "Notes on Kempe's Paper" (MSS 711: 2–5, 712: 2, 712s: 2–3, 713: 2–3), and "Kempe Translated into English" (MS 715: 2), as well as "Notes on Kempe's Paper on Mathematical Form" (MS 714), "Reply to Mr. Kempe" (MS 708: 2–19), and "Notes on A. B. Kempe, 'On the Relation Between the Logical Theory of Classes and the Geometrical Theory of Point' " (MS 1584: 17–24), and his notes in his copy of *Kempe 1886*.

In his next paper, "The Subject Matter of Exact Thought" (*1890*), Kempe expands upon the axiom (K5) of continuity, according to which no entity is absent from the system which can consistently be present in the system, thus introducing what is apparently a bivalent truth-theoretic aspect for equations of his system; literally "truisms" and "falsisms" are understood to be equations having the same algebraic value as other equations of the system. In fact, however, as Stokes (*1900*: 7) pointed out, Kempe was primarily concerned here with the consistency of his system, and "truism" and "falsism" turn out to be the disguised laws of identity and contradiction respectively.

Peirce expressed particular concern for what he considered Kempe's careless use of the difference between *distinguished* and *undistinguished* terms (e.g. in MSS 709–711). In MS 709: 3–4, Peirce attempted to help Kempe out by supposing that he had meant that the claim that $ab \cdot c$ and $ba \cdot c$ are indistinguishable merely asserts that a and b have exactly the same relationship to the system, not that $a = b$. If in fact Kempe had meant that $ab \cdot c$ and $ba \cdot c$ are indistinguishable means that $a = b$, then, said Peirce, c must fail to exist; however, (K5) asserts that c does exist. (K5) "would seem to imply that there always is such an 'entity' " (MS 709: 4), although then it must be asked (MS 709: 3): "But what has all this to do with 'the distribution of linear triads through the system,' which is all these laws [(K1)–(K5)] profess to define?" Kempe's problem in this matter is traced to a problem with his misuse of commutativity, and to his failure to consider the construction of ordered pairs and triples (see, for example, MS 708: 6–8; 12–13). Some additional confusions in the rules [(K1)–(K4)] defining the distribution of linear triads through a system are noted (MS 713); in particular, a contradiction occurs with respect to the meaning of $ab \cdot c$, which, by (K4), asserts that $c = a$ or a function of $a \cdot b$, though clearly $a \cdot b \neq a$. Similar difficulties arise with respect to applications of (K2) and (K3) by the definition of $ab \cdot c$ (MS 713: 3).

For Peirce, the concept of the triple relative or triad is basic, since, according to Peirce (in the 1882 paper "Brief Description of the Algebra of Relatives" (W4: 328–333)), as we have already seen, dyadic relations can be expressed in terms of triples, while tetradic relations and higher-degree polyadic relations can be reduced to triples. In his unpublished and undated "Brief Account of the Principles of the Logic of Relative Terms" (MS 531), which I judge to be one of Peirce's earlier attempts to account for triple relatives, Peirce (MS 531: 2) defined a *conjugative term* as a term which represents its object "as a medium or third which brings two (or more) things into relation." This definition is rather ambiguous, since it is not completely clear here, at least in the expression of the definition, whether the third term is the product of a pair of terms (that is, with $\langle A, B, C \rangle$ defined as $(A{:}B) = C$), or a distinct term in a triple which is not definable in terms of a dyadic relative taken as an atom together with a monad or atom. The rules which Peirce presents, and in particular his appeal to the transitivity principle for relative triples, however, seems to me to suggest that he is already thinking of relative triples as products of two binary relations between logical atoms (that is, where, if $(A{:}B) = C$ and $(X{:}Y) = Z$, then $\langle C, Z, T \rangle$ is just $(C{:}Z) = T$), albeit not yet very clearly. The more interesting aspect of the *Brief Account* manuscript, however, is that it provides a formal proof of transitivity rather than assuming transitivity as an axiom. Indirectly, this can be interpreted as a proof that transitivity is expressible within not less than a three-variable calculus.

An equally informal statement appears in the manuscript "Simplification for Dual Relatives" (MS 419: 2–3) arguing that quantified terms of relational equations can be reduced to quantifier-free terms which appear as dual relatives, and that nonrelative terms in such equations can be rewritten as dual relatives by the simple expedient of taking such terms to express identities, so that any nonrelative term N is rewritten as the dual relative $(N{:}N)$, and that polyadic relatives ("plural relatives") "can be decomposed," following a device developed by Kempe (*1886*), "into dual relatives as follows":

> Take A gives B to C. Call this individual act of giving D. Then the triples relation is compounded of these three dual relations, A is the agent of D, B is the direct object to D, C is the indirect object of D.

Peirce then presents the mathematics intended to show how this is to be done, with special attention to quantified relations. Simplifying the mathematics that Peirce gave to show how all dyadic and all polyadic relations can be rewritten as triples of dyadic relations, we have in effect the equation

$$(A{:}B){:}C = A{:}(B{:}C) \equiv (A{:}D){:}(D{:}B){:}(C{:}D)$$

where

$$\{[(A{:}D){:}(D{:}B){:}(D{:}C)] = [A{:}(D{:}D){:}(D{:}D){:}B]{:}(D{:}C)] =$$
$$[A{:}(D{:}D){:}(D{:}B){:}D]{:}(D{:}C) = \{[(A{:}D){:}D]{:}[(D{:}B){:}(D{:}C)]\}\}$$

and each (X, Y) is commutative, for any X and any Y. (As we shall see, a careful working out of a "proof," or more accurately a sketch for an actual proof of the reduction of polyadic equations to combinations of triadic, dyadic, and monadic relations, is presented in MSS 543 and 544: 33–34 (see also NEM 3: 832–833). The inverse of this technique can of course also be used to show how polyadic relations can be reduced to triadic relations. The techniques for quantifier elimination are fairly familiar, appearing as a version of the techniques found in *Löwenheim 1915* and taken directly from Schröder, where indices of quantifiers are abbreviations for multiple quantifiers, and quantifiers are eliminated by interpreting their indexed terms as relative products and relative sums.

The unpublished manuscript "The Logic of Relatives" (MS 534), which begins with a very brief mention of De Morgan's and Kempe's contributions to the logic of relations, is obviously a more mature work, which must have been written no earlier than 1887. It uses the classifications of relations presented in the manuscript "On the Formal Classification of Relations"

(MS 533). In the classification there are various ways in which terms of relations permute within the relation—in terms of the identity relation (A:A), for various types of combinations between dyadic, triadic, polyadic finite relations, and among infinite relations. (This paper was probably Peirce's first, tentative, step in a series of papers written in the decade of the late 1890s to about 1905, and designed to classify relations by both their arity and by the permutations of their terms.) Much of MS 534 is concerned with the classification of dual relatives, according to which there are two principal types, identity, having the form (A:A) and that having the form (A:B), where $A \neq B$. Another manuscript entitled "The Logic of Relatives" (MS 547), which also begins with a brief mention of contributions of De Morgan, Kempe, as well as some results attributed to Peirce's Johns Hopkins student O. H. Mitchell, and therefore probably dates from the late 1880s to late 1890s, is a detailed analysis of deductive rules for the calculus of relations, extending deductive rules of propositional calculus, and thus utilizing technical results from his published papers of the period 1880–1897, including the application of quantifiers. It does not, however, deal specifically with the question of the adequacy of a dyadic relation as compared with that of a triadic relation.

The manuscript "Dual Relatives" of 1889 (MS 536) may serve as a crucial link between MSS 531 and 419. In "Dual Relatives," Peirce makes a step in clarifying his earlier definition, given in "A Brief Account of the Principles of the Logic of Relative Terms" (MS 531), of a *conjugative term* as a term which represents its object "as a medium or third which brings two (or more) things into relation." In "Dual Relatives," it was argued (MS 536: 2–5) that a pair of atoms tied into a collection is itself a new, third, "higher-order" atom, itself connected to no other atoms than those which comprise the pair taken as a single atom, and according to which the resultant of a pairing of atoms (the pairing of two "second-order" atoms), both of which were themselves resultants of a pairing of ("first-order" or simple) atoms, is itself a "third-order" atom. This provides clear evidence that Peirce was thinking of relative triples as products of two binary relations between atoms. Moreover, this hierarchical structuring of dyadic relations allows propositions to be taken as terms of dyadic relations, with their connecting "third" defined in terms of the truth values, according to whether the logical connection between the propositions is conjunction or disjunction. Thus, the truth-value of the "third" in the relation is determined by the truth values of the two propositions which it connects. In effect, Peirce (MS 536: 4–6) has created a truth-value semantic for the calculus of relations and has simultaneously permitted his atoms to be represented as truth-values. Much of MS 536 (7–18) is another endeavor to classify dual relatives by both their arity and by the permutations of their terms. The manuscripts "Division and Nomenclature of Dyadic Relations" (MS 538), "Nomenclature and Divisions of Dyadic Relations" (MS 539), and "Class of Dyadic Relations" (MS 542) are principally concerned with this endeavor, along with a study of the laws which apply to the various permutations, including symmetry and transitivity, and are consequently largely terminological. A similar endeavor was at least begun for triadic relations in the manuscript "Nomenclature and Division of Triadic Relations" (MS 540). "A Proposed Logical Notation," summarizing the combinatory results, asserted (MS 530: 109) that there are 2^3-many ways to connect a dyad with any one triad, and 3^2-many ways of connecting a triad with any one dyad. It becomes necessary, however, to distinguish the relative product obtained from counting the permutations involved in connecting a dyad with a triad from the case in which there are n–many terms transitively connected in a chain. The number of combinations of the dyad obtained by composition, or nonrelative multiplication, involved in a connection with only the first term of an n-ad whose individual terms are dyads for the second term of the n-ad, whose terms are triads of the n-ad, . . . which in turn are $(n–1)$-ads of the $(n–1)^{th}$ term of the n-ad, whose terms are n-ads of the n-ad, that is, until the n-ad is exhausted, is by itself only 2^n (compare MS 530: 111). A relation is called *cyclic* (MS 536: 18) if is consists of a (finite) chain or "line" of terms such that, if the there are n-many terms $t_1, t_2, \ldots, t_n$ in the chain, there is an ith and a jth term $(1 \leq i < j \leq n)$, then there is a subchain $t_1 \prec \ldots \prec t_j \prec t_i \ldots$; that is, "a *cyclic* . . . relation is one of which some line is inexhaustible by returning into itself." It seems clear that Peirce had in mind an analogy with cycles in group theory. In "Nomenclature and Divisions . . . ," Peirce expanded upon the definition of *cycle*. After defining the *seed* of an existential relation as an existential individual (that is, an atom which is neither an empty sign nor a logical law) which has a reciprocal relation to some other existential atom, which bears the same relation to another atom as that atom bears to it, and the *spike* as a collection of seeds (MS 539: 5), Peirce refined his earlier (MS 536: 18) definition of cycle, albeit less elegantly, by defining a *cyclic spike* (MS 539: 22) as a spike in which every atom individually stands in the same relation to one term of the spike as the spike itself stands to that particular term, and in the converse relation to only one atom of the spike. Cyclic spikes may be either singletary (single spikes), or finite, or infinite, although infinite spikes may be regarded as acyclic. An example of infinite spikes which may be regarded as acyclic include series of the type

$$\sum_{i=-\infty}^{\infty} a^i, \quad i \in Z$$

where each successive a^i in the series is defined as a successor of a^{i-1} by a successor function such as '$\prec$'. An open question for Peirce is whether such series, having left-sided limits of zero and right-sided limits of infinity, can properly be regarded as cyclic (MS 539: 22), since it is not obvious that there are any j^{th} and k^{th} terms in the series ($0 \le j < k < \infty$), defining a subchain $t_j \prec \ldots \prec t_k \prec t_j \ldots$ of the series (although there is a promise to return to the question in the course of lectures on the subject). The next section (MS 539: 24–28) is another summary of Kempe's "Memoir," the major theme here being an expression of a preference for Schröder's terminology and the Grassmannian matrix-theoretic notation over Kempe's terminology and tabular arrangement.

Over all, in his reactions to Kempe's work Peirce was concerned with a discussion of some technical considerations concerning the problem of symmetry and asymmetry of combinations of relations. This was reflected in some of Peirce's notes on Kempe's papers, in his criticisms of Kempe, both published and private. Thus, in his letter to William James of August 1905 (L 224: 40–76), the published version of which appears in NEM 3: 809–835, Peirce referred to an error by Royce which mimics an error found in Kempe's "Memoir." In particular, Peirce was responding to claims by Royce (*1905* [*1951*]: 388) that "the principles of logic can be developed solely in terms of a polyadic symmetrical relation" (NEM 3: 821). But as Peirce noted (NEM 3: 821), the most fundamental relations of the calculus of relations are asymmetrical, including, for example that between antecedent and consequent (MS 708: 16–17) and that of $A{:}B + B{:}C$ and $A{:}C + C{:}A$, whose product $B{:}C$ is unsymmetrical (NEM 3: 821). In fact, since Kempe's rules (K3) and (K4) define "betweenness" of triads, and are therefore closer to Peirce's own connective '$\prec$', Peirce's criticism was correct to the extent that it argued against the purported symmetricality of Royce's fundamental relation, but misdirected to the extent that he felt it necessary to prove that the fundamental relations for such systems must be asymmetrical. The "proof" (NEM 3: 821–822) involves the argument that the combination of two unsymmetrical relations will yield a symmetrical relation, whereas the combination of two symmetrical relations will never yield an unsymmetrical relation. As Peirce remarked (NEM 3: 822–823), Royce's argument closely resembles Kempe's assertion in the "Memoir" (1886: 13) that, given an unsymmetrical relation, its graphical representation can be presented by simple (symmetrical) connecting "links," *provided additional terms (atoms) are introduced* into the graph. While Peirce disagreed with Kempe on the nature of the necessity or absence of additional links between asymmetrical pairs (see NEM 3: 435, 822–823, and Peirce's marginal annotations in his copy of *Kempe 1886*: 13), he agreed with Kempe, and so disagreed with Royce, that the relation of an unsymmetrical pair is unsymmetrical. Royce's error, said Peirce (NEM 3: 823) resulted from an unfa-

miliarity with matrix multiplication: "the relation expressed by relative multiplication . . . is asymmetric. That is to say, it is asymmetric unless the relations multiplied are converses of each other." At NEM 3: 824, Peirce recalled that in either *1883c*—which he called "Note B of 1882"—or in *1870*
he "expressly mentioned and extensively applied the fact that a dyadic relation could be regarded as a triadic relation by formally adding a correlate
common to all dyadic relations . . . ," although "it is not pretended, and is
not true . . . that a dyadic relation is a triadic relation." He added (NEM 3:
824) that, although a class-name is not a relative term, if one considered
an individual atom of a class of atoms, where the property of the class was
constant while some aspect of that property might vary for each member
of the class, then the atom could be interpreted as a relative term; that is,
"you convert non-relative terms into relative terms" by considering a variable property of the non-relative terms. Similarly, a relative term is interpretable as non-relative by applying a choice function to the correlate of
the term and interpreting the relative term as a "field" or representative of
a property for an entire class-name.

The heart of Peirce's case was made in reply to Royce's claim that there
is no real difference between dyadic and triadic relations. Peirce (NEM 3:
825) responded by presenting and defending the propositions

(P1)   *No dyadic relation can be composed of non-relative factors alone*
(P2)   *No triadic relation can be composed of dyadic and monadic factors alone*
(P3)   *Every tetradic and higher relation can be composed of monadic, dyadic,
       and triadic factors.*

The proof of (P1) consists simply of the argument (NEM 3: 825–826)
that monads can produce dyads in one of two ways, either a single monad
splits into two monads, in which case it was a dyad from the outset, or two
monads combine. If three monads are combined and then one subtracted,
so that the result is a dyad, then one has shown only that "twoness" is inseparable from "thirdness." Finally, an irreducibly symmetrical relation or
"degenerate dyad," such as (*A:A*), is just a monad treated as though it were
a dyad. Furthermore, nothing (except a degenerate dyad) can be in relation
to itself, although a whole may be in relation to its parts. But a collection
and its members are different entities. □

The proof of (P2), it is claimed (NEM 3: 826), might run the same way.
But instead, an alternative, informal, "proof," using function-theoretical
concepts, is offered.

It is argued (NEM 3: 826) that a unary function stands in a dyadic
relation to its variable. And while it is true that fixing the value of the variable transforms the function into a constant, that does not entail that functionality is just the same as the value (of the variable), since, if it did entail

that functionality is the value, that would contradict (P1). (We know of course that fixing the value of the variable in $f(x)$ does not transform f into a constant, but that $f(a)$ identifies a constant; but this does not affect Peirce's argument.) In the case of a binary function such as $x + y\sqrt{-1}$ it is argued (NEM 3: 826) that if either variable were fixed, with either $x = a$ or $y = b$, then the result would be a unary function, either $a + y\sqrt{-1}$ or $x + b\sqrt{-1}$. But "that does not prove that the *combining* power" of a binary function can be made into a unary function. Thus, given $a + b\sqrt{-1}$, one may first substitute either $a + y\sqrt{-1}$ or $x + b\sqrt{-1}$ and then $x + y\sqrt{-1}$, freeing first one variable, then the other.

Peirce comments (NEM 3: 826) that it is only in respect to the nature of the substitutions, taken from a nontechnical perspective, that he can understand how Royce might have concluded that the difference between unary and binary functions, or between dyadic and triadic relations, is "superficial."

Next (NEM 3: 826) it is argued as "self-evident" that binary functions or triadic relations combine the effects of different variables, whereas unary functions or functions of a (single) function cannot combine the effects of different variables. But neither can a dyadic relation. (Here, Peirce must clearly have meant that a dyadic relation cannot combine the effects of monadic *or of triadic functions* alone. Otherwise, the proof by counterexample that he presents must fail.)

The set-theoretic counterexample which is then offered (NEM 3: 826), and which Peirce believed would be the strongest possible argument in favor of Royce's claim that there is no real difference between dyadic and triadic relations, relies upon Cantor's diagonal argument. Any positive rational number $^a/_b$ is a function of two positive integers a and b. Royce would claim, according to Peirce, that all elements of $\mathbb{Q}$ are functions of a single variable. We know, however, that $\mathbb{Q} \cong \mathbb{Z}$. But Peirce (NEM 3: 826–830) applies a geometric variant to Cantor's diagonal argument to show that, although $\mathbb{Q}^+ \cong \mathbb{Z}$, the integers are ordered and definable in terms of ordinal numbers, we nevertheless cannot "reduce the values of fractions to their ordinal place" in the mapping from the integers to the rationals. This argument, Peirce contended, would fail to show that rational values are functions of a single variable; on the contrary, the proof shows that positive rational fractions are functions of two positive integers. Finally, a part of the proof states that the ordering of the rationals requires three terms for its completion, namely, that, for any a, b, c, $d \in \mathbb{Z}^+$, and for $^a/_c$, $^b/_d$. This part of the proof is clearly required in order to make the case for the difference between dyadic and triadic relations; for with it Peirce has shown that ordering among rationals is a triadic relation, while a dyadic successor relation is enough to order the integers, and, without it, he would have shown only that a binary function is sufficient to order and define the rationals, since the piece of the argument that was implicit, but otherwise missing, was that the binary function $f(x, y) = z$ expresses a triadic relation $R(x, y, z)$. □

Before Peirce began his proof of (P3), the reduction conjecture, which was the centerpiece of his theory of the algebra of triadic relations, he eliminated possible misinterpretations that might arise (NEM 3: 829–832). He also suggested (NEM 3: 832) that (P3) could be considered to be axiomatic, rather than a theorem.

To substantiate the "self-evidence" of his proposition, that "in all cases a proposition concerning a set of objects is equivalent to a special proposition that something combines a truth concerning an object described in terms of any part of that set and an object described in terms of the rest of the set" (NEM 3: 832), Peirce offered several examples, taken from the proof-theoretic nature of geometry where, for example, by combining two propositions, a third, new, proposition can be derived. This shows that the combination of two terms creates a triadic relation.

Peirce frequently asserted (P3) (see *Herzberger 1981*: 56–58 for a list of these claims), and frequently alluded to a proof supposed actually to have been carried out. Nevertheless, the one that appears in (NEM 3: 832–833) has so far gone apparently undetected (leaving aside momentarily the question of whether it is correct).

The proof itself (NEM 3: 832–833) is by contradiction. It is, in effect, a detailed account of the work merely suggested in "Simplification for Dual Relatives" (MS 419: 2–3).

Suppose there is a tetradic relation R, and let A, B, C, D be the four elements which, taken in the order presented, are members of R. Let R be such that the equation $\mathscr{E}_R$ fails to imply that A, B, C, D are in relation R in any order. Then there are four distinct places in the relation (call them α, β, γ, δ). We interpret $\mathscr{E}_R$ then as

$$\mathscr{E}_{R^4} = (A : \mapsto \alpha \wedge B : \mapsto \beta \wedge C : \mapsto \gamma \wedge D : \mapsto \delta) = (\alpha A \wedge \beta B \wedge \gamma C \wedge \delta D)$$

where, for any term X and any position χ, $\mapsto$ assigns the relational term X to the relational position χ. In that case χX is in a triadic relation to χ and X.

Now let $\mathscr{E}_{R^1}$ be the subformula of $\mathscr{E}_R$ according to which the relation $\alpha A \wedge \beta U \wedge \gamma V \wedge \delta W$ is true, where U, V, and W are variables; let $\mathscr{E}_{R^2}$ be the subformula of $\mathscr{E}_R$ according to which the relation $\alpha A \wedge \beta B \wedge \gamma V \wedge \delta W$ is true, and V and W are variables; and let $\mathscr{E}_{R^3}$ be the subformula of $\mathscr{E}_R$ according to which the relation $\alpha A \wedge \beta U \wedge \gamma C \wedge \delta W$ is true, and W is a variable. In this case, $\mathscr{E}_{R^1} \subset \mathscr{E}_{R^2} \subset \mathscr{E}_{R^3} \subset \mathscr{E}_{R^4}$, and each antecedent $\mathscr{E}_R{}^{n-1}$ is less well-defined than its consequent $\mathscr{E}_{R^n}$. If the relation R between A, B, C, D defined by the order $\mathscr{E}_{R^1} \subset \mathscr{E}_{R^2} \subset \mathscr{E}_{R^3} \subset \mathscr{E}_{R^4}$ is equivalent to another relation R' which differs from R only in the order of the $\mathscr{E}_{R^i}$, then this merely adds a new relation to the one given as $\mathscr{E}_{R^1} \subset \mathscr{E}_{R^2} \subset \mathscr{E}_{R^3} \subset \mathscr{E}_{R^4}$ by R. The argument is that this shows that a tetradic relation can be reduced to a triadic relation. Peirce claims (NEM 3: 833) that this conclusion is now "self-evident," and adds that it also holds, *mutatis mutandis,* that all polyadic relations greater than the tetradic can likewise be reduced to some combination of triadic, dyadic, and monadic relations.

Peirce's proof is at this point unclear, since we cannot determine whether Peirce meant a new relation $R^+ = (R:R')$ or a new relation $(\mathscr{E}_{R^1} \subset \mathscr{E}_{R^2} \subset \mathscr{E}_{R^3} \subset \mathscr{E}_{R^4}):R'$ or a new relation $\mathscr{E}_{R^1} \subset \mathscr{E}_{R^2} \subset \mathscr{E}_{R^3} \subset \mathscr{E}_{R^4} \subset \mathscr{E}_{R^i} \subset \mathscr{E}_{R^j} \subset \mathscr{E}_{R^k} \subset \mathscr{E}_{R^\lambda}$ (where each of the equations $\mathscr{E}_{R^i}, \mathscr{E}_{R^j}, \mathscr{E}_{R^k}, \mathscr{E}_{R^\lambda}$ expresses a relation in which each of the four terms A, B, C, D are uniquely mapped to one of the positions α, β, γ, δ by some other mapping function than $\mapsto$ such that, if $\lrcorner$ is the mapping function for R', $x \lrcorner Y$ (not necessarily $X = Y$, $x = y$) for each x, y of α, β, γ, δ and each X, Y of A, B, C, D, or whether all of these alternatives are equivalent.

Suppose, then, that R' is the relation such that $\alpha \lrcorner B$, $\beta \lrcorner C$, $\gamma \lrcorner D$, and $\delta \lrcorner A$. If $R = R'$, then clearly

$$(\alpha A \wedge \beta B \wedge \gamma C \wedge \delta D) = (\alpha B \wedge \beta C \wedge \gamma D \wedge \delta A)$$

In that case, we may understand Peirce to interpret the tetradic relations

$$((\alpha{:}A){:}(\alpha{:}B)){:}(\alpha A{:}\alpha B), \qquad ((\beta{:}B){:}(\beta{:}C)){:}(\beta B{:}\beta C),$$
$$((\gamma{:}C){:}(\gamma{:}D)){:}(\gamma C{:}\gamma D), \qquad ((\delta{:}D){:}(\delta{:}A)){:}(\delta D{:}\delta A)$$

as yielding the triadic relations $(\alpha{:}A){:}\alpha A$, $(\alpha{:}B){:}\alpha B$, $(\beta{:}B){:}\beta B$, $(\beta{:}C){:}\beta C$, $(\gamma{:}C){:}\gamma C$, $(\gamma{:}D){:}\gamma D$, $(\delta{:}D){:}\delta D$, and $(\delta{:}A){:}\delta A$, to obtain the equivalences $\alpha A = \alpha B$, $\beta B = \beta C$, $\gamma C = \gamma D$, and $\delta D = \delta A$, such that $((\alpha{:}A){:}(\alpha{:}B)){:}(\alpha A{:}\alpha B)$, $((\beta{:}B){:}(\beta{:}C)){:}(\beta B{:}\beta C)$, $((\gamma{:}C){:}(\gamma{:}D)){:}(\gamma C{:}\gamma D)$, and $((\delta{:}D){:}(\delta{:}A)){:}(\delta D{:}\delta A)$ are all expressible as binary relations between triadic relations (but not, as Peirce suspects Royce would conclude, as binary relations between dyadic relations). $\square$

There remain some fundamental questions, however. There remains in particular the question of the correctness of this constructive proof of the reduction theorem, which Peirce himself, as noted, considered to be axiomatic, and "self-evident," rather than a theorem. The answer to this question is at least partially tied to Peirce's disagreement with Kempe as to whether triads $ab{\cdot}c$ and $ba{\cdot}c$ are distinguishable or indistinguishable, and, if indistinguishable, whether a and b are identical or simply stand in the same relationship to the entire system. Unfortunately, Peirce did not address this question in his presentation of the proof. Even so, it is not clear how indistinguishability in Peirce's sense justifies reduction of tetradic relations to triadic relations, unless there is a hidden identity lurking in the systematic equivalences $\alpha A = \alpha B$, $\beta B = \beta C$, $\gamma C = \gamma D$, and $\delta D = \delta A$. The part of the proof that would respond to this worry is completely missing, although it is reflected in Peirce's notes and papers on Kempe. The "proof" of (P3) is precisely the heart of Peirce's argument. Even if Peirce's relation algebra is fundamentally equivalent to the one studied by Tarski, however, can we agree that Peirce's results are not necessarily fully invalidated by the counterexamples presented by Korselt, Löwenheim, Tarski, and others of irreducible four-variable statements, since they, Tarski and Tarski's colleagues, are concerned with the expressive power of equations, and not with the irreducibility of the relations which those statements express. Nevertheless, we can reiterate that there is some justification in Quine's assertion (*1935*: 292–293) that Peirce's attempts to reduce all tetradic relations to triadic relations will fail to yield the desired result, in view of the proof presented at (*Tarski 1941*: 89) that the sentence expressing the existence of four elements is not expressible in the calculus of relations. The addition of the proviso is warranted, lest there be misunderstanding, that the work of Tarski and his colleagues indicates, not that polyadic relations are irreducible, but that equations about an important class of equations concerning polyadic relations are not translatable into equations of three-variable fragments of first-order logic.

While Peirce's arguments in the letter to James (NEM 3: 809–835/L 224) are certainly not "proofs" in the sense of Bourbaki, they are Peirce's informal arguments or sketches of proofs of the three propositions, and in particular for his reduction thesis (P3). They would almost certainly, if correct, count as proofs in the nineteenth century style of mathematics. Before dealing with these "proofs" too harshly and from the perspective of hindsight, it ought to be noted that, after all, the formal proof by Tarski and Givant for the generalization of the main theorem (of the equipollence of $\mathfrak{L}^+{}_3$ and $\mathfrak{L}^\times$) as the Main Mapping Theorem of $\mathfrak{L}^+$ to $\mathfrak{L}^\times$ required fully fourteen pages (*Tarski and Givant 1987*: 110–124). Nevertheless, it is absolutely essential to determine whether Peirce's analysis, as we have sought to understand it, implies that there are some statements about tetradic and polyadic relations which are translatable into an equation in the algebra of relations which is expressible in three variables. If it does, then we may suppose Tarski's Main Mapping Theorem to be related, if not equivalent to (P3), and that Peirce's analysis, therefore, indeed obtains for all equations about tetradic relations requiring four variables in FOL, and in particular for those which Tarski and his colleagues have proven are not translatable into $\mathfrak{L}^+{}_3$. In case such a connection can indeed be established, then the equations presented in *Tarski and Givant 1987* clearly constitute counterexamples to Peirce's "proof," unless it can be shown that Peirce's bonding algebra, as depicted here, is not equivalent to either $\mathfrak{L}^+{}_3$ or to any system

$\mathfrak{L}_3$. Both Herzberger (*1981*), who has presented a formalization of Peirce's bonding algebra, and Burch (*1991*), who has created a system combining the Herzberger formalization of Peirce's bonding algebra with Peirce's graph-theoretical representation of his algebra, have undertaken to show that the reduction conjecture is provable within their respective systems. Nevertheless, it remains to show that either or both of these versions of the Peirce bonding algebra are representable relation algebras. The correctness of Herzberger's and Burch's reduction theorem proofs must yet be determined, even apart from the question of whether Tarski's irreducible sentences count as counterexamples to their proofs.

As has already been noted, very little work was done in algebraic logic for nearly fifty years, in the interval between the publications of Peirce, Schröder, Kempe, and Royce, and the publication of *Tarski 1941.*

Finally, continuing along lines initially set out by Peirce and Kempe, but working directly within the developments set forth by Tarski and his colleagues, and without any evidentally direct contact with the work of Peirce and Kempe, Lyndon (*1950*) developed the concept of cycles as triples of relations which, when subject to certain specified conditions, and together with a full survey of the basis of a relational algebra, permits the complete characterization of that algebra. Lyndon (*1950: 709*–710) defined a *basis B* of a complete relation algebra, or relation algebra which is a complete Boolean algebra, as the set of all minimal elements, where every element is uniquely expressible as a union of minimal elements, and noted that a relation algebra is completely determined by a knowledge of the minimal elements contained in the identity relation of the algebra, that is, its *units*, of the converse of each minimal element, and of the product of each two minimal elements. In this case, a minimal element M is a unit if and only if $MM^{\cup} = M$, for the converse $M^{\cup}$ of M. The complete relation algebra is fully characterized by its basis when the mapping $M \rightarrow M^{\cup}$ of B onto itself is specified and when a complete set of "incidence relations" $M \subset LN$ for triples of elements of B are specified. The axioms characterizing proper relation algebras (**PRA**) can be expressed in terms of basis axioms, as follows:

PRA axioms	Basis axioms
$R\,\mathrm{Id} = R$	
$R^{\cup\cup} = R$ ...	B2. $M^{\cup\cup} = M$
$(RS)^{\cup} = S^{\cup}R^{\cup}$ ...	B3. $(M^{\cup} \subset LN) \rightarrow (M \subset N^{\cup}L^{\cup})$
$(RS \cap T^{\cup} = 0) \rightarrow (ST \cap R^{\cup} = 0)$ ...	B4. $(M^{\cup} \subset LN) \rightarrow (L^{\cup} \subset NM)$
$(RS)\,T = R(ST)$ ...	B5. $((T \subset ML) \,\&\, (U \subset TN)) \rightarrow$
	$(\exists W)((W \subset LN) \,\&\, (U \subset MW))$
$(R \neq 0) \rightarrow (1\,(R1) = 1)$	

(The translation, however, is not easy. Additionally, these same **PRA** axioms characterize relation algebras (**RA**) generally; and the same translation into basis axioms holds.)

It follows from (B2)–(B4) that the conditions

$$M^\cup \subset LN, \; L^\cup \subset NM, \; N^\cup \subset ML, \; M \subset N^\cup L^\cup, \; L \subset M^\cup N^\cup, \; N \subset L^\cup M^\cup$$

are all equivalent. Lyndon (*1950*: 710) then defined a *cycle* as a triple of relations which satisfies these six conditions, and presented (*1950*: 715–716), by cycles, a finite relation algebra which is not isomorphic to any proper relation algebra, and hence is not representable; moreover (*1950*: 717–718), he created an infinite algebra which he uses to prove that the class of all representable relation algebras is not characterizable by algebraic axioms. But his proof had an error in it (see *Lyndon 1956*).

I do not, however, know of any use of cycles, other than that appearing in *Lyndon 1950*, for the analysis of relation algebras. Nor is there any evidence from Lyndon that he had either Peirce's relative triples or Kempe's linear triads in mind when using cycles to determine the characteristics of representable relation algebras. It is difficult, therefore, to assert conclusively that Lyndon's cycles were direct descendants of relative triples or linear triads—although they certainly appear to be related algebraically. There is likewise no published evidence that Lyndon was familiar with Kempe's base algebra when he presented his own work on the basis of complete relation algebras, whatever the similarities, apparent or real. Nevertheless, it appears from the axioms of MS 548: 20–22 that, for Peirce, the product of two minimal elements is always minimal, although we must also add that Peirce's system likewise appears to satisfy Lyndon's definition (*1950*: 709–710) of a *basis*, the basis axioms (B2)–(B4), and the conditions $M^\cup \subset LN, \; L^\cup \subset LN, \; N^\cup \subset ML, \; M \subset N^\cup L^\cup, \; L \subset M^\cup N^\cup,$ and $N \subset L^\cup M^\cup$ (*Lyndon 1950*: 710) required for cycles. (It is fairly easy to check that Peircean cycles satisfy the requirements for being Lyndon cycles.)

Tarski (*1941*) presented two methods for constructing the foundations of the calculus of relations. One used equational logic and yields the theory of **RA**s. The more significant method, from our present perspective, is an extension of FOL which augments FOL with operations on relations and with axioms defining these new operations. This permitted him to prove that every equation in the calculus of relations is equivalent to an FOL-formula which contains only three distinct variables and which does not use any operations on relations. He also proved that any equation which does not use more than three distinct variables can be translated into an equation of the calculus of relations. This is clearly a major step towards the development of Tarski and Givant's formalism $\mathfrak{L}_3$ as a three-variable fragment of first-order logic in *Tarski and Givant 1987*. As a converse to the result that any equation not using more than three distinct variables is translatable into an equation of the calculus of relations, Tarski noted that there are equations of the classical (that is, unquantified) Boole-Schröder calculus which cannot be translated into FOL-formulae. As Dipert (*1984*: 50, 52–53) correctly noted, this result shows only that algebraic logic without quantifiers is less expressive than the Peirce-Schröder calculus, the algebra of relations with quantifiers. It does not, however, show that the Peirce-Schröder algebra is necessarily less expressive than FOL. It is only with *Tarski and Givant 1987*: 54 that we arrive at the result that the addition of finitely many more logical operations to $\mathfrak{L}^\times$ is still not sufficient to guarantee that every FOL-formula can be expressed in $\mathfrak{L}^\times$. But this is a much more specific and precise claim than the informal supposition, attacked by Dipert (*1984*) in reply to Tarski (*1941*: 86–89), that the calculus of relations is incomplete compared with FOL, for example as presented by Frege's *Begriffsschrift* or by Whitehead and Russell's *Principia*. In particular, Tarski (*1941*: 86–89) proved the precise metalogical theorem of Schröder (*1895*: 153), *that every sentence of the calculus of relations can be transformed into an equivalent equation of the form "R = S," and even of the form "T = 1,"* a theory, that is, corresponding to restricted FOL, and then shows (*Tarski 1941*: 87–89) that there are several separate and quite precise issues, specifically: (a) whether every sentence of the calculus of relations which is true in every domain of individuals is derivable from axioms of the extension of FOL which augments FOL with operations on relations and with axioms defining these new operations; (b) whether every model of the axiom system of the calculus of relations is equipollent with a class of binary relations containing relations absolute (Boolean) zero Ø, absolute (Boolean) unit **1**, relative (Peircean) zero Ø', and relative (Peircean) unit **1'**, and closed under all operations of that calculus, that is, whether the relation algebra is representable; (c) whether every equation of the theory of relations expressible in FOL is expressible in the calculus of relations; and (d) whether there is an algorithm for determining which equations about relations are translatable from FOL into the calculus of relations. In *Tarski 1941*, (a), (b), and (d) are open questions, while (c) is shown to have a negative solution, and it is conjectured that the solution to (d) is also negative. *Tarski and Givant 1987* is devoted to understanding, and insofar as possible answering, these open questions. The question of whether every sentence provable in equational logic is provable in the extension of FOL which augments FOL with operations on relations and with axioms defining these new operations is a generalization of (a) which makes it false.

In a related work, McKinsey (*1940*) presented a set of postulates for the calculus of binary relations which permitted him to prove the completeness of the theory in the sense that any two realizations of the postulates having the same (finite or infinite) number of homogeneous atoms are equipollent; the result is also obtained that every realization of the postulates is isomorphic to a complete atomic relational realization.

Some additional results, based on *Tarski 1941* and leading to *Tarski and Givant 1987*, include Lyndon's proofs (*1950*) that there exists a finite algebra satisfying all of the axioms of *Tarski 1941* but which is not isomorphic to a proper relation algebra, and hence is not representable, and the more general but false result that there is no set of finitary axioms sufficient to characterize the class of representable relation algebras; in *Lyndon 1956* the error is corrected and proofs are given of the theorems that every equation in the set of all equations

$$\cup_n^m = (\alpha, \beta), \quad 1 \leq m \leq \omega$$

is equivalent to a universal equation in the elementary language of relation algebra, that every equation in the set $\cup_n^m$ is valid in any proper relation algebra, and, most importantly, that a relation algebra is equipollent to a proper relation algebra if and only if it satisfies the set of equations. In general, Lyndon (*1950, 1956*) presented the set of conditions which finite algebras must satisfy in order to be isomorphic to proper relation algebras, and thus to be representable. Lyndon (*1961*) gave an example of a non-representable relation algebra, in particular an integral relation algebra derived from a projective geometry of dimension 1 and finite order $n \geq 3$, that is, whose projective plane is order n. In addition, Lyndon proved that any projective plane whose order is not the power of some prime generates an integral relation algebra which is nonrepresentable. Lyndon borrowed the techniques for construction of relation algebras from projective planes from Jónsson (*1959*). Monk (*1965*) proved the nonfinite axiomatizability of the class of representable three-dimensional cylindric algebras, Monk (*1969*) proved the nonfinitizability of representable cylindric algebras, and in a related paper, Johnson (*1969*) proved the nonfinite axiomatizability of representable polyadic algebras. *Henkin 1971* is a generalization of results obtained by Tarski (*Tarski and Thompson 1952; Tarski 1952*); Henkin proved in particular that if η is any initial transfinite ordinal, and D is the underlying set of an η-dimensional cylindric algebra $\mathcal{D}$ which is dimensionally complemented, then there exists an η-dimensional proper cylindric algebra $\overline{\mathcal{D}}$ and a mapping μ of D onto $\overline{D}$ which is an isomorphism between $\mathcal{D}$ and $\overline{\mathcal{D}}$. In *McKenzie 1970*, another example is given of a nonrepresentable integral relation algebra which is a model of Tarski's axioms (*Tarski 1941*) which is not representable. Following Lyndon (*1950*: 709), an algebra is an *integral relation algebra* if it has no zero divisors and its identity relation is minimal (an atom). Maddux (*1985*), extending and detailing some results of McKenzie (*1970*), gave additional examples of nonrepresentable integral relation algebras; in particular, he showed that for sufficiently large finite n there are at least $2^{(1/7)n^3}$–many nonrepresentable isomorphism types.

Another step was the category-theoretic characterization of nonrepresentable relation algebras by Comer (*1983*). A survey of decidability results for equational theories, along with proofs, was presented by McNulty (*1976a; 1976b*; see also *Anellis 1989*), while Maddux (*1983*) used proof theory to obtain foundations of a classification theory for nonrepresentable relation algebras and relation algebraic equations. He was able to classify several equations of binary relations; following Chin and Tarski (*1951*) (and possibly known also to Jónsson in the 1940s), Maddux was able to prove in sequent calculus, with equations using four but not three variables, the property of binary relations that *the relative product of complement of equivalence relation with itself is again an equivalence relation*; beyond that, Maddux was able to show that there is a natural hierarchy of classes MA_n ($3 \leq n \leq \omega$) in which relation algebras are shown to be MA_4, that is, those algebras which satisfy all equations provable in the sequent calculus using no more than four variables. Meanwhile, Németi (*1985*) gave a comprehensive list of decidable cases of infinite- and finite-dimensional varieties of cylindric algebras, and chapter 3 of *Henkin, Monk, and Tarski 1985* presents a lengthy discussion of representable and nonrepresentable cylindric algebras. (The first part of chapter 3 of *Henkin, Monk, and Tarski 1985* is an up-dated summary of *Henkin et al. eds. 1981.*) More recently, Andréka (*1988*), following the work of Maddux, has shown how to generate cylindric and polyadic algebras by their relation algebra reducts and has studied the conditions of the representability of those cylindric and polyadic algebras generated by representable relation algebras. More importantly, it has now been shown that it does not follow from a theory being weakly representable (is a **wRRA**) that it is therefore representable. Andréka (*1990*) has shown that a relation algebra (A, $\cup$, $\cap$, $\setminus$, 0, 1, ;, 1′) is weakly representable if it is representable as an algebra of binary relations where all of the operations (except perhaps $\cup$, and $\setminus$) have their natural set-theoretic meanings. Since weak representable relation algebras are such that **wRRA** $\subseteq$ **RRA**, she has been able to give a negative reply to Jónsson's (*1959*) question of whether every **wRRA** is representable such that every operation, including $\cup$ and $\setminus$, is standard (is an **RRA**) or not. Moreover, she has proven that **RRA** is not finitely axiomatizable over **wRRA**, meaning that for no finite set Ax of first-order formula is **RRA** = $\text{Mod}(Ax) \cap$ **wRRA** true.

There are, of course, a large number of works in recent years which present representability results for various specific theories, as well as works which generalize some of these results or offer new proofs of earlier results. We shall not, however, present an exhaustive list, since the most important of these results are already considered in *Tarski and Givant 1987* or elsewhere in the literature cited. Some of the most recent results have been summarized by Givant (*1988*) and Thompson (*1988*). In addition, some questions concerning representation of reducts of Tarski's relation algebras which still remain open are presented by Schein (*1991*). *Tarski and Givant 1987* has been reviewed by Bacon (*1989*) and by Monk (*1989*). There is the very important special issue of *Studia Logic* (vol. 50, nos. 3/4, 1991), edited by Willem J. Blok and Don Pigozzi, devoted to algebraic logic and its applications. A summary of the work of Tarski, Lyndon, Jónsson, and McKenzie on representability is found in this issue (*Maddux 1991b*, 438–449, and especially 444–449); the paper by I. Németi in the same issue (*1991*) surveys all of algebraic quantification theory from the perspective of the unifying principle of an algebra of sets of consequences, and helps make problems of axiomatization and representation the dominant theme of his survey.

As an alternative to cylindric algebras as three-variable algebraic fragments of FOL, Everett and Ulam (*1946*) developed projective algebras as two-variable algebraic fragments of FOL, without either identity or the substitution operation. A projective algebra is a Boolean algebra whose operations are the projective operations on coordinate axes, and whose relations are subsets of the plane ^2U. The indicated Boolean algebra is closed under projection and $\square$-product, where $X \square Y$ is a direct product, or class $[X; Y]$ of all pairs $\langle x, y \rangle$ in the set of points X and Y. The aim of this work was to find the set of axioms necessary and sufficient for determining which projective algebras are representable. Everett and Ulam (*1946*: 87) proved the representation according to which *a projective algebra defined on all subsets of a set s of points P is isomorphic to a projective algebra of certain subsets of a direct product of [X;Y] having unit preservation (i.e. s → [X;Y]), union, intersection, complementation, projection, and $\square$-product.* The task of classifying representable projective algebras was undertaken by McKinsey (*1948*).

An historically interesting question concerns the basis for any possible connections between the work of Peirce and Kempe on the one hand and of Tarski and his colleagues and followers on the other. In particular, it is striking that so much of the work in algebraic logic of Tarski, his colleagues, and followers, from the 1940s through the 1980s, should be devoted to determination of the representability of relation algebras and to some extent to the determination of which relation algebras are capable of expressing all equations concerning relations in a three-variable fragment of FOL, even if it should be that the modern algebraic forms of these questions turn out not to be equivalent to or directly connected with the modern equivalent of Peirce's claim or conjecture that all polyadic relations could be reduced to combinations of monadic, dyadic, and triadic relations.

Since Tarski referred in his own writings to so few of Peirce's works (and none of Kempe's or Royce's), we cannot ascertain whether his concern with representability was triggered by Peirce's conjecture, and if so, in which of Peirce's writings he may have read the conjecture. Nevertheless, we may reasonably assume that he came across the conjecture in one or more of the writings in the Harvard Edition (for example, 4.12–20), or other published papers of Peirce in which it appears. Our conjecture here is based upon our knowledge (for which I am indebted to Prof. Hiż in a private communication of 1989) that Tarski had become fully acquainted with Peirce's published papers while still a student in Poland. (A listing of many of the formulations of the reduction thesis found in the Peirce corpus appears in *Herzberger 1981*: 56–58.) Neither can we ascertain whether Peirce's reduction claims in fact provided the impetus for the work on representability undertaken by Tarski, his colleagues, and followers, or whether that work assumed such an extensive role in research in algebraic logic during the last fifty years because of Peirce's claims. It would seem even less likely, however, that the specific question of the reducibility of sentences about relations requiring four variables to equivalent three-variable equations undertaken by Tarski and others would have been inspired by Schröder's more general question of the expressibility of all equations about relations in the relation algebra, in particular since Schröder was concerned almost exclusively with dyadic relations.

OPEN QUESTIONS

1. Determine whether Peirce's *vids* are equivalent to Lyndon's *units*.

2. Determine whether Lyndon's cycles and Peirce's and Kempe's linear triads or relative triples are related algebraically, and, if they are, determine whether or not they are algebraically equivalent.

3. Explore the connections, if any, between Peirce's cycles and those of Lyndon. Do Peirce's cycles satisfy the conditions presented by Lyndon whereby a basis is shown to be a cycle?

4. Determine whether Herzberger's formalization of Peirce's bonding algebra (HBA) is a representable relation algebra (**RRA**). If not, is it a weakly representable relation algebra (**wRRA**)?

5. Is HBA isomorphic with $\mathfrak{L}^{\times}$; with $\mathfrak{L}^{+}_3$; with $\mathfrak{L}_3$?

6. Determine whether Burch's formalization of Peirce's bonding algebra (BBA) is a **RRA**. If not, is it a **wRRA**?

7. Is BBA isomorphic with $\mathfrak{L}^{\times}$; with $\mathfrak{L}^{+}_3$; with $\mathfrak{L}_3$?

8. Using the axioms of 548: 20–22, determine whether in Peirce algebras the product of two minimal elements is always minimal, or is minimal only in case the basis forms a group.

Acknowledgements. An early version of this paper was read and commented upon by Roger Maddux and Steven Givant. The author, however, assumes sole responsibility for any mathematical errors. Nathan Houser assisted by lending his historical and bibliographical expertise regarding Peirce.

17.

NEW LIGHT ON PEIRCE'S ICONIC NOTATION FOR THE SIXTEEN BINARY CONNECTIVES

Glenn Clark

§1. Introduction

It is remarkable that ninety years after Charles Sanders Peirce's notations for the sixteen binary connectives first appeared in his manuscripts they are still so little known, even among Peirce scholars. One would have thought that such a fundamental set of signs, created by a pioneer in the field of symbolic logic and "one of the founders, perhaps even *the* founder, of modern semiotic," (*Fisch 1978*: 131) would have received more attention.

No doubt one reason for this situation was the widespread popularity of the Frege-Peano-Russell line in logic following the publication both of Russell's *The Principles of Mathematics* in 1903 and of the more detailed *Principia Mathematica* by Whitehead and Russell in 1910–13. Furthermore, since Peirce's death his notations have been the victims of a number of unfortunate circumstances. First, it may be that, in working with those manuscripts that lead up to and include "The Simplest Mathematics," which is one chapter in the "Minute Logic" (1902), scholars have tended to rely on the relatively short and typed version in MS 429 (127 pages) rather than on the earlier but much longer and handwritten MS 431 (402 pages). Nearly half of the pages in MS 431 are early excursions that led to rewrites (431B), and one such lengthy excursion contains, in comparison with MS 429, a much clearer and more complete picture of what Peirce was doing and thinking when he constructed his iconic signs. In addition, some significant pages which really belonged with MSS 430, 431, and 429 were, until recently, functionally lost to scholars, "buried" elsewhere in the Peirce collection. One such page, recently found and identified, plays an important role in this paper. What happened to "The Simplest Mathematics" is just one example of the misfortunes that have befallen the Peirce manuscripts (*Houser 1989a*).

Another reason for the scant attention to Peirce's signs grew out of what happened to MS 429 when it was about to be published. The decision to include this manuscript in the *Collected Papers* provided an opportunity for making information on Peirce's signs more widely known. This opportunity, however, was squandered on two counts: (1) by a lengthy omission and (2) by the replacement of Peirce's signs by other symbols. Between 4.261 and 4.262 the editors left out nearly seven pages of typed manuscript. These pages are precisely the ones in which Peirce introduced his sixteen signs, chiefly those in cursive form, along with comments on certain modifications which he adopted for historical reasons. Thus, at this first opportunity in print, the reader of the *Collected Papers* is deprived of seeing Peirce engaged in applying the principles he believed should be used in sign creation. Fortunately the omitted pages have since been published by Carolyn Eisele at NEM 3: 272–275n. Also, further injustice was done in 4.268–273, where "a more conventional symbolism was substituted by the editors" (4.261n.). In contrast to Peirce's signs, the replacement set lacked any attempt at iconicity.

Thus, even after the publication of most of "The Simplest Mathematics," Peirce's manuscripts remained the only source of a full presentation of his sixteen signs. This situation, in view of the above comments about the manuscript collection, has delayed the emergence of publications dealing with these signs. Eisele's contribution in 1976 helped to change the pattern of neglect, and recent preparation for the publication of MS 431A, and some of MS 431B, in *Writings of Charles S. Peirce* has led to continued attention to these signs.

In the footnote to 4.261, the reason given for omitting the sixteen signs from the *Collected Papers* was that "Peirce below abandons these signs" for a sign of logical multiplication ($\cdot$) and a sign of logical disjunction (Ψ). It is true that those two signs, together with one for logical equivalence, are frequently used after MS 431A: 100, but the others are not completely abandoned after their introduction.[1] To the contrary, all or some of the sixteen cursives are used or referred to in MS 431A on pages 70–74 (MS 429: 45–47) and 86–95. In particular these signs are used very heavily in the stretch of pages MS 431A: 87–95 (MS 429: 62–68), where Peirce examined several different logical forms in a search for what he called "propositions necessarily true." More will be said below about pages 87–95, since an analysis of them is the prime purpose of this paper, but first a description of Peirce's notations for the binary connectives is in order.

Max Fisch has remarked that "No other logician compares with Peirce in attention to systems of notation and to sign-*creation*" (*Fisch 1982* [1980–1982]: 132). Such a strong statement invites an example in detail. The following account of Peirce's systems of notation for the sixteen binary connectives will not only show how he applied some of his principles of sign-creation but will also make it possible to exploit the properties of these systems in his algebra of logic.

§2. Notations for the Binary Connectives

What has been called the box-X notation was described in 1902 in MS 431A: 53–54, which became MS 429: 34–35, but was not fully displayed there. In those pages (4.261 has the first part) *x* and *y* represent "two quantities," each having one of two values. The value assigned to a letter was indicated by a line below or above the letter. This gave rise to four "possible states of things, as shown in this diagram." Then followed our Figure 17.1, in which the expression in the left quadrant is for $\underline{x}\bar{y}$ and the expression in the right quadrant is for $\bar{x}\underline{y}$.

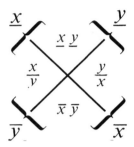

Figure 17.1

In terms of the truth values T and F, Peirce assigned the pairs TT, TF, FT, and FF to the top, left, right, and bottom quadrants, respectively. He then observed that "in each possible assertion" about the values of *x* and *y* "each of these states of things is either admitted or excluded; but not both." Thus each of the sixteen possible assertions may be represented by using the X frame and by "closing over the compartments for the excluded sets of values." At this point 4.261 ends except for the footnote mentioned above.

In what immediately follows in MS 431A, Peirce showed how to close the compartments in only five cases; he did the closing by inscribing the X frame in a circle (to close all four compartments) or in a semicircle (to close two adjacent compartments). While this notation, with a partially or totally closed X, may well be called "box-X," the idea of a box is more strongly conveyed if the compartments are closed by straight line segments. Peirce used this method as early as January 1902 (MS 431: 338, 341) but he did not, in MS 431, list all sixteen box-X signs in one place. He did include such a list on pages 27–28 of MS 530, "A Proposed Logical Notation," written about 1904 and as yet unpublished. Before that, at various times he closed compartments in different ways: shading, circular arcs, line segments. He even tried closing one *angle* when three compartments were to be left open—"making the blackened quadrant shrink to nothing and representing it by the single line of its two boundaries thus brought into coincidence" (MS 431B: 332).[2] For example, to indicate that only the left quadrant (TF) is to be closed, as for the "if-then" connective, the X-frame would be reduced to —≺. This "claw" is similar to and has the same meaning as the claw that Peirce introduced in 1870 (W2: 360).

Peirce's approach in MS 530 is worth noting because he also arrived at the X-frame in a different way. He began by representing two propositions *A* and *B* by two side-by-side intersecting circles, forming an Eulerian diagram in which he thickened the lower intersection. He then discarded the rest of the diagram and straightened out the intersecting arcs to form an X-frame in which the quadrants represented combinations of truth values of *A* and *B*. In the next step, for each of the sixteen possible "conjunct propositions" he obtained a representation by closing, with line segments, all quadrants representing combinations *excluded* by the proposition. He then listed these sixteen propositions, in sentence form and with their box-X signs. Thus $A \boxtimes B$ represents "*A* is not true but *B* is true," while $A \times B$ represents "If *A* is true, so is *B*." Peirce included $A \boxtimes B$ and $A \times B$ in his list, although he called the former "Everything is impossible (absurd)" and the latter "Meaningless." In another version of this system, at MS 530: 126, he suggested that, instead of closing the *excluded* quadrants, a dot be placed in each of the *included* quadrants, thus anticipating in detail the notation proposed by Warren S. McCulloch in 1942.[3]

After Peirce devised his set of signs featuring a basic X frame, he described how they could be modified, both in MS 431A: 54–64 (reproduced at NEM 3: 272–275n) and in MS 530: 28–34. His purpose was to create signs which were more easily written and which conformed to a two-part ethical principle expressed often in his writings. In the version at the beginning of MS 530, this principle declared it to be the right and obligation of the introducer of a concept into science to prescribe a terminology and notation for it. Moreover, this prescription should be respected by others unless it has serious disadvantages which cannot be removed by slight modifications.

For example, he wrote the box-X sign $\times$ in the cursive form forms which was further modified to forms in recognition of William Stanley Jevons's introduction of the corresponding operation and of the sign ⊣· for it. Peirce thought that the Jevons sign was liable to be mistaken for three signs but that in forms the two dots and the upright part preserved the essential features of Jevons's sign, while the loop at the bottom recalled the closed bottom quadrant of the X-frame and thus the exclusion of the FF combination.

Peirce gave similar reasons for other modifications. Recorde, Descartes, De Morgan, Harriots, and especially Peirce's former student, Christine Ladd-Franklin, were credited with ideas or signs leading to some of these modifications. In accordance with his ethical principle, Peirce claimed that his own sign for inclusion, —≺, introduced in 1870, should be recognized in a cursive form of $\times$, and he felt that the new sign ∝ did justice to his claim.

In 1882, a full twenty years before "The Simplest Mathematics," Mrs. Franklin used the sign ∨ for substantially the same purpose as Peirce's later ⋈. Peirce reduced his form to her form by discarding the three closed quadrants. Moreover, in MS 431A: 56 (or NEM 3.272n) he wrote: "It ought to be remembered that it was Mrs. Franklin who first proposed to put the same character into four positions in order to represent the relationship between logical copulas, and that it was part of her proposal that when the relationship signified was symmetrical, the sign should have a left and right symmetry." Peirce applied the first part of her proposal three times; more specifically, his cursives include each of ∧, ∝, and ◺ in four positions. Also eight of his signs have left-right symmetry and eight have up-down symmetry. Those with left-right symmetry signify relationships that are also symmetrical. In each of these eight signs the left and right quadrants are both open or both closed.

In their discussions and correspondence on notation Peirce and Mrs. Franklin did not always agree. On 8 January 1902, while working on MS 431, Peirce drafted a five-page letter intended for her (L237: 174–178). In it he reviewed his thinking and revealed some indecision about the eight signs having one compartment or three compartments closed. He also suggested that, if he and Mrs. Franklin could come to an agreement, they should publish in collaboration so that "our combined weight might do something toward agreements." By 20 January the indecision of 8 January was mostly resolved, and on that date his output included the material in 4.261 and in Eisele's long footnote, as well as the draft of a one-page letter to Mrs. Franklin (L237: 186) listing the sixteen signs which "I have pretty much decided to adopt." Thirteen of these preferred signs are in the third row of our Fig. 17.2b, and later he adopted the other three signs in the same row. Also, the second Ladd-Franklin letter has an open rhombus for FFFF.[4]

In Figure 17.2, part (a) shows the binary connectives in sixteen columns, each containing four entries. These entries indicate the quadrants that are closed (F) or left open (T) in Peirce's X-frame, with the entries in rows one, two, three, and four applied, respectively, to the top, left, right, and bottom quadrants of the frame. The (b) part of Figure 17.2 contains three rows that show Peirce's iconic signs, first his box-X forms, then his *original* cursives, and then his *modified* cursives.

§3. A Search for Tautologies

When Peirce wrote "The Simplest Mathematics," he had a highly productive day on Saturday, 25 January 1902. On that day he showed how to obtain thousands of tautologies by substituting his iconic signs in six different logical forms. The complexity of these forms varied all the way from

1	2	3	4	5	6	7	8	9	10	11	12	13	14	15	16
F	F	F	F	T	T	T	T	F	F	F	F	T	T	T	T
F	F	F	T	F	T	F	F	T	T	F	T	F	T	T	T
F	F	T	F	F	F	T	F	T	F	T	T	T	F	T	T
F	T	F	F	F	F	F	T	F	T	T	T	T	T	F	T

(a)

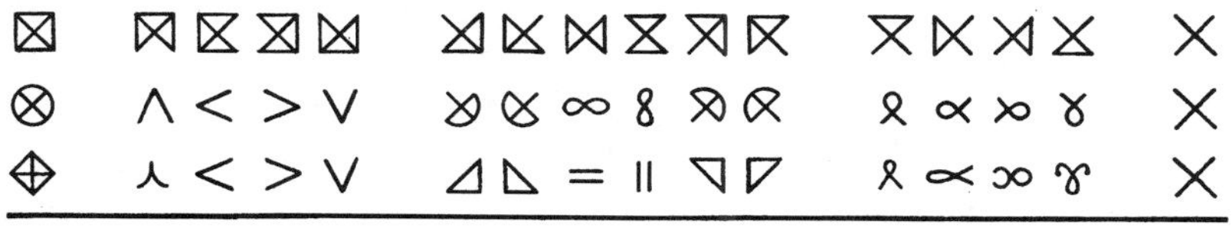

(b)

Figure 17.2

one letter and one binary connective to three letters and five binary connectives.

These tautologies are described in MS 431A: 87–93; the total number of them is 25,988. It seems likely that the actual spade work, which included the construction of a special four-in-one 16 by 16 table entered by four pairs of sequences in the margins, was not all done on 25 January. This table shows how deeply and how profoundly Peirce activated the properties of his iconic signs. It does not appear in the *Collected Papers* nor in MS 431. It was recently found at MS 839: 262, with many entries blurred, and also at MS 1537: 17, with 240 clear entries but with one row missing and with just one pair of marginal sequences. In this paper we offer a reconstruction of Peirce's thinking in building this table and in stating his conclusions by means of it. On page 94 of MS 431A, he gave a four-part rule for finding 680 more tautologies of a form involving two letters and three binary connectives, and, on page 95, he gave a table showing 256 sets of choices for these connectives.

The six logical forms that Peirce considered are listed below in the order in which we shall examine them. They are designated here (but not by him) as Forms A-E and G. The letters x, y, and z stand for statements, and the symbols O, ♦, ♥, ♠, and ♣ stand for binary connectives.

Form A:	x O x
Form B:	x ♦ $(x$ O $x)$
Form C:	$(x$ O $x)$ ♥ x
Form D:	$(x$ ♦ $x)$ O $(x$ ♥ $x)$
Form E:	$(x$ ♦ $y)$ O $(x$ ♥ $y)$
Form G:	$(x$ ♠ $y)$ ♥ $[(y$ ♣ $z)$ ♦ $(x$ O $z)]$

Form G is actually a representative of a collection of *twelve* such forms, to be shown below, which Peirce considered as a single set. Each form in this set has three letters and five binary connectives, but these forms differ from one another in the order of the letters and in the placement of the brackets. Peirce actually dealt with Form G before the much less complicated Form E.

Each of these forms, A-E and G, has one or more unspecified binary connectives, indicated by one or more of the five symbols O, ♦, ♥, ♠, ♣. For each of these symbols, Peirce imagined the substitution of all 16 of his signs for the binary connectives, giving 16 "propositions" (as he called them) in the form A, 16^2 in each of the forms B and C, 16^3 in each of the forms D and E, and 16^5 in each of the twelve forms represented by G. He was interested in determining which of these propositions are true for every possible assignment of truth value assignments to the letters x, y, and z.

Here, since we are expounding Peirce's manuscript, we employ his terminology without defending it. In MS 431A: 87–94 he used "proposition" for a logical expression in which the binary connective(s) were specified even if the component statement(s) were not. Thus, in his usage, $x \infty x$ is "a proposition of the form x O x," and it is "a proposition necessarily true" (we would say "a tautology") if it is true, "whatever be the value of the single letter" it contains. Likewise, in his usage, $(x \propto y) \times (x \vee y)$ is a proposition of the form $(x ♦ y)$ O $(x ♥ y)$; it is necessarily true since the $\times$ assures that it is true for every assignment of truth values to x and y. Thus in treating Forms A-E and G, he was seeking to specify the binary connectives in such a way that the resulting propositions, as he called them, would be necessarily true. To do this, he had to take into account the two 1-place truth values for x in Forms A-D, the four 2-place combinations of truth value assignments for x and y in the Form E, and the eight 3-place combinations of truth values for x, y, and z in the twelve variations of Form G.

It should be emphasized that in this investigation Peirce was operating at a level of abstraction that was unusual for his time. Instead of restricting himself to those binary connectives that represented a relatively few relations then in common use between two propositions, he insisted on making general statements that apply to the full replacement set of all of the connectives that are theoretically possible. Even though he might call $x \otimes z$ an "absurd" proposition, as he did in MS 431A: 96, he immediately pointed out that it "is often valuable." For him, since $2^4 = 16$, there are sixteen binary connectives, no more and no less. At a time when many of these were being ignored, he was using all of them to construct 16 by 16 tables to help answer his questions about Forms E and G above.

In basing his systems of signs on an X-frame having horizontal, vertical, and diagonal symmetry, Peirce brought a strong two-dimensional geometric element into his system. Also, when his iconic signs were substituted for the connective variables (O,♦,♥,♠,♣), he used a procedure similar to the substitution of numerical values in an algebraic expression. Thus, in his treatment of logical expressions, such as Forms A-E and G, some features of both geometry and algebra were involved in the design and in the use of his system of signs.

Peirce did not justify his choice of connectives for any of the logical forms that he treated. He told us how to select the connectives, but he did not prove that the method of selection was valid. In this paper we take an added step, going beyond "how" to "why" for all of the forms that he discussed.

§4. Form A: x O x

Peirce began with the simple expression x O x. There are 16 possible substitutions for O; that is, the substitution set for O is the set of 16 binary connectives. The substitutions may be made by using one of his three sets of symbols in Figure 17.2b. In the pages of the manuscript under discussion, he favored the modified cursives on the third row of Figure 17.2b, but with = and ‖ sometimes replaced by ∞ and 8 respectively. We shall sometimes write ⊗ for ⊕.

The substitution set for O in x O x readily separates into four subsets. Since x O x becomes T O T or F O F, conditions are imposed on only the top and bottom quadrants of the X-frame for whatever symbol is substituted for O. It is easy to see that, whatever the truth value of x, a proposition of the form x O x is *necessarily* true when O (that is, the X-frame for O) is open in both the top and bottom quadrants, the TT and the FF quadrants. Since there is no restriction on the left and right quadrants, there are four possibilities for O, namely ✕, ∝, ∞, and ∞. Similarly, x O x is *necessarily* false if O is closed in the top and bottom quadrants, as in ⊗, >, <, and 8. Moreover x O x agrees in value with x when O is open at the top and closed at the bottom (ᛦ,◺,◿,∨) and agrees in value with $\bar{x}$ ("not x") when O is closed at the top and open at the bottom (⋏,◹,◸,⋊). Thus a proposition of the form x O x is necessarily true, is necessarily false, agrees in value with x, or agrees in value with $\bar{x}$, according as O is in the set N = {✕,∝,∞,∞}, in the set S = {⊗,>,<,8}, in the set W = {ᛦ,◺,◿,∨}, or in the set E = {⋏,◹,◸,⋊}.

Peirce noted the above properties associated with N and S but did not mention those associated with W and E. The elements in each of these four sets (with ⊗ in S replaced by ⊕) appear at the corners of one of the four quadrants in Peirce's page 87 diagram, reproduced here as Figure 17.3. In the oval in the top quadrant of the diagram, the signs ◿ and ◺ should be interchanged. Directions for use of the diagram are included in 4.269.

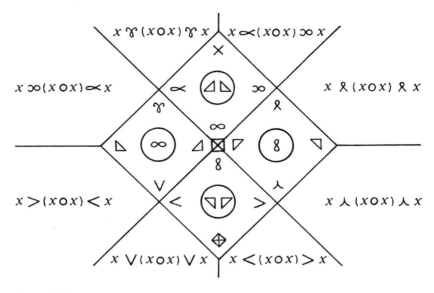

Figure17.3

§5. Form B: x ♦ (x O x) and Form C: (x O x) ♥ x

Peirce next considered Forms B and C, in which the diamond and heart are two of the four playing-card symbols that he introduced to stand for binary connectives. On his pages 88–89, he treated these two forms separately, even though Figure 17.3 shows them combined and with particular choices for ♦ and ♥.

The complete substitution set for each of Forms B and C consists of the $16^2 = 256$ ordered pairs of binary connectives, namely, those replacing (♦,O) in B or (O,♥) in C. In both cases Peirce designated, without proof, the subsets which would produce propositions necessarily true or necessarily false. The subsets he found are precisely those arrived at below.

Since Form B includes x O x, the information we now have on Form A will be useful in finding, for each O, the proper choices for ♦. There are four considerations. (1) If O is in N, Form B becomes x ♦ T.[5] Now x ♦ T is necessarily true if ♦ is open in both the TT and FT (top and right) quadrants, as in ⨯, ℽ, ∝, and ◺. (2) If O is in S, Form B becomes x ♦ F, which is necessarily true if ♦ is open in both the left and bottom quadrants, as in ⨯, ∞, ৪, and ◹. (3) If O is in W, Form B becomes x ♦ x, which is Form A and so is necessarily true if ♦ is any one of ⨯, ∝, ৪, or ∞. (4) If O is in E, Form B becomes x ♦ x̄, which is necessarily true if ♦ is open in both the left and right quadrants, as in ⨯, ℽ, ৪, and 8. For each of (1)–(4) there are 4 choices for O and then 4 for ♦, or 16 for the (O,♦) pair. In all 64 of these cases, taking the complement of ♦ opens the closed quadrants and vice versa. This changes the above propositions from necessarily true to necessarily false.

To devise an easy passage from Form B to Form C, let y stand for x O x, so that the two forms become x ♦ y and y ♥ x. For example, take ♦ as <. Since < is open in the right quadrant only, $x < y$ is true if and only if x is false and y is true. Then $x < y$ is logically equivalent to y ♥ x if the sign for ♥ is open in the *left* quadrant only, since that quadrant is for y true and x false. Thus, $x < y \equiv y > x$, and we write $<_h \, = >$ to indicate that a horizontal (left-right) flip applied to < yields >. Since interchanging x and y amounts to interchanging the left and right quadrants of the X frame, we have, in general, that x ♦ $(x$ O $x) \equiv (x$ O $x)$ ♦$_h$ x. Thus to convert any (♦,O) pair used in Form B to a (O,♥) pair that will yield the same truth value in Form C, use the same O and give ♦ an h-flip to get ♥. In this manner 64 tautologies in Form C are produced.

Peirce constructed an ingenious diagram to help him record the proper substitutions for the pairs (♦,O) and (O,♥) in his Forms B and C. This diagram, on his page 87 and shown above as Figure 17.3, exhibits the pairs that lead to the propositions that are necessarily true and to those that are necessarily false. Partial lists of such propositions, with substitution of specific binary connectives, appear in MS 430B: 408–410, so these pages really belong with MS 431A.

§6. Form D: (x ♦ x) O (x ♥ x)

On his pages 90–91 Peirce treated propositions of the form P = $(x$ ♦ $x)$ O $(x$ ♥ $x)$. With 16 choices for each binary connective, the complete substitution set consists of $16^3 = 4096$ ordered triples replacing (♦,O,♥). Of the 4096 propositions in Form D, Peirce considered only those which are necessarily true. Obviously, this includes the 256 with O as ✕, which will be absorbed in the following discussion, and excludes the 256 with O as ⊗.

To simplify Form D we write Q = x ♦ x and R = x ♥ x. Then the parentheses may be omitted, the result being P = Q O R. Here Q and R are repeats of Form A already discussed, and thus an analysis can be made in terms of the sets N, S, W, and E. Specifically, we may replace Q by T, F, x, or $\bar{x}$ according as ♦ is in N, S, W, or E; likewise, there are four possibilities for R, and thus there are 16 cases to consider. For example, if ♦ and ♥ are both in N, P becomes T O T. This expression is true if O is open in the top quadrant, without restriction on the other quadrants. Thus there are eight choices for O, as noted in line 1 of Table I. Similarly, in each of the next three lines of this table, O must be open in only one specified quadrant, again without any restriction on the other quadrants.

In lines 5 and 6, P is in Form A, so O is taken in N. In lines 7 and 8,

Table I

Sets for ◊♡	P	Quadrants necessarily open for o	Number of choices for o
1. N N	T o T	Top	8
2. S S	F o F	Bottom	8
3. N S	T o F	Left	8
4. S N	F o T	Right	8
5. W W	x o x	} Top and bottom	4
6. E E	$\bar{x}$ o $\bar{x}$		4
7. W E	x o $\bar{x}$	} Left and right	4
8. E W	$\bar{x}$ o x		4
9. N W	T o x	} Top and left	4
10. N E	T o $\bar{x}$		4
11. S W	F o x	} Right and bottom	4
12. S E	F o $\bar{x}$		4
13. W N	x o T	} Top and right	4
14. E N	$\bar{x}$ o T		4
15. W S	x o F	} Left and Bottom	4
16. E S	$\bar{x}$ o F		4
Total			80

whatever the truth value of x, P becomes T O F or F O T, so O must be open in the left and right quadrants. In lines 9 and 10, P becomes T O C, where C is "contingent," so O must be open in the top and left quadrants, whereas, in lines 11 and 12, P becomes F O C and O must be open in the right and bottom quadrants. Similarly, with P as C O T or C O F, O must be open in the top and right quadrants in lines 13 and 14 and open in the left and bottom quadrants in lines 15 and 16.

In each row of Table I, there are 4 choices for ♦ and 4 for ♥, so there are 16 choices for the pair (♦,♥). Hence the number of tautologies represented by each row is 16 times the number of choices for O in that row. Thus the number of tautologies is 128 for each of the first four rows and

64 for each of the others, giving 1280 necessarily true propositions of the
Form D. Peirce gave the number as 1100, counting 844 in addition to the
256 resulting from taking O as ✕. However, his own explanation of the
three-color diagram on his page 90 (4.270), which he says exhibits the 844
"not very elegantly," leads to 1024, for a total of 1280. Thus it turns out that,
of the full substitution set of 4096 elements (ordered triplets of signs), 1280
were of interest to Peirce. Subjecting Form D to a similar procedure shows
that a like number of propositions, namely 1280, are necessarily false, and
thus the remaining 1536 are contingent.

§7. Form E: (x ♦ y) O (x ♥ y)

Peirce treated propositions of Form E on his pages 94–95, dated 26
January 1902. The substitution set for (♦,O,♥) again has $16^3 = 4096$ members; that is, there are 4096 propositions of Form E, and the 256 with O as
✕ are among those which are necessarily true. Since Form E is similar to

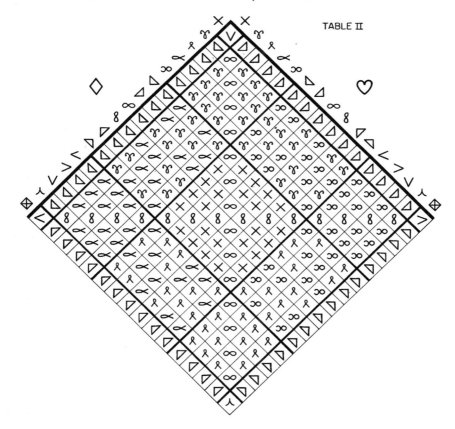

TABLE II

Form D in the pattern of connectives but has stricter conditions for tautology (TF and FT for xy in E do not occur in D), the number of tautologies in Form E cannot exceed 1280, the number in Form D; it turns out that the number in Form E is 680.

In identifying tautologies of the Form E, Peirce simply gave, without explanation, a four-part rule for telling which quadrants, if any, must be open in the box-X form of O when ♦ and ♥ have been prescribed. Then he stated that the "principal results" of the rule are shown in a 16 by 16 table on his page 95, shown here as our Table II. (The < and > at the left and right corners were incorrectly interchanged in Peirce's table.) This table is given in the *Collected Papers* at 4.273 with different notation and with 24 cells blank. We read ♦ from the left margin in the upper half of this table and ♥ from the right margin, with a value of O found in the cell determined by the two coordinates. Let us see how Pierce's four-part rule for filling in the table might be derived.

To simplify Form E we write Q = x ♦ y and R = x ♥ y. Then the parentheses may be omitted, leaving Q O R. The four parts of the rule relate in turn to the TT, TF, FT, and FF quadrants of the X frame for O. In particular, a quadrant must be open if the corresponding truth-value pair can occur as a pair of values of Q and R, resulting from the choice of ♦ and ♥. Note that TT occurs, at least once, as a pair of values of Q and R, if and only if there is at least one pair of values of x and y for which x ♦ y and x ♥ y are both true. In effect there is at least one quadrant (top, left, right, or bottom) in which the X-frames for ♦ and ♥ are *both* open. In that case, in order for Q O R to be a tautology, T O T must have the value T, which means that the *top* quadrant of the X frame for O must be open. This is the substance of the first part of Peirce's rule.[6]

Similarly, if there is any quadrant (one or more) in which the X-frame for ♦ is open and that for ♥ is closed, then TF occurs as a pair of values of Q and R. It follows that the *left* quadrant in the X frame for O must be open. This is the second part of the rule. Likewise, any quadrant closed for ♦ and open for ♥ calls for an open *right* (FT) quadrant for O, and any quadrant closed for both ♦ and ♥ calls for an open *bottom* (FF) quadrant for O. This completes the four-part rule.

In the X-frame for O, as Peirce observed, any quadrant that is not required by the rule to be open may be either open or closed. Thus if n quadrants are required to be open (n = 1, 2, 3, 4), then 4-n quadrants may be either open or closed, and hence there are 2^{4-n} choices for O. For example, when ♦ and ♥ are both taken as ✕, so that Q and R are necessarily true, the X-frame for O is required to be open only in the top quadrant; thus there are $2^3 = 8$ choices for O, the one with three quadrants closed being ∨. This ∨ is the entry at the apex of Table II. In that table, each recorded entry is the one obtained by closing all quadrants not required to be open; the other possible choices for the entry may then easily be written. It has been shown that if all possible choices are counted for every entry, the total is 680.

The construction of the preceding table does not require that the 256 entries in it must be obtained one at a time by using Peirce's four-part rule. Some entries are obtainable in batches of 14 and others in pairs. There are 6 sets of 14 identical entries in a straight line, and the table is symmetric about each of its diagonals. This symmetry depends in part on the *iconicity* of Peirce's system of signs, so that (for example) ♈ and ♈ are symmetric about the horizontal diagonal (each reflects into the other), as are ∞ and (again) ∞. Similarly, about the vertical diagonal there are the pairs ∝ and ∞ as well as ♈ and ♈. We regard the pairs ∝, ∞ and ♈, ♈ as exhibiting the symmetry described, since their box-X forms do.

§8. Form G: $(x \spadesuit y) \heartsuit [(y \clubsuit z) \blacklozenge (x \bigcirc z)]$

Statement of the problem. Form G gives us an excellent opportunity to see what Peirce had in mind for his iconic signs. On his pages 92–93 he was bold enough to take on the complexity in a collection of twelve related forms, each involving three distinct letters and five binary connectives. He used the letters x, y, and z for statements, and he added two more playing card symbols, a spade and a club, to his diamond, heart, and circle to denote arbitrary binary connectives. Actually he used an inverted spade, with the stem at the top. (Note that when these twelve related forms appeared in the *Collected Papers* at 4.271 some of the notation was changed.)

Here are Peirce's twelve forms in full, with his labels, eleven Greek letters and an F. In (F) Peirce wrote z for the last letter, but it clearly should be y.

$$(\alpha) \quad (x \spadesuit y) \heartsuit [(y \clubsuit z) \blacklozenge (x \bigcirc z)]$$
$$(\beta) \quad (x \spadesuit y) \heartsuit [(x \bigcirc z) \blacklozenge (y \clubsuit z)]$$
$$(\gamma) \quad (y \clubsuit z) \heartsuit [(x \spadesuit y) \blacklozenge (x \bigcirc z)]$$
$$(\delta) \quad (y \clubsuit z) \heartsuit [(x \bigcirc z) \blacklozenge (x \spadesuit y)]$$
$$(\varepsilon) \quad (x \bigcirc z) \heartsuit [(x \spadesuit y) \blacklozenge (y \clubsuit z)]$$
$$(F) \quad (x \bigcirc z) \heartsuit [(y \clubsuit z) \blacklozenge (x \spadesuit y)]$$

$$(\xi) \quad [(x \spadesuit y) \blacklozenge (y \clubsuit z)] \heartsuit (x \bigcirc z)$$
$$(\eta) \quad [(x \spadesuit y) \blacklozenge (x \bigcirc z)] \heartsuit (y \clubsuit z)$$
$$(\theta) \quad [(y \clubsuit z) \blacklozenge (x \spadesuit y)] \heartsuit (x \bigcirc z)$$
$$(\iota) \quad [(y \clubsuit z) \blacklozenge (x \bigcirc z)] \heartsuit (x \spadesuit y)$$
$$(\kappa) \quad [(x \bigcirc z) \blacklozenge (x \spadesuit y)] \heartsuit (y \clubsuit z)$$
$$(\lambda) \quad [(x \bigcirc z) \blacklozenge (y \clubsuit z)] \heartsuit (x \spadesuit y)$$

Since the labels have no special significance we shall, for convenience, use lower-case Roman letters a, b, c . . . l.

To simplify form (a), which is also our Form G, we first write P = x ♠ y, Q = y ♣ z, and R = x O z. Then the original parentheses may be omitted, and thus the brackets may be replaced by parentheses, the result being P ♥ (Q ♦ R). When these simplifications are adopted in all twelve of the forms in Peirce's set, the forms become:

(a)	P ♥ (Q ♦ R)	(P ♦ Q) ♥ R	(g)
(b)	P ♥ (R ♦ Q)	(P ♦ R) ♥ Q	(h)
(c)	Q ♥ (P ♦ R)	(Q ♦ P) ♥ R	(i)
(d)	Q ♥ (R ♦ P)	(Q ♦ R) ♥ P	(j)
(e)	R ♥ (P ♦ Q)	(R ♦ P) ♥ Q	(k)
(f)	R ♥ (Q ♦ P)	(R ♦ Q) ♥ P	(l)

Peirce set himself the task of finding "some propositions necessarily true" in the above forms. That is *some*, not *all* as in Forms A-E. Since P, Q, and R involve the ultimate components x, y, and z as well as the connectives ♠, ♣, and O, a necessarily true proposition in the form (a) is one, in Peirce's usage, in which the five connectives are so specified that the resulting expression is true for all eight combinations of truth values for x, y, and z. A similar statement holds for each of the other eleven forms above.

The set of all possible substitutions for the binary connectives in each of the twelve forms consists of ordered quintuplets of signs that replace (♠,♥,♣,♦,O). In each form this yields 16^5 or 1,048,576 possibilities. Peirce passed over what he probably took as too obvious to mention, namely, the 16^4 or 65,536 tautologies that are obtained in each of the twelve forms by taking ♥ as ✕. This deliberate omission was part of his "some but not all" decision. He further reduced his task by considering only the tautologies of a particular type, now to be described. By means of a complex procedure, including a special set of directions that tell us how to use the recently found 16 by 16 table, he specified 2048 tautologies for each of the twelve forms, which gives a total of 24,576. By then he was ready to rest from his labors, observing that "these are all of this class that it seems worth while to give." Careful study, both of his 16 by 16 table and of his directions for using it, makes it possible to specify the main parts and the key steps in his thinking.

Peirce's approach to his task was based on the different roles played by the triple (♠,♣,O) and by the pair (♥,♦). The signs in the triple act directly on x, y, and z to produce what we have called P, Q, and R. In contrast, the signs in the pair connect P, Q, and R in the twelve ways shown in forms (a)-(l). These roles, though different, are related because P, Q, and R are involved in both. A particular choice for the triple determines a list of distinct combinations of truth values for P, Q, and R. This list then controls the possible choices for the pair.

For example, if the ordered triple (♠,♣,O) is taken as (≺,𝒴,◺), it is easy to verify that the *eight* different combinations of truth values for x, y, and z produce just *five* different combinations of truth values for P, Q, and R. These combinations are TTT, TTF, FTT, FTF (three times), and FFT (twice). The combinations TFT, TFF, and FFF do not occur; they are excluded by the choice of ≺,𝒴, and ◺ for ♠, ♣, and O.

The exclusion of three truth-value combinations for P, Q, and R in this example reduces the requirement on the (♥,♦) pair. To make a proposition of form (a) necessarily true when the triple is (≺,𝒴,◺), we need only to choose ♥ and ♦ so that P ♥ (Q ♦ R) becomes true for the five combinations of truth values just noted. We shall see later how such choices of ♥ and ♦ can be made, but at this point we merely observe that it is easy to verify that P ∞ (Q ⅄ R) is true for all five combinations of truth values for P, Q, and R listed above. Thus we may confidently assert that the substitution of the ordered quintuplet (≺,∞,𝒴,⅄,◺) for (♠,♥,♣,♦,O) in form (a) yields a tautology. Now we need to devise a systematic procedure for identifying triples (♠,♣,O) and pairs (♥,♦) that will result in more "propositions necessarily true," to use Peirce's phrase.

An outline of this procedure can be obtained from observing that Peirce stated his conclusions about forms (a) − (l) under four headings. This division into what we shall call four "cases" arose from his approach to the choice of triples. He did not pick a triple and then determine what truth-value combinations of P, Q, and R would be excluded, as we did for the triple (≺,𝒴,◹). Instead, he found that he could pick any one combination and then determine many different triples that would exclude that combination. There are eight possible combinations, and of these he used four, one at a time, the four with R false. When these four *cases* are considered along with the twelve *forms* (a)-(l), the (♠,♣,O) *triples*, and the (♥,♦) *pairs*, we arrive at a four-part outline that is partitioned according to the following formula, expressed first in words and then in numbers, two of which will be justified later:

pairs	×	cases	×	forms	×	triples	yields	tautologies
2	×	4	×	12	×	256	=	24,576

The presentation that follows will proceed in reverse order, starting in the first part with the triples and ending in the fourth part with the pairs. The full product, which comes to 24,576, is the number of tautologies for which Peirce gave proper triples and pairs, although he misstated the total as 24,376. The above factors show that the number of tautologies per form is 2 × 4 × 256 = 2048, while the number per case is 2 × 12 × 256 = 6144.

Part 1: 256 triples for form (a) in Case 1. Each of the 4096 possible substitutions for (♠,♣,O) will produce from one to eight different combinations of truth values for P, Q, and R. From those 4096 substitutions Peirce first identified 256 that were so chosen that the combination FFF for P, Q, and R could not occur. This exclusion of FFF is what we will call Case 1. We wish to see how he chose (♠,♣,O) so as to exclude FFF and then to see later how he chose (♥,♦) so that form (a) is true for *the other seven* combinations of truth values for P, Q, and R.

When Peirce sought to exclude FFF for P, Q, and R, he could have chosen ✕ for ♠, for ♣, or for O, but this would have been an easy way out. He found instead that, whenever he picked ♠ and ♣ arbitrarily (256 ways), he could then for each such choice determine O so that any combination of truth values of x, y, and z that made P and Q false would make R true. For example, with ⊏ for ♠ and ℧ for ♣, he chose ◺ for O; this gives the combination FFT twice but not FFF.

To see how he could have arrived at ◺ for O in this example consider the following table.

Table III

x	y	z	P $x \lessdot y$	Q $y\, ℧\, z$	R $x\, O\, z$
T	T	T	F	T	
T	T	F	F	T	
T	F	T	F	T	
T	F	F	F	F	T
F	T	T	T	T	
F	T	F	T	T	
F	F	T	F	T	
F	F	F	F	F	T

On the two lines where P and Q are both false we want R, or $x\,O\,z$ to be true. On these lines the truth value combinations for x and z are TF and FF, so R will be true if the sign for O is open in the left and bottom quadrants. Since there is no restriction on the value of R on the other six lines, there are four possibilities for O, namely, ◺, ◹, ∞, and ✕. When Peirce had more than one choice for O, he consistently favored the sign with quadrants that were open only where they were required to be open, and thus in this case he chose ◺.

Peirce made 256 choices of O, one for each of his choices for the or-

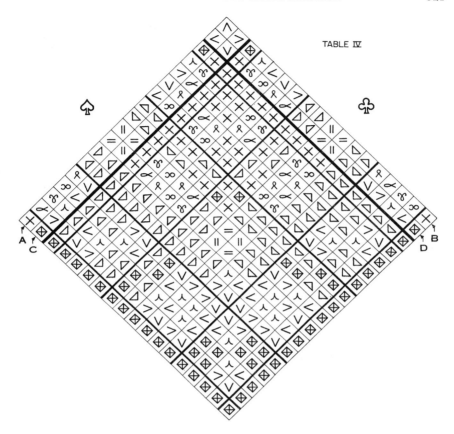

TABLE IV

dered pair (♠,♣). He did not reveal his method for determining which quadrants of O must be open, whether by a truth table or by some other means. He simply recorded his choices for O in the body of a 16 by 16 table, the same one already mentioned as only recently found and identified. Although this table is referred to twice in the *Collected Papers* at 4.271, the editors gave no indication that they realized it was missing. They may have assumed that Peirce was referring to a different table, the one that appears in 4.273 (from MS 429: 68, now our Table II). In fact, the table mentioned in 4.271 was not found with MS 431 or MS 429, and it has not previously been published. In 1983 Christian Kloesel and Shea Zellweger coordinated their efforts at the Peirce Edition Project in Indianapolis, where they located two imperfect versions of this table. A recent reconstruction of it now appears as our Table IV, where we have added the sequence labels A, B, C, and D.[7] The original (imperfect) version from MS 839: 262 is reproduced as our Table V.

When Peirce used the prototype of Table IV to exclude the combina-

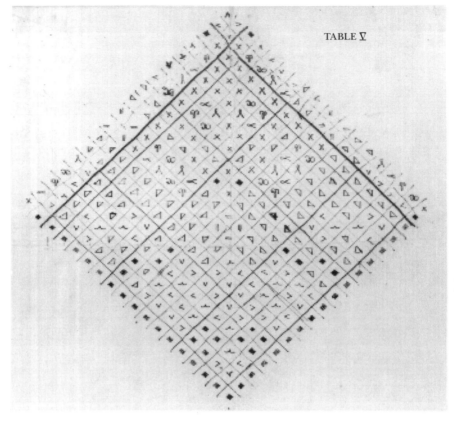

tion FFF for P, Q, and R, he read ♠ in the left margin in the A sequence, ♣ in the right margin in the B sequence, and O in the cell determined by those two coordinates. As he expressed it, "those parts of the table are used which intersect in ∧." The symbol ∧ used here is, evidently by intent, the same as that for one of Peirce's binary connectives. In Case 1, the one just mentioned, it is associated with "P and Q false implies R true." Then since "P ∧ Q true" implies "P and Q are false," we have "P ∧ Q implies R". Presumably, this accounts for Peirce's use of ∧ to indicate the intersection of the outer marginal sequences. Similarly the use of <, >, and ∨ at the peak of Table IV can be accounted for in the other three cases below.

Now recall that Peirce was looking for *some* of the possible substitutions that would produce tautologies in the twelve forms (a)-(l). It has been shown that if he had counted all admissible choices for O instead of just one in each cell, the total number of choices would have been 1699 instead of 256.

Although the entries in Table IV can be found by the truth table method exemplified above, much of this work can be eliminated by using certain generalizations and relationships *not* mentioned by Peirce. Of course the validity of these relationships would have to be proved before they could be used in the construction of the table. These short-cuts are not essential and not all of them will be mentioned, but we do call attention to the easily observed fact that the table is symmetrical with respect to its vertical diagonal.

Part 2: 12 forms. The 256 triples found in Part 1 were "for form (a) in Case 1." However, these triples were chosen solely for their ability to exclude the situation "P, Q, and R all false." This exclusion is not affected by writing P, Q, and R in a particular order or by grouping two of them in parentheses in a certain way. Consequently the same 256 triples that "work" (exclude FFF) for form (a) also work for forms (b)-(l). It follows that, for each of the twelve forms, we have the same 256 choices for (♠,♣,O), obtained from Table IV by using its marginal sequences A and B. At this point we have $12 \times 256 = 3072$ potential tautologies in Case 1, with the (♥,♦) pair(s) remaining to be found for each of them. But first we take note of some other cases.

Part 3: 4 cases. In addition to the exclusion of FFF in Case 1, Peirce went on to consider three more combinations for exclusions. He did not say that he was doing this, but a study of his use of four pairs of marginal sequences in Table IV shows that, out of the eight possible truth-value combinations of P, Q, and R, he was dealing with only four cases. We have numbered these cases in the order in which he reported his conclusions about them.

Case 1. "P, Q, and R all false" is excluded;
Case 2. "P false, Q true, and R false" is excluded;
Case 3. "P true and Q and R false" is excluded;
Case 4. "P and Q true and R false" is excluded.

In each case, the exclusion of the designated combination is accomplished by properly choosing (♠,♦,O), and in each case 256 choices for this triple are found by using Table IV. In Case 2 the exclusion of "P false, Q true, and R false" can be achieved by using the (A,D) marginal sequences to enter Table IV, by reading ♠ in the A sequence as before, but this time, by reading ♣ in the D sequence in the right margin. The D sequence is obtained by taking the negations, or complements, of the adjacent signs in the B sequence. Thus if any sign in the D sequence makes Q true, that is, makes $y \clubsuit z$ true, the adjacent sign in the B sequence makes Q false. The body of the table, where O is read, is unchanged. It is readily seen that any triple (♠,♣,O) chosen by using the A and D sequences, "which intersect in <," leads to this result: P false, Q true, and R false cannot occur. If this combination did occur, the sign in the B sequence which is adjacent to the one used in the D sequence would have resulted in an occurrence of "P, Q, and R all false" in Case 1, which is contrary to the way the entry for O in the table was determined.

In Case 3, the exclusion of "P true and Q and R false" can be accomplished by entering Table IV through the C and B sequences, "intersecting in >." In the C sequence, each sign is the complement of the adjacent sign in the A sequence. Any triple (♠,♣,O) chosen by taking ♠ from the C sequence, ♣ from the B sequence, and O from the body of the table, has this result: P true and Q and R false cannot occur. The argument is like that in the preceding paragraph.

By now it should be clear that in Case 4 the exclusion of "P and Q true and R false" can be accomplished by reading ♠ and ♣, respectively, in the C and D sequences, "intersecting in ∨."

We have just seen how to obtain four sets of 256 triples. Each of the four cases will make use of just one set, the one obtained by entering Table IV through the pair of marginal sequences appropriate for that case. We now assert that, in each case, the 256 triples so found work in all twelve forms. The reasoning is similar to that used for Case 1 in Part 2. Neither the different orders for P, Q, and R in the various forms nor the presence of parentheses affects the choices for the (♠,♣,O) triples, although these differences will play a crucial role in the choices for the (♥,♦) pairs. Thus for each of the twelve forms, we have the same four sets of 256 choices for (♠,♣,O), obtained from Table IV by using its four pairs of marginal sequences.

Part 4: 2 pairs. Part 3 gave us $4 \times 12 \times 256 = 12{,}288$ potential tautologies, with substitutions for ♥ and ♦ still to be decided in each of them. Fortunately, we can deal with these expressions in batches of 256. The strategy here is to group together those which have the same excluded truth-value combination for P, Q, and R, for it is that excluded combination which controls the choice(s) for the (♥,♦) pairs. Thus we may turn our attention away from specific (♠,♣,O) triples and consider how the exclusion of each of the four truth-value combinations operates to identify (♥,♦) pairs for each of the twelve forms.

We shall show two ways to obtain valid (♥,♦) pairs for the 48 combinations of 4 cases times 12 forms. The first way is to treat these 48 situations individually by a direct procedure, "from scratch," for each one. We shall go through this procedure twice, for (a) in Case 1 and for (i) in Case 3. This should show how the procedure could be used 48 times, so that the problem is theoretically solved. The second way to get valid pairs is to use the direct procedure to obtain some of the pairs and then to use these selected pairs to obtain the remaining ones. In this second way, we will be aided both by some logical equivalences involving negation and by partitioning the 48 case-form combinations into 14 sets in such a way that a

($\heartsuit$,$\diamondsuit$) pair for one element of a set works for all of the elements of that set. The second way will receive the lion's share of attention in what follows, and results for it will be given in detail. First consider the direct procedure.

In form (a) in Case 1 we suppose that any one of the 256 valid substitutions for ($\spadesuit$,$\clubsuit$,O) has been made by using the (A,B) sequences in Table IV, so that P, Q, and R cannot all be false. Specifically, suppose ($\spadesuit$,$\clubsuit$,O,) is ($<$,Υ,$\triangledown$). Then we will have a tautology *if* $\heartsuit$ and $\diamondsuit$ are chosen *so that* P $\heartsuit$ (Q $\diamondsuit$ R) is converted into a false statement when P, Q, and R are all false and into a true statement otherwise. In other words, we wish to choose $\heartsuit$ and $\diamondsuit$ *so that*, in the truth table with four columns headed P, Q, R, and P $\heartsuit$ (Q $\diamondsuit$ R), in that order, the last column has an F on the one line where the first three columns all have Fs, but the last column has Ts on the other seven lines. This condition (after "*so that*") can be restated as follows: F $\heartsuit$ (F $\diamondsuit$ F) is false but becomes true if any one or more of the letters are negated (replaced by T). The condition can be stated even more compactly as "F $\heartsuit$ (F $\diamondsuit$ F) is barely false," where the meaning of "barely false" comes from what follows the colon in the preceding sentence.

The requirement that F $\heartsuit$ (F $\diamondsuit$ F) be barely false imposes conditions on both $\heartsuit$ and $\diamondsuit$. For $\diamondsuit$ (that is, for its box-X form) the bottom (FF) quadrant must differ from the other three quadrants as to closure. For if not, then F $\diamondsuit$ F has the same truth value as at least one of T $\diamondsuit$ T, T $\diamondsuit$ F, and F $\diamondsuit$ T, and thus any choice of $\heartsuit$ that makes F $\heartsuit$ (F $\diamondsuit$ F) false also makes false at least one of the expressions F $\heartsuit$ (T $\diamondsuit$ T), F $\heartsuit$ (T $\diamondsuit$ F), and F $\heartsuit$ (F $\diamondsuit$ T). This violates the "barely false" condition for F $\heartsuit$ (F $\diamondsuit$ F). Thus, in modified cursive form, $\diamondsuit$ has two possibilities. First, if $\diamondsuit$ is taken as Υ, so that F $\diamondsuit$ F is false, then we have F $\heartsuit$ (F). Consequently (from the barely-false requirement), for $\heartsuit$, the bottom quadrant is closed and all others are open, so that $\heartsuit$ is also Υ. Second, if $\diamondsuit$ is taken as $\curlywedge$, so that F $\diamondsuit$ F is true, then we have F $\heartsuit$ (T). It follows that for $\heartsuit$, the right quadrant (the FT quadrant) is closed and all others are open, so that $\heartsuit$ is ∞. This gives us two pairs for (a) in Case 1, namely, (Υ,Υ) and (∞,$\curlywedge$). Combining either of these pairs with the triple ($<$,Υ,$\triangledown$) produces a tautology in the original Form G.

The same type of procedure can be used for (i) in Case 3, with rather simple modifications. Since in Case 3 "P true and Q and R false" is excluded, we want (F $\diamondsuit$ T) $\heartsuit$ F to be barely false. Hence, for $\diamondsuit$, either the right (FT) quadrant is closed and all others are open or the right quadrant is open and all others are closed. Thus $\diamondsuit$ has two possibilities. First, if $\diamondsuit$ is taken as ∞, so that F $\diamondsuit$ T is false, then we have (F) $\heartsuit$ F. Consequently, for $\heartsuit$, the bottom quadrant is closed and all others are open, so that $\heartsuit$ is Υ. Second, if $\diamondsuit$ is taken as $<$, so that F $\diamondsuit$ T is true, then we have (T) $\heartsuit$ F. It follows that, for $\heartsuit$, the left quadrant is closed and all others are open, so that $\heartsuit$ is $\propto$. This also gives us two pairs for (i) in Case 3, namely, (Υ, ∞) and ($\propto$, $<$).

The technique demonstrated in the two foregoing examples shows how

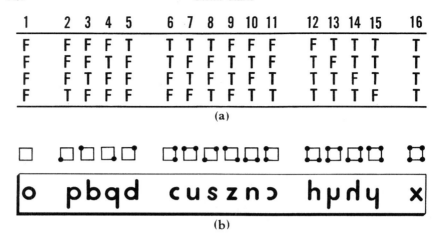

1	2	3	4	5	6	7	8	9	10	11	12	13	14	15	16
F	F	F	F	T	T	T	T	F	F	F	F	T	T	T	T
F	F	F	T	F	T	F	F	T	T	F	T	F	T	T	T
F	F	T	F	F	F	T	F	T	F	T	T	T	F	T	T
F	T	F	F	F	F	F	T	F	T	T	T	T	T	F	T

(a)

o   pbqd   cuszn⊃   hυny   x

(b)

Figure 17.4

Peirce *could* have found two (♥,♦) pairs for each case-form combination and so $2 \times 48 \times 256 = 24{,}576$ tautologies. Thus the discussion of Form G could be ended rather quickly at this point, but to do so would ignore the possibility that he obtained some of his results by the direct method (as in the two examples) and that, from these, he derived others by using one or more relationships among the binary connectives. It would seem to be virtually certain that he was aware of the conversion rule already discussed for our Form C and also included in the list of six rules that appear below as logical equivalences. These rules incorporate some of the symmetry properties of Peirce's iconic signs, and they suggest an alternate route to his final conclusions.

Three of the rules just referred to were suggested by their equivalents in the "logic alphabet," a notation recently devised by Shea Zellweger. Since an important clue to Peirce's choices for ♥ and ♦ was also based on the logic alphabet, it is appropriate to explain the framework for this notation, to list its sixteen signs, to indicate its relevance to the choices of ♥ and ♦, and to state a few symmetry rules based on it. Then, after stating the corresponding rules in Peirce's notation, we shall outline the steps for combining these rules both with the direct "from scratch" technique for finding (♥,♦) pairs and with the grouping of barely-false expressions.

The logic alphabet is a notation for the sixteen binary connectives, a notation devised by Zellweger independently of Peirce but sharing with Peirce's the features of *symmetry* and *iconicity*, the shape of each sign indicating its meaning (*Zellweger 1982*: 17–54 and his "Untapped Potential" paper in the present volume). The essential features of this notation are shown in Figure 17.4, where part (a) is repeated from Figure 17.2 and part

(b) refers to the logic alphabet. The signs in the last row are "letter shapes", twelve lower-case letters as they are commonly used in print and four instances in which the ordinary c and h shapes are placed in a new orientation.

The system is derived from an all-common basic square labeled externally at the corners: TT, TF, FT, FF in the order upper right, lower right, upper left, lower left. For each binary connective, a quadruplet of Fs and Ts, listed vertically in part (a), provides internal labels for the same four corners, each T determining a "stem" in the corresponding letter-shape. Thus the "and" symbol, determined by TFFF, must have a stem at the upper right and nowhere else and so is **d**. The "or" symbol, determined by TTTF, has three stems; it is **ɥ**, which has the same shape as the letter **h** but a different orientation. In part (b) the location of each stem is also shown by a heavy dot at the corresponding corner on the basic square.

Zellweger considers his notation to be a direct continuation not only of the approach taken but also of the subject matter developed by Peirce. Their notations can be interchanged by switching from "compartments" to "stems" or vice versa. The logic alphabet has been helpful both in revealing Peirce's use of the "barely-false" technique to choose (♥,♦) pairs by the direct method and also in suggesting his possible use of symmetry properties to derive other (♥,♦) pairs.[8]

The clue that cracked the code on ♥ and ♦ is interesting. Recall that Peirce's first two recorded choices for this pair were (𝒳,𝒳) and (∞,⅄), found in Case 1 for form (a). In the logic alphabet these pairs become (ɥ,ɥ) and (ᒷ,p), and they lead to the expressions P ɥ (Q ɥ R) and P ᒷ (Q p R). By the use of previously constructed tables, these expressions were both reduced to the form (ɥ x)(P Q R). Here ɥ x is a seven-stemmed ternary connective that indicates a false statement for P, Q, and R all false and indicates a true statement for each of the other seven truth-value combinations of P, Q, and R. The strategy suggested by this result was repeatedly confirmed when it was found that, in the logic alphabet and for all four cases, each of Peirce's choices for (♥,♦) always led to a seven-stemmed ternary connective, with the blank corner always corresponding case by case to the same excluded combination of truth values for P, Q, and R, and such that the iconic pattern of stems always displayed the barely-false requirement.

In the logic alphabet notation, when one of the two statements in (x O y) is negated, the letter-shape symbol at O is given a left-right flip (Nx) or an up-down flip (Ny). Corresponding negations in Peirce's box-X notation also involve flips; they are made about the diagonals of the X frame. To formulate and extend these rules, let r, u, and w be any statements. Let ♥ₘ be the sign obtained by flipping the box-X form of ♥ about its main diagonal (upper left to lower right); for example, ⟋ is transformed into ⟍. Let ♥s be the sign obtained by flipping the box-X form about its other (secondary) diagonal; for example, ⨯ is transformed into ⋉. Let $♥_h$ be the sign obtained by a left-right flip of ♥. Finally, let the negation of a statement or a connective be indicated by a horizontal bar above its sign. Then it may easily be shown that:

Rule 1. r ♥ $u \equiv r$ $♥_m$ $\overline{u}$;
Rule 2. r ♥ $u \equiv \overline{r}$ ♥s u;
Rule 3. r $\overline{♦}$ $u \equiv \overline{r ♦ u}$;
Rule 4. r ♥ $u \equiv u$ $♥_h$ r.[9]

Rules with all three letters are obtained by combining Rule 1 or Rule 2 with Rule 3:

Rule 5. r ♥ $(u$ ♦ $w) \equiv r$ $♥_m$ $(u$ $\overline{♦}$ $w)$;
Rule 6. $(u$ ♦ $w)$ ♥ $r \equiv (u$ $\overline{♦}$ $w)$ ♥s r.

As we shall see, Rules 4–6 are useful within any one case, but the grouping of forms by barely-false expressions can cut across cases. For example, the barely-false statement T ♥ (F ♦ F), abbreviated as T:FF and required for form (c) in Case 2, is precisely the one required for form (d) in Case 2 and also for forms (a) and (b) in Case 3. Similarly, we want T ♥ (F ♦ T),

Table VI

			Rule 5					Rule 4	
Forms	♡	◇	$♡_m$	$\overline{◇}$	Forms	$♡_h$	◇	$(♡_m)_h$	$\overline{◇}$
1a, b, c, d, e, f	ϒ	ϒ	ꝏ	⅄	1 j, l, h, k, g, i	ϒ	ϒ	∝	⅄
2a, f; 3c, e	ϒ	∝	ꝏ	>	2 j, i; 3h, g	ϒ	∝	∝	>
2b, e; 3d, f	ϒ	ꝏ	ꝏ	<	2 l, g; 3k, i	ϒ	ꝏ	∝	<
2c, d; 3a, b	∝	ϒ	ጰ	⅄	2h, k; 3j, l	ꝏ	ϒ	ጰ	⅄
4a, c	∝	∝	ጰ	>	4 j, h	ꝏ	∝	ጰ	>
4b, d	∝	ꝏ	ጰ	<	4 l, k	ꝏ	ꝏ	ጰ	<
4e, f	ϒ	ጰ	ꝏ	V	4g, i	ϒ	ጰ	∝	V

or T:FT, to be barely false for forms (b) and (d) in Case 4. It is readily seen that, by such a process, the 24 combinations of the six forms (a)-(f) in the four cases are partitioned into seven sets of 6, 4, 4, 4, 2, 2, and 2 combinations. All of the case-form combinations in any one of these seven sets will have the same valid (♥,♦) pairs. The forms and barely-false statements in these seven sets are as follows, where, for brevity, each form is indicated by a number (for one of the four cases) and a letter (for one of the six forms):

1a-f (F:FF)

2a,f; 3c,e (F:TF)	2b,e; 3d,f (F:FT)	2c,d; 3a,b (T:FF)
4a,c (T:TF)	4b,d (T:FT)	4e,f (F:TT)

With these groupings and Rules 4–6 before us (Rules 1–3 were needed to obtain Rules 5–6), we can readily devise a plan for getting all of Peirce's results, although we are not certain how *he* got those results. The plan is as follows.

1. Use the direct method to obtain just one (♥,♦) pair for each of the seven key (representative) forms 1a, 2a, 2b, 2c, 4a, 4b, and 4e. These seven pairs are shown in Table VI under the heading ♥ ♦, with each pair written on the line that has the corresponding key form at the far left.

2. Use Rule 5 to obtain another pair for each of the seven key forms. For example, having found (Υ,$\propto$) for 2a, we know by Rule 5 that (Υ_m,$\overline{\propto}$), or (∞,$>$), is also a valid pair. The seven additional pairs found in this manner are shown in Table VI under the heading ♥$_m$$\overline{♦}$.

3. Use Rule 4 to obtain two pairs for each of the seven forms 1j, 2j, 2l, 2h, 4j, 4l, and 4g on the right side of Table VI. For example, it follows by Rule 4 that P ♥ (Q ♦ R) ≡ (Q ♦ R) ♥$_h$ P, and thus the (♥,♦) and (♥$_m$,♦) already recorded for 1a, 2a, and 4a are transformed into (♥$_h$,♦) and (($♥_m$)$_h$,$\overline{♦}$) respectively. Similarly, this procedure of giving ♥ a left-right flip and retaining ♦ enables us to pass readily from 2b to 2l, 4b to 4l, 2c to 2h, and 4e to 4g. The fourteen pairs found in this manner are shown on the right side of Table VI.

4. (Optional). Rule 6 could be used to get from the first to the second pair in each line on the right side of the table or to check the second pair.

5. Use the grouping of barely-false statements to copy in the remaining 17 (24–7) case-form combinations on the left side of the table. For example, when we enter at 2a, write 2f, 3c, and 3e.

6. Use the Rule 4 pairing of forms within each case (a-j noted above, b-l, c-h, d-k, e-g, and f-i) to copy in the remaining 17 case-form combinations on the right side of the table. For example, 2f, 3c, and 3e on the left side lead to 2i, 3h, and 3g on the right.

The 96 (♥,♦) pairs, the (2 × 48) pairs, also the (2 × 4 × 12) pairs, obtained by this process yield a total of 21 distinct pairs plus duplicates. These pairs as they are grouped in Table VI agree exactly with the substitutions for ♥ and ♦ listed by Peirce. His system for recording his choices of ♥ and ♦, though not his notation, may be seen in the *Collected Papers* at 4.271. He gave no idea of his procedure and did not attempt to justify his choices, but it seems likely that he at least used the idea in Rule 4. Although he did not state and may not have known Rule 1 or Rule 2, he did know what the asterisk should be in both of the equivalences r ♥ u ≡ r $*$ $\overline{u}$ and

r ♥ $u \equiv \bar{r}$ ✳ u, when ♥ is 𝒴, ∝, ∞, or 𝒳. He included these eight results, and much more, in tables occupying two entire pages, MS 431A: 99–100, omitted from the *Collected Papers* (4.274). Again, he did not explicitly relate any of this to his choices for the (♥,♦) pairs.

Two additional comments are relevant following the above discussion of Cases 1–4. First, Peirce could have found suitable (♠,♣,O) triples in the other four cases, the ones that exclude in turn the other four combinations of truth values for P, Q, and R (FFT, FTT, TFT, TTT). He would have needed another table like Table IV, except that the entry in each cell would be the complement of the corresponding entry in Table IV. If he had done this, he would still have found $2 \times 12 \times 256 = 6144$ tautologies per case, but $8 \times 6144 = 49{,}152$ altogether. When the other four cases are treated by our methods, it turns out that the seven barely-false statements encountered in forms (a)-(f) include the last six of the seven already listed. The only new one is T:TT, associated with all six forms (a)-(f) in what may be called Case 8. The (♥,♦) pairs for 8a-f are (∝,𝒳) and (𝒳,∨). When the study is thus extended to include all eight cases, each of the eight barely-false statements is associated with exactly six forms, repairing the imbalance in Table VI. As representative of these eight barely-false statements, and thus as key forms, we may take 1a, 2a, 3a . . . 8a.

Second, it is possible that what we have called Case 4 was actually the first case to be considered by Peirce, although it was certainly the last to be reported by him on his pages 92–93. The imperfect (and earlier?) version of Table IV found at MS 1537:17 has only one pair of marginal sequences, basically our C and D. With ♠ read in the left margin, ♣ in the right margin, and O in the interior, the result is that if P and Q are true then R is true. Thus "P and Q true and R false" is excluded. In this connection, it is worth noting that on Peirce's pages 96–97 he discussed a technique involving Euler diagrams and illustrated it by deducing that if x 8 y and y ∞ z are true then x 𝒳 z is true. However, when he did this, he made no reference to any table.

Table VI displays in compact form all of the (♥,♦) pairs found by Peirce, and thus it complements his own Table IV, which displays the (♠,♣,O) triples. As both an exercise and a review, the following examples illustrate the four-fold use of Table IV and the selection of options from Table VI. The pairs of letters on the right indicate the pairs of marginal sequences used in the various cases to obtain these tautologies.

Case 1. (a)	$(x < y)$ 𝒴 $[(y \infty z)$ 𝒴 $(x △ z)]$	(A,B).
Case 2. (c)	$(y \infty z)$ 𝒳 $[(x < y)$ 𝚲 $(x ∝ z)]$	(A,D).
Case 3. (h)	$[(x < y) > (x ⊗ z)] ∝ (y \infty z)$	(C,B).
Case 4. (l)	$[(x ▽ z) \infty (y \infty z)] \infty (x < y)$	(C,D).

§9. Conclusion

It is hoped that the preceding exposition of Peirce's iconic signs for the binary connectives, together with the exploitation of their properties in the solution of some problems in logic, will draw attention to his outstanding achievement, one that deserves to be more widely recognized. Those who to date have not been familiar with these signs will now have an opportunity to learn more about them. In this paper we have begun to see some of the properties that have all the while been latent in Peirce's iconic signs. The six rules stated herein express a property that Zellweger calls "relational iconicity," a property that is also inherent in his own "logic alphabet" (*Zellweger 1982*: 19–24, 31–37, 44–45, and *1987*: 377–378).

This is not to say that Peirce was fully aware of the potential of his signs for the binary connectives, but he surely realized that, for a complete set of signs, iconicity was essential *if logical relations and operations on these relations are to be mirrored in the properties of the signs*. Actually Rules 1, 2, and 4 are only three of a set of eight rules related to the eight rigid motions of the X-frame, or of a square, into itself, these eight motions constituting a nonabelian group called the octic group, or the dihedral group, D_4. One of the subgroups is a Klein 4-group generated by the m-flip and the s-flip. The octic group results from combining *negation*, as in Rules 1 and 2, with *conversion*, as in Rule 4, and includes not only the flips mentioned in those rules but also rotations of 90, 180, and 270 degrees. Adding *complementation* (Rule 3) to the mix doubles the number of transformations to which all sixteen signs can be subjected. These remarks introduce both a larger context and a very interesting topic that must await exposition at a later time.[10]

The full 16-set of iconic signs that Peirce devised in 1902–04 represented his mature judgment about a matter that first engaged his attention no later than 1880. It was probably in the fall of that year that Christine Ladd-Franklin, then one of his students at Johns Hopkins University, wrote some notes on his paper, "On the Algebra of Logic," which had recently appeared in the *American Journal of Mathematics* (now at W4: 163–209). In that paper there was no discussion of a full set of signs, but in her notes Ladd-Franklin found fifteen "possible worlds" (later sixteen "propositions") based on two classes.[11] Although one of the fifteen was represented by an X-frame and the others by plane regions created by one or more lines, she did not have a complete system of signs based on a common frame. Her own paper entitled "On the Algebra of Logic" included two wedges to which Peirce later gave recognition in his cursives (*Ladd-Franklin 1883*: 25–27). Another one of Peirce's students, O. H. Mitchell, included a 16-line "Table of Propositions" in his paper, "On a New Algebra of Logic" (*Mitchell 1883*: 75). Still further indications of things to come appeared in Peirce's "Lecture on Propositions" (W4: 490–492 [1883]), in his "Lecture on Types of

Propostions" (W4: 493–500 [1883]), and in his "How to Reason" (or "Grand Logic") of 1893–95 (see MS 415: 02–20).

Eventually, in January 1902, Peirce applied himself seriously to the task of devising a complete system of suitable signs for the binary connectives. By that time he had developed and was continuing to work on his system of existential graphs, and it is quite likely that his predilection for iconic representation, as exemplified by this system of graphs, was influential in his creation of the box-X notation (see n. 2).

At any rate, in MS 431A: 87–95 Peirce gave a virtuoso performance on his newly-invented logical instrument. The crescendo, of course, appears in Table V, on that one amazing page, which once was lost but now is found— identified at "Fragments" MS 839: 262. This paper provides an extended comment on these pages, and it should be especially helpful when MS 431A, "The Simplest Mathematics," appears in *Writings of Charles S. Peirce.*

N O T E S

1. Tables, figures, and quotations from the Peirce manuscripts are by permission of the Department of Philosophy of Harvard University. I am grateful to the Peirce Edition Project in Indianapolis for access to their copies of some of the manuscripts, and especially to Max Fisch and Nathan Houser for their interest in this paper.

2. The idea of closing a space to denote exclusion appeared earlier in Peirce's treatment of existential graphs, where enclosure in a circle or oval indicated negation.

3. This date has been confirmed by McCulloch in personal correspondence with Shea Zellweger.

4. Figures 17.2 and 17.4 are taken from a composite figure designed by Shea Zellweger, who has kindly permitted its use here. During the preparation of the present paper Professor Zellweger made a great many helpful suggestions for its improvement and supplied several of the references. He, along with Donald Ray and Debbie Eglin at Mount Union College, saved me much typing time by their use of word processors.

All of the art work for the tables and figures in this paper, except Table V, was done by Mr. Warren Tschantz, Artec Incorporated, Alliance, Ohio. The playing card symbols in Tables II and IV have been added to the reconstruction of Peirce's drawings, as have the marginal sequence labels A, B, C, and D in Table IV.

5. Peirce often used expressions like $T \vee F$, and occasionally ones like $x \curlywedge T$, where T and F, respectively, represent a true proposition and a false proposition.

6. At 4.272, the X-frame is replaced by $: V :\text{-}$, and thus the references to "the quadrants" of the sign are meaningless.

7. Although this table was constructed by the writer by first checking all legible entries in Peirce's two incomplete versions (MS 839: 262 and MS 1537: 17) and

correcting an error, it is of course Peirce's work and appears here by permission of the Department of Philosophy of Harvard University.

8. Experience with the logic alphabet has also provided independent confirmation of the number of tautologies in Peirce's Forms D and E. Zellweger has constructed sixteen 16 by 16 tables showing, for each choice of the triad ◆, O, and ♥, the unique binary connective ⁂ which satisfies $(x ◆ y)$ O $(x ♥ y) \equiv x ⁂ y$. In Form D, with y identified with x, we want ⁂ to be in the set N established for Form A; so we ask for the total number of occurrences of the connectives s, ᵽ, ᴅ, and x in the 4096 cells of Zellweger's tables. The answer is 1280. For Form E the number sought is the number of occurrences of x in the tables, namely 680. These tables have not yet been published. Karl Menger (1962), using a quite different notation, developed a number of algebraic formulas for finding the ⁂ above, but he told Zellweger that he did not construct any of the 16-by-16 tables.

9. Rules 1–3 were suggested by their logic alphabet equivalents: R2, R3, and R5. Zellweger (*1982*: 28–29) applied all eight combinations of them (each rule active or inactive) to all sixteen binary connectives, displaying the results in an 8 by 16 table and from it obtaining "two waves of generalization" of De Morgan's Laws.

10. Zellweger (*1982* and *1987*) deals with much of this material, except for conversion, when it is treated not only as a logical operation but also as a symmetry operation. Also see Table IV of his "Untapped Potential" paper in this volume. [A more recent brief exposition can be found in *Clark and Zellweger 1993*.]

11. Fortunately these notes, with comments by Peirce written in the margins of some of the pages, are preserved with the Peirce manuscripts at L237: 1–54. On enlarging the set of propositional forms, see also W4: 169–173, the lengthy note thereon in W4: 569–571, and the paragraph from Nathan Houser's introduction at W4: xlix.

18.

UNTAPPED POTENTIAL IN PEIRCE'S ICONIC NOTATION FOR THE SIXTEEN BINARY CONNECTIVES

Shea Zellweger

§1. Remembrance of Things Past (Setting the Stage)

Once in a while it happens. Patience in the archives produces a letter that will serve as an ideal introduction to a paper.[1] Such a letter was written to Christine Ladd-Franklin. This letter follows in full; it will set the stage for what lies ahead. The date is 1898, at a time when Peirce was already developing his Existential Graphs (1896–1906), when he was only four years away from devising his notation for the sixteen binary connectives (1902–1904), and when he was a like number of years from an early excursion into his triadic logic (1902, 1909).[2]

Avebury
Newton Abbot
15th August 1898

My Dear Madam

I have to thank you most sincerely for your kind letter and for the promise of further communication which I shall be very glad to receive.

I think I ought to explain to you how it is that I became interested in your paper of 1883. I have been in occasional correspondence with Miss E. E. C. Jones of Girton College, Cambridge, whom I have never seen, but whom I wrote to some time ago with reference to a book she published. This was the beginning. Early this year she kindly sent a paper in "The Paradox of Logical Inference" . . . and in thanking her for this I explained that from various causes I had let logic slide lately, but that I still thought that the logical problems suggested by Jevons' logical alphabet (with which my interest in logic first began) *ought* to be more completely solved though I could not see of what practical use this would be. I asked her, therefore, not having easy access to a library here, if she could inform me how far these investigations had now been carried. Upon this she consulted Mr W. E. Johnson who replied that "with regard to the question of *terms* what Jevons began has been more fully carried out [in *your* paper of 1883] and that the other and much more difficult question (about relation of *classes* I think) has been worked out by Clifford in a paper in his Essays. Mr Johnson says the former question is quite possible to work out for four terms, but the latter he thinks not possible and that Clifford has gone as far as it is likely that anyone can go in attempting it."

Now I have Cliffords paper, and I want to see yours because I do not understand clearly the distinction between "terms" and "classes," but presume it has to do with what Miss Jones calls "The intermittent logical controversy as to Predication and Existence." [3] I am inclined to agree (as opposed to "Lewis Carroll" in his 'Game of Logic') that we should assume attributes to be compatible till they are proved incompatible—"Welsh Hippopotami" can certainly exist in our imagination, but if it can be proved that "Welsh" and "Hippopotami" are incompatible terms we need not trouble ourselves as to whether they are "heavy" or not.

In my view of things there are as follows:

[with "*Terms*" "*Marks*"] *Cases to be accounted for.*

0	1	2	The 2 "marks" for 1 term
1	2	4	are A (positive)
2	4	16	a (not-A) negative
3	8	256	and the 4 "Cases" are
4	16	65,536	

or (n being the No. of Terms)

n	2^n	2^{2^n}	

	1	2	3	4
	A	A	-	-
	a	-	a	-

Case 1 - No limiting relation

Case 2 - "Denial of Universe"

I shall be glad to see how the subject presents itself to you.

With many thanks
Yours very truly
Fredk. Binyon

P. S. I presume that you find no great crowd of people interested in this subject. This is my experience.

This letter is in the papers of Ladd-Franklin, where we find that she took courses in logic with Peirce at Johns Hopkins (1879–82), that she contributed with him to "Studies in Logic" (*1883*), and that she collaborated with him on Baldwin's *Dictionary* (*1901*).[4] The letter cuts nicely into the community of logicians: E. E. C. Jones, W. E. Johnson, Jevons, Clifford, Lewis Carroll, and of course, Ladd-Franklin. It nicely tags the issues that were struggling for clarification: terms, classes, predication, existence, the presence of the denied universe. Not only all of this. It also ends with a clear presentation of a triple series of numbers: n, 2^n, $2^{(2^n)}$.

Today, as a basic part of 2-valued logic and the propositional calculus, we start with the same triple series of numbers: n is for the number of atomic propositions, 2^n is for the 1st-order truth tables, and $2^{(2^n)}$ is for the 2nd-order truth tables. Most of what lies ahead in the present paper will focus on letting what I call *negation tables* add a 4th series of numbers to the above, namely, $2^{(n+1)} \times 2^{(2^n)}$. Formalizing this series in detail, giving it a good syntax, is where we make contact with the untapped potential in Peirce's notation for the sixteen binary connectives. The clue to all of this lies in a careful reading of Binyon's letter. It lies in W. E. Johnson's reply to E. E. C. Jones (uppercase added): "the other and much more difficult question [is] about RELATION of *classes*. . . . " Clifford (*1879*) described types of 1st-order relations in $2^{(2^n)}$, when $(n = 4)$, but as we will see, the 4th Series takes another step, one that leads into 2nd-order relations.[5]

Before we pick up on Peirce and make our way to the 4th Series, we need to say more about Ladd-Franklin. As a matter of historical perspective, it is important to recognize that she had a significant influence on Peirce when he laid the ground-work for what lies ahead. Gradually a run of main characters added some momentum to her efforts: Ladd-Franklin (*1880, 1883, 1890*), Schröder (*1892, 1898a*), Peirce (the 1902 "Minute Logic"), Davis (*1903*), and Shen (*1927, 1929*). But all too few were ready to listen— least of all after the full sweep in the tidal wave that was set in motion by Whitehead, Russell, and the *Principia*.

In reference to the present paper, it will become evident that in 1880 Ladd-Franklin was as much Peirce's mentor as she was his student. Her main idea here is contained in a golden quote that will for us become a primary precursor of all that lies ahead. This quote is from her 1880 paper (p. 218):

> There is great advantage in having every symbol of such a form that its opposite can be indicated by actually turning it over, as Mr. Peirce's symbol for the copula, which is such that $A \prec B$ means A *is wholly contained in B*, and $A \succ B$ means A *wholly contains B*. It is an advantage possessed by many of the letters of the alphabet, but not yet made use of.

Ladd-Franklin continued to pursue this idea right up to her last years. Her papers at Columbia show that well into the 1920's, along with her influence on Shen, she continued to experiment with the "mirror" (p. 218) symmetry properties of the letters of the alphabet. Unfortunately, although she played a significant role in establishing the count for the sixteen binary connectives (also *1880*; see n. 26), the notation that she devised got stuck on only half of the connectives (*1890*), those that we will call the eight pawns of a chess set (see Fig. 18.1 in §3).

When we cross the Atlantic and go to England, we come to the writings of Hugh MacColl. From 1877 to 1897 he published seven installments in the *Proceedings of the London Mathematical Society*, all of them under the title

"The calculus of equivalent statements." From 1880 to 1906 he had eight entries in *Mind*, all of them under the title "Symbolic reasoning." When we continue with the year 1880, we come to the first paragraph of the first paper in *Mind 5*: 45–60. This paragraph has a ring that is very modern and it could serve as a preamble for what lies ahead, especially when it is looked at in the light of the computer age. With the italics added, it reads as follows:

> Symbolical reasoning may be said to have pretty much the same relation to ordinary reasoning that machine-labour has to manual labour. In the case of machine-labour we see some ingeniously contrived arrangement of wheels, levers, &c., producing with speed and facility results which the hands of man without such aid could only accomplish slowly and with difficulty, or which they would be utterly powerless to accomplish at all. In the case of symbolical reasoning we find in an analogous manner some regular system of rules and formulae, easy to retain in the memory *from their general symmetry and interdependence*, economising or superseding the labour of the brain, and enabling any ordinary mind to obtain by simple mechanical processes results which would be beyond the reach of the strongest intellect if left entirely to its own resources.

When we continue with 1880 and return to the same volume in which the quote from Ladd-Franklin appeared, we see that the *American Journal of Mathematics* carried two more papers that significantly set the stage for what lies ahead. The first paper is by Peirce (*1880*: 15–57). "On the Algebra of Logic" (W4: 163–209) lays the groundwork for what came close to serving as a complete basis for logic. This was followed by another version in the same journal (*1885*: 180–202). "On the Algebra of Logic: A Contribution to the Philosophy of Notation" (W5: 162–190) combined his revised theory of signs with more of his improvements in logic, so that notation in logic was cast in terms of his three main kinds of signs, there called icons, indexes, and tokens. The second paper is by Stringham (*1880*: 1–14). "Regular Figures in *n*-dimensional Space" not only became a jump start that spurred the study of forms in higher dimensions, it also included many drawings of 4-dimensional forms. Even more to the point and very much in our favor, this paper included what may be the first drawing of a 4-cube.

What we have seen so far indicates that the closing decades of the nineteenth century was a time rich with absorbing the challenge of the impact of Boole and De Morgan. Taking full aim at what lies ahead, the five examples so far have called attention to Binyon and his clear statement of the triple series of numbers that underlies the 2-valued propositional calculus, to Ladd-Franklin and her focus on the symmetry of letter-shapes, to Mac-Coll and his notational overview of the connection between rule-governed systems of formulae and better reasoning, to Peirce and his semeiotic approach to the algebra of logic, and to Stringham and his lead off display in his drawing of a 4-cube. All of this at least raises the suggestion that what lies ahead would have been much more at home at the beginning of our century than at the end of it.

§2. One Part of One Part of the Minute Logic (1901–1902)

This part centers on Peirce. In July 1901 he outlined a plan for a book with one word in the title, "Logic" (MS 1579: 2–3). Later it became "Minute Logic." In November 1901 he outlined a chapter for that book with one word in the title, "Mathematics" (MS 250). Later it became "The Simplest Mathematics." This chapter was handwritten (MS 431) and then typed (MS 429; most of it at 4.227–323). The handwritten part consists of over 400 pages; the second half is rewrite pages. His notation took shape in several efforts in the rewrite pages at MS 431: 315–363 (January 1902), especially at MS 431: 330. At MS 430: 289 Peirce wrote "the mathematics of two values only, which is the simplest possible mathematics . . . "[6]

The iconic notation for the sixteen binary connectives that Peirce devised is still very much alive today. The untapped potential that it contains deserves to be examined very carefully. This potential, in fact, is robust enough so that, as an exercise at least, it will force us to *rethink* the propositional calculus. This could lead to a watershed clarification.

Such a strong claim, if it can be substantiated, does indeed give us pause to wonder about what dynamic has been afoot so that this part of Peirce has managed to stay submerged right up to his one hundred and fiftieth birthday. Whatever we come up with as an explanation for this long-standing inattention,[7] let me suggest that there is more at stake here than we may be inclined to suppose. There is a strong possibility that what we have before us is something that goes much deeper than, something that is far more significant than, just another exercise in passing. Some say that, as a minimum, it will simplify and consolidate what we already know.

The best preparation for what follows can be found in three places. The first is Max Fisch's paper on "Peirce's General Theory of Signs" (*1978*). The second is Glenn Clark's "New Light" paper, which is the immediately preceding paper in this volume. Clark gives us a detailed description of how Peirce constructed his notation and how he used it. I show what can be done with it, when it is carried beyond itself, when a direct continuation of it opens up new territory. The third is Schmeikal-Schuh's symmetry approach to logic (*1993*).

If you already know about Peirce, his general theory of signs is a matter of review. All thought is in signs. Cognition is minimally three-termed or triadic. Operations upon symbols, including substitution, are present in all logical thought. Logic calls for a mixture of icons, indices, and symbols. All three are needed, as is also the case for a perfect notation for logic. Even though central importance belongs to the logic of relations and even though logic is not the whole of semeiotic, logic in full is logic that is wholly semeiotic.

If you do not know about Peirce, look at what lies ahead from a historical point of view. Also from an evolutionary point of view. Consider the tides of change in the use of signs across the millennia, especially since the advent of writing. Some perspective here would include Gelb on writing (*1963*), Diringer on alphabets (*1968*), Ogg on the letters of the Roman alphabet (*1961*), and more recently, Sampson (*1985*) and Coulmas (*1989*), also on writing.

A less ambitious effort would use the history of numbers as a mental guide. Consider in particular what took place during the several centuries that it took Europe to go from the use of Roman numerals to the use of Arabic numerals (1000–1450). Much of this story can be recounted by considering the evolution of three things: number words, number symbols, and number operations. See Menninger (*1969*), Wilder (*1968*), Ifrah (*1985*), and Rotman (*1987*). In a parallelism that cuts across history and subject matter, most of the present paper is a replay, in uncanny detail, of going from Roman numerals to Arabic numerals. What lies ahead can be thought of in terms of the same three things: logic words, logic signs, and logic operations.

In brief, in what looks like backward history, the main thrust in what lies ahead will follow a plan that calls for a full reversal: logic operations, logic signs, and logic words. Start by putting the main emphasis on what is needed to make it easy to perform *logic operations*. Then build the *logic signs* so that they will fit, and cooperate with, these operations. After that, very much toward the end, come along with *logic words* that will reflect the ground already covered, not only for the logic operations but also for the good-fit logic signs.

What lies ahead is both disturbing and delightful. How this much can come pouring out of Peirce at this late date is disturbing. The economy of expression, the elegance, the perspicuity in this approach is delightful. Notice especially how small are the changes, how small are the departures from Peirce that are needed to activate the untapped potential in his notation. This brings up another moment of disquiet. If those changes are all that small, it follows that all the while we have been very close, just a step away—and still all that potential continues to go unnoticed.

§3. The Main Ingredients (2 4 16 128)

Acknowledge that the full story of the steps and stages in which Peirce engaged in sign-creation for logic has not been told. Looking at a few highlights of what happened during the evolution of his "claw" will serve as an illustration:

1870	1893–1895	1897	1902
$(x \prec y)$	$(A \propto B)$	$(l \propto b)$	$(x \propto y)$

His famous paper on the logic of relatives (1870) introduced the "claw" as a sign for inclusion (W2: 360). This is the same sign given in the golden quote from Ladd-Franklin. After that he chided Descartes for using ∞ for equality, since its "dissymmetry" far better qualifies this shape to serve as the copula for inclusion (MS 594: 54). At that time he also added, "Perhaps I might have suggested reversing it, . . . if the happy thought had not occurred to me twenty years and more too late." Later he placed it between "loved" and "benefactors" in his algebra of dyadic relatives (MS 513: 62). Still later, which brings us to MS 431 (1902), the same shape was modified as it now appears in Column 13 in Fig. 18.1c. This put legs on the Greek

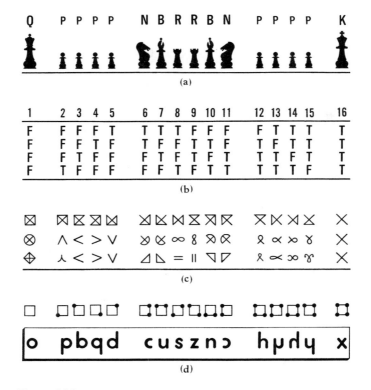

Figure 18.1

letter *alpha*, which for Peirce was shorthand for *and*. Peirce became his own best example when he wrote, "Indeed, a *mind* may, with advantage, be roughly defined as a *sign-creator in connection with a reaction-machine*" (MS 318 [1907]: 208).

Start with the sixteen binary connectives. This already allows entries in the first three series of numbers, when ($n = 2$), namely, two atomic propositions (A,B), four 2-place truth-value combinations (TT, TF, FT, FF), and the sixteen 4-place columns in Fig. 18.1b. This also introduces 128 as the corresponding number in the 4th Series. We will build the present paper around that number. The algebra that generates the progression to 128 can be easily expressed as follows:

n	2^n	$2^{(2^n)}$	$2^{(n+1)} \times 2^{(2^n)}$
2	4	16	128

The 128 will become the leading example, and it will serve as a model for the general case. Later, as the 4th number in the Ladder of Products (2 4 16 128), it will be treated as a particular case in the 4th Series.

Realize that school children could easily learn a yes-no sing-song for the first three numbers (2 4 16 -). Finger Logic runs a repeat on a point made by Dantzig (*1967*: 9). He tells us that "wherever a counting technique, worthy of the name, exists at all, *finger counting* has been found to either precede it or accompany it." Look at the obvious, the natural partitions in our anatomy, and notice that we have two thumbs (A,B), four pairs of corresponding fingers (TT-index, TF-middle, FT-ring, FF-little), and sixteen ways that the two fingers in each of the four pairs can (T)ouch or (F)ail-to-touch each other (the 4-place columns in Fig. 18.1b). The fun part is more advanced. Generating the 128 in terms of finger logic turns into a game in which the four pairs of fingers dance along to music written for the sixteen 4-place Touch-Fail combinations.

Bring in the iconic notations that Peirce devised, as shown in Fig. 18.1c, namely, his box-X system of signs, his original cursives, and his modified cursives. It is important to see that what he did in 1902 captures and stabilizes all of the information in the first three numbers (2 4 16 -). It would be a mistake to abstract away from, to leave any of it out. All of this iconicity, every bit of it, is needed for the 128.

Make a clear distinction between a *notational primitive* and an *axiomatic primitive*. For example, "8" is a notational primitive in the decimal system but it is not a primitive in the axiom systems for natural numbers, such as the one presented by Peano (*1889*), and earlier by Peirce (*1881*).[8] It follows that the signs in Fig. 18.1c are notational primitives (1902, 1904), even though, again back to 1880, Peirce knew well before Sheffer (*1913*) that all of the sixteen connectives could be expressed in terms of a single connec-

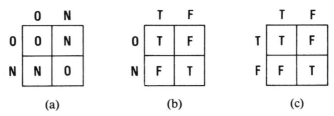

Figure 18.2

tive (W4: 218, 575 [1880]).[9] The lesson here is that elegance in building axiomatic systems should not overdetermine notation.

Introduce what I call *negation tables*.[10] These tables look just like truth tables. However, instead of having (T,F) stand for True and False, let (O,N) stand for Ogation and Negation, when Ogation stands for the *absence* of Negation. The corresponding 2-place table contains four negation pairs: OO, ON, NO, NN. Likewise, the 3-place table contains eight negation triplets: OOO, OON, ONO, NOO, ONN, NON, NNO, NNN. And so forth, for all combinations of (O,N) in the 2^n *n*-tuples in *n*-place negation tables.

Notice that Figs. 18.2a and 18.2c are full-fledged 2-groups in the algebra of abstract groups.[11] Notice especially that Fig. 18.2b is a mixed case, a hybrid that behaves internally just like an ordinary 2-group and one that will take on an extremely important role in what lies ahead. The advantage of negation tables is that they separate explicitly, they avoid pileups and run togethers, among the differences in Figs. 18.2a, 18.2b, and 18.2c. Even more important, a negation table is the *one multiplier that in one step goes from the 3rd Series to the 4th Series.* For example, the 8 in (8×16), which gives us the 128, when $(n = 2)$, $(n + 1 = 3)$, and $(2^3 = 8)$. As we will see, negation tables lead to group operations in symbolic logic that Whitehead (*1901*) was unaware of when he compared Boolean algebra to the inertness of argon (*Royce 1913*: 293–296).

Consider forms. Appreciate the level of abstraction already fully present in the six logical forms, the six general forms, the six algebraic forms, in which Peirce placed his signs and for which he identified *substitution sets of tautologies.* Along with the exposition that appears in the preceding paper by Clark, these forms and the number of substitutions in them are repeated in Table I. It follows that, in 1902, Peirce reached for a level of abstraction that went beyond "propositional variables" (x, y, z). These forms also contain "connective variables" (♠, ♥, ♣, ♦, O). The number at Form G is in parentheses; it covers the subset of (T)autologies in which Peirce was interested. Table II lists the forms that will be developed in the present paper. It continues with the level of abstraction found in Peirce but, in what leads to some surprising results, it looks for substitution sets of (E)quivalences.

Table I

Form A:	x O x	T	4
Form B:	x ♦ (x O x)	T	64
Form C:	(x O x) ♥ x	T	64
Form D:	(x ♦ x) O (x ♥ x)	T	1280
Form E:	(x ♦ y) O (x ♥ y)	T	680
Form G:	(x ♠ y) ♥ [(y ♣ z) ♦ (x O z)]	T	(24,576)

Table II

(1)	(A ♦ B) O (A ♥ B)	T	680
(2)	(A ♦ B) O (A ♥ B) ≡ (A × B)	T(E)	680
(3)	(A ♦ B) O (A ♥ B) ≡ (A ✳ B)	E	4096
(4)	(A ✳ B)	–	0
(5)	(A ✳ B) ≡ (A × B)	E(T)	1
(6)	(A ✳ B) ≡ (A ✳ B)	E	16
(7)	(A ✳ B) ≡ (?A ? ✳ ?B)	E	128
(8)	(?A ? ✳ ?B) ≡ (A ✳ B)	E	128
(9)	(?A ? ✳ ?B) ≡ (?A ? ✳ ?B)	E	1024
(10)	(?A ? ✳ ?B) ≡ (?B ? ✳ ?A)	E	2048
(11)	(?B ? ✳ ?A) ≡ (?A ? ✳ ?B)	E	2048
(12)	(?B ? ✳ ?A) ≡ (?B ? ✳ ?A)	E	4096
(13)	(A ✳ B) ≡ (?B ? ✳ ?A)	E	256

Consider tables. Take special notice of the two (16 × 16) tables that helped Peirce identify the substitution sets that he found for his Form E and his Form G. This is a real treat. At last, ever so belatedly, we have full access to the iconicity in these tables, in Clark at Table II and Table IV (this volume, pp. 315, 321). This presents a double opportunity: to take full measure of what Peirce was doing and, as we will see, to give central importance to a new (16 × 16) table.

Consider models. Recognize that Peirce did not have any hand-held models for his notation, especially those determined by that part of logic where he was clearly in the forefront, namely, the *logic of relations*. As we will see, many relations-theory-generated models are possible.

Now call upon analytic geometry as a primary analogy. The algebra of logic is not enough. The same goes for the geometry of logic. The same goes for using them side by side. Instead, giving a full account of the 128 will force us to combine and coordinate them at their best. Consequently, be prepared to use analytic geometry as method and to adapt it to logic.[12]

In the next three sections we will race across an analytic exposition of a triple isomorphism in the semantics of truth tables. First, we will isolate an algebraic form that covers 256 substitutions. Next, we will let the new (16×16) table display these substitutions. Then, after we cut back to the top half of this table, we will build a geometric model that contains 128 vertices. This makes the 128 a subset of the 256. All of this will ease into a fourth isomorphism, when we let transformational truth tables lead the way to some good syntax and a much better notation.

The form, the table, the model, and then notation, will play into a direct continuation of what we find in Peirce. Along the way and with great care, we will introduce *five small changes* that are modifications and extensions of what we find in Peirce. All of this will lead to a perfect match between a carefully designed notation and an analytic treatment of the 4th Series, a match that will also put an emphasis on iconicity, symmetry, and the logic of relations.

§4. Isolating the One Form ($2 \times 8 \times 16$ Substitutions)

When we look at what Peirce did in (1) in Table II, which is a repeat of his Form E, we see that he used a (16×16) table to help him find 680 triplets, all of them replacements for ($\blacklozenge$,O,$\heartsuit$) and all of them specifying a like number of tautologies.[13] In (2), another term that contains his sign for tautology has been added on the right; this expresses more explicitly how he used (1). In (3), continuing with how he used the card symbols, the sunburst asterisk on the right ($\ast$) introduces another algebraic mark that stands for any one of the sixteen binary connectives. All triples on the left side of (3) cover a set of 4096 equivalences, which include the 680 tautologies in Peirce's (1) and in our (2). This also introduces the *first small change*. Peirce was looking for substitution sets of *tautologies* (T), whereas (3) is looking for a substitution set of *equivalences* (E).

Now do the same thing at (4). It allows sixteen substitutions and it allows zero equivalences. Substituting in (5) is so trivial that, hardly worth any notice from Peirce, it allows only one tautology; nevertheless, also one equivalence. Algebraically, as a matter of pure form, (6) is like (3) because it is also at a higher level of abstraction. It too is a *master equivalence*, one that covers sixteen like-term identity-term equivalences.

Let the question marks in (?A ? $\ast$?B) stand for all combinations when Negation operates on the (A, $\ast$, B) parts of ($A \ast B$): OOO; OON, ONO, NOO; ONN, NON, NNO; NNN. This gives us eight negation triplets (1 3 3 1), the same ones that constitute the 3-place negation table $2^{(n+1)}$, when (n) is for the two propositions (A,B) and when the (+1) tells us that negation is also operating on the connective itself (N$\ast$). More explicitly, when we count the number of elements in (A,B + $\ast$), (2 + 1) becomes the exponent that generates the eight entries in the 3-place table. As a second example, let conversion (c) refer to the operation of converting from ($A \ast B$) to ($B \ast A$).[14] Hereafter, "OOOO" will stand for combinatorial identity, and "4-tuples" will refer to the sixteen combinations of (???c), as they are listed along the left side of Tables III and IV.

All equivalences from (6) through (13) are such that, when one of the asterisks contributes a factor of sixteen, the other asterisk is determined. It contributes a factor of one. Each question mark and the letter c contribute a factor of two. It follows that, in the present approach, these forms are sufficiently abstract so that they go beyond "propositional variables" (A,B) and "connective variables" $(*, *)$.

These forms also contain "operation variables" (???c). In other words, what makes (6) interesting is what happens to this general form when it is subjected to *logical operations*. This is something Peirce did not do, and it introduces the *second small change*. This adds (7), (8), and (9) to the picture. Notice that N is not allowed in (6), that it is *attached* on the right in (7), that it is *detached* from the left in (8), and that it is present in all combinations on both sides in (9). Even though (9) is without conversion, it still stands as a master equivalence that covers 1024 atomic equivalences.

Now bring in conversion and see that this adds (10), (11), and (12) to the picture. Here the 4-tuples (???c) operate on $(A * B)$ to obtain $(B * A)$. It follows that (12) allows all combinations of negation and conversion on both sides. It also follows that, along with including the 1024 in (9), (12) is a master equivalence that covers a special set of 4096 atomic equivalences. This master form fits together in such a way that, when all of the primary parts are continued in the same system, it unifies three subsystems: propositional variables, connective variables, and operation variables.

What we have seen so far shows that it takes only two operations (N,c) to explode the simplicity of (6) into the complexity of (12). What follows will stay manageable if we cut back to operations on only one side. This cuts back to (13) and to sub-form (7). This centers not only on the 256 (both operations) but also, as promised, on the 128 (negation alone).

These numbers count the entries in certain substitution sets of operation-driven *2nd-order relations*, the same ones that specify equivalence relations ($\equiv$) between the binary connective relations ($*$) between (A,B).[15] These sets of 2nd-order relations go a step beyond Clifford (*1879*) and his types of 1st-order relations in $2^{(2^n)}$, when $(n = 4)$. These sets become small systems that exist as total systems, when the equivalence interrelations are considered not one at a time, and perhaps trivial, but collectively. These systems are loaded, in layers, with many configurations of abstract structure. Those that involve negation alone constitute the 4th Series.

§5. Displaying the One Table (2 × 8 × 16 Cells)

We already mentioned the (16 × 16) table that Peirce used to help him find the 680 tautologies in his (1) and in our (2). A complete table for (3) calls for a 3-dimensional block of cells. The (16 × 16 × 16) entries in

Table III

(A * B)	1	2	3	4	5	6	7	8	9	10	11	12	13	14	15	16
OOOO	1	2	3	4	5	6	7	8	9	10	11	12	13	14	15	16
OONO	1	3	2	5	4	6	10	9	8	7	11	13	12	15	14	16
NOOO	1	4	5	2	3	11	7	9	8	10	6	14	15	12	13	16
NONO	1	5	4	3	2	11	10	8	9	7	6	15	14	13	12	16
NNNO	16	12	13	14	15	6	7	9	8	10	11	2	3	4	5	1
NNOO	16	13	12	15	14	6	10	8	9	7	11	3	2	5	4	1
ONNO	16	14	15	12	13	11	7	8	9	10	6	4	5	2	3	1
ONOO	16	15	14	13	12	11	10	9	8	7	6	5	4	3	2	1
OOOC	1	2	4	3	5	7	6	8	9	11	10	12	14	13	15	16
OONC	1	4	2	5	3	7	11	9	8	6	10	14	12	15	13	16
NOOC	1	3	5	2	4	10	6	9	8	11	7	13	15	12	14	16
NONC	1	5	3	4	2	10	11	8	9	6	7	15	13	14	12	16
NNNC	16	12	14	13	15	7	6	9	8	11	10	2	4	3	5	1
NNOC	16	14	12	15	13	7	11	8	9	6	10	4	2	5	3	1
ONNC	16	13	15	12	14	10	6	8	9	11	7	3	5	2	4	1
ONOC	16	15	13	14	12	10	11	9	8	6	7	5	3	4	2	1

($\spadesuit$,O,$\heartsuit$) are placed along three axes, and the connectives for the asterisk are in the corresponding cells. In other words, along with containing the 680 in (1) and (2) as a subset, this block of 4096 cells has three degrees of freedom. Algebraically, specify all three in ($\spadesuit$,O,$\heartsuit$) and the (*) will be determined, in order to honor the equivalence sign (see n. 13).

Table III is the new (16 × 16) table that we will highlight. The (2 × 8 × 16) has been cut in two: (c × ???) is for the (2 × 8) 4-tuples down the side and (*) is for the 16 connectives along the top. The first row in this table is for (6), when all positions in the identity element (OOOO) operate on all substitutions for the asterisk (*). The first four rows are for all pairs of (O,N), when they operate on (A,B). The first eight rows are for (7) and (8), when the triplets of (O,N) operate on (A * B). All sixteen rows are for (13), when the 4-tuples (???c) operate on the primitive places in (B, *, A). It follows that the 256 cells in Table III are isomorphic to the 256 substitutions in (13).

The first eight rows of Table III have become a lopsided hybrid with respect to Fig. 18.2b. This (8 × 16) half table continues with conversion on hold, it is restricted to negation, and it readily partitions itself into two (8 × 8) sub-tables. Fig. 18.3 is a complete simplex that displays the interconnectivity in the 8-set of connectives that have an *odd* number of T's in their 4-place truth-value combinations (- 4 - 4 -). Fig. 18.4 is a degenerate simplex that does the same for the *even* T's (1 - 6 - 1). These diagrams will come back again.

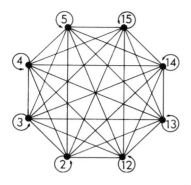

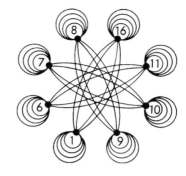

Figure 18.3

Figure 18.4

One way to obtain the 1024 in (9) is to extend the top half of Table III. This yields another 3-dimensional block of cells (8 × 8 × 16): the negation triplets (???) by the negation triplets (???) by the connectives ($A * B$). We need two such blocks (2048) for (10), and also for (11). Now double this for the 4096 in (12). A one-shot display of (12) not only contains all of the above as subsets but it also starts outright with a special block of cells (16 × 16 × 16): the 4-tuples (???c) by the 4-tuples (???c) by the connectives ($A * B$). This master form includes both departures from Peirce. It is for *equivalences* and two of the axes in it are for *logical operations*.

Another way to obtain the 4096 in (12) is to extend the columns in Table III. This gives rise to sixteen *super equivalences*, each of them consisting of sixteen mutually equivalent terms. Go to the top cell in each column to find the first term, the identity term, in each super equivalence. Then go down each column and combine all of the 4-tuples along the left with the connectives in the corresponding cells. This leaves all of the terms in each column equivalent to the one at the top and thus to one another.

For example, the second column starts with (A 2 B), also called the Peirce connective (Not-A and Not-B). The first four terms are as follows:

$$(14) \quad (A\ 2\ B) \equiv (A\ 3\ NB) \equiv (NA\ 4\ B) \equiv (NA\ 5\ NB) \equiv \ldots$$

Repeating the sixteen terms of this super equivalence *along both axes* of still another (16 × 16) table yields a total of 256 ordered pairs. Each pair specifies the elements in a 2-term equivalence, so that the 256 pairs consist of sixteen self-pairs in the main diagonal and two sets of 120 unlike pairs, all pairs in one set standing in commuted order to those in the other set.

Sixteen of these (16 × 16) tables, one for each column in Table III, add up to the 4096 atomic equivalences in (12). This view of this master equivalence puts the super equivalences in pairs, one pair for each substitution in the sunburst asterisk. Sixteen rows across the top line up with and repeat the sixteen columns along the left. This gives us two copies of Table III that are now serving as two matching (16 × 16) border faces. All of the 4096 ordered pairs become 2-term equivalences that appear full blown in the (16 × 16 × 16) block of cells.

When any triplet (???) operates on this *total block of cells*, the same triplet acts on both sides of all of these 2-term atomic equivalences. When all of the triplets are activated, the eight big blocks (8 × 4096) could be placed at the vertices of a very large 3-cube. The great beauty here is that all of the cells in all of the blocks continue to contain valid 2-term equivalences.

§6. Building the One Model (2 × 8 × 16 Vertices)

We started with a number of forms, all of them toward the simple side, and then we cut back to (13). We called on a matching set of tables, and then we centered on Table III (the 256), the one that goes with (13). This time, amid all those possibilities, we will go directly to the one model. It will go with the top half of Table III (the 128), and also with the corresponding substitutions in sub-form (7). It will complete the triple isomorphism.

This model will continue with what we said about backward history. Start with logic operations *and then* look to the needs of logic signs and logic words. But we know that (N,c) explodes (6) into the thousands of interrelations in (12). It will help to cut back again, this time to one operation on one side, namely, to negation alone and the 128. More specifically, back to sub-form (7), which also puts conversion on hold. With respect to generalizing, the 128 is still a small number, one that comes early in the 4th Series.

Before we continue with the top half of Table III, we need to prepare more deeply for the part played by the eight negation triplets (???). When the triplets are placed at the vertices of a 3-cube, it introduces a Boolean lattice of length three, one that has (OOO) as the least element and (NNN) as the greatest element. When the origin of the 3-cube (OOO) is treated as the identity element, the triplets constitute an Abelian 8-group, the one obtained from a recursive direct product of three 2-groups (O,N), namely, $(C_2 \times C_2 \times C_2)$ or $(C_2)^3$. When, as in crystallography, three coordinate mirror planes act on a nearby point, such as the origin of the 3-cube (OOO), it generates the same 8-group. When this Abelian 8-group is extended to a vector space over the field of integers modulo 2 (Z_2), this vector space is 3-dimensional and has $\{(NOO), (ONO), (OON)\}$ as a basis. All of this and more to come, just because we let the O-letter in Ogation stand for the absence of Negation.

We can now say that the top half of Table III, along with the 128, comes from taking the Cartesian product of two Boolean lattices, when the 3-place negation table (1 3 3 1) operates on the connectives (1 4 6 4 1). Fig. 18.5

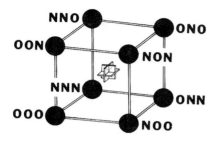

Figure 18.5

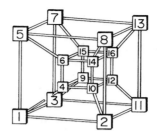

Figure 18.6

puts the first lattice (1 3 3 1) at the vertices of the 3-cube. The origin is at (OOO), and (ONO), also expressed as (N ✳), passes through the center of the cube. Specifying this part of this lattice in this way forces duality (NNN) to become a mirror reflection. It follows that, along with being treated as a third dimension, duality is acting from front to back in Fig. 18.5.

The large circles in Fig. 18.5 are now ready to become containers for the top half of Table III. It is not very helpful, however, to make a vague heap every time the numbers (1–16) are placed in one of these circles. This brings the second lattice to the rescue. Fig. 18.6 puts the connectives (1 4 6 4 1) at the vertices of a 4-cube. When it becomes the identity 4-cube and when it is put inside of the OOO-circle in Fig. 18.5, two things happen. Variations of Fig. 18.6 appear in the other seven circles, and these variations are in keeping with the other seven rows in the top half of Table III. The drawback, however, and there is an improvement just ahead, is that these variations are being determined by Table III and not by the mirror reflections.

Notice the mismatch in dimensions, when Fig. 18.5 acts on Fig. 18.6. Something 3-dimensional is acting on something 4-dimensional. One way out of this mismatch is to do something about Fig. 18.6. This brings Fig. 18.7 to the rescue. This is the 4-cube when it is casting a shadow of itself in 3-space. It has become a "solid shadow," seen here as a 3-skeleton of a 4-cube, and it takes the shape of a rhombic dodecahedron.[16] In keeping with the quotation from MacColl, it also puts us one step away from a model that will let the mirrors do the work.

When Fig. 18.5 acts on Fig. 18.7, the mismatch has gone away. Something 3-dimensional is acting on something 3-dimensional. This puts all of Fig. 18.7 inside of the OOO-circle and then the negation triplets operate on Fig. 18.7. The mirrors themselves generate the top half of Table III. The eight rows generated by the 3-place negation table (???) now inhabit variations in the eight dodecahedra in the circles in Fig. 18.5.

Fig. 18.8 gives a better view of this model. The identity dodecahedron

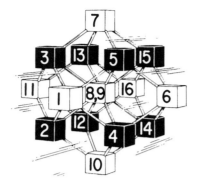

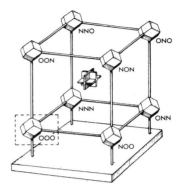

Figure 18.7   Figure 18.8

(OOO), now seen as a "logical garnet" and with all of the detail in Fig. 18.7, is inside of the broken rectangle. This completes the triple isomorphism. The 128 vertices in Fig. 18.8 correspond one for one with the 128 cells in the top half of Table III and with the 128 substitutions in sub-form (7): the three entries in each vertex-cell-substitution convey the same information. This also completes our analytic exposition. All of it followed from doing no more than letting a carefully selected negation table (???), seen as $2^{(2+1)}$, operate on a carefully prepared arrangement of the sixteen binary connectives $(A * B)$, seen as $2^{(2^2)}$. All of it lays out what we have in mind for the 4th Series, for the 128, when $(n = 2)$.

Meanwhile, more about the dimensionality of Fig. 18.8, which is indeed a very special logical structure. The 8-set of logical operations in Fig. 18.5 is also the 8-group $(C_2)^3$. The 16-set of logical connectives in Fig. 18.7 is also a 3-shadow of a 4-cube (Z_3^4). When Fig. 18.5 acts on Fig. 18.7, also seen as (8×16), it generates the *product* of $(C_2)^3$ and (Z_3^4). It should have been obvious: add exponents. Fig. 18.8 is a model that is 7-dimensional.

Lalvani (*1989*), with reference to logic or not, would treat the under-lying structure activated by Fig. 18.8 so that it would appear as a 2-flat draw-ing of the skeleton of a 7-dimensional cube (128 vertices). The 3-cube in Fig. 18.5 would be highlighted as a sub-frame of lines on the 2-flat drawing. The 4-cube in Fig. 18.6 would appear eight times, all of them identified as position-specific sub-skeletons on the 2-flat drawing. The variations in the vertices of the eight sub-flat 4-cubes occur with respect to both the 2-flat drawing taken as a total frame and the 3-cube sub-frame on it.

My way of doing the same thing is to see Fig. 18.8 as a custom-designed adaptation in which the mirror reflections in $(C_2)^3$ generate a configura-tion of eight "solid shadows" of a 4-cube (the dodecahedra), each of them displayed as a (Z_3^4). This treats Fig. 18.8 as a 7-dimensional object, one that could be called a $(C_2)^3(Z_3^4)$. In keeping with a row by row partition of the top half of Table III, this view activates but does not visually display all of the lines in the 7-dimensional 2-flat mentioned above. This view, instead, emphasizes that Fig. 18.8 has been designed to display the direct connections in the *transformational 3-dimensionality* that shows up when the three mirrors that embody the 3-place negation table (???) generate the 8-set of logical garnets.

Both of these 7-dimensional models call attention to the clarifications that must be mastered to make transformational peace with the 128. The challenge lies in capturing the supersymmetry in the logical garnet. (i) Each connective at the vertices of Fig. 18.7 is 2-dimensional. It will become obvious that all of them are bound by the symmetries of a square. (ii) The total count of the connectives is 4-dimensional. They are at the vertices of a 4-cube. (iii) Relocations of the connectives are 3-dimensional. They obey the mirrors in the 3-cube. All of this all at the same time: *3-dimensional mirrors are operating on a 3-dimensional arrangement of a 4-dimensional count of 2-dimensional connectives.* The consequences are severe. The standards for a good notation are going up because it will be necessary to capture and stabilize all of this to arrive full force at the 4th Series.

§7. The View from Where We Are (By the Number Names)

We have made good on the semantics of the 128. The negation table (???) became the multiplier that operated on a 16-set of truth-value combinations $(A * B)$. This led to the 128, also seen as a substitution set of 2nd-order interrelations, namely, the negation-driven equivalences that inhabit these truth-value combinations. Analytically exposing the 128 led to a triple isomorphism; more explicitly, to what we found for sub-form (7), for the top half of Table III, and for Fig. 18.8. All of this was done to get our logic straight, clear and precise, *before* we make any case at all for notation and syntax.

We have, meanwhile, fallen upon some good fortune. It follows from the level of abstraction (L75: 404–406) that Peirce brought to the six algebraic forms listed in Table I. This is a lot like making statements that cover whole classes of numbers, instead of making statements for a few specific cases. The same thing follows from our 3-way analysis of the 128. The connectives themselves ($*$), the connectives in general ($*$), have become abstract objects that show up in the mathelogical forms that cover the layers of substructure that exist in whole classes of logical expressions.

For example, from an analytic point of view, when we go back to the product of the two Boolean lattices, we see that Equivalence (15) does not focus on any connective in particular:

$$(15) \quad (\text{NOO}) \ [(???) \times (A * B)] \equiv (???) \times (A * B)$$

It treats Fig. 18.8 as a transformational model of the set (8×16) of all substitutions in $(???) \times (A * B)$. When the vertical front-to-back A-negation mirror plane (the NA in NOO) that passes down through the center of Fig. 18.8 operates on this *total structure*, the left four and the right four dodecahedra are reflected into each other. This pushes 128 connectives in one push; two sets of 64 vertices trade places. Later we will use a notation that will show that the right side of (15) must be a repeat of what is in brackets on the left.

The abstract structures that are coming to the fore are a fundamental part of the algebra of logic. They reaffirm what Peirce was doing and they put in place the long and sustained preparation up to this point. They have also been reduced to a minimum. The connectives, as we have expressed them, are down to nothing but number names (1–16). "By the number names," we might say, such that all substitutions, all cells, all vertices, as we have them, will not change one whit in the next hundred and fifty years.

At the same time, however, all of this logic by the number names is just one question away from making contact with the deep issues that show up in aesthetics, ethics, and pragmatics. This question could take many forms. For example, what kind or type of notation, not what notation in particular, *should* we bring into common use, when it is built to fit and connect precisely with the triple isomorphism in the semantics of truth tables? The answer lies in the untapped potential.

§8. Easing into the Fourth Isomorphism (Return to Peirce)

We need to replace the number names (1–16) with a special set of signs. The requirements are such that the standards should grow out of the 128. The signs should embody, reveal, and participate in all of the abstract structure that we have described for the 128. We will settle for nothing less than a perfect match. But this is only a small part of what lies ahead. What works for the 128 must also serve as groundwork that will generalize to any number in the 4th Series. This can be done and the place to start is with Peirce.

We know that Peirce activated the *elements* in the two Boolean lattices because the eight negation triplets and the sixteen connectives were two cognitive frames that repeat many times in his writings. For example, he used the triplets in his 1880 paper (W4: 172, 203), and his 1880 class in logic struggled to establish the count on the sixteen (W4: 569). In 1902 he took a big step when he put the sixteen connectives in eight columns.[17] This called on only half of his signs. Each column was in terms of negation and the same sign. To date, however, we have no evidence that he ever took the *product* of the two Boolean lattices, as we have it, for the 128 and the top half of Table III.

Peirce's iconic notation contains enough good clues to specify the *class of notations* that will meet the requirements. A notation that meets them in a particular way will now be described in detail. This notation repeats what we find in Arabic numerals, which is an abacus carried to the mental level. It will consist of a special set of *notational primitives* that behave like small Venn-Carroll diagrams, when they are carried to the mental level. This notation also gives full rein to the quote from Ladd-Franklin. It will consist of *phonetic letter-shapes* that also possess some carefully determined algebraic and geometric symmetry properties.

I call it the "Logic Alphabet." This notation should not be confused with Jevons's "logical alphabet," which was also mentioned in Binyon's letter. His alphabet is an early version of 1st-order truth tables: large letters for true $(A,B,C\ldots)$ and small letters for false $(a,b,c\ldots)$.[18] This stops at the 2nd Series; it falls far short of the requirements for the 4th Series. The logic alphabet that I have in mind will now be described in three parts, first the genesis, then the anatomy, and then the physiology of the letter-shapes.

§9. Genesis of the Letter-Shapes (Separate the Box)

All of the letter-shapes are placed inside of an *all-common basic square.* The construction of this square repeats what Peirce did when he treated two "quantities" as dimensions (a,b), such as we see in Fig. 18.9a.[19] This also repeats how Carroll (*1958*) used the same two lines, as he did in 1886. Peirce drew these lines as early as 1880 (W4: 569); again in 1902 (MS 431: 330), before he switched on the same page to Fig. 18.9b. His cursives *close and call false* all combinations of four regions (Fig. 18.1c). In 1904 he not

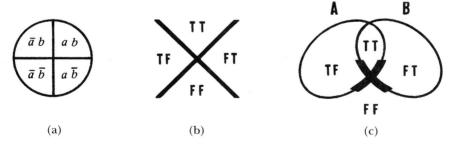

(a)　　　　　　　(b)　　　　　　　(c)

Figure 18.9

only added the enclosing sides of a box to the X-part of Fig 18.9c. He also, like McCulloch (*1965*), tells how to put a dot in a quadrant (MS 530: 126).

Now add all combinations of *enlarged dots* to the corners of the basic square. This goes back to the first row of Fig. 18.1d. As we will see, it is astonishing how much follows from, how rich these (dot, no dot) corners will become, when the squares are treated as adapted matrices, in this case (2×2) matrices. These dot-squares *separate the box* from Peirce's box-X, and this introduces the *third small change*. This change is similar to Boll (*1948*: 133), who placed the same combinations of dots inside and not far from the corners of a 16-set of squares. This change is also similar to Wiscamb (*1969*), who darkened all combinations of four-dot Cartesian lattice-corners in precisely the same square pattern used in the present paper.

This change reminds us of the special act that isolates the key characteristic that made our alphabet possible, invented only once, as told by Diringer (*1968*: vol. 1, 435), and based on using letters for sounds alone and not for things, ideas, or syllables. This change also contains a full repeat of the key distinction that we find in Arabic numerals and not in Roman numerals. Expressing this distinction is easily made in terms spurred by Chomsky (*1957*) and used by Skemp (*1982*). When the digits are all alike in what we write in the surface structure, as in "222," the differences are carried in the deep structure, namely, "(H)undreds, (T)ens, (U)nits." The mental abacus works because we think "HTU" and we write "222." But Peirce could not separate his all-common X-frame from what he wrote, every time for every sign. The logic alphabet has no difficulty on this point. It thinks dot-squares and it writes stem-shapes.

§10. Anatomy of the Letter-Shapes (Code the Stems)

All of the letter-shapes belong to a *system of stems*. This introduces the *fourth small change*. Peirce centered on the four open regions that surround his X-frame. The logic alphabet puts the main focus on the four stems that constitute the **x**-letter. Reducing the code space to short dot-oriented stems is what makes it possible to separate the dot-squares. The all-common square, like "HTU," stays in the deep structure, in the unwritten think part. This abandons Peirce's Fig. 18.9b and returns to his Fig. 18.9a, which continues to dove-tail with analytic geometry and its use as a primary analogy.

All of the letter-shapes are shown in Fig. 18.10, where they have been placed at the sixteen stations of a convenient (4×4) diamond-shaped frame.[20] The labels given to the stations read like a "clock-compass." The letter-shapes have also been placed inside the dot-squares. The stems correspond one for one with the dot-corners. Consequently, the (stem, no

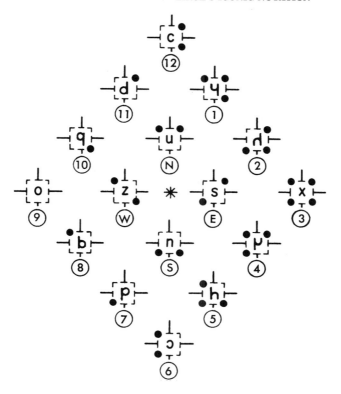

Figure 18.10

stem) corners of the letter-shapes yield a matching set of adapted (2 × 2) matrices.

The anatomy of the letter-shapes contains a *stem test* that determines the positions that are true. Put the four pairs in finger logic (TT, TF, FF, FT) at the corners of the basic square, clockwise from the upper-right. Now, like (T)ouch, let a s(T)em stand for (T)rue, and consider the d-letter located at 11 o'clock in Fig. 18.10. As illustrated in Equivalence (16), this phonetic letter-shape sign takes on triadic meaning:

(16) $(A \text{ d } B) \equiv (A \ \Box \ B) \equiv (A \ \$\$ \ B) \equiv (A \text{ and } B)$

It is an *icon* when it stands for the dot-square at 11 o'clock because, like one dendrite joined across a synapse, it has one s(T)em close to, leading from, and determined by one enlarged dot in the upper-right corner (T - - -) and nowhere else (- FFF). It is an *index* not only when it is treated as a 4-fold truth table combination ($\vdots\ \vdots$) but also when it substitutes for (A TFFF B). It is a *symbol* when it abbreviates for the word "an(d)" in (A and B).

And so forth, in Fig. 18.10 and like (16), for all of the letter-shapes, all of them in triadic parallel with the corresponding dot-squares, truth table 4-folds, and words for the connectives. This is where we repeat the special act that gave us a sound alphabet, one that, within limits, makes what we read a phonetic copy of what we say. Now we have a carefully crafted, generic, shape alphabet. It comes from recasting Peirce's X-frame when it is combined with an organic, built-in version of Leśniewski's spoke system (see n. 19). It carries a context specifiable triadic copy of the intended meaning.

The anatomy also contains a *size test* that partitions the letter-shapes into a pair of half sets (8,8). The first half is "tall." All of them are pulled long in the vertical (**p b q d h ꟼ ꓷ ꓴ**); this tells us that they have an *odd* number of stems (- 4 - 4 -). The other half is "squat." All of them are pushed from above to match their width (**o c u s z n ꓛ x**); they have an *even* number of stems (1 - 6 - 1). As we will see, this partition (8,8) plays a fundamental role in the underlying structure of the binary connectives (1 4 6 4 1).

When the stem condition and the size condition interact, the anatomy maintains an orderly *stem/circle proportion* that runs the length of the Boolean lattice (1 4 6 4 1). The least-stems-element **o**-letter (1 15) is all circle and no stems, and the most-stems-element **x**-letter (15 1) is no circle and all stems. Organically, while the stems emerge one by one along the (**o-x**) lattice direction of change, the curved section of a circle decreases. In reverse (**x-o**), while the stems drop off, the point at the cross-center of the **x**-letter grows until it becomes a circle. The **z**-letter is the only exception to this orderly exchange between the number of stems and the curved section of a circle. It could be replaced by a reverse **s**-letter.

The anatomy also contains a *diagonal test* that partitions the letter-shapes into another pair of half sets (8,8). This test finally comes around to *conversion* and the bottom half of Table III, when ($A * B$) changes to ($B * A$). Start with the dot-squares in Fig. 18.10 and the diagonals that pass from upper right to lower left. Letter-shapes in dot-squares that are symmetrical are convertible (**p d h ꓴ o z s x**). The other half is not convertible (**b q ꟼ ꓷ c u n ꓛ**). This diagonal test is what has become of Ladd-Franklin's suggestion that a sign should have left-right symmetry when the *relationship* to which it refers is symmetrical. Accepting her suggestion implies that Peirce must set aside his Fig. 18.9a and opt for his Fig. 18.9b (MS 431: 330). The duality at work here is such that locating the (A,B) sign-frame axes in either figure shifts the diagonal-test axis into the other position.

Along with this joint differentiation with respect to stems, size, proportion, and diagonals, the internal anatomy of the letter-shapes must also meet a vital test that looks for the presence of a *super truth table*. This test has gone unnoticed in the approach suggested by the quote from Ladd-Franklin. This test can easily go unnoticed by someone who has been using the logic alphabet a long time. It follows that, with respect to this test, a "blind user" could have *logica utens* without having the corresponding *logica docens.*

In any case, the conditions required by this test must be present in the internal anatomy of the letter-shapes, exactly, or the system will not work. These conditions (self-flippable, self-rotatable) are determined by the symmetry requirements of the letter-shapes because the letter-shapes are also *topological icons*, at least to the extent that they must bend and twist and stem in just the right places.

This test puts all of the letter-shapes into four levels of symmetry-asymmetry (2 2 4 8). Two of the letter-shapes (**o x**) are two-way self-flippable, with reference to (x,y) axes, and they are self-rotatable, with reference to 180 degrees and the flat of the page. Two (**s z**) are not self-flippable but they are self-rotatable. Four (**c u n ɔ**) are one-way self-flippable but they are not self-rotatable. Eight (**p b q d h ꓕ ꓒ ꓩ**) are neither self-flippable nor self-rotatable.

Now place the top row of Fig. 18.1 into correspondence with the bottom row. It so happens that the power of chess pieces can be put into the same four levels (2 2 4 8). The Queen and King stand over (**o x**), the Rooks over (**s z**), the Knights and Bishops over (**c ɔ**)(**u n**), and the pawns over (**p b q d h ꓕ ꓒ ꓩ**). This correspondence vividly exposes the logic alphabet as a total system.[21]

Four letter-shapes (2 2 - -) subdivide into two kinds of central symmetry (**o x**)(**s z**). Consequently, in Flatland and in Figs. 18.11 and 18.12, four chess pieces (Q K)(R R) will compete for the symmetry combinations that pile up at the center where the (A,B) axes intersect. The heavy chess pieces are over the even-stemmed squat half, which contains all of the *special* symmetry letter-shapes (2 2 4 -). They stand inside of the circles in Fig. 18.11. The pawns are over the odd-stemmed tall half, which contains all of the *general* symmetry letter-shapes (- - - 8). They stand outside of the circles in Fig. 18.11, away from the (A,B) axes but over the corresponding signs in Fig. 18.12.

At this point it is helpful to sound a precaution, one that Menninger tells us about in an interesting historical parallel (*1969:* 322–327). By best account, when Gerbert (940–1003) traveled south to Spanish border towns (967–970), he learned how to write West Arabic numerals. But by the time he became Pope Sylvester II in 999, he and his disciples still did not know how to make written computations. Not the existence but the essence of Arabic numerals had eluded him. In other words, we will come to the same dead end should we stop short at the anatomy of the letter-shapes.

With or without this precaution, what we have said about the anatomy

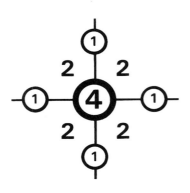

Figure 18.11

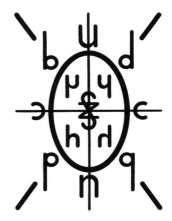

Figure 18.12

of the letter-shapes underscores the fact that all of the binary connectives are now full-fledged members in a democratic community, one that, as paradoxical as it may seem, has also become a specialized and a highly differentiated system. All of them must be present and ready for action. The need to have easy access to all of the meaningful partitions among the letter-shapes will become obvious when we examine some well chosen physiology.

§11. Physiology of the Letter-Shapes (Flip the Signs)

All of the letter-shapes *stand in new positions*, when they participate in logical operations. Now that they have the needed anatomy and now that they populate a community of dot-squares, all of them dance the same steps to make the same logical operations. We will use both the front sides and the back sides of the letter-shapes, when we *flip* them. This gives center stage to the quote from Ladd-Franklin, and it introduces the *fifth small change*. We have no evidence that Peirce used diagonal flips on his box-X forms. Some of his cursives do not work for these symmetry operations.

The trick is to treat negation as a *transformational mirror*, one that reflects or flips 180 degrees. The consequences in the algebra of logic are deep and systematic because *anytime that a negation table enters into a product set with a truth table it activates transformational symmetry groups*, of the hybrid kind, as in Fig. 18.2b. In reference to (16) and the iconic truth tables in Fig. 18.10, logic operations become symmetry transformations that act on the notational 4-folds (⁚ ⁚) obtained from the adapted (2 × 2) matrices.

We have, in effect, come to the moment when the letter-shapes, built to exist at several levels of symmetry-asymmetry (2 2 4 8), will find themselves trapped in the same system of symmetry transformations (???c). Like Arabic numerals, the payoff lies in the *rules that fit* the calculations.

The four symmetry rules, also called flip-mate-flip and flip, are as follows, when (*A* d *B*) serves as an example. (R1) The (NOOO) in (N*A* ✳ *B*) is an *A*-flip negation that *flips* all of the letter-shapes from left to right, when the N*A*-flip for (N*A* d *B*) becomes (*A* b *B*). The ready made reminder here is that it just so happens that the shape of the *A*-letter itself has left-right symmetry. (R2) The (ONOO) in (*A* N ✳ *B*) is a counterchange negation that *mates*, or reverses, the presence and absence of all stem positions, when the (N ✳)-mate for (*A* Nd *B*) becomes (*A* h *B*). (R3) The (OONO) in (*A* ✳ N*B*) is a *B*-flip negation that *flips* from top to bottom, when the N*B*-flip for (*A* d N*B*) becomes (*A* q *B*). The shape of the *B*-letter also has up-down symmetry. (R4) The (OOOc) in (*B* ✳ *A*) is a conversion operation that *flips* along the diagonal from upper right to lower left, when the diagonal c-flip for (*A* d *B*) remains (*B* d *A*) but when the one for (*A* b *B*) becomes (*B* q *A*). Notice that the (NONO) in (N*A* ✳ N*B*) is a double flip, one that also rotates 180 degrees, when (N*A* d N*B*) becomes (*A* p *B*).

The four symmetry rules, also abbreviated as (f-m-f and f), grow out of what we see in Fig. 18.13. The negation table mirrors (OO, ON, NO, NN) are operating on the 1st-order truth table (TT, TF, FT, FF) in the upper right 4-fold (⁚ ⁚). Negation is differentiated: (N*A*) and (N*B*) are flips but not across the same dimension. The four enlarged one-dot corners are from the 2nd-order truth table matrices (TFFF, FTFF, FFTF, FFFT), when the (R1,R3) mirrors operate on the same 4-folds. The (R2) mate rule is a simple (dot, no dot) reversal. The (R4) diagonal conversion axis is anchored along the alike-value truth-table pairs (TT, FF). In effect, and in continuity with Chomsky (*1957*), these rules are based on regularities that exist in what I am calling *transformational truth tables*.

The deed is done. The letter-shapes do not stay stuck to the page. Three negation rules (flip-mate-flip) cover the top half of Table III (the 128) and the conversion rule (flip) brings in the whole table (the 256). These symmetry rules belong to the syntax of the letter-shapes, with respect to the semantics of the forms, the tables, and the models. This syntax, built to fit the logic operations, becomes grammar by design, specialized in such a way that it also leads to some transformational logic.

§12. More about Transformational Logic (Going All the Way)

Forward history indicates that numbers were used for centuries before they were looked at in terms of abstract structures, such as the group properties of addition and multiplication. In other words, Arabic numerals al-

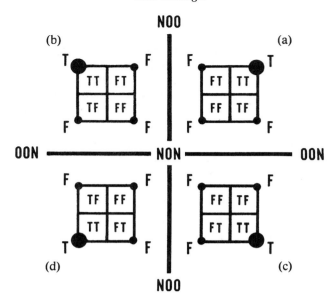

Figure 18.13

ready had those properties long before anyone paid attention to them. But where are the logic signs, just waiting to be noticed, that already have the required group properties? Backward history puts this in reverse. Start with logic operations and then do what is needed to construct logic signs and logic words.

What we have done is unbelievably simple. We have let the symmetry group properties of the transformational truth tables become the syntax in a notation that has been designed to mimic these properties. This looks again at the quote from MacColl. The signs have been constructed so that, to bypass unnecessary thinking, logic operations are being introduced in such a way that they will act on the signs themselves. This reclaims the underlying structure; this repeats the interrelational structure. It carries that structure into the notation. From now on, in one sweeping replacement, we will no longer go by the number names. We will, instead, go by the letter-shapes. Several examples will show what happens.

After we go to the second column of Table IV and rewrite (14), let us take another look at the analytic correspondence between (15) and the 3-cube of logical garnets in Fig. 18.17. The observation here is very simple and extremely subtle. In Fig. 18.17, before the front-to-back (NA) mirror plane is activated, all of the 128 letter-shapes are in flip-mate-flip orientation. When the (NA) mirror operates, two 4-sets of logical garnets are reflected into each other; 64 letter-shapes are carried to the right and 64 to the left. *At that moment all of the 128 letter-shapes do a left-right flip.* After the (NA) mirror operates, Fig. 18.17 continues as it was at the outset, a logical structure that looks exactly as it did before anything happened.

In one stroke this example jumps far ahead of the number names in Table III and in Fig. 18.8. This brings us back to the quote from Ladd-Franklin. "By actually turning it over," how does one give a number name a left-right flip? The same goes for any sign in any notation no better than a number name.

In like manner, repeatedly and rapidly attaching N's to (A) drives any letter-shape into a left-right spin. For example, the simple form in (17) has become a Law of Multiple Negation:

$$(17) \quad (\dots \text{NNNNNNNNNNNNNA} * B)$$

Detaching the N's from (A) could be seen as reversing the direction of the spin. If the number of N's is finite, the spin comes to a stop. If not, an unending spin shows up as a Law of Nonstop Negation. These developments, in conjunction with Fig. 18.16 and with many of the symmetry point groups in crystallography, bring us very close to asking about the differences, if any, between the nature of logic and the logic of nature.

Table IV is rock-bottom basic. It replaces Table III and it reinstates the ground already covered. Nowhere in terms of any signs that we care to choose has this table, or the top half of it, been found. This new (16×16) table is a showcase example of the potential that lies untapped in Peirce's notation. In many ways it is treating the connectives as if they were numbers. To carry along an important distinction, the letter-shapes could be called "connectals" (not numerals). This table of the binary connectals will do for logic what a multiplication table does for arithmetic.

The fact that the (R1,R2,R3,R4) symmetry rules operate directly on the letter-shapes themselves leads to a revealing question. What would happen to the simplicity, to the compactness, to the perspicuity of Table IV, if we would try to fill in all those cells with any notation now in common use? Such an effort would force us to see that, as much as we let Arabic numerals set the standard for number symbols, that much like Roman numerals are the logic symbols that we continue to use for the binary connectives. Such an effort would also make it obvious that the construction of Table IV has become a lesson in what a symmetry notation can do for logic.

Let Tables V and VI replace the top half of Table IV, after it has been partitioned into the two sub-tables, one for the odd-stemmed tall letter-shapes and the other for the even-stemmed squat letter-shapes.[22] The letter-shape transformations in both of these sub-tables are driven by the symmetry properties of the 8-group $(C_2)^3$. This includes the seven 4-subgroups,

Table IV

A*B	1 o	2 p	3 b	4 q	5 d	6 c	7 u	8 s	9 z	10 n	11 ɔ	12 h	13 ʮ	14 ɹ	15 ɥ	16 x
OOOO	o	p	b	q	d	c	u	s	z	n	ɔ	h	ʮ	ɹ	ɥ	x
OONO	o	b	p	d	q	c	n	z	s	u	ɔ	ʮ	h	ɥ	ɹ	x
NOOO	o	q	d	p	b	ɔ	u	z	s	n	c	ɹ	ɥ	h	ʮ	x
NONO	o	d	q	b	p	ɔ	n	s	z	u	c	ɥ	ɹ	ʮ	h	x
NNNO	x	h	ʮ	ɹ	ɥ	c	u	z	s	n	ɔ	p	b	q	d	o
NNOO	x	ʮ	h	ɥ	ɹ	c	n	s	z	u	ɔ	b	p	d	q	o
ONNO	x	ɹ	ɥ	h	ʮ	ɔ	u	s	z	n	c	q	d	p	b	o
ONOO	x	ɥ	ɹ	ʮ	h	ɔ	n	z	s	u	c	d	q	b	p	o
OOOC	o	p	q	b	d	u	c	s	z	ɔ	n	h	ɹ	ʮ	ɥ	x
OONC	o	q	p	d	b	u	ɔ	z	s	c	n	ɹ	h	ɥ	ʮ	x
NOOC	o	b	d	p	q	n	c	z	s	ɔ	u	ʮ	ɥ	h	ɹ	x
NONC	o	d	b	q	p	n	ɔ	s	z	c	u	ɥ	ʮ	ɹ	h	x
NNNC	x	h	ɹ	ʮ	ɥ	u	c	z	s	ɔ	n	p	q	b	d	o
NNOC	x	ɹ	h	ɥ	ʮ	u	ɔ	s	z	c	n	q	p	d	b	o
ONNC	x	ʮ	ɥ	h	ɹ	n	c	s	z	ɔ	u	b	d	p	q	o
ONOC	x	ɥ	ʮ	ɹ	h	n	ɔ	z	s	c	u	d	b	q	p	o

the seven 2-subgroups, and their cosets, in this 8-group, as they combine to form hybrids, as in Fig. 18.2b. The interconnectivity in these sub-tables is also repeated in Figs. 18.14 and 18.15, where the general symmetry shapes from Fig. 18.12 occupy the nodes of a complete simplex (- - - 8) and where the special symmetry shapes also from Fig. 18.12 occupy those of a degenerate simplex (2 2 4 -).

The Flatland 2-views in Figs. 18.11 and 18.12 now appear in Fig. 18.16 as the 3-skeleton of a 4-cube. Here the nodes from Fig. 18.14 go to the black 3-rotor vertices of a cube (pawns) and those from Fig. 18.15 go to the white 4-rotor vertices of an octahedron (heavy chess pieces). All of this forces Fig. 18.16 to stand in as a special rhombic dodecahedron. It has become a logical garnet, in which a cube of tall letter-shapes interpenetrates with an octahedron of squat letter-shapes.

In other words, there is another way to interpret (8,8) the two sets of

Table V

(???)	2	3	4	5	12	13	14	15
	P	b	q	d	h	P	d	4
OOO	P	b	q	d	h	P	d	4
OON	b	P	d	q	P	h	4	d
NOO	q	d	P	b	d	4	h	P
NON	d	q	b	P	4	d	P	h
NNN	h	P	d	4	P	b	q	d
NNO	P	h	4	d	b	P	d	q
ONN	d	4	h	P	q	d	P	b
ONO	4	d	P	h	d	q	b	P

Table VI

(???)	1	6	7	8	9	10	11	16
	o	c	u	s	z	n	ɔ	x
OOO	o	c	u	s	z	n	ɔ	x
OON	o	c	n	z	s	u	ɔ	x
NOO	o	ɔ	u	z	s	n	c	x
NON	o	ɔ	n	s	z	u	c	x
NNN	x	c	u	z	s	n	ɔ	o
NNO	x	c	n	s	z	u	ɔ	o
ONN	x	ɔ	u	s	z	n	c	o
ONO	x	ɔ	n	z	s	u	c	o

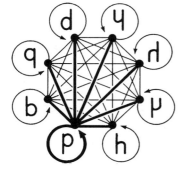

Figure 18.14

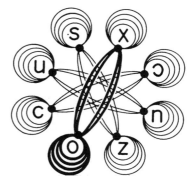

Figure 18.15

vertices in Figs. 18.14 and 18.15. Remember that (4,4) two 3-dimensional tetrahedra can be inscribed in a 3-cube so that they interpenetrate, and then recognize that (8,8) two 4-dimensional cross-polytopes, called 16-cells, likewise inscribed in a 4-cube also interpenetrate (*Coxeter 1991*: 31). The interesting point here is that, when a 4-cube is collapsed along one of its major diagonals (**sz**), thereby shadowing it into 3-space (Z_3^4), the two 16-cells are changed into two new forms that *continue to interpenetrate*. The 16-cells, now seen in Fig. 18.16 as black vertices (- 4 - 4 -) and white vertices (1 - 6 - 1), show up so that (8,8) they are in keeping with one of the defining properties of both a rhombic dodecahedron and the logical garnet: the tall-eight 3-cube interpenetrates with the squat-eight 3-octahedron (**sz** at the cocenter). Uncovering all of this analytic detail makes it easy to let Fig. 18.16 stand in as a 3-view iconic stick figure of a 4-dimensional hyper-snowflake (*Miyazaki 1986*: 98). Capturing all of this interconnectivity displays the great beauty of this approach.

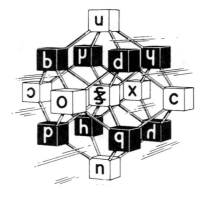

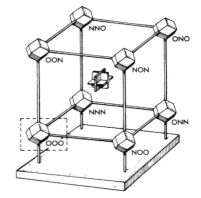

Figure 18.16   Figure 18.17

How closely this approach follows Peirce is shown in Fig. 18.18. At MS 431: 79, "in their most iconic shapes," Peirce arranged his cursives into six subsets (1 4 2 4 4 1), so that, as repeated in Fig. 18.18a, those with the same symmetry shape would be those in the same subset. As displayed in Fig. 18.18b, the five face-on top-down vertical slices that pass from the front to the back of Fig. 18.16 give almost the same view of the logic alphabet (1 4 6 4 1). The left side of Fig. 18.18a starts with tautology whereas, like Fig. 18.1, Fig. 18.18b starts with contradiction. Like Peirce, in the mnemonic at n. 21, and in keeping with her quote, the logic alphabet follows Ladd-Franklin three times (- 4 - 4 4 -), when she proposed that the same character be placed in four different positions (MS 431: 56, NEM 3: 272, *Fisch 1986*: 335).[23] These 4-sets (**p b q d**) (**c ∪ ∩ ⊃**) (**h ꟼ ꓒ ɥ**) occupy planar 4-folds that radiate away from the four that have central symmetry (**o, s z, x**), the same ones that line up (1 - 2 - - 1) along the front-to-back central axis of Fig. 18.16.

Peirce would have been delighted. Continuing with the level of abstraction that he favored and taking a deeper look at (12) tells us that Equivalence (18) has become a Law of Symmetry Equivalences:

$$
(18) \quad \overset{\text{R1} \quad \text{R2} \quad \text{R3}}{(\overset{\bullet}{\mathbf{A}} * \underset{\text{R4}}{\underbrace{\overset{\bullet \quad \bullet}{\mathbf{B}}}})} \equiv \overset{\text{R1} \quad \text{R2} \quad \text{R3}}{(\overset{\bullet}{\mathbf{A}} * \underset{\text{R4}}{\underbrace{\overset{\bullet \quad \bullet}{\mathbf{B}}}})}
$$

This master equivalence does not fall short. It adds enough algebra to the algebra of logic so that, fully functioning as three systems in one, it contains fourteen nodes of substitution (4 2 8), seven on each side of the equivalence sign (2 1 4). Two A's and two B's cover the propositional variables. Two asterisks cover the connective variables. Six N-carrying over-dots and two c-carrying under-arcs cover the operation variables.

Good fortune prevails because, when (18) is cast in terms of the four symmetry rules (R1,R2,R3,R4), the logic alphabet becomes a notation that covers *distinctly and uniquely* all and exactly all of the well-formed atomic equivalences (4096) that can be written under this master equivalence. This happens because, along the axes in Table IV, the four symmetry rules under combinatorial identity are *robust enough*, and the sixteen letter-shapes are *configurated enough*, so that these two conditions *covary enough* to fit what is needed to cover the atomic equivalences. This includes De Morgan's laws and many of the standard tautologies, such as are listed in *Reichenbach 1947*: 38–39.

One precaution about conversion and (R4). This operation may not be commutative when it combines with negation. How it combines with negation is the work of Glenn Clark. In Table IV and in the algebra of abstract groups, he has shown that the first four and the third four rows activate another 8-group, the octic group (D_4), the one that includes diagonal flips and quarter turns.[24] This non-abelian 8-group is obtained from the eight symmetries of a square, which in this case are applied one by one to all of the dot-squares in Fig. 18.10. At this point it is easier if the scope of (R1) is extended so that it acts on whatever (A or B) is on the left side of the asterisk, and likewise for (R3) on the right.

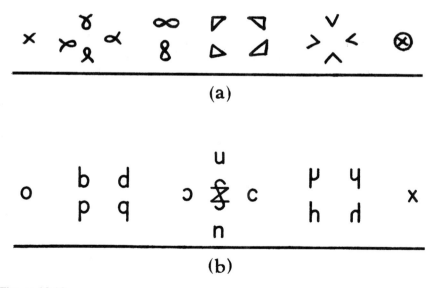

(a)

(b)

Figure 18.18

An interesting reference never seen by Clark comes from Renton (*1887*: 8) and how early he published the same (8 × 8) group table for (D₄) to describe what happens when the eight combinations of (negation, negation, conversion) act on (subject *a*, predicate *b*). Clark came to the same 8-group when he let the eight triplets from (N-Nc) act on (*A* ✳ *B*). How (N) and (c) combine in Fig. 18.16 and in Fig. 18.17 will remain an exercise.[25]

It is clarifying to identify the plane of conversion that cuts through Fig. 18.16. This symmetry plane is the diagonal front-to-back internal mirror that cuts through (**p o d**), the same mirror plane that contains the eight letter-shapes that allow conversion. Even more interesting, the internal mirror that cuts through (**b o q**) is the contrapositive plane, the same one that also contains eight letter-shapes and the same one (N*B*,N*A*) that is (R1,R3) rotated a half turn from (**p o d**). Cutting along both planes at the same time is like passing through Fig. 18.16 head-on from front to back with a knife in the shape of large X-letter. This complex slice passes along both (2 4 2) diagonals in Fig. 18.11. It follows that four dot-squares, along with the stem corners of the corresponding letter-shapes (**o z s x**), have just the right symmetry (self c-flippable and self-rotatable) so that they allow both conversion and the contrapositive.

Another view of conversion comes from building another model, one that partitions the top half of Table IV into columns instead of rows. Instead of a large 3-cube of logical garnets, as in Fig. 18.17, this model yields a large logical garnet of 3-cubes, one at each of the sixteen vertices (two at the cocenter). The 128 vertices are doubled (all of Table IV), when a front-to-back diagonal conversion mirror plane *is placed external to* the large logical garnet taken as a total unit. This generates another view of (2 × 8 × 16).

At this point it would be interesting to see what happens when the same transformational approach carries the same letter-shapes into logical forms that are more complex, such as the (16 × 16 × 16) block of cells for (3) and how it combines with Peirce's Form E (see n. 13). For example, exactly how are the 680 in (1) and (2) distributed throughout the 4096 in (3)? More fascinating for the beginner is to lay out symmetry models not only for the 256 vertices in (13) but also for the 1024 vertices in (9) and for the 4096 vertices in (12) and (18).

The physiology of the transformational letter-shapes is now active on such a grand scale that, all at the same time, logical operations are leaving their impact on single signs, on tissue collections of sign structures, and on muscle sets, all of this in large and complex models that are being treated as body configurations. This puts the emphasis not on small pieces, not on fragmentism, not on narrow-gauge axiomatics, but on unfolding patterns, evolving structure, and the gestalt advantages of a full-fledged systems approach.

§13. Climb Carefully on the Cognitive Ladder (I/PPRR/P)

We have made good on the syntax of the 128 but we have not looked at what made this possible. Becoming aware of what happened calls for a close analysis of the four numbers for (A,B), namely, (2 4 16 128). Once this analysis specifies the hurdles that the logic alphabet had to cross to meet the requirements of the 128, it will be easy to generalize to the 4th Series.

Analysis of (2 4 16 128) shows that *any notation* that is going all the way will give special attention to the delicate interplay that exists in five carefully coordinated properties. Notice that the first two numbers (2 4 - -) are for conditions that are fixed. The logic alphabet uses (i), *dimensions*, which repeats Peirce, stabilizes the first number, and assigns coordinates to (A,B). The logic alphabet has (ii), *frame consistency*, which repeats Peirce, stabilizes the second number, and uses the corners of a square for the 1st-order truth table. So much for what remains fixed.

Notice that the second two numbers (- - 16 128) are for conditions that are subject to change. The logic alphabet has (iii), *visual iconicity*. It is a property of a single sign. This repeats Peirce, captures the variations covered by the third number, and uses matching 16-sets of dot-squares and stem-shapes to stabilize the 2nd-order truth table.

Right here we need to go beyond Peirce because we are looking ahead to the 128. The three properties so far need to be brought into cooperation so that the logic alphabet will have (iv), *eusymmetry*. The letter-shapes have good symmetry because, as a special set of cursives, all of them are capable of obeying the symmetries of the all-common basic square. This makes it possible for all of them to participate in the same system of flips and rotations.

The four properties so far are still working on the 128. The logic alphabet has (v), *relational iconicity*. This key property is more abstract than visual iconicity. It is a property of a system of signs. When the symmetry motions in the 3-place negation table for $(A * B)$ operate on the letter-shapes (8×16), it captures the variations in the 128. We now have a perfect match between the negation-driven equivalence relations among the connectives and the rule-governed syntax relations among the letter-shapes.

The five properties are now ready to serve the primary purpose for all of this sign engineering and notational architecture. These properties lead the way to the fourth isomorphism, in such a way that it also arrives at the 4th Series. *A symmetry operation that does a logical calculation becomes a symmetry calculation.* This brings logic closer to mathematics, so that together they reach into deeper common ground.

Giving the logic alphabet the five properties, along with the consequent syntax, has cleared the way so that this notation can be carried all the way to the top of a cognitive ladder, one with *six levels* under it, also abbreviated as (I/PPRR/P). Start with Propositions (-/- - - -/P); they are in the 1st Series. Jump across the 2nd Series and treat the connectives as Relations; they are in the 3rd Series. Specify the negation-driven equivalence Relations between the connective Relations (-/- - R R/-); they are in the 4th Series. Establish the Patterns in the abstract graphs for all of the preceding; this is revealing. Determine the Properties in these Patterns (-/P P - -/-); this is surprising at first and then obvious. Now Isolate the Properties so that they can be built into a good notation. Going the full distance carries with it the prescription that we will be able to Isolate the Properties in the Patterns in the Relations between the Relations between the Propositions.

Another view of the cognitive ladder puts the four numbers into four layers of transparencies (2 4 16 128). Two lines that look like a plus sign cross on the bottom. Four vertices are one layer up at the corners of a square. Sixteen combinations of enlarged dots are another layer up, as they appear at the stations of the clock-compass. And 128 flip-mate-flip symmetry motions are in the top layer, when all combinations of the three negation table mirrors operate on the system of dot-squares and the corresponding stem-shapes.

Another view of the cognitive ladder puts the logic alphabet in a top row that contains sixteen dot-cornered front-back flip cards. Leśniewski's hub-spokes notation goes one row down because it does not become explicit enough about the back sides of the cards. Łukasiewicz's upper-case alphabet (Polish notation) is another row down because it leaves out visual iconicity. The standard Peano-Whitehead-Russell notation (dot, vee, horseshoe) belongs in the bottom row because it is confined to a small subset of the cards. Climbing all the way back to the top means that we start with all of the connectives, that we add visual iconicity to the signs, and that we let the back sides of the flip cards activate the relational iconicity.

The use of flip cards again calls attention to another replay of what happened when Europe entered the age of Arabic numerals (*Menninger 1969:* 319–388). Writing numbers was no longer separated from the act of making computations by placing coin-like counters on decimal-lined counting boards. The act of writing Arabic numerals became an integral part of making the computations. Likewise when we elect to write what happens to the flip cards. Writing the changes in the positions of the letter-shapes not only performs the logical operations. It also lets the mirrors do the thinking. In passing, note that if we are clever about placing four of the letter-shapes (**s z**)(**u n**) on the back sides of the flip cards, no letter-shape will ever be carried into positions that make it look backward or upside down.

Peirce set the stage in 1902 when he stabilized the first three numbers (2 4 16 -), the same numbers that are listed in Binyon's 1898 letter. As we have seen, it takes only five small changes to carry his *cursives* to the top of the cognitive ladder, in such a way that they will stabilize the fourth number, namely, the 128. The grand irony in Fig. 18.1 is that Peirce came ever so close on his own. As already noted, Pope Sylvester II also came very close, in regard to the full use of Arabic numerals. Although it would have been very awkward, all Peirce had to do was flip his box-X forms along the diagonal lines of his X-frame, to activate the full force of his notation.

§14. Generalization to the Fourth Series (Taking the Big Step)

We are now prepared to reduce all of the foregoing to a simple routine. All we have to do is run repetitions on what we did for (A,B). We can expect that, for every finite instance of $(A,B,C \ldots n)$, there will be a corresponding display in the quadruple isomorphism. The routine will repeat not only for the substitutions, for the cells, and for the vertices, in the semantics of the transformational truth tables, but also for the corresponding symmetry rules, in the syntax of the letter-shapes. This last step, already recognized by Messenger (*1986*), is not all that difficult, if we allow the logic alphabet appetite enough to live and move about in n-dimensional geometry.[26]

The first three series build up to the 4th Series, so the main emphasis will be on what happens to $2^{(n+1)} \times 2^{(2^n)}$. Let n equal 2, 3, and 4. We already know that the letter-shapes in (19) call for the symmetries of the all-common basic square:

(19) $2^{(2+1)} \times 2^{(2^2)} \rightarrow (8 \times 16) \rightarrow (???)(A * B)$

In (20) it will be the symmetries of a 3-cube and in (21) it the symmetries of a 4-cube:

(20) $2^{(3+1)} \times 2^{(2^3)} \rightarrow (16 \times 256) \rightarrow (????)(* *)(ABC)$

(21) $2^{(4+1)} \times 2^{(2^4)} \rightarrow (32 \times 65,536) \rightarrow (?????)(::)(ABCD)$

In general, each 1st-order truth table is placed at the 2^n vertices of an n-cube and the corresponding 2nd-order truth table becomes a $2^{(2^n)}$ set of dot n-cubes. The 4th Series appears when all combinations of $(n+1)$ mirrors operate on these sets of dot n-cubes.

In (20), the sum of the ternary connectives for (A,B,C) is 256 dot 3-cubes, one for each 8-place truth table combination. These dot 3-cubes are placed at the 256 vertices of a large 8-cube. Four negation table mirrors generate sixteen of these 8-cubes. This gives us still another set of 4096 equivalence interrelations in the corresponding (16×256) table. Then we shadow the 8-cubes into 4-space to match the number of symmetry mirrors. In other words, when $(n = 3)$, 4-dimensional mirrors are operating on a 4-dimensional arrangement of an 8-dimensional count of 3-dimensional connectives.

In (21), the sum of the quaternary connectives for (A,B,C,D) is 65,536 dot 4-cubes, one for each 16-place truth table combination. These dot 4-cubes are placed at the 65,536 vertices of a large 16-cube. Five negation table mirrors generate 32 of these 16-cubes. This gives us 2,097,152 equivalence interrelations in the corresponding $(32 \times 65,536)$ table. Then we shadow the 16-cubes into 5-space. In other words, when $(n = 4)$, 5-dimensional mirrors are operating on a 5-dimensional arrangement of a 16-dimensional count of 4-dimensional connectives.

And so forth, for both sides of (22), in such a way that the right side of (22) becomes a custom designed recasting of the 4th Series:

$$(22) \quad 2^{(n+1)} \times 2^{(2^n)} \rightarrow (C_2)^{n+1} (Z_{n+1}^{2^n})$$

The right side of (22) generates the *transformational symmetry model* for n propositions. The $(C_2)^{n+1}$ is for the n-case negation table, when the recursive direct product of the 2-group (C_2) operates on the unit negation mirror (O,N). The $(Z_{n+1}^{2^n})$ *is for the* n-case identity garnet. For (19), as in Fig. 18.17, three mirrors generate an 8-set of (Z_3^4)'s. For (20), four mirrors generate a 16-set of (Z_4^8)'s. For (21), five mirrors generate a 32-set of (Z_5^{16})'s. And so forth, for all positive integers.

Obviously, the number of equivalence interrelations becomes explosive when n increases. But, and this is not in Binyon's letter, we have found an easy way to keep up with the exponential runaway in $2^{(2^n)}$. The clue lies on the right side of (22), in the growing difference between the superscript 2^n and the $(n+1)$. Every time we *add* one more mirror to $(n+1)$.

Fig. 18.19 lays out how the logic alphabet will keep pace with (22). This large adapted matrix is one way to express the connective generalized [∗]. It absorbs all possible cases in the 3rd Series. It tells us how to *compound the letter-shapes*, and it could be counted as the *sixth departure* from Peirce (MS 431: 65). More specifically, demarcated subsets of asterisks will be filled in with selected combinations of letter-shapes.

Fig. 18.19 shows how to write the *n*-ary connectives, when they are laid flat on a *spread sheet of hyperdimensionality*. It starts in the upper right corner and moves in steps that build to the left and to the bottom. Steps to the left cover rectangles of letter-shapes, when n is odd; those to the bottom cover squares of letter-shapes, when n is even. The sixteen connectives for

(A,B) are carried by single letter-shapes in the upper-right corner (✳). The 256 connectives for (A,B,C) are carried by all side-by-side pairs of letter-shapes (✳ ✳); C on the right and not-C on the left. The 65,536 connectives for (A,B,C,D) are carried by all four-folds of letter-shapes (⁛⁛); D above and not-D below. And so forth on the spread sheet, for all 2-flat displays of all 2nd-order truth-value combinations that define all 2-valued n-ary connectives.

Fig. 18.19 also shows how to operate on the n-ary connectives. The same flip-mate-flip rules (NA, N ✳ , NB) continue to operate on all of the asterisks that carry letter-shapes. The NC negation rule is a *translation* that slides 2-fold halves past each other (✳ ✳), when (**o d**)(ABNC) becomes (**d o**)(ABC). Be careful, because (**o d**) stands for the *ternary* conjunction of (ABC). It is not the same as writing a quick string of binary conjunctions, as in (A **d** B **d** C). The ND negation rule is also a translation, one that slides 4-fold halves past each other (⁛⁛), so that (**o** d-above, **o o**-below) becomes (**o o**-above, **o** d-below). And so forth, for all (NC,ND,NE . . . Nn) translations past each other, in halves of demarcated subsets, *across* the solid line parts in Fig. 18.19. The larger the n universe and the closer to the C-end in the $(C,D,E . . . n)$ series the more the dotted lines in Fig. 18.19 are included in these translations.

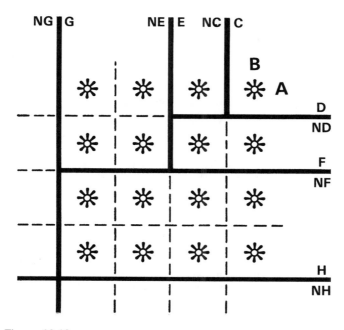

Figure 18.19

The series of models that keep up with the generalized logical garnet in (22), with the doubling of the asterisks in Fig. 18.19, and with the squaring of the vertices in the 3rd Series, behaves like the self-similar repetitions found in a fractal, this time applied to $(A,B,C \ldots n)$ increasing one by one. To go up one proposition, start with just enough copies of a given garnet so that one copy of itself can be put at each one of its own vertices. More specifically, to go to $(n + 1)$, each vertex of the n-garnet repeats itself until the repeats at each vertex are equal to, and then they are paired with, exactly all of the vertices of another copy of itself, the n-garnet. When this happens, things get thick in the middle. The garnet that starts the series contains the (**s z**) cocenter.

This series of models reaches for another sequel, in the same way that Abbott's *Flatland* in the early 1880's was followed by Burger's *Sphereland* in the early 1960's. This sequel enters into a *Shadowland* that lays open a fascinating landscape that includes a series of hypershadows. For (A,B), the front vertex of the garnet for the 16 vertices in (Z_3^4) starts out with an **o**-letter. For (A,B,C), the front vertex of the front garnet for the (16×16) vertices in (Z_4^8) carries the **oo**-pair of letter-shapes. For (A,B,C,D), the vertex most to the front for the (256×256) vertices in (Z_5^{16}) carries the 4-fold (a double pair) of **o**-letters. The next one would carry an 8-fold of **o**-letters, also seen as a pair of side-by-side 4-folds. And so forth, for (22), for Fig. 18.19, and for what happens to the n-garnet at $(n + 1)$.

Consequently, everything is in place. The matrix of letter-shapes at each vertex is located so that, in particular and within the system, it cooperates precisely with the relational iconicity in the symmetry properties of the transformational truth tables, each time in the series of models increasing by one proposition in $(A,B,C \ldots n)$.

Four examples will stay on the simple side. In (23) and (24), binary connectives come between the quaternary singularies for (A,B,C,D):

$$(23) \quad \begin{pmatrix} c & c \\ c & c \end{pmatrix} d \begin{pmatrix} u & u \\ u & u \end{pmatrix} d \begin{pmatrix} o & x \\ o & x \end{pmatrix} d \begin{pmatrix} x & x \\ o & o \end{pmatrix} \equiv \begin{pmatrix} o & d \\ o & o \end{pmatrix}$$

$$(24) \quad \begin{pmatrix} c & c \\ c & c \end{pmatrix} ч \begin{pmatrix} u & u \\ u & u \end{pmatrix} ч \begin{pmatrix} o & x \\ o & x \end{pmatrix} ч \begin{pmatrix} x & x \\ o & o \end{pmatrix} \equiv \begin{pmatrix} x & x \\ ч & x \end{pmatrix}$$

In (23), the four singularies are placed one on top of another to form a compound binary conjunction, which leaves one stem *in common* in the upper right corner (TTTT) and which yields quaternary conjunction for $(ABCD)$. In (24), the binary disjunctive superimposition of the same singularies yields quaternary disjunction, with one position *empty* in the lower-left corner (FFFF). In (25), as expected, the mate of the quaternary (p)eirce connective also yields quaternary disjunction:

$$(25) \quad N\begin{pmatrix} \text{o} & \text{o} \\ \text{p} & \text{o} \end{pmatrix} \equiv \begin{pmatrix} \text{x} & \text{x} \\ \text{ч} & \text{x} \end{pmatrix}$$

In (26), when a ternary conjunction is placed in front of three quaternary 4-folds, the stem that is in common in all three 4-folds also yields quaternary conjunction:

$$(26) \quad (\text{od})\left[\begin{pmatrix} \text{o} & \text{d} \\ \text{d} & \text{d} \end{pmatrix}\begin{pmatrix} \text{ч} & \text{ч} \\ \text{o} & \text{ч} \end{pmatrix}\begin{pmatrix} \text{x} & \text{x} \\ \text{x} & \text{o} \end{pmatrix}\right] \equiv \begin{pmatrix} \text{o} & \text{d} \\ \text{o} & \text{o} \end{pmatrix}$$

Look how easy it is to write *n*-ary conjunctions, which again reminds us of what we take for granted when we write decimal numbers. When the powers of 10 increase one by one, the same digit in a line of zeros is repeatedly pushed to the left (10, 100, 1000 . . .). Likewise for a given *n*-ary conjunction. Always start with a demarcated frame of **o**-letters and then insert a **d**-letter, the same one in the same place, always in the upper-right corner. That same **d**-stem takes on more and more truth value when the *n*-universe increases one by one (TT, TTT, TTTT . . .).

Another point, just to flag the riches that abound. It follows from another property of the logic alphabet. *Any upper-case letter (A,B,C . . .) in a n-universe can always be replaced by its n-singular.* In (23), we built up the quaternary conjunction by first replacing (A,B,C,D) with the corresponding 4-universe singularies and then obtaining the binary superimposition of these singularies. Likewise, the (A,B) in (A **d** B) can be replaced by the corresponding 2-universe singularies, as in (**c d u**). Notice that, at the mental level and like the overlap in two Venn circles, **c** on top of **u** leaves a **d**-stem in common. When this **d**-stem is written as a **d**-letter, it not only stands alone, without (A,B). It still stands for (A **d** B). But this comes right back to where we started, to the **d**-letter in (A **d** B), before (A,B) was changed to the (**c,u**) in (**c d u**).

Without debating whether to call this a "definition" of conjunction or a "primitive operation" which yields conjunction, there is more than one way to think about the (**c,u**) in (**c d u**). The (**c,u**) could be *propositions*, when the stems are for truth table positions that are true. They could be *classes*, when the stems are for subclasses that are occupied. They could be *connectives*, when, unlike (6) which repeats what is in the middle (✳), (**c** ✳ **u**) becomes a master form that always reduces to what is in the middle (✳). In all three cases, depending on the universe under consideration, the stem-tip of the **d**-letter puts a dot in the overlap of the **c**- and **u**-letters.

This multiple use of the letter-shapes brings us back to Binyon's letter, where he is concerned about the *distinction* between "terms" and "classes." We can only hope that he read Ladd-Franklin's 1883 paper. Also O. H. Mitchell's, also in *Studies in Logic.* And Peirce praised both papers many times for their attention to what they called a "universe of discourse."

As we have just seen, in reference to the deep *commonality* that runs through propositions, classes, and connectives, the emphasis is on what universe we start with, on what sub-universe is under consideration, on what operation is acting at what scope in that sub-universe, and especially, on what part-universe, including the whole universe, is activated, when that operation goes into effect. By now all that potential is upon us. Any notation in the same class as the logic alphabet will easily earn its keep.

§15. A Few Remarks about a 5th Series ($6 \times 8 \times 16 = 768$)

We started with an emphasis on writing a paper that would build up to and include the 4th Series. In the process, however, new possibilities came rushing in from many directions. At this point, without trying to characterize the outer limits of the potential before us, we will settle for one more excursion that feeds into a larger context.

Glenn Clark introduced the possibility that we might add a 5th Series to the Ladder of Products developed in the present paper. He suggested that the next multiplier could be $n!$. This would run *permutations* on the n-axes in any set of n-ary dot-cubes. When ($n = 2$), the extended series would become (2 4 16 128 256). In (27), when ($n = 2$), the logic operation of conversion becomes a simple case, the least case, of changing from one permutation to another:

$$(27) \quad n! \times 2^{(n+1)} \times 2^{(2^n)} = 2 \times 8 \times 16 = 256 \rightarrow \text{Table IV}$$

This comes around to the same 256 that now includes the bottom half of Table IV. In the logic alphabet this is the same diagonal c-flip that interchanges the positions of the (A,B) axes in all of the dot-squares and the corresponding stem-shapes.

Although we did not know it at first, when Clark introduced the 5th Series, he made contact with some abstract structure that comes up large in the wake of the present approach. The Cartesian product of all combinations and all permutations of n elements, also expressed as ($2^n \times n!$), is called a "wreath product"; see *Coxeter 1991*: 31. All cells in a table of a wreath product, one obtained from combinations in rows and permutations in columns, constitute a formal roster of the symmetry elements of an n-dimensional cube. An interesting spin-off here is that introducing this product reaches deeply into a confluence with crystallography.

Then came some surprises in logic and notation. Suppose we continue with (???c), which means continue with Table IV and the 16-group ($D_4 \times C_2$). Now permute the asterisk in ($A * B$). This includes ordinary infix ($A * B$), regular Polish prefix ($* A B$), and reverse Polish postfix ($A B *$); see *Katzan 1975*: 282. This also adds to what Clark started. It goes beyond (27), to a (28) that is not shown, because ($n!$) becomes ($n + 1$)!, when the permutations are determined *after* the connective generalized [$*$] has been added to ($A,B,C \ldots n$). Doing this for the six permutations of the three elements in ($A * B$) puts the emphasis not on ($2 \times 8 \times 16$) but on ($6 \times 8 \times 16$); that is, on (48×16), which equals 768 and not the 256 (or the 128).

Without knowing what we had done, Clark on the algebraic side and I on the tabular side, we had bumped into a wreath product for three dimensions. Some time later Coxeter called our attention to his reference, which describes a wreath product for four dimensions.

The cross-disciplinary infusion from crystallography comes rushing in when we let the (6×8) in the wreath product act on the 16 connectives. This calls on the 5th Series in full, all of (28). It follows that, with (R5) under ($A *$) and (R6) under ($* B$), the ($6 \times 8 \times 16$) becomes (permutations $\times$ combinations $\times$ connectives). Algebraically, with the emphasis on a system that is unified, we have elected to express all of this as (under-arcs $\times$ over-dots $\times$ asterisk). This comes back to the 768 seen as (48×16), when the wreath product lies in the 48 expressed as (under-arcs $\times$ over-dots).

When the 48 symmetry operations act on the 16 connectives, it generates the highest symmetry point group, the one called (m3m), the same one in which we find the gemstone diamond. This abstract structure is rich enough so that the 48-group ($S_4 \times C_2$) contains the 16-group ($D_4 \times C_2$) which in turn contains the 8-groups (D_4) and (C_2)^3 as subgroups. In effect, when we start with only one term, namely, with no more than ($A * B$), and when we let a logical garnet become the identity motif, one that replaces the asterisk, this 48-group generates a model that includes and adds to the three mirrors and the eight logical garnets in Fig. 18.17. This model is a marvel to behold! A sunburst of 48 logical garnets is located at the 48 vertices of a greater rhombicuboctahedron (also called a truncated cuboctahedron).

This overall shape not only surrounds a point center of symmetry. It has also been reduced to nothing but the 48 vertices that are now occupied by 48 asterisks. These are the markers suspended in midair that locate the overall shape of the greater rhombicuboctahedron, the same overall shape that outlines the sunburst of 48 logical garnets. All at the same time the 48 logical garnets subdivide into six sets of eight, eight sets of six, and twelve sets of four, such that 48 o-letters face out and 48 x-letters face inward, not directly at the point center, where nine mirrors intersect, but inward along (xyz) planes. See *Coxeter 1988*: 74 and *Loeb 1988*: 114.

Instead of having Einstein ride on a beam of light, the viewer of this model rides along on the symmetry operations that change the position of the asterisk. The viewer rides from one fix-specific perspective to another. This partitions the 48 logical garnets into three sets of sixteen, so that the 768, seen here as $(3 \times 16 \times 16)$, becomes (3×256). Infix $(A * B)$ is viewed from front to back, face on in the ordinary reading position; prefix $(* A B)$ from right to left, face on along the A-axis; and postfix $(A B *)$ from top to bottom, face on along the B-axis. When the viewer looks along the tri-fix directions (xyz), the viewer shifts position to isolate the three 16-sets of logical garnets (3×256). This gives us three copies of Table IV, one for each fix.

Each fix is located at 16 vertices, each fix subdivides into (8,8), and each fix is found at one of the three pairs of near-far octagonal faces of the greater rhombicuboctahedron. For example, in reference to the main thrust of the present paper $(2 \times 8 \times 16)$, ordinary infix $(A * B)$ is a single fix that is often treated as standard form. It appears in (27) and on both sides in (18), it is analytically specified by the (1×256) cells in Table IV, and the (???-) 8-group half of it is displayed in Fig. 18.17. It includes the (???c) 16-group of symmetry operations $(D_4 \times C_2)$ that both generate and act on the two octagons of logical garnets that are located at and that are framed by the sixteen vertices (8,8) that demarcate the front-back octagonal faces of the greater rhombicuboctahedron. In brief, to obtain the three 16-sets of logical garnets, namely, the 16-sets for $(A * B)$, $(* A B)$, and $(A B *)$, start with the 48-group $(S_4 \times C_2)$ and isolate the 16-element cosets for $(D_4 \times C_2)$. It follows that the three pairs of (8,8) near-far octagonal faces run full fit into *the* perfect 3-coloring of a 3-cube.

It also follows that it is very easy to take the road that leads to the cross-disciplinary infusion. In order to obtain the 48 symmetries of a logical garnet, instead of letting a crystallographer subject the Miller symbol (hkl) to a wreath product, here defined as $(2^n \times n!)$ when $(n = hkl = 3)$, which yields the (8×6) Cartesian product of all combinations and all permutations, let a logician start at (4) and do exactly the same thing to the three elements in $(A * B)$. The surprise is that so much logic and crystallography fall out of each other, and so easily. Carrying this back to (18) and subjecting one term and then the other term of this fundamental master equivalence to a wreath product generates a second-order sunburst. For more about these developments, see *Zellweger 1994, 1995*.

So, as a matter of historical perspective, Jevons brought in the 2nd Series (*1880:* 181), Ladd-Franklin in Peirce's 1880 logic class listed the sixteen binary connectives, thereby making contact with the 3rd Series (W4: 569–

571), and Peirce in 1902 devised his box-X notation for those connectives (MS 431: 315–363). As it turns out, this notation has precisely enough visual iconicity in it to put us in a strong position to take the next jump step. When we call on sign engineering to take full advantage of the central importance of symmetry and relational iconicity, it becomes apparent that, as an excellent first effort, Peirce's notation easily makes contact with the formal, the relational, and the interrelational structures contained in the 768. This brings us around to another 3-dimensional block of cells ($6 \times 8 \times 16$), when all of them belong to variables that act on only one term ($A * B$): the under-arcs by the over-dots by the connectives.[27]

"Almost trivial, hardly worth a moment's notice, just another set of sixteen signs" is what we might hear. But there is much more to it than that, when we examine carefully what Peirce was doing when he devised his notation. As we have just seen, for example, in a world ready to continue in the tradition tagged by what is in the first section of the present paper, looking at what follows when Clark introduced the 5th Series comes down to a leading question. What are variations in the forms, the tables, and the models that line up with the symmetry operations contained in a wreath product of a logical garnet?

§16. Comments on the Ground Covered (What Lies Ahead)

Many slow changes in signs and sign systems have accumulated in the tides of evolutionary notation. Notations in recent millennia have gradually become more differentiated, so that, for example, we now have specialized notations for reading (a,b,c . . .), for counting (1,2,3 . . .), and for singing (doh, ray, me . . .). Someday logicians and the lay public alike will settle upon a simple, a common, and a specialized notation for the sixteen binary connectives. What we see in Fig. 18.1, for Peirce and for the logic alphabet, are a couple of notches along the way on that journey. In keeping with the quote from MacColl, the case for (o,p,b . . .) is a contribution to the evolution of cognitive economy.

All by itself the repeated appearance of iconic notations for the binary connectives carries a message. Examples are found in Peirce's box-X (1902 Minute Logic), Leśniewski's hub-spokes (*1929*), Gonseth's windowpanes (*1937*), McCulloch's jot-X (*1965* [1942]), Parry's trapezoid (*1954*), and Frazee's dots-lines (*1988, 1990*).[28] All of these efforts are strong on interrelations but none of them goes all the way. More efforts along this line will continue to appear, until we settle upon a notation that embodies precisely the property of relational iconicity.

The pragmatics of the 128 runs full face into the profound significance of and the amazing consolidating power of the concept of symmetry. Peirce made much of it in his iconic notation. The logic alphabet is not only truth table sensitive and user friendly. It goes all the way. Negation becomes a work slave and symmetry is the primary muscle. Group theory and negation table mirrors carry the same emphasis into a whole family of hand-held models.

Some of these models show up in the present paper, when another set of letter-shapes is placed on the back sides, in see-through orientation. The bottom row of Fig. 18.1 becomes a front-back flipstick. Fig. 18.10 becomes a set of logic blocks. It also becomes a flip-mate-flip clock-compass, when all of it is constructed as a single unit. Fig. 18.12 takes the shape of a logic bug. Fig. 18.16 stands in as a logical garnet. Other examples are diagrammed and described in *Zellweger 1981*. In contrast, not at all like Arabic numerals and how we use an abacus or a slide rule, we do not even ask ourselves, "How come the standard notations for the binary connectives do not have any hand-held models?"

These models have implications for cognitive development and education. These models become important at the stage when children are learning to count on their fingers. They supply the pupil's brain-mind-self with valuable developmental underframes of sensori-ideo-motor experience. Preschoolers at play with a set of hand-held front-back dot blocks is an easy introduction to such things as Finger Logic, Lazy Logic, and the game of Flip-Mate-Flip and Flip, long before they know anything about abstractions, flip card logic, and mind-held models. The pedagogical merit of these models is also enhanced by the ease with which it is possible to control a wide range of increasing complexity that reaches in steps and stages all the way from childhood to graduate school.

What happens in school again calls attention to a marked tendency in our time. We continue to expect so much more in the world of numbers than is the case for logic. We take it for granted that a high school student in ordinary geometry will learn about (x,y) coordinates. It is routine that a straight line goes with one equation and a circle with another. But, from the same point of view, what about all possible substitutions in (18)? It is very odd indeed that in our time we seem to be so disinclined to use a little bit of analytic geometry on this fundamental master equivalence.

These models have implications for cognitive ergonomics and computers.[29] Here we look at the mental work needed to use a notation, when we call on our engineering and architectural skills, in the act of designing signs and sign systems so that they will make efficient use of our mental activity. Thinking *about* signs and using them to think *with* are very different kinds of mental activity. This emphasis carries over to computers, especially to those that have been given a nervous system that is connected to large complex symmetry components, so that they will be able to make the best use of a notation such as the logic alphabet. Feeding the letter-shapes into the hexadecimal system and letting optical symmetry mirrors activate the negation table mirrors would be a small step in that direction. In reply to the quotes from both Ladd-Franklin and MacColl, mirror logic is logic in which the mirrors do the thinking.

Right here, also loaded with untapped potential, the same letter-shapes could be used as notational primitives in another font. In arithmetic and in computer languages, the same letter-shapes would become number symbols (0, 1, 2, 3, 4 . . . 8 . . . 15, 16, 17 . . .). They would all at once become all three, alphanumeric, binary stem-positional, and hexadecimal (o, d, q, c, b . . . p . . . x, d o, d d . . .). In effect, along with being computer friendly, number symbols and logic symbols for binary connectives would become identical twins but if the output dresses them differently we should have no difficulty telling them apart.

These models have implications for what happens when we consolidate the logic of propositions, classes, and connectives. The classical square of opposition extends to Peirce's cube-torus of the De Morgan eight and then to the letter-shape sixteen (see n. 17). De Morgan's laws generalize to a master equivalence that covers 4096 symmetry equivalences, and then to the 4th Series. The standard list of tautologies is built into the syntax of the letter-shapes. Modus ponens along with conjunctive and disjunctive routines is subsumed under relational iconicity. Whatever works for truth tables also works for what happens when we operate on the letter-shapes themselves. All of this, and more, will organize what we put on the inside covers of our logic text books.

Sad indeed it is, for a world that might have been. It is extremely unfortunate that Peirce did not have an eager graduate student, say about 1905–1907. Call him OTTO. This name, of course, is symmetrical; the spelling comes from (O)ne, (T)wo, (T)hree. He tells us that it stands for the Grand Unity (- - - O) that comes from seeing Everything in terms of Firstness, Secondness, and Thirdness (O T T -). As we have shown, with this much going for him, a few small changes in just the right places would easily have been enough to activate the untapped potential in Peirce's iconic notation for the sixteen binary connectives.[30]

OTTO would have found more than logic and mathematics in "The Simplest Mathematics." A couple of pages past the beginning of the section on "Trichotomic Mathematics," he would have seen that, in order to take on the philosophical importance of the generative potency of the number three, "It will be convenient to begin with some *a priori* chemistry" (MS 431: 165–168). This leads into the concept of valency and the sufficiency of triads, when the types of atomic combinations become increasingly complex. Elsewhere in the "Minute Logic," when Peirce again takes up the classification of the sciences, he repeatedly uses crystallography as an example with respect to its relations with mathematics, chemology, mineralogy, and chemistry (MS 427).

But where do we put the likes of the logical garnet? If Fig. 18.16 belongs in the "Crystallography of Logic," then the present paper brings into the same mathelogical frame what are often counted as two primary experiences in Peirce's mental development, namely, his taking up the study of chemistry when he was eight and his reading of Whately's logic when he was almost twelve.

N O T E S

1. The present paper is the culmination of a long apprenticeship that has enjoyed continued encouragement. I am especially grateful to Max Fisch, Nathan Houser, André De Tienne, and Christian Kloesel. Also for two full-time sabbaticals at the Peirce Edition Project. At the University of Toronto, to H. S. M. Coxeter. At the University of Chicago, to Milton Singer. At North Dakota University, to Tim Messenger. At Princeton University, to John Puterbaugh, doctoral student in composition and computer music. At Mount Union College, to Mary Ellen Nurmi, George Thomas, William Carter, and Donald Ray. And to William Glenn Clark, faculty colleague emeritus, longstanding co-worker, correspondent for 15 years, whose patient and clarifying influence has left its mark on many pages in the present paper.

I am also grateful to the Philosophy Department at Harvard University for permission to include Peirce's notations, as they appear in Fig. 18.1c. See *Zellweger 1982*: 44–45 for an earlier display of these notations. Some of this is also in print not in the *Collected Papers* at 4.261 but in the long footnote at NEM 3: 272–275.

An integral part of the present paper is the art work for the tables and figures. All but the first three tables were done by Mr. Warren Tschantz, Artec Incorporated, Alliance, Ohio. Tables V, VI, and Figures 3, 4, 8, 10, 16, and 17 are repeats from *Zellweger 1982*.

Many background references are in *Zellweger 1982*. Several areas of interest that are not developed in the present paper come to the fore when we have a notation such as the logic alphabet. One looks at networks of logical structure: *Keynes 1884 [1894]*: 113, *Gottshalk 1953*, *Blanché 1957*, *Menger 1962*, *Roberts 1973*, *Hacker 1975*, and *Lehmann 1992*. A second looks at cognitive development: *Baldwin 1906–1915*, *Piaget 1957*, *Adler 1968*, *Commons et al. 1984*, and *Wallace et al. 1989*. A third looks at mind and machines: Peirce MS 1101/831 [1900], *McCulloch 1965*, and *Sowa 1984*. A fourth looks at crystallography: *Buerger 1963*, *Yale 1968*, and *Boisen and Gibbs 1990*. The core concern with the coordinate use of a hierarchy of systems, especially the long span conditions at the level of sign systems, comes from *Von Bertalanffy*

1868 and *Laszlo 1972.* The central emphasis that I have given to transformational truth tables should leave no doubt about the influence of *Chomsky 1957.* The section on "Cognitive Ladder" makes a special case of what we find in Plato's divided line (*Wood 1991*) and in Korzybski's structural differential (*1933*). Also see *Morris 1955, Kent 1987,* and *Anellis and Houser 1991.*

Stern (*1988*) and Freytag Löringhoff (*1967, 1985*) are in a league all by themselves. Stern is strong on calculation, he expresses all of the binary connectives in terms of (2×2) matrices of zeros and ones, and he then uses matrix algebra as the computational engine that performs logical operations, which makes contact with many of the interrelations considered in the present paper. Freytag Löringhoff is strong on structure, he understands the role of symmetry, he has a diagram like our Fig. 18.16, and he makes much of the interrelations among the eight corners of a cube, which is a full return to what we see in Peirce's drawing at (MS 415: 004); see n. 16. and n. 17. Neither Stern nor Freytag Löringhoff are strong on notation, to make the best case for what they are doing. Note that I did not see *Stern 1988* until 22 October 1990, likewise for *Freytag Löringhoff 1967* until 7 January 1992, which was long after I submitted the present paper. Fortunately the editorial process permitted the inclusion of these important references.

Very much at the last moment, namely, on 8 September 1995, I received a copy of *Schmeikal-Schuh 1993* from Fritz Lehmann. Paleolithic, anthropological, and social notions of space are related to what are taken to be genetically based commonalities that have emerged in human cognition. The author starts with patterns of orientation in physical and social space and then finds the same patterns in mental space, namely, in (some of) the symmetry groups of logical operations contained in classical logic and Boolean algebra. All of this places prime importance on the symmetries that can be applied to a quartered circle. This geometric form is the lead icon in the construction of Peirce's notation and it is repeated at Fig. 18.9a in the present paper.

2. Peirce continued to be a creative logician into and across the first decade of the twentieth century. In the three examples that I mention, he called on iconicity to help him devise expressions that would enhance logical analysis and expose abstract structure. In 1902 he was already experimenting with 3-valued logic (MS 431: 279–283); add this to *Turquette 1972* and *Zellweger 1991.* For a full sweep of the perspective from which Peirce viewed diagrammatic notations, see "A Comparative and Critical Outline of the Useful Systems of Logical Representation" (MS 283: 346–347 [1906]). It is interesting that several of Peirce's students contributed to logic studies that also involved iconic forms, geometric representations, or abstract relational patterns: Ellery W. Davis (*1903*), Benjamin I. Gilman (*1892, 1923*), Christine Ladd-Franklin (L237, *1880, 1883*), Allan Marquand (*1881, 1883, 1886*), O. H. Mitchell (*1883*), and Henry Taber (*1890*).

3. See "Predication and Existence," in *Jones 1890:* 86–102. Also see "Implication and Existence in Logic," in *Ladd-Franklin 1912,* which post-dates Binyon's 1898 letter but Ladd-Franklin is the person to whom it was written.

4. I am grateful for permission to repeat this letter in full, and the letter in part at n. 9. They are retained in the Ladd-Franklin Papers, Rare Book and Manuscript Library, Columbia University. So far I have been unable to identify Frederick

Binyon. The ellipsis in the second paragraph replaces Binyon's bracketed reference: "offprinted from 'Mind' Vol vii N. S. No 26." (See references for this volume.)

5. Notice that Kempe (*1886*: 70) also refers to Clifford's paper and the case when ($n = 4$). Also see n. 11.

6. The story here is more complicated. Peirce went for hundreds of pages when he wrote "The Simplest Mathematics." In fact he wrote it twice: first MS 430, which contains about 180 mainline pages and 230 rewrites; and then MS 431, which was followed by typed MS 429. On 19 January 1902, while he was writing MS 431, he was about to set aside a long run of pages, including those in which he devised his notation. On that day he wrote to Royce, "I am now writing, with great labor, that chapter in my Logic book which is to treat of the two subjects of the Logic of Mathematics and the Mathematics of Logic. You cannot conceive the difficulty I am having in putting it into a form that will satisfy me" (L385: 27). See *Grattan-Guinness 1988* and Section 4 of his paper in this volume. On 8 January and 20 January Peirce also wrote to Ladd-Franklin first to collaborate with her and then to favor his own notation (L237: 174–178 and 186).

The above details are from the research that I did during my first full-year sabbatical at the Peirce Edition Project (1982–83) when, along with identifying pages and tables located elsewhere in the Peirce Papers, I drew tree diagrams that reconstruct the order in which Peirce wrote the 950 pages, including rewrites, that constitute the three versions of "The Simplest Mathematics" (MSS 430, 431, and 429). During my second sabbatical (1989–90) I did the same for almost 500 pages in the seven versions of Peirce's "Carnegie Proposal" (L75) to write the "Minute Logic" as a series of memoirs. This work is in preparation for a later volume of *Writings of Charles S. Peirce* and is on file at the Peirce Edition Project.

7. With respect to MSS 431 and 429, see Clark's explanation in the introduction to his paper in this volume. Also see *Houser 1989a* for more about the archival past and the present condition of the Peirce Papers.

8. See W4: xliv, 1, 222, 299, and Paul Shields, this volume.

9. Again at (MS 431: 70–71). On the first page of the Introduction to the Second Edition of the *Principia* (1927), the impact of the Sheffer connective ("*p* and *q* are incompatible") is noted as "the most definite improvement resulting from work in mathematical logic during the past fourteen years". The emphasis there is on axiomatic primitives and not on notational primitives. Ladd-Franklin also took notice of the same use of the Sheffer connective. This appears in a letter to Whitehead dated 20 November 1928 (see n. 4 above and Box 8 in her papers; also n. 26 below). The first part of this letter reads as follows: "This is something that I have been intending to ask you: Why, in changing in the second edition of your book from '*a* is all *b*' to 'no *a* is *b*' (or from '*p* entails *q*' to '*p* is-incompatible-with *q*'—the simple and the compound are the same thing, you know, in my logic) as the foundation-stone of your system, do you attribute to *Sheffer* the discovery of the vast superiority of the universal-negative-symmetrical form of proposition (u s n) to the universal-affirmative-nonsymmetrical (u s̄ a) form? You told me (when I had the happiness of seeing you) that you were very familiar with my paper on symbolic logic in the Studies in Logic by Members of the Johns Hopkins University—you told

me in fact (much to my pleasure) that you kept the book always on your study-table."

10. See n. 4 in *Zellweger 1982*.

11. See *Adler 1967, Budden 1972*, and *Weyl 1952* for more about the algebra of groups. Also look under *groups* in NEM and include the footnote at 3: 954. The extent to which Peirce knew about groups *is also shown* in his complimentary copy of Kempe's "Theory of Mathematical Form" (MS 1599; author signed, 18 November 1886), which carries a full list of tables and diagrams in a section called "Groups containing from one to twelve Units" (pp. 37–43). Peirce annotated the (C_2) 2-group, heavily annotated the $(C_2)^2$ Klein 4-group, and rediagrammed the $(C_2)^3$ 8-group. These three groups are central to the development of the present paper. They are, in fact, another way of expressing what happens when *negation tables* operate on themselves, as in Fig. 18.2a. Again, see n. 4 in *Zellweger 1982*.

12. What a full-fledged analytic geometry of logic would be like deserves more attention. *Whitehead 1901, Royce 1913*, and *Tarski 1941* saw the merits and the difficulties of combining Boolean algebra and group theory. Tarski is a good example here because he used both an algebraic method and a geometric method to treat binary relations. Also see Ladd-Franklin at n. 26. Note especially that Peirce is the same person who devised *both* the existential graphs and the iconic notation under discussion. At what end-combination these two would have run parallel, if at all, had Peirce continued his work indefinitely long, is an interesting question.

13. See *Menger 1962*, and Clark, n. 8, this volume.

14. Standard vocabulary is confusing. We use "commutative" with two of the binary connectives (and, or) but we use "conversion" with a third connective (if, then), in the sense of "All *A* is *B*." In the present paper, however, (✳) is a generalized connective, and we will say that $(B ✳ A)$ is obtained by taking the converse of the relation $(A ✳ B)$. In this sense, see the quotes from De Morgan and Peirce in *Bocheński 1961*: 375–379.

15. Do not forget to paint the paint brush. Relations apply to many things, including relations between relations. Relations may even be applied to relations. See Burch, this volume.

16. Even more accurately, because this rhombic dodecahedron is a 3-shadow of a 4-cube, it is called a *shadow rhombic dodecahedron*. It can be dismantled into all of the component parts of a 4-cube. Internally it contains an 8-spoked cocenter such that two of the vertices (8,9) pile up and fuse at the exact center of Fig. 18.7. The same diagram, *without this conceptual context and without a matching notation*, also shows up in *Freytag Löringhoff 1967*: 75. See *Coxeter 1948 [1973]*: 256 for how a 3-shadow of a 4-cube becomes the zonohedron (Z_3^4); *Loeb 1976*: 131, *Engel 1980, Miyazaki 1986*: 88, *Edmondson 1987*: 182, *Banchoff 1990, Robbin 1992*, and *Coxeter 1995* for more about this geometric form; and *Shephard 1974* for the general case of the zonotope (Z_r^n). It is also interesting, if not ironical, that the entry for "dodecahedron, rhombic" in the *Century Dictionary* was submitted by Peirce.

17. In this context, see *Zellweger 1993*. Early on and later on, in continuation with *De Morgan 1966*, Peirce and his students had different ways of treating half of the connectives (Fig. 18.3), also called the eight pawns (Fig. 18.1). Peirce was superb on the lines and planes of interrelations among them, both at the eight vertices

of a 3-cube and on the surface of a torus (MS 415 [1893]). Davis gave them an 8-group analysis (*1903*); the same 8-group of negations $(C_2)^3$ operated on the same eight connectives, as this appears more recently in *Bachtijarov 1979*. Ladd-Franklin had a 3-bit 3-set of (4–4) conditions, universal, symmetrical, and negation, that acted directly on her own 8-set of signs (*1890*; *Shen 1929*). A version similar to Fig. 18.3 appears in *Hacker 1975*; see his connection with *Parry 1954*. Much beyond any of this, in 1902 Peirce put the sixteen connectives in eight columns (MS 431: 99-100; MS 429: 71–72; omitted in the *Collected Papers* at 4: 274). Each column, along with variations in negation, was expressed in terms of the same sign. This limited the eight columns to only half of his cursives (pawns).

18. See *Jevons 1880*: 181, and Chapter 5 on Jevons in *Gardner 1958*: 96. Also in 1880, again seen as a windfall year, Venn had five publications on geometrical diagrams, mechanical representations, and various notations. The same emphasis was continued in his book (*1881*), with a second edition in 1894.

19. Peirce worked with *more than one* dimensionalized notation. Hawkins (*1981*) considers the notational attitudes behind Peirce's existential graphs and Frege's system of branching conditionals. Along with McCall (*1967*) and Hiż (this volume), see Luschei (*1962*), who shows a significant parallel between Peirce in our Fig. 18.1 and Leśniewski's ideographic system of hub-spokes. Chapter 8 in Luschei, especially pp. 292–297, also stands as a strong precursor of how the logic alphabet uses the letter-shapes. Ladd-Franklin (n. 4, Box 39) had her own 2-dimensional ("magic") notation. Also be sure to compare Peirce and the (x,y) lines in Fig. 18.9a with the claims made for the same 4-fold diagram in *Schmeikal-Schuh 1993*: 119.

20. Peirce at MS 431: 87 had a "4-square 4-fold." See Clark's Fig. 17.3 (this volume) and *Scharle 1962*. Also see n. 25 in *Zellweger 1982*, which mentions why the letter-shapes in the non-print positions (ɔ, μ, ꓒ, ꓕ) could be called *ray, mif, rif,* and *yor*. Where I use the expression *letter-shapes*, Schmeikal-Schuh has vocabulary that is very Peircean. He calls them *relational icons*.

21. A chess game could be played with two sets of letter-shapes that have different colors. The tall letter-shapes (- - - 8) would be the pawns in the front row and the more symmetrical squat eight (2 2 4 -) would become the ranking classes appropriately placed in the back row. Another flight of fantasy here is worth a good mnemonic. Let the symmetry plane that separates the two halves of a chess board become the bite plane between the upper and lower jaws, the same bite plane that separates a full set of adult human teeth, sixteen above and sixteen below. We could even plead for natural partitions, alike in both jaws, so that (2 2 4 8) becomes (2 canines, 2 wisdom molars, 4 remaining molars, 4 premolars and 4 incisors). This (- - 4 4) also lines up with "orbits," as in n. 22, and with the three times that the logic alphabet places the same shape in four different positions.

22. As an application of group theory, the "orbits" in Tables IV, V, and VI, as well as in Figs. 18.2, 18.14, 18.15, and especially 18.18a, can be used to illustrate the Polya-Burnside Theorem (*Yale 1968*: 37–38). See the models in *Verheyen 1996*.

23. When (R1,R3) operates on the same character in four positions, such as (p b q d) and (h μ ꓒ ꓕ), it activates the $(C_2)^2$ Klein 4-group, as in n. 11. After enlarging (o,x) and after shifting four of the tall eight (p b q d), Fig. 18.12 is spread out like a pancake. It comes from a front-to-back polar 2-flat projection of Fig. 18.16.

24. Again, see *Schmeikal-Schuh 1993*: 125 for more about the same octic 8-group (D_4) when it is related to logic and the symmetries of a quartered circle. Also note that the subtlety of what we find in Table IV can easily remain inaccessible to the logician as logician, at the same time that it goes unnoticed by the mathematician trained in the algebra of abstract groups. Nor are the close connections likely to be recognized by the crystallographer. Glenn Clark has shown that, when the 4-tuples along the left side of Table IV are repeated along the top, as in Fig. 18.2a, and when each of these 4-operations is replaced by its "net effect" (the compound outcome in each case), it leads to the 16-group called ($D_4 \times C_2$); see *Budden 1972*: 291. When the 4-tuples (???c) operate on the four levels of symmetry-asymmetry in the letter-shapes (2 2 4 8), it leads to Table IV as we have it.

25. Find the front-to-back diagonal planes that contain the convertible connectives, and then, for Fig. 18.17, specify the two ways of locating one symmetry mirror so that it will generate the bottom half of Table IV. Note that the display value of Fig. 18.17 is superior because *external* mirrors keep the eight reflections separate, whereas Fig. 18.16 and *internal* mirrors would send all combinations of 3-directional halves across and into each other.

26. Again, recalling Ladd-Franklin's paper in the same journal in the same year, see *Stringham 1880* and his paper on regular figures in *n*-dimensional space. Again in the same year, this time in Peirce's class, Ladd-Franklin considered the possibility of introducing "an *n*-dimensional logic" (L237: 50). This was part of her conceptual struggle, along with "imaginary logic" and "non-Euclidean logic," to build a universe of discourse that would include contradiction (FFFF) as one of the sixteen binary connectives. In 1880, in her geometric model, it was difficult for her to admit "a region of space in which all things which exist are things which do not exist and all things which do not exist are things which exist" (L237: 47–48). See n. 3 and W4: 569–571. Also see Tarski at n. 12 and *Green and Hamberg 1968*.

27. The one sentence that nicely expresses the approach taken in the present paper is also the last paragraph in *Kiss 1947*: 315. He tells us that "Mathematics and logic are essentially devoted to the construction of abstract structures and to the study and use of their *symmetries*" (italics added).

28. Frazee is recent and interesting. But standards are going up. He has no general forms, no tables of substitutions, and alas, no models. His references are in one world and what he does is in another. For example, he ends up including the paper by Davis (*1903*), who like Ladd-Franklin was one of Peirce's students, but he makes no mention at all about the algebra of abstract groups. Furthermore, he tells us nothing about the desirability of building signs that embody the group structures contained in *transformational truth tables*. Even more serious, as Kauffman (*1991*) would remind him, a lot happens when we do something as simple as cross two lines. Unlike Peirce, such as we see in Fig. 18.9, and unlike Carroll, Leśniewski, Gonseth, McCulloch, and Parry, Frazee does not cross the two lines for (*A,B*).

29. In the summer of 1988, while attending a meeting at Yale University, Ralph E. Kenyon, Jr., suggested that "Cognitive Ergonomics" would be a better label for what in turn I had been calling "Man-Sign Engineering" and "Sign Factors Engineering."

30. The deep continuity that exists between the main thesis of the present paper and the many historical ties on which it stands suggests that, without the post-1907 references, this paper could pose as the dissertation that OTTO might have submitted. Also allow the luck of the draw to be such that Peirce, Ladd-Franklin, Royce, and perhaps Binyon, served on his dissertation committee.

19.

A DECISION METHOD FOR EXISTENTIAL GRAPHS

Don D. Roberts

§1. Introduction: Anticipations

Quite a few of us who have built Peirce into our careers have discovered important ways in which he has anticipated later investigators. This, and similar discoveries involving other thinkers, have pretty well convinced me that if we try hard enough, we might find anticipations of almost anything, somewhere. The idea is not a new one. Alfred North Whitehead expressed it this way: "Everything of importance has been said before by somebody who did not discover it" (*Newman 1956:* 1: 75). Even this way of putting it is not new. Sometime between 185 and 159 B.C. the Roman dramatist Terence wrote "Nothing is said that has not been said before" (*Byrne ed. 1988*: 70).

This paper reports a quasi-anticipation. It involves Peirce's system of Existential Graphs (EG), which he sometimes presented as though it consisted of several parts. The part that has only two special symbols—the sheet of assertion (the surface on which the graphs are drawn) and the cut (a self-returning finely drawn line that serves as the sign of negation)—he called the *alpha* part; it is a complete propositional calculus. The part that adds the third special symbol—the thickly drawn line of identity—he called *beta*; it is a complete first order functional calculus with equality. A third part, *gamma*, was apparently intended in part for people who wanted to say something that had *not* been said before.

One of the topics we all cover when we teach elementary logic is the decision problem, which is the problem of finding an *effective* test by means of which it can always be determined whether or not a given formula is a theorem of some system or other. By "effective test" is meant a test that is a "matter of direct observation, and of following a fixed set of directions for concrete operations with symbols" (*Church 1960* [1983]). The decision problem for the propositional calculus can be solved by reduction to normal form, by the method of truth tables, by some version of the tree method, and, I am pleased to announce, by Charles Peirce's 1903 rule for determining whether any alpha graph is possible or not. I call this a "quasi-anticipation" because Peirce had already, in 1885, given the first statement of the truth table method as a *general* decision procedure (3.385–389). In addition, his 1903 work antedated the 1921 publications of Łukasiewicz and Post that have been the source of much of the recent development of the truth table method (see *Church 1956:* 162).

§2. The Decision Problem

2.1. *Background.* When I first stumbled onto Peirce's work on a graphical decision procedure I was working with Max Fisch on the manuscripts at Harvard in the early 1960's. Among the things we found puzzling was the first sentence of several drafts of the third Lowell Lecture of 1903, each of which referred to a method for determining the truth value of an alpha graph. Here is one example: "I explained at the end of the last lecture how to ascertain speedily and certainly whether a graph is absurd or not after the junctures [lines of identity] are got rid of" (MS 457; see also MSS 462 and 464). Repeated searches of the collection failed to find any such "last lecture" containing any such alpha procedure. The puzzle was finally solved in 1969 when a substantial number of Peirce papers were discovered in table drawers at Harvard's Houghton Library; included were six pages which had been cut from MS 455, six pages which present the alpha procedure. The story of these "lost and found manuscripts" is told by Richard Robin in the introduction to his catalogue of this material (*Robin 1971*).

Peirce's method had two parts: the first part was a rule for eliminating the lines of identity from beta graphs, in order to make them subject to the second part of the method, which was the alpha decision procedure. If Peirce had successfully developed both parts, he would have solved the decision problem for the first-order functional calculus. Each Lecture Three draft begins to describe the first part of the method; none of them contain a finished exposition, although they provide interesting insights into the way Peirce worked with EG. Peirce apparently ran into difficulties which forced him to abandon this part of his project, at least temporarily. Nevertheless, he remained confident that "it would be easy enough" to describe a rule which *would* do the job by reducing *any* beta graph to an alpha graph "while preserving its entire meaning" (MS 457: 6, and see MS 464: 6), at which point the method for alpha graphs could be applied. He had no illusions about the convenience of such a rule: "the performance of it," he said, "would be utterly impracticable owing to the stupendous complications that it would lead to" (MS 464: 8).

We now know that the decision problem of the first order functional calculus is unsolvable; this was proved by Alonzo Church in 1936, and it is one thing Peirce seems not to have anticipated. Still, although Peirce's reduction rule cannot do all that he wanted it to, it may turn out to be an alternative to the extended truth table method presented by Kleene and others (see *Kleene 1967:* 86ff for instance). This possibility is worth examination, but in the present paper I am concerned only with the second part, the successful part, of Peirce's method, his graphical solution to the decision problem for the propositional calculus.

2.2. *Peirce's Intentions.* There is no doubt that Peirce knew what he was doing when he invented this method; that much is clear from his presentation of it in his Lowell Lectures of 1903.

He clearly states the major reason for having such a method. He knew that to show that a graph is a theorem it is sufficient to derive it by using the transformation or inference rules of the system, and that a graph is impossible if one can show that applying the rules to it results in a contradiction (in the case of EG this is shown by obtaining the empty cut); but he also knew that the inability to *find* such proofs does not mean that no such proofs exist (MS 457: 1 and MS S32: 3, quoted below). Peirce was fully aware of the distinctions between tautologies, contingencies, and contradictions, and he coined a set of terms to express these and related distinctions: alpha possible (and possibility), beta possible, gamma possible; alpha, beta, and gamma necessary; alpha, beta, and gamma contingent.[1]

These points are simply made in a short passage from Peirce's second lecture:

> Now we may have a complicated graph and after transforming it according to the rules, may fail to find it either beta impossible or beta necessary; and yet we may be unable to make quite sure that there is not some mode of transforming it that we have not thought of which would prove one or the other. We thus stand in need of a rule as simple as may be, which shall determine whether or not the rules of transformation of graphs would in any way result in the canceling of every area on which that graph should be scribed. If this effect would be produced the graph is *beta impossible;* if not it is *beta possible.* In the latter case, we examine its negative, and if that be *beta impossible* the graph itself is *beta necessary.* But if both the graph and its negative are *beta possible,* they are both *beta contingent.*[2]

There are other passages which reveal more of Peirce's thought. In the first draft of his third lecture he gives an account of how he developed the method, in order (he said) to give his audience "some rough idea of the kind of reasoning required" in the logic business (MS 457: 1,6). This account amounts to a proof that the five operations are justified by the EG rules of transformation.[3] We could show that the method is adequate for the propositional calculus by means of a proof by mathematical induction on the number of letters and cuts in a graph, but we do not do it here.

One thing more: Peirce *knew* his alpha method was effective, he *designed* it that way. He wrote that it would be "easy to devise a machine that would perform" his decision procedure, because the method is "a comprehensive routine" (MS 462: 6) that calls for the exercise of "little intelligence" (MS 464: 6). This claim may be tested in the near future by a computer expert at Watcom International Corporation of Waterloo.[4]

2.3. *Peirce's Method.* Let us now turn our attention to the method itself as Peirce developed it for the alpha part of EG. For this purpose it will be convenient to have in front of us not only a concise statement of the five operations which constitute the method, but also the alpha conventions and rules of transformation. I remind the reader that alpha has only two proper symbols, the surface on which the work is done (a sheet of paper, a blackboard, etc.) called "the sheet of assertion," and the cut, which is a finely drawn self-returning line. Cuts do not intersect, but may be made within other cuts, creating "nests" of cuts. A nest of two cuts, called a scroll, is particularly important; four varieties are shown in Fig. 19.1. A scroll which has no graph (other than the second cut of the scroll itself) on its outer, or once-enclosed area, is called a double cut. Figs. 19.1(a) and 19.1(d) illustrate the double cut.

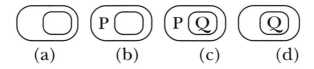

| (a) | (b) | (c) | (d) |

Figure 19.1

We now list the conventions, which present the principal interpretation of the system (see *Roberts 1992* for a brief introduction to the full system, and *Roberts 1973* for a comprehensive treatment):

C1. The sheet of assertion (SA) is a graph expressing whatever is taken for granted at the outset to be true of the universe of discourse.

C2. Whatever is scribed (Peirce's term for making a graph) on SA is asserted to be true of the universe represented by SA.

C3. Graphs scribed on different parts of SA are *all* asserted to be true; hence juxtaposition on SA is the sign of conjunction.

C4. The scroll is the sign of the (material) conditional: Fig. 19.1(c) means "if *p* then *q*."

C5. The empty cut (Fig. 19.2) is an "always false" proposition (some-

Figure 19.2

times called the "pseudograph"); and the cut precisely denies its contents.[5]

The following five rules of transformation yield a complete elementary logic:

R1. *Erasure*: Any evenly enclosed graph may be erased.

R2. *Insertion*: Any graph may be scribed on any oddly enclosed area.

R3. *Iteration*: A graph which already occurs may be scribed again within the same or additional cuts.

R4. *Deiteration*: Any graph whose occurrence is, or could be, the result of iteration, may be erased.

R5. *Double cut*: The double cut may be inserted around or removed (where it occurs) from any graph on any area.[6]

Use of the decision procedure supposes the existence of a proposition whose status is to be determined. Let *X* represent the graph of such a proposition. There are two preliminary steps: (1) construct the graph of the conditional sentence (a familiar theorem) "If *X* then *X*," as shown in Fig. 19.3; (2) separate the area of the consequent (the twice-enclosed area) into an upper and a lower region, as shown in Fig. 19.4. The following five operations are to be applied (again and again, until they can no longer be performed) to the graph which occurs in the lower region of this twice-enclosed area:

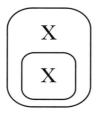   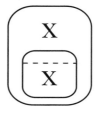

Figure 19.3   Figure 19.4

Op. 1. Cancel any enclosure (cut plus contents) containing an empty cut.

Op. 2. Cancel any double cut.

Op. 3. Transfer one letter that stands otherwise unenclosed in the lower region to the upper region, and cancel it wherever it occurs, however enclosed, in the lower region.

Op. 4. For each letter which occurs as the sole contents of an unenclosed cut in the lower region, substitute an empty cut throughout that lower region, and insert in the upper region a cut containing that letter and nothing else.

Op. 5. When the first four operations can no longer be applied, if graphs remain in the lower region, they will consist of letters enclosed in various nests of cuts. Select, as next to be dealt with, a letter which remains, however enclosed, in the lower region. Then (a) iterate (copy) on the once-enclosed area (alongside *X* representing the antecedent) every enclosure which occurs there; this will double the number of enclosures on that area. (b) In the upper region of each of the original enclosures scribe the letter selected, canceling that letter throughout the lower region, as in Op. 3. (c) In the upper region of each of the copies scribe an enclosure containing only the letter selected, and throughout the lower region substitute an empty cut for that letter, as in Op. 4.

When no further operations can be performed, no graphs will remain in any lower region. The graph in each upper region, consisting of letters or single letters inside cuts. shows a truth value combination of the letters which makes the original graph, *X*, true. This graph, says Peirce (MS S32: 006), "is to be understood as referring to the universe of alpha possibility." If, however, no enclosures remain alongside *X*, so that no upper regions remain in which letters or single letters within cuts can be found, then *no* combination of truth values to its components will make the original graph true: *X* is alpha impossible.

2.4. Application of the Method. As might be expected, the five operations which constitute Peirce's decision procedure for alpha are most easily and quickly understood in connection with specific examples.

(A) Let us first use the method to determine whether or not a given graph is alpha consistent, and to show, if it is, what the truth values of its component propositions must be to make the graph true. Consider the set of propositions "It rains," "if it rains it does not pour," and "it does not pour." Expressed in EG the conjunction of these propositions looks like Fig. 19.5, where we use the lowercase letters '*r*' and '*p*' to represent, respectively, "it rains" and "it pours." The two preliminary steps of our method yield the graph of Fig. 19.6.

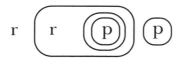

Figure 19.5

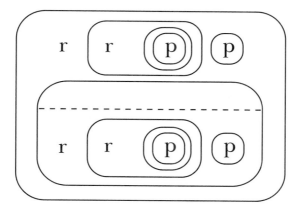

Figure 19.6

In the figures which follow, we could continue to scribe the antecedent (everything on the outer, or once-enclosed area) of our graph just the way it is given in Fig. 19.6, but since all the work of Peirce's operations will be done with the consequent of the graph of Fig. 19.6 (on the twice-enclosed area), there is no need to write out the antecedent in detail; hence, following Peirce's practice we abbreviate it by the letter *X* as in Fig. 19.7.

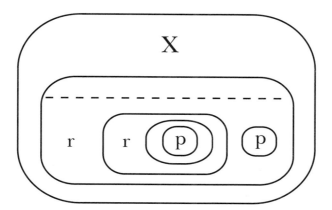

Figure 19.7

Consider Fig. 19.7: Op. 2 directs us to cancel the double cut; Op. 3 directs us to transfer the r to the upper region (above the separator) and cancel its other occurrences in the lower region (below the separator); this work is indicated in the graph of Fig. 19.8 with the help of some cross-out lines. You will notice that what remains in the lower region of Fig. 19.8 (the cross-

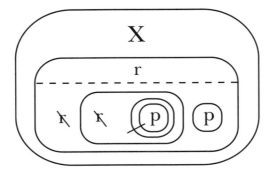

Figure 19.8

outs represent erasures) are two occurrences of cut-*p*. Op. 4 can now be
applied: an empty cut is substituted for each *p*, and cut-*p* is scribed in the
upper region, as indicated in Fig. 19.9. Op. 2 can again be applied to the
lower region of Fig. 19.9, and this empties the region.

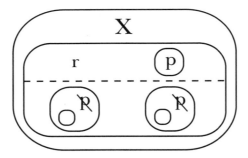

Figure 19.9

The resulting graph can be read as an ordinary material conditional:
If *X* then *r* and not-*p*. The interpretation of this is that when *X* (the conjunction we started with) is true, *r* must be true and *p* must be false. In
other words, our original graph is alpha possible, and we have discovered
the values that *r* and *p* must have in order to make *X* true.

(B) The method may also be used to determine the validity of an Alpha
argument by testing the premises of the argument together with the *denial*
of its conclusion. The argument is valid if and only if the method assigns
the value *false* to this set of propositions. This method is called "indirect
proof" in some contemporary introductory texts. Peirce's student Christine

Ladd-Franklin introduced a generalized form of it in her theory of the syllogism; she called it the "inconsistent triad," or the "antilogism."

The three propositions of our first illustration comprise an argument in the form *modus ponens*, as the reader has probably noticed. The denial of that conclusion is double-cut-*p*, and the two preliminary steps yield the graph of Fig. 19.10, in which two double cuts have already been marked for erasure. (Keep in mind that the *X* in Fig. 19.10 represents the new set

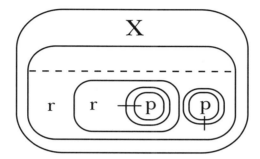

Figure 19.10

of graphs, with the negated conclusion; it does not represent the set of Figs. 19.7–9.) Op. 3, applied as indicated in Fig. 19.11, then yields Fig. 19.12, which is marked for a final application of Op. 3. Notice that the result will leave *r* and *p* in the upper-region, and an empty cut in the lower region. This sets up an application of Op. 1, which will leave *X* as the sole content of a cut; and this means that *X*, our graph of the argument with its conclusion negated, has the value *false*; it is alpha-impossible. Hence our original argument is valid.

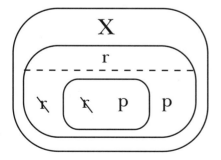

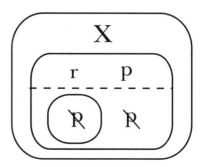

Figure 19.11   Figure 19.12

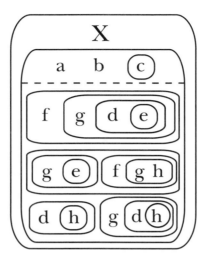

Figure 19.13

(C) To illustrate the use of Op. 5 we will consider one of Peirce's own examples, given in Fig. 19.13. Since the lower region of the two-times enclosed area contains no empty cuts, no double cuts, no unenclosed letters, and no cases of a single letter as the sole contents of an unenclosed cut, none of the first four operations can be applied. Op. 5 tells us to select some letter from the lower region to work with—let us choose the letter *g*; before we do anything with *g* we must, according to Op. 5(a), duplicate the large enclosure (the one containing the separated regions), making a copy of it on the same area as the original, alongside of *X*. This gives us Fig. 19.14.

Let us call the enclosure on the left the original, and the enclosure on the right, the replica. Following the instructions in 5(b) we scribe *g* in the upper region of the original enclosure, and cancel every occurrence of *g* in the lower region; and following 5(c) we scribe cut-*g* in the upper region of the replica enclosure, substituting an empty cut for each occurrence of *g* in the lower region. These operations are shown in Fig. 19.15, which also indicates (by means of stroke marks) impending cancellations of resulting double cuts (by Op. 2) and of enclosures which end up containing empty cuts (by Op. 1). A little reflection should convince the reader that these operations set up new opportunities for the use of Ops. 3 and 4, and so on.

A reader who continues to apply Peirce's method to Fig. 19.15 will discover that the original enclosure and its copy (the "replica") will both be canceled: the copy because the presence of *f* and cut-*f* in the lower region

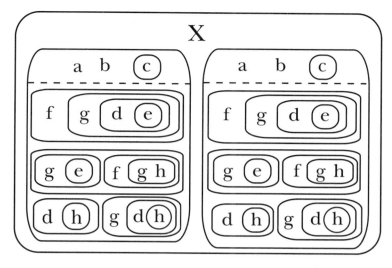

Figure 19.14

will yield an empty cut, the original because h and cut-h will appear, also yielding an empty cut. What will be left at the end is cut-X, which tells us that the original graph is alpha impossible.

Peirce's fifth operation closely resembles the elegant method developed by Quine in early versions of his *Methods of Logic*, in a section entitled

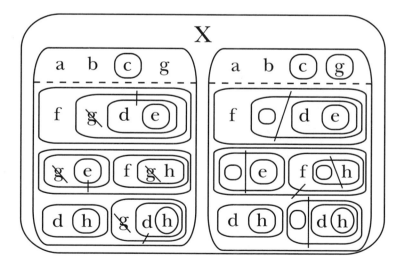

Figure 19.15

"Truth-value analysis." Quine replaced sentence letters with markers (elongated T's supplanting truths and upside-down T's supplanting falsehoods), and formulated "rules of resolution" whose applications progressively reduced the number of cases to be considered. The rationale is the familiar one in truth-value analysis generally: first consider what is the case if a given letter is true; then consider what happens if it is false. Both methods, Peirce's and Quine's, and tree methods generally, have the advantage of progressive simplification over a standard comprehensive truth table which becomes cumbersome when more than a few distinct letters or variables are involved. A complete truth table analysis of Peirce's graph in Fig. 19.13 would require a table of 256 rows.

§3. Concluding Remarks

Logic intrigues me. The sheer difficulty of it acts as a challenge; its continuing rich harvest of practical and theoretical results promises the excitement of discovery; and there is a beauty in the relationships it reveals. Even more important, perhaps, is the feature which gave rise to the notion of effectiveness: the insistence on clarity of ideas and argument, and on the conception of truth as something public. Howard DeLong (*DeLong 1971:* 164) has argued that the notion of proof itself

> was invented by such Pre-Socratics as Pythagoras and Zeno [in order] to make knowledge no respecter of persons, to take it out of the theological realm and into the secular. Thus a proof, in order to be a proof, had to be recognizable *as* a proof by anyone.

I consider this an inseparable part of the belief that philosophy, like the other sciences, "is essentially, not accidentally, a shared inquiry—an inquiry in which it is pointless to offer answers until the reader has become completely engaged in the questions that they answer" (*Brumbaugh 1989:* 39).

Imagine my surprise when I learned that Robert Nozick was propounding the thesis that we ought *not* to try to persuade people by rational argumentation: to do so is to "force" someone to believe something he may not *want* to believe, and this is wicked, because it is not "a nice way to behave toward someone" (*Nozick 1981:* 13). Was this something new, a Ms Manners approach to philosophy, or a philosophical concession to political correctness? Then I remembered an old comic strip character created by Al Capp, and realized that Nozick's recommendation is a kind of application of L'il Abner's moral philosophy, "Good is better than evil, 'cuz it's nicer."

Here is more evidence in favor of the claim that everything has been said before. Actually, it is better to say that *many* things, both true and important, have already been said. For I do not believe that an intelligent person can hold to the extreme form of this thesis. Gödel's result and the related limitative theorems are new, although the notion that there are limitations to knowledge can be found in the pre-Socratics. What is new is the precision with which these logically unsolvable problems are stated and the relative simplicity of the problems themselves (see *DeLong 1971: 209*). The knowledge that Neptune has at least eight moons is new. Quantum theory is new, for while the ancients had the idea of a kind of atomism, to push this "anticipation" any further is to succumb to what Richard Feynman aptly called "Cargo Cult Science" (*Feynman 1985: 338–346*).

A wonderful inversion of this theme occurs in the account of Pierre Menard, by the late Argentinean writer Jorge Luis Borges. Menard's fascination with *Don Quixote* stimulated—perhaps overstimulated—his own creative energy. "Every man should be capable of all ideas," wrote Menard (*Borges 1962:* 54), who certainly practiced what he preached: without for a moment denying the temporal priority of Cervantes, Menard himself authored "the ninth and thirty-eighth chapters of Part One of *Don Quixote* and a fragment of the twenty-second chapter" (*Borges 1962:* 48). Menard too had predecessors. In Part Two of his *Discourse,* Descartes reports that in his college days he "discovered" that "nothing can be imagined which is too strange or incredible to have been said by some philosopher" (*Descartes 1985:* 1: 118). Fourteen years later Hobbes discovered the same thing, but he discovered it in Cicero, who wrote "Somehow or other no statement is too absurd for some philosophers to make" (*De Divinatione* ii.58; see *Hobbes 1947:* Part I, ch. v).

The possibility that our interaction with great writings will stimulate our own creative activity lies behind the characterization of that interaction as a "Great Conversation."[7] Reflection on conversation in this extended sense can enrich our understanding of another Peircean favorite of mine, the notion that meaning is virtual; rather than diminish the importance of a truth or discovery, this latter thesis makes successive generations of scholars seem like a community of investigators. This brings me to the third part of EG, which Peirce called the Gamma part. Most of his own work on this part of EG dealt with modalities, and scholarly accounts of it largely follow Peirce's lead. Nevertheless, one 1903 manuscript makes it clear that Gamma was not viewed as having a fixed agenda in the way Alpha and Beta do: "*The Gamma Part* supposes the reasoner to invent for himself such additional kinds of signs as he may find desirable" (MS 693: 282). For a while, Peirce's expectations were modest, as indicated in a March 9, 1906 draft of a letter to Lady Welby (L 463):

> The system of Existential Graphs (at least, so far as it is at present developed) does not represent every kind of Sign. For example, a piece of con-

certed music is a sign; for it is a medium for the conveyance of Form. But
I know not how to make a graph equivalent to it. So the command of a
military officer to his men: " 'Halt!'" "Ground arms!" which is interpreted
in their action, is a sign beyond the competence of existential graphs. All
that existential graphs can represent is *propositions*, on a single sheet, and
arguments, on a succession of sheets, presented in temporal succession.

Apparently this letter was never posted, perhaps because when Peirce wrote
it he was feverishly developing his tinctured graphs which soon *would* have
a way to analyze commands, hypotheses, and questions (4.554). The tinc-
tures were designed, in short, to analyze "all that ever could be present to
the mind in any way or any sense" (MS 499(s)). At the height of his enthusiasm Peirce meant EG to apply to the whole of the proper business of logic:
"All our thoughts are signs"; "it is in their character as signs that logic is
applicable to them"; "logic is applicable to all signs whether they are directly
mental or not." (MS 425, 1903; the "Minute Logic.")

    This is very ambitious. It reminds me of a remark by Whitehead about
the future of his own "first love," symbolic logic (*Whitehead 1948:* 99):

> When in the distant future the subject has expanded, so as to examine
> patterns depending on connections other than those of space, number,
> and quantity—when this expansion has occurred, I suggest that symbolic
> logic, that is to say, the symbolic examination of pattern with the use of
> real variables, will become the foundation of aesthetics. From that stage
> it will proceed to conquer ethics and theology.

Neither Whitehead's nor Peirce's expectations have yet been realized.
Still, recent work on deontic and erotetic logic, though none of it (as far as
I am aware) owes anything to EG, shows that they were on the right track
in some respects, at least. Perhaps some qualifications are necessary, both
for symbolic logic in general and EG in particular; Susanne Langer's treatment of the arts as "non-discursive and non-systematic symbolic expression"
of an artist's "knowledge of feeling" seems to be a significant advance
(*Langer 1967:* xv). Perhaps digital recording techniques will turn out to be
relevant to broad questions about representation; perhaps continuing work
in artificial intelligence, and in semiotics, will make a difference here. A
good example of what can be done in connection with EG—by no means
the only example—is the work of John Sowa in adapting the graphs to information processing.

    I mention all this in connection with the graphs, but the point is more
general: it is to reinforce the message of Peirce, Whitehead, Menard, and
others that we should be experimental in our work, everywhere; we should
keep the conversation going.

N O T E S

1. MS S32, especially pages 2–4; MS 457: 2f.; MS 462: 2, 8f., 26f.; MS 464: 2, 6f., 18, 26–28.

2. MS S32: 3. Peirce mentions complications involved in the beta part of the rule at MS 457: 6; he elaborates on this at MS 462: 6ff. and MS 464: 6fff. For statements that his rule was probably capable of improvement, see the continuation in MS S32, as well as MS 457, MS 462, and MS 464.

3. See MS 457: 4–6 for an explanation of Operations 4 and 5. A later draft of Lecture III, MS 464, marked "to be used" by Peirce, begins with a statement of regret that "there is not time" for presenting "the train of thought" which yielded the rule.

4. This expert, Dan Pronovost (*BMath 1989, Waterloo*), also has some ideas about computerizing a little decision procedure for simple beta graphs I invented about fifteen years ago, or whenever it was that the method of trees was all the rage in Waterloo. I call the method "weeds."

5. There are four more conventions, having to do with the line of identity:

C6. The scribing of a heavy dot or the line of identity on SA denotes the existence of a single, individual object in the universe of discourse.

C7. The line of identity asserts the numerical identity of the individuals denoted by its two extremities.

C8. A branching line of identity with n number of branches will be used to express the identity of the individuals denoted by its n extremities.

C9. Points on a cut shall be considered to lie outside the area of that cut.

6. These same five rules are adequate for the Beta part of the system also; their application to lines of identity might require some explanation.

7. Robert Hutchins (*1952*) gives an inspirational treatment of this notion.

20.

PEIRCE AND PHILO

Jay Zeman

Charles Peirce, logician and philosopher, contributed notably to the theory of the conditional. Actually, from his perspective and in his terminology it is better, as we shall see, to link his work on the *conditional* with his discussions of the *hypothetical* proposition. Peirce spoke often of the *consequentia de inesse*,[1] the concept of which is intimately linked with the material, or "Philonian" conditional; indeed, we shall see him calling himself a Philonian. And it is not uncommon to hear Peirce—at least prior to the last decade of his life—declared a Philonian, whose fundamental analysis of the conditional was essentially the same as that of Philo (and of more modern types like Russell and like Quine).

In this paper, I intend first to examine Peirce's understanding of "Philonian"; I will then look at the Philonian or "*de inesse*" conditional in the context of his overall logical thought. It is commonly held that Peirce in his early years held to the "nominalistic" Philonian conditional, and only later "surrendered" it in favor of a more "realistic" view; the study we are here undertaking will indicate that this does not adequately reflect his position.

As we shall see, Peirce at one time or another called himself a Philonian. He was, of course, also quite aware that Philo is historically paired with his teacher Diodorus. That pairing, in fact, is so integral to the meeting of "Philonian" that without consideration of it, "Philonian conditional" becomes a somewhat pretentious term for "truth-functional implication."

When he discusses the historical pairing I have mentioned, Peirce employs "Diodoran" as the adjective derived from the name of Diodorus. Our considerations here are complicated by contemporary discussions; an index of the complication is the fact that the usual contemporary form of the adjective is "Diodorian." Arthur Prior, one of the rediscoverers of the possible worlds approach to modal logic,[2] used Diodorus as a take-off point in his work (see *Prior 1955, 1957*); Prior, and modal logicians following him, employed this latter form of the relevant adjective. Prior at first thought that what he called the "Diodorian modal system" supplied semantics for S4; this misapprehension was rather quickly dispelled, however, and 'Diodorian modality' gravitated to its contemporary usage: generally, it means the class of modal logics with reflexive, transitive, and *linear* (total-ordered) relational frames (that is, the system containing S4.3); specifically, it refers to one of these systems, "D," the system in whose frames "possible worlds" are "discrete" (see *Zeman 1973*).

Peirce's discussions of the Philo-Diodorus debate are among those in which he anticipates the discovery of possible-worlds modal semantics; we must not, however, conclude that his term "Diodoran" means anything like the contemporary "Diodorian." We find him saying that

> Cicero informs us that in his time there was a famous controversy between two logicians, Philo and Diodorus, as to the signification of conditional propositions. Philo held that the proposition "if it is lightening it will thunder" was true if it is not lightening or if it will thunder and was only false if it is lightening but will not thunder. Diodorus objected to this. Either the ancient reporters or he himself failed to make out precisely what was in his mind, and though there have been many virtual Diodorans since, none of them have been able to state their position clearly without making it too foolish. Most of the strong logicians have been Philonians, and most of the weak ones have been Diodorans. For my part, I am a Philonian; but I do not think that justice has ever been done to the Diodoran side of the question. The Diodoran vaguely feels that there is something wrong about the statement that the proposition, "If it is lightening it will thunder," can be made true merely by its not lightening. (NEM 4: 169; this is MS 441 [1898])

Although Peirce comments elsewhere that "the Diodoran view seems to be the one which is natural to the minds of those, at least, who speak the European languages" (3.441), he takes (as is suggested above) a rather dim view of those who have advocated this position. He contrasts the Philonian and Diodoran positions thus:

> According to the Philonians, "It is now lightening it will thunder," understood as a consequence *de inesse*, means "It is either not now lightening or it will soon thunder." According to Diodorus, and most of his followers (who seem here to fall into a logical trap), it means "It is now lightening and it will soon thunder." (3.442)

Far from being connected with the contemporary "Diodorian" modal logic (which, from an important perspective, fits perfectly within Peirce's own theory of possibility), the Diodoran position *as understood by Peirce* seems to require a sort of "existential import" for the antecedent condition (in the above case, "It is now lightening"). A function with the meaning ascribed above by Peirce to the conditional of "Diodorus, and most of his followers," is, of course, not really a conditional, but is much closer in meaning to a *conjunction.* This won't do, but Peirce is unwilling simply to reject the "Diodoran" strategy. Presumably, this approach expresses *something* about the conditional that can be taken as common-sense wisdom about it; perhaps Peirce himself can come to the aid of the inept Diodoran, and can

> fit him out with a better defense than he has ever been able to construct
> for himself, namely, that in our ordinary use of language we always under-
> stand the range of possibility in such a sense that in some possible case
> the antecedent shall be true. (NEM 4: 169)

Taken thus, the "Diodoran" position is not really in competition with the Philonian, but might be considered *complementary* to it, reflecting something to do with ordinary uses of language.

This sense of "Diodoran" stands in contrast to the technical modal "Diodorian" of the last couple of decades: Peirce's reading of Diodorus is quite different from Prior's. And Peirce's understanding of Philo/Diodorus also differs from another way in which these Ancients may be compared. It is possible to see Philo *vs.* Diodorus as the preliminary bout of a card of fights which has more recently matched C. I. Lewis with the Russellians and W. V. O. Quine with Ruth Marcus. The most visible issue in the disputes I mention here is that of the nature of modal propositions, and in particular, of the relationship of "implication" to modality. In the terminology of Lewis, this is the question of "material" *vs.* "strict" implication. And in terminology often used in discussing broad movements in western philosophy, it is the matter of "nominalism" *vs.* "realism." Peirce's understanding of Philo *vs.* Diodorus does not seem, as our quotes indicate, to reflect this contrast. I emphasize this so that we won't unconsciously read current approaches back into Peirce.

Now, it is generally accepted that a development can be traced over the years in Peirce's thought. As one commentator has put it,

> Peirce's philosophy is like a house which is being continually rebuilt from
> within. Peirce works now in one wing, now in another, yet the house stands
> throughout, and in fact the order of the work depends upon the house
> itself since modification of one part necessitates the modification of an-
> other. And although entire roofs are altered, walls moved, doors cut or
> blocked, yet from the outside the appearance is ever the same. (*Murphey
> 1961: 4*)

Well, perhaps; I think that it is well to examine carefully the extent to which "modifications" actually occur, however. The richness of a later position may not at all mean that its earlier forebears have been abandoned. It may be that these earlier positions are, in fact, so well established as to be unproblematic with respect to the later position; three decades after the *Illustrations of the Logic of Science* series, for example, Peirce was writing expansive foot-notes on those papers—but hardly rejecting Pragmatism as laid out there.

Now, coming near matters which will be near to our concern here, in tracing "Peirce's Progress from Nominalism to Realism," another author tells us that

> Since the proof of pragmaticism, and thereby of realism, can be most co-gently stated in existential graphs, the series [of *Monist* articles beginning in 1905] proceeds to a fresh exposition of the graphs (4.530–572). But the *Monist* printer's ink is scarcely dry on that when Peirce hits upon an improvement to enable them to represent different kinds of possibilities; and, confronted by this improvement, his other self, his nominalist self, surrenders his last stronghold, that of Philonian or material implication. (*Fisch 1986:* 196) [3]

Actually, a fairly careful study of material available from the period in which this "surrender" is supposed to have taken place doesn't give much evidence of it; in fact, there is at this period very little testimony of concern with the conditional as such. Now I certainly don't believe that a dearth of discussions specifically regarding the conditional indicates that Peirce had lost interest in it. I believe, rather, that Peirce's theory of the conditional had taken its basic form long before, and that while there would be elabo-rations and amplifications of that theory, that basic form—which was inte-gral to his thought—would maintain itself intact.

My own investigations lead me to believe that Peirce's relationship to the Philonian conditional is quite complex; although there was an in-creased emphasis on the reality of thirds and firsts in the later Peirce, his view of the conditional *de inesse* sees it as second to a larger third from about the days of the *Illustrations of the Logic of Science*, anyway.

I note that in many locations Peirce speaks not of *conditional* proposi-tions, but of *hypotheticals*. The term "hypothetical" suggests a strong link between mathematical logic and philosophy for Peirce; we note its role in the thought of Kant, for example, who was a major influence on Peirce, while on the other hand, seeing that Peirce's discussion of "hypotheticals" are almost always located in the context of his symbolic logic. Let's look at what he says in the well-known *Philosophy of Notation* articles of 1885:

> To make the matter clear, it will be well to begin by defining the meaning of a hypothetical proposition, in general. What the usages of language may be does not concern us; language has its meaning modified in technical logical formulae as in other special kinds of discourse. The question is what is the sense which is most usefully attached to the hypothetical propo-sition in logic? Now the peculiarity of the hypothetical proposition is that it goes out beyond the actual state of things and declares what *would* hap-pen were things other than they are or may be. The utility of this is that

it puts us in possession of a rule, say that "if *A* is true, *B* is true," such that should we hereafter learn something of which we are now ignorant, namely that *A* is true, then, by virtue of this rule, we shall find that we know something else, namely, that *B* is true. There can be no doubt that the Possible, in its primary meaning, is that which may be true for aught we know, that whose falsity we do not know.[4] The purpose is subserved, then, if throughout the whole range of possibility, in every state of things in which *A* is true, *B* is true too. The hypothetical proposition may therefore be falsified by a single state of things, but only by one in which *A* is true while *B* is false. States of things in which *A* is false, as well as those in which *B* is true, cannot falsify it. If, then, *B* is a proposition true in every case throughout the whole range of possibility, the hypothetical proposition, taken in its logical sense, ought to be regarded as true, whatever may be the usage of ordinary speech. If, on the other hand, *A* is in no case true, throughout the whole range of possibility, it is a matter of indifference whether the hypothetical be understood to be true or not, since it is useless. But it will be more simple to class it among true propositions, because the cases in which the antecedent is false do not, in any other case, falsify a hypothetical. This, at any rate, is the meaning which I shall attach to the hypothetical proposition in general, in this paper. (3.374)

I note especially in the above Peirce's speaking of "a proposition true in every case throughout the whole range of possibility"; the notion of quantification over a range of possibility is central here. And we see him in 1902 saying that

the quantified subject of a hypothetical proposition is a *possibility*, or *possible cause*, or *possible state of things*. In its primitive state, that which is *possible* is a hypothesis which in a given state of information is not known, and cannot certainly be inferred, to be false. The assumed state of information may be the actual state of the speaker, or it may be a state of greater or less information. Thus arise various kinds of possibility. (2.347)

The notion of quantification over a range of possibilities is a basic theme in Peirce's work; in 1902, he himself sees his understanding of hypotheticals in these terms as going back even earlier than the *Philosophy of Notation* paper:

In a paper which I published in 1880, I gave an imperfect account of the algebra of the copula. I there expressly mentioned the necessity of quantifying the possible case to which a conditional or independential proposition refers. But having at that time no familiarity with the signs of quantification which I developed later, the bulk of the chapter treated of simple consequences *de inesse*. Professor Schröder accepts this first essay as a satisfactory treatment of hypotheticals; and assumes, quite contrary to *my* doctrine, that the possible cases considered in hypotheticals have no multitudinous universe. This takes away from hypotheticals their most characteristic feature. (2.349)

Peirce probably refers here to the well-known paper of 3.154 ff.; he notes there that

> De Morgan, in the remarkable memoir with which he opened his discussion of the syllogism . . . has pointed out that we often carry on reasoning under an implied restriction as to what we shall consider as possible, which restriction, applying to the whole of what is said, need not be expressed. The total of all that we consider possible is called the *universe* of discourse, and may be very limited. One mode of limiting our universes by considering only what actually occurs, so that everything which does not occur is regarded as impossible. (3.174)

And a universe so limited would be, of course, the universe of the *de inesse*, it seems clear to me, even without Peirce's 1902 testimony, that even in 1880 he considers this realm just a limiting case of a broader domain which cannot be ignored; in fact, in another 1880 paper, we find a harbinger of the thought which was to be made explicit with the development of notations for quantification:

> To express the proposition: "If S then P," first write
>
> $$A$$
>
> for this proposition. But the proposition is that a certain conceivable state of things is absent from the universe of possibility. Hence instead of A we write
>
> $$B \; B^5$$
>
> Then B expresses the possibility of S being true and P false. Since, therefore, SS denies S, it follows that (SS,P) expresses B. Hence we write
>
> $$SS,P;SS,P$$

(4.14)

This is close to the time of the "*Illustrations of the Logic of Science*"; it might be argued that, with that series of essays and with the significant contributions to symbolic logic dating from this period, we have the beginnings of Peirce's maturity as a philosopher-logician. And his theory of the conditional—or, more generally, of the *hypothetical*—is consistent from this point on, and is integral to his thought as a whole.

Examination of Peirce's later work in logic strongly supports this position. A decade after the *Philosophy of Notation* paper, Peirce had changed notations. The algebraic notations we have been examining were successful vehicles for his deductive logic; for a logic, however, it is not enough that a notation be mathematically correct. The purpose of a logic is not efficient calculation. The purpose is to provide an appropriate representation of the process of necessary deductive reasoning; one of the measures of appropriateness, Peirce tells us, is *iconicity* (*Zeman 1986*: 13). And Peirce felt that his Existential Graphs were a more appropriate representation of the subject matter of logic. The Graphs provide an alternative and arguably more iconic (than the algebras) representation of a number of features of deductive logic. The feature we are interested in at this point is the relationship between the *de inesse* conditional and the "hypothetical."[6]

The most basic of the signs of the Existential Graphs is the "Sheet of Assertion" (SA); SA is itself a graph, which represents whatever is true about the universe of discourse; effectively, it represents that universe of discourse. And

> A proposition *de inesse* relates to a certain single state of the universe, like the present instant. Such a proposition is altogether true or altogether false. But it is a question whether it is not better to suppose a general universe, and to allow an ordinary proposition to mean that it is sometimes or possibly true. Writing down a proposition under certain circumstances asserts it. Let these circumstances be represented in our system of symbols by writing the proposition on a certain sheet. (4.376)

It seems natural to employ a given SA to represent what Peirce calls the "quasi-instantaneous" state of the general universe; Peirce sometimes employs SA in this manner, and sometimes suggests other representations (or at least that there are other representations; see *ibid.*, we shall eventually note at least two specific cases in which Peirce sets separate "quasi-instantaneous" states upon the same sheet). In 1903 Peirce tells us that

> If a system of expression is to be adequate to the analysis of all necessary consequences,[7] it is requisite that it should be able to express that an expressed consequent, *C*, follows necessarily from an expressed antecedent, *A*. The conventions hitherto adopted do not enable us to express this. In order to form a new and reasonable convention for this purpose we must get a perfectly distinct idea of what it means to say that a consequent follows from an antecedent. It means that in adding to an assertion of the antecedent an assertion of the consequent we shall be proceeding upon a general principle whose application will never convert a true assertion into a false one. . . . But before we can express any proposition referring to a general principle, or as we say, to a "range of possibility," we must first find means to express the simplest kind of conditional proposition, the *conditional de inesse*, in which "If *A* is true, *C* is true" means only that, principle or no principle, the addition to an assertion of *A* of a assertion of *C* will not be the conversion of a true assertion into a false one. That is, it asserts that the graph of Fig. 20.1, anywhere on the sheet of assertion, might be transformed into the graph of Fig. 20.2 without passing from truth to falsity.

a   a c

Figure 20.1   Figure 20.2

This conditional *de inesse* has to be expressed as a graph in such a way as distinctly to express in our system both *a* and *c* and to exhibit their relation to one another. To assert the graph thus expressing the conditional *de inesse*, it must be drawn upon the sheet of assertion, and in this graph the expressions of *a* and *c* must appear; and yet neither *a* nor *c* must be drawn upon the sheet of assertion. How is this to be managed? (4.435)

Peirce goes on to develop the typical alpha-graph representation of the conditional. Significantly, he stresses iconicity:

In order to make the representation of the relation between [antecedent and consequent] iconic, we must ask ourselves what spatial relation is analogous to their relation. (*ibid.*)

His solution is the representation in Fig. 20.3:

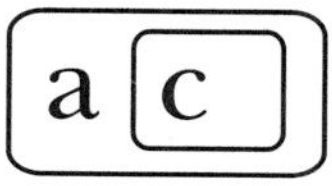

Figure 20.3

I have remarked that Peirce will give us ways of dealing with more than one "quasi-instantaneous state" on a given sheet of assertion; he also, however, describes sheets themselves as representing such states:

in the gamma part of the subject all the old kinds of signs take new forms. . . . Thus in place of a sheet of assertion, we have a book of separate sheets, tacked together at points, if not otherwise connected. For our alpha sheet, as a whole, represents simply a universe of actual existent individuals, and the different parts of the sheet represent facts or true assertions concerning that universe. At the cuts we pass into other areas, areas of conceived propositions which are not realized. In these areas there may be cuts where we pass into worlds which, in the imaginary worlds of the outer cuts, are themselves represented to be imaginary and false, but which may, for all that, be true, and therefore continuous with the sheet of assertion itself, although this is uncertain. You may regard the ordinary blank sheet of assertion as a film upon which there is, as it were, an undeveloped photograph of the facts in the universe. I do not mean a literal picture, because its elements are propositions, and the meaning of a proposition is abstract and altogether of a different nature from a picture. But I ask you to imag-

ine all the true propositions to have been formulated; and since facts
blend into one another, it can only be in a continuum that we can conceive
this to be done. This continuum must clearly have more dimensions than
a surface or even that a solid; and we will suppose it to be plastic, so that
it can be deformed in all sorts of ways without the continuity and connection of parts being ever ruptured. Of this continuum the blank sheet of
assertion may be imagined to be a photograph. When we find out that a
proposition is true, we can place it wherever we please on the sheet, because we can imagine the original continuum, which is plastic, to be so
deformed as to bring any number of propositions to any places on the
sheet we may choose. (4.512)

So the "alpha sheet . . . represents simply a universe of actual existent individuals, and the different parts of the sheet represent facts or true assertions concerning that universe." We seem to have "gates"[8] into other universes: "At the cuts we pass into other areas, areas of conceived propositions
which are not realized."[9] The model for the relationship between the universe of *all possibles* and the actual universe is given geometrically (actually,
topologically). Further, Peirce is not interested only in "the" universe of
actual existent fact, but as well in certain other subsets of the realm of all
possibles—he is interested in those subsets which *could* themselves constitute existential universes.

Now, *qualities* are not, properly speaking, individuals. . . . Nevertheless,
within limitations, which include most ordinary purposes, qualities may
be treated as individuals. At any rate, however, they form an entirely different universe of existence. It is a universe of logical possibility. As we
have seen, although the universe of existential fact can only be conceived
as mapped upon a surface by each point of the surface representing a vast
expanse of fact, yet we can conceive the facts [as] sufficiently separated
upon the map for all our purposes; and in the same sense the entire universe of logical possibilities might be conceived to be mapped upon a surface. Nevertheless, if we are going to represent to our minds the relation
between the universe of possibilities and the universe of actual existent
facts, if we are going to think of the latter as a surface, we must think of
the former as three-dimensional space in which any surface would represent all the facts that might exist in one existential universe. (4.514)

So moving from the topological icon of the basic Sheet of Assertion (considered to be a surface) as representing the universe of actual existent fact,
we take an appropriate space (that "surface" plus another dimension) as
representing the overall realm of possibles; other surfaces within that space
would then represent other possible existential universes. The "book of
separate sheets, tacked together at points, if not otherwise connected" of
4.512 is an approximation of this "possibility space"; somehow the cuts
(which are intimately associated with the conditional) are means of passing

from one possible existential universe to another. It seems clear to me that the model here proposed for "the relation between the universe of possibilities and the universe of actual existent facts" is also the model for the relation between the *hypothetical* (which is concerned with the universe of possibles in general) and the conditional *de inesse* (which focuses on conditions at just one "quasi-instantaneous" state).

Peirce does not at this point see it possible to exploit this model adequately:

> In endeavoring to begin the construction of the gamma part of the system of existential graphs, what I had to do was to select, from the enormous mass of ideas thus suggested, a small number convenient to work with. It did not seem to be convenient to use more than one actual sheet at one time; but it seemed that various different kinds of cuts would be wanted. (*ibid.*)

He suggests the "broken cut" as a way of entering the broader universe of possibility; thus

Figure 20.4

> does not assert that it does not rain. It only asserts that the alpha and beta rules do not compel me to admit that it rains, or what comes to the same thing, a person altogether ignorant, except that he was well versed in logic so far as it [is] embodied in the alpha and beta parts of existential graphs, would not know that it rained. (4.515)

He sketches out rules for this cut; our purpose at this moment is not to explore these rules in detail except insofar as that exploration helps us to investigate the model for the relationship of the possible to the realm of the *de inesse* as Peirce tries to lay it out. The broken cut is a "possibly not" operator, which Peirce uses in conjunction with the standard cut (simple negation) to define the usual modal operators. Apropos of the aim of this paper, we note that he employs this notation along with some different new signs (4.517):

> It must be remembered that possibility and necessity are relative to the state of information.
>
> Of a certain graph *g* let us suppose that I am in such a state of information that it *may be true* and *may be false*, that is I can scribe on the sheet of assertion Figs. 20.5 and 20.6.

Figure 20.5

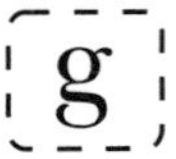

Figure 20.6

Now I learn that it is true. This gives me a right to scribe on the sheet Figs. 20.5, 20.6, and 20.7.

g

Figure 20.7

But now relative to this new state of information Fig. 20.6 ceases to be true; and therefore relatively to the new state of information we can scribe Fig. 20.8.

Figure 20.8

Presumably, Fig. 20.6 was "scribable" at one state of our information, and ceased being so at a later state; Peirce will now suggest within the graphs ways of representing such states (as we noted earlier, he had been talking about such states and about quantification over them since the 1880s); some kind of organized representation is needed, since

> You thus perceive that we should fall into inextricable confusion in dealing with the broken cut if we did not attach to it a sign to distinguish the particular state of information to which it refers. And a similar sign has then to be attached to the simple *g*, which refers to the state of information at the time of learning that graph to be true. I use for this purpose cross marks below, thus:

Figure 20.9

These selectives are very peculiar in that they refer to states of information as if they were individual objects. They have, besides, the additional peculiarity of having a definite order of succession, and we have the rule that from Fig. 20.10 we can infer Fig. 20.11.[10]

Figure 20.10   Figure 20.11

These signs are of great use in cleaning up the confused doctrine of *modal propositions* as well as the subject of logical breadth and depth. (4.518)

Note that Peirce refers to the new "cross marks" as *selectives*. This is a little unusual, since selectives are ordinarily letters of the alphabet. It is clear, however, that he wishes these signs to be (implicitly) quantified variables, which is precisely what selectives are. These selectives are in appearance much like lines of identity, which is Peirce's preferred form of implicitly quantified variable in the graphs. I suggest that he thinks of them as selectives because they show the "order of succession" he refers to, and show it in a way that may not be quite as iconic as he would prefer; in the last two figures, the order of succession is indicated by the single or the double nature of the crossmark selectives. As we shall see, he also comes up with a representation of this situation employing lines of identity. Before we examine that however, we take note of another element that Peirce just mentions, the *state of reflection*:

There is not much utility in a *double broken cut*. Yet it may be worth notice that Fig. 20.7 and

Figure 20.12

can neither of them be inferred from the other. The outer of two broken cuts is not only relative to a state of information but to a state of reflection. The graph asserts that it is possible that the truth of the graph *g* is necessary. (4.519)

Peirce is here exploring a theory of modality, and his exploration is continuous with the views on the conditional/hypothetical which had been part of his approach to logic since about 1880, anyway. He had then written about states of information (or of "things," or of "the speaker"), but had not had a formal mathematical mechanism especially for such states—he *had* held that they were capable of being represented in his quantified logic, as we have seen. With the Gamma Graphs, he has a wealth of signs to help him deal with these "states"; as he goes on:

> It becomes evident . . . that a modal proposition is a simple assertion, not about the universe of things, but about the universe of facts that one is in a state of information sufficient to know. [Fig. 20.6] without any selective, merely asserts that there is a possible state of information in which the knower is not in a condition to know that the graph *g* is true, while Fig. 20.8 asserts that there is no such possible state of information. Suppose, however, we wish to assert that there is a conceivable state of information of which it would not be true that, in that state, the knower would not be

Figure 20.13

in condition to know that *g* is true. We shall naturally express this by Fig. 20.13. But this is to say that there is a conceivable state of information in which the knower would know that *g* is true. [This is expressed by] Fig. 20.10.

Now suppose we wish to assert that there is a conceivable state of information in which the knower would know *g* to be true and yet would not know another graph *h* to be true. We shall naturally express this by Fig. 20.14. Here we have a new kind of ligature, which will follow all the rules of ligatures. We have here a most important addition to the system of graphs. There will be some peculiar and interesting little rules, owing to

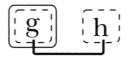

Figure 20.14

the fact that what one knows, one has the means of knowing what one knows—which is sometimes incorrectly stated in the form that whatever one knows, one knows that one knows, which is manifestly false. (4.520–521)

So Peirce applies ligatures—systems of lines of identity—to the states of information. He makes suggestions that go even further, introducing a sign specifically relating such states of information; this is effectively a special "spot" whose 2 "hooks" are marked by the A and the B below (4.522):

> The truth is that it is necessary to have a graph to signify that one state of information follows after another. If we scribe,

Figure 20.15

> to express that the state of information B follows after the state of information A, we shall have

Figure 20.16

This last is a version—employing lines of identity ("ligatures")—of the rule of necessitation done with "selectives" in Figs. 20.10–20.11.

So we see that Peirce from about 1880 on, anyway, was interested in what we recognize as "possible worlds" logic. He understood the conditional as appropriately interpretable within such a context; he never really "abandoned" the *de inesse* conditional, but through this entire period he understood this "material implication" as a **second** to the larger **third** of the "hypothetical," which cannot be understood outside this context of possible states of information, states which can be "quantified over"; he alludes to this quantification early on (in 1885, anyway). When he has, in the Existential Graphs, what he considers an adequate notation for dealing with such matters, he discusses and gives examples of such quantifications, as illustrated in the text and illustrations we have been presenting. I note that in this material, Peirce is very close to our notion of possible worlds semantics, and, indeed, may have something to teach *us* about it!

N O T E S

1. Which term Peirce credits to none other than:

Duns Scotus, who was a Philonian, [and who] as a matter of course, threw considerable light upon the matter by distinguishing between an ordinary *consequentia*, or conditional proposition, and a *consequentia simplex de inesse*.

A *consequentia simplex de inesse* relates to no range of possibilities at all, but merely to what happens, or is true, *hic et nunc.* But the ordinary conditional proposition asserts not merely that here and now either the antecedent is false or the consequent is true, but that in each possible state of things throughout a certain well-understood range of possibility either the antecedent is false or the consequent true. So understood the proposition "If it lightens it will thunder" means that on each occasion which could arise consistently with the regular course of nature, either it would not lighten or thunder would shortly follow. (NEM 4: 169 [1898])

2. The original *discoverer* was Peirce (see *Zeman 1986:* 9); Kripke is often given the credit.

3. Fisch at this point refers to 4.580 (1906) and 4.546 (1906, "*Prolegomena*"). Although Peirce in these passages does indeed advert to complications in the determination of the truth values of conditionals, I find it difficult to interpret the passages' actual text as evidence of what we could call a "surrender of his last [nominalist] stronghold," especially in the light of what I will argue about the early Peirce's view of the conditional. We might take as a summary of the arguments of these paragraphs the conclusion of 4.580:

Some years ago [1903 is suggested by the editors of CP] when . . . I was led to revise [the] doctrine [that a mere possibility is an absolute nullity], in which I had already found difficulties. I soon discovered, upon a critical analysis, that it was absolutely necessary to insist upon and bring to the front, the truth that a mere possibility may be quite real. That admitted, it can no longer be granted that every conditional proposition whose antecedent does not happen to be realized is true, and the whole reasoning just given breaks down.

4.546 is even more difficult to see as the "surrender of [a] stronghold," although its emphasis is on the reality of possibles (see my discussion of the EG rule given in 4.569; the "suiciding wife" example is connected to this, and it is further discussed in 4.580).

4. The editors of CP at this point suggest looking at 3.527. This is an 1896 location which (especially in the light of Peirce's own 1908 comments on it) must be regarded as transitional to the view of possibility developed by Peirce over the next decade in the Existential Graphs.

5. The propositional forms here are the result of this being Peirce's presentation of "A Boolian Algebra with One Constant," namely, "neither-nor."

6. We are about to examine some of Peirce's work in the logic of "Possible Worlds." At this point I will note that his 1906 work in and about the "*Prolegomena to an Apology for Pragmaticism*" (4.530 ff.) contains much potentially fruitful material in this direction, in the form of his "Tinctured Existential Graphs." This is the topic of a paper in itself; it takes somewhat different directions than those I am emphasizing here, and I will, in this paper, just note the existence of the "Tinctures," and promise an investigation of them at an early opportunity.

7. According to Peirce, in the language of logic "consequence" does not mean that which follows, which is called the *consequent,* but means the fact that a consequent follows from an antecedent.

8. As in contemporary science fiction!

9. This fits with his discussion of 4.435: "one should observe that the consequent of a conditional proposition asserts what is true, not throughout the whole

universe of possibilities considered, but in a subordinate universe marked off by the antecedent. This is not a fanciful notion, but a truth."

10. As the editors of CP point out, this is effectively a rule of necessitation. It would appear that implicit in this rule is that the first figure appears on the Sheet of Assertion by logical rule rather than by contingency.

21.

MATCHING LOGICAL STRUCTURE TO LINGUISTIC STRUCTURE

John F. Sowa

Systems of logic with equivalent expressive power may have very different structure. Short, simple statements in one system can often be expressed only by awkward circumlocutions in another. During the last quarter of the nineteenth century, three complete systems of first-order logic were developed: Frege's Begriffsschrift, Peirce's linear form of predicate calculus, and Peirce's existential graphs. This paper compares the structures of propositions stated in these systems to one another and to the underlying semantic structures of language. Of the three, existential graphs have the most direct translations to language. Remarkably, they are isomorphic to the discourse representation structures that were independently developed over 80 years later. They also form the logical foundation for conceptual graphs, which combine Peirce's logic with research on semantic networks in artificial intelligence and computational linguistics. Although Peirce's linear notation has proved to be a powerful tool for foundational studies in mathematics, his existential graphs are better suited to studies of language.

§1. Search for a Natural Logic

Versions of logic with equivalent expressive power may have very different mappings to and from language. As a basis for semantics, an ideal version of logic should not only have the expressive power of natural language, it should also have the simplest and most direct mapping to linguistic structures. Although defining a system of logic is not an empirical matter, the complexity of the mapping from language to different systems is an empirical issue that has been addressed by various linguists, including Keenan (*1972*), Lakoff (*1972*), and Jackendoff (*1983*). To show how different systems of logic represent the same features of language, consider the sentence "Every ball is red." In Frege's *Begriffsschrift* of 1879, it would be represented by the following diagram:

The cup containing the variable x represents the universal quantifier ($\forall x$), and the hook represents material implication $\supset$. As a representation for the English sentence, it is isomorphic to the corresponding formula in predicate calculus:

$$\forall x(\text{ball}(x) \supset \text{red}(x)).$$

Since there is one-to-one mapping between the features of Frege's diagram and the formula in predicate calculus, the differences in the way they are drawn are stylistic matters that have no relevance to their naturalness as systems of logic. Neither of them is especially natural, since they represent the simple sentence "Every ball is red" by an awkward paraphrase of the form, "For all x, if x is a ball, then x is red." Yet since both systems represent the sentence in the same way, they would be equally natural or unnatural according to all empirical tests. A more significant difference appears with the sentence "There exists a red ball," which could be represented by the following formula in predicate calculus:

$$\exists x(\text{ball}(x) \wedge \text{red}(x)).$$

This formula corresponds to the sentence "There exists an x, x is a ball, and x is red." Since Frege's primitives included only the universal quantifier, implication, and negation (represented by a short vertical line), that formula would be more complex in the *Begriffsschrift*:

This diagram is the equivalent of the English sentence "It is false that for all x, if x is a ball, then x is not red." The clumsiness of this paraphrase is much more than a stylistic issue. It is strong evidence that Frege's operators by themselves are insufficient to support a natural semantic representation for language. Although predicate calculus also distorts English sentence structure, it does not distort it as much as Frege's form.

Frege, however, was not concerned about natural languages. He even took pride in the way his notation diverged from language. In the preface of the *Begriffsschrift*, he said "I found the inadequacy of language to be an obstacle; no matter how unwieldy the expressions I was ready to accept, I was less and less able, as the expressions became more complex, to attain the precision that my purpose required." Later he continued, "These deviations from what is traditional find their justification in the fact that logic has hitherto always followed ordinary language and grammar too closely. In particular, I believe that the replacement of the concepts *subject* and *predicate* by *argument* and *function*, respectively, will stand the test of time." That replacement has indeed stood the test of time for representing an important body of mathematics. Yet Frege's argument is directed at the subject-predicate model of language, which is so oversimplified that it is almost a caricature of sentence structure. A functional representation is just as oversimplified, but in a different way. Relations, which include functions as special cases, are more general; and Peirce's long studies of relations led him to a more felicitous form of logic.

In searching for a better logic, Peirce experimented with graph notations as early as 1882. In his *relational graphs*, he could represent the sentence "There exists a red ball" in the simplest possible way:

$$red\text{——}ball$$

In this notation, the bar means "There exists something," and the two predicates attached to it represent the facts "It is red" and "It is a ball." That is far simpler than Frege's form, and it is also simpler than the predicate calculus, which requires a conjunction $\wedge$, three occurrences of a variable symbol x, and a number of parentheses to express a statement that requires no conjunctions, variables, or parentheses in English. As evidence against a system of logic, syntactic features like parentheses are of minor significance; semantic features such as the proliferation of variables and conjunctions are stronger evidence against the predicate calculus as a natural logic. A point in favor of Peirce's graph is that its three structural features correspond exactly to the three words in the English phrase "a red ball." Feature counting is one kind of evidence for a natural logic; but a more detailed analysis is needed, since the linguistic relationship implicit in word order must also be considered.

Although Peirce's relational graphs were an excellent notation for representing conjunctions and existential quantifiers, they could not show the scope of quantifiers and Boolean operators. With those early graphs, Peirce could negate a single predicate or relation to say There exists a nonred ball or "There exists a red nonball." But the graphs could not express operators whose scope was larger than a single predicate. Meanwhile, Peirce developed his linear notation of 1883, which with a change of symbols by Peano became the modern predicate calculus. With minor changes, that notation evolved into an outstanding tool for studying the foundations of mathematics. Yet its structure diverges from natural languages in a number of important areas:

- Multiple sorts. For mathematics, a logic with only one sort of individuals, such as the integers or real numbers, is adequate for many purposes. But the world contains an enormous variety of things, which lead to vocabularies of thousands of words. Versions of multisorted logic exist, but theoreticians prefer to work with the simpler logics, and practical programmers ignore logic altogether.
- Quantification. In predicate calculus, quantifiers range over a single universe of discourse. In natural languages, the range is explicitly stated either by a single word, as in "every dog," or by a phrase or relative clause, as in "every dog that may have eaten two meatballs on some Wednesday afternoon." This feature requires not only a sorted logic, but a logic in which sorts can be dynamically defined by arbitrarily complex expressions.
- Variables. Pronouns in natural languages correspond roughly to variables in predicate calculus. Yet the predicate calculus usually has many more variables than an equivalent sentence in ordinary language. Variables tend to scatter the references to each individual throughout a formula; graph or network systems tend to link all references to an individual to a single node.
- Functional notation. Predicate calculus puts the emphasis on predicates, and the objects they apply to are buried inside the arguments of the predicates. Graphs turn the formulas inside out: they provide separate nodes for objects and predicates. This transformation makes it easier to link new information to those nodes in order to handle context and background knowledge.
- Plural nouns. English can switch from singular to plural simply by adding "s" to a noun. Set notation in logic is more cumbersome and requires a major restatement of the formula to transform a statement about one individual into a statement about more than one.
- Context. The meaning of a sentence depends heavily on context to resolve pronouns, tenses, definite articles, deictics, and indexicals. Yet predicate calculus makes no provision for context or context-dependent references.

These arguments concern practical questions of readability as well as theoretical issues about the underlying structures of language and logic. With a poor choice of logic, linguists must go through needless contortions to work around the notation, knowledge engineers have no guidelines for mapping specifications from language to logic, and students are systemati-

cally misled by the mismatch between language and logic. Although predicate calculus *lets* them do everything, it doesn't *help* them do anything.

Further examples may illustrate the issues more clearly. Consider the
sentence "A cat chased a mouse." Most logic texts would represent it by a
formula like the following:

$$\exists x\, \exists y (\mathrm{cat}(x) \wedge \mathrm{mouse}(y) \wedge \mathrm{chased}(x,y)).$$

This formula is already more complicated than the English sentence, but
it does not represent all the information as clearly as the sentence. The
predicate "chased(x)," for example, implicitly includes the tense. But the
time should be explicit instead of having separate predicates for "chase,"
"chased," and "will chase:"

$$\exists x\, \exists y\, \exists t (\mathrm{cat}(x) \wedge \mathrm{mouse}(y) \wedge \mathrm{time}(t) \wedge \mathrm{chase}(x,y,t) \wedge \mathrm{succ}(t,\text{s-time})).$$

Here the symbol s-time indicates a special speech time, which is a successor
of the time t of the chasing. The speech time is not really a constant, but a
contextually bound reference, which is already outside the realm of standard logic. Yet time governs the entire context, not just the verb. The cat and
the mouse should also be inside the scope of the variable t, since one of
them (perhaps the mouse) might not exist after time t:

$$\exists t (\mathrm{time}(t) \wedge \mathrm{succ}(t,\text{ s-time}) \wedge \mathrm{ptim}(t, \exists x\, \exists y (\mathrm{cat}(x) \wedge \mathrm{mouse}(y) \wedge \\ \mathrm{chase}(x,y)))).$$

The predicate ptim(t,p) shows that a proposition p is true at the point in
time t. But this formula is no longer first-order, since it contains a proposition nested inside a predicate.

Although the formula is already much more complex than the original
sentence, further refinements are necessary to show other linguistically relevant features. The cat and the mouse, for example, each have a unique
variable. Further references to them could use the corresponding variable
x or y. But the verb "chase" should also have its own variable, since the
speaker might want to refer back to it. In the sentence "The chase lasted
39 seconds," the noun "chase" refers to the event previously introduced by
a verb. The next formula assigns an *event variable* z to the act of chasing:

$$\exists t (\mathrm{time}(t) \wedge \mathrm{succ}(t,\text{ s-time}) \wedge \mathrm{ptim}(t, \exists x\, \exists y\, \exists z (\mathrm{cat}(x) \wedge \mathrm{mouse}(y) \wedge \\ \mathrm{chase}(z,x,y)))).$$

In this formula, the predicates corresponding to nouns have only one argument, but the predicate corresponding to the verb has three. Linguists
eliminate that asymmetry by introducing special dyadic predicates to show
the *case roles* or *thematic relations* that hold between the noun phrases and
the verb. For this sentence, the predicate agnt (z,x) would show that the cat

is the agent of chasing, and ptnt(z,y) would show that the mouse is the patient or the one who is chased:

$$\exists t(\text{time}(t) \land \text{succ}(t, \text{s-time}) \land \text{ptim}(t, \exists x \exists y \exists z(\text{cat}(x) \land \text{mouse}(y) \land$$
$$\text{chase}(z) \land \text{agnt}(z,x) \land \text{ptnt}(z,y)))).$$

To use a formula this complex to say "A cat chased a mouse" is a *reductio ad absurdum*. It indicates that something is wrong with predicate calculus as a linguistic representation.

Many expert systems require sets and structures, and expressing them is also awkward in the predicate calculus. Consider the sentence "Every trailer truck has 18 wheels."Before that sentence could be expressed in the predicate calculus, it would have to be paraphrased in the form, "For all x, if x is a trailer truck, then there exists a set s with 18 elements, and for all y, if y is a member of s, then y is a wheel and x has y as part:"

$$\forall x(\text{trailer-truck}(x) \supset \exists s(\text{set}(s) \land \text{count}(s,18) \land \forall y(y \in s \supset (\text{wheel}(y) \land$$
$$\text{part}(x,y))))).$$

This example has only one plural noun. With two plural nouns, the formulas are even worse. Consider the sentence "Some blocks are each supported by 3 pyramids."That could be paraphrased as "There exists a set s, where for every x in s, x is a block and there exists a unique set r with 3 elements, where for every y in r, y is a pyramid and y supports x:"

$$\exists s(\text{set}(s) \land \forall x(x \in s \supset (\text{block}(x) \land \exists!!r(\text{set}(r) \land \text{count}(r,3) \land$$
$$\forall y(y \in r \supset (\text{pyramid}(y) \land \text{support}(y,x))))))).$$

Uniqueness is an awkward relationship to express with the standard quantifiers. In this example, the quantifier $\exists!!r$ indicates that there exists a distinct set of three pyramids for each block. The unique existential quantifier $\exists!!$ corresponds to the operator E!! that Whitehead and Russell (*1910–1913*) introduced for representing similar dependencies between sets. The quantifier with one exclamation point $\exists!$ was introduced by Kleene (*1952*) with the following definition:

$$\exists!x\, P(x) \equiv \exists x(P(x) \land \forall y(P(y) \supset x = y)).$$

The quantifier $\exists!$ indicates that there exists exactly one object that satisfies the condition. But for the distributive interpretation of plurals, the $\exists!!$ quantifier is necessary to show that for each block, the pyramids that support it are different from the ones that support any other block. That quantifier is defined by the following formula:

$$\forall x\, \exists!!y\, P(x,y) \equiv \forall x\, \exists y(P(x,y) \land \forall z(P(x,z) \supset y = z) \land \forall w(P(w,y) \supset x = w)).$$

This definition takes into account the dependency of y on x. If y depends on several variables $x1, x2, \dots$, the definition can be generalized by treating x and w as vectors of variables. Such extended quantifiers are necessary to simplify logical expressions, as Whitehead and Russell realized in 1910.

Sorted logic can simplify formulas that deal with the variety of things in the real world. For the two examples at the beginning of this section, "Every ball is red and There exists a red ball," the formulas in sorted logic are distinctly shorter:

$$(\forall x{:}\text{ball})\,\text{red}(x), \qquad (\exists x{:}\text{ball})\,\text{red}(x).$$

These two formulas may be read as the English sentences "For every ball x, x is red" and "There exists a ball x, where x is red." In sorted logic, the extra implication "⊃" and conjunction "∧" have been eliminated, and the number of variable symbols has been reduced. Sorted logic is therefore an important step towards a more natural logic, but it is not sufficient by itself to make all formulas more English-like. Following is a sorted-logic formula for "Every trailer truck has 18 wheels":

$$(\forall x{:}\text{trailer-truck})(\exists s{:}\text{set})(\text{count}(s,18) \wedge (\forall y \in s)(\text{wheel}(y) \wedge \text{part}(x,y))).$$

And following is the formula for "Some blocks are each supported by 3 pyramids":

$$(\exists s{:}\text{set})(\forall x \in s)(\text{block}(x) \wedge (\exists!!\,r{:}\text{set})(\text{count}(r,3) \wedge (\forall y \in r)(\text{pyramid}(y) \wedge \text{support}(y,x)))).$$

As these formulas illustrate, English plurals cannot be represented easily in predicate calculus, whether sorted or not. Besides a better representation for plurals, a natural logic should also have a better way of handling contexts and context-dependent expressions, such as pronouns, indexicals, tenses, and deictics.

Logicians make excuses for predicate calculus by saying that a precise formalism is bound to be longer than an informal language like English. Yet many notations used in artificial intelligence are more compact, and they can be formalized. Conceptual graphs (*Sowa 1984*) are an extension of Peirce's existential graphs designed to accommodate linguistically relevant features. For the examples in this section, the conceptual graphs are consistently shorter than the predicate calculus, yet they are just as formal, and they capture the same information. In fact, the only examples where predicate calculus is more succinct than conceptual graphs are mathematical statements where predicate calculus is also more succinct than English.

§2. From Predicate Calculus to Existential Graphs

After a dozen years of working with his linear form of logic, Peirce developed his *entitative graphs*, which took the universal quantifier, disjunction, and negation as primitives. Except for the graph notation it had a strong resemblance to the clausal form used in resolution theorem provers. But in 1896, Peirce abandoned the entitative graphs in favor of the dual form, the *existential graphs*, with the existential quantifier, conjunction, and negation as primitives. One reason for switching from the entitative to the existential form is the problem of mapping language to logic: for most sentences, the most common quantifier is the existential ∃, and the most common Boolean operator is the conjunction ∧.

Peirce presented existential graphs in three steps: Alpha, Beta, and Gamma. Alpha is equivalent to standard propositional logic. Beta extends Alpha to a complete theory of first-order logic. Gamma goes beyond Beta to include modal logic, higher-order logic, and propositional attitudes. In Alpha, conjunction is shown by drawing two graphs in the same context. Negation is shown by a closed curve that separates the outer positive context from the inner negative context. If p, q, and r are any three graphs that represent propositions, the following diagram is an existential graph that corresponds to the formula $(p \wedge \sim(q \wedge \sim r))$:

Since $\sim(q \wedge \sim r)$ happens to be equivalent to $q \supset r$, this graph may also be read as "p and if q then r." The nest of two ovals is the standard Alpha form for implication.

Peirce's system Beta by itself corresponds to his relational graphs of 1882. In fact, propositions that contain only conjunctions and existential quantifiers would be identical in Beta and in relational graphs. Fig. 21.1 shows a graph for the sentence "A farmer owns and beats a donkey." This

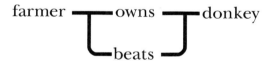

Figure 21.1. Graph for "A farmer
owns and beats a donkey."

graph could be either a relational graph or an existential graph using the Beta conventions. Fig. 21.1 contains two linked sets of bars, which Peirce called *lines of identity*. The bars on the left represent an existential quantifier (∃x) for the farmer, and the bars on the right represent another quantifier (∃y) for the donkey. The graph as a whole corresponds to the following formula in predicate calculus:

$$\exists x \, \exists y (\text{farmer}(x) \wedge \text{donkey}(y) \wedge \text{owns}(x,y) \wedge \text{beats}(x,y)).$$

Although the formula takes less space on the printed page, Fig. 21.1 has fewer symbols. Even a logician with many years of experience with the predicate calculus can see relationships more quickly with a diagram like Fig. 21.1 than with the formula in predicate calculus.

Without negative contexts, Beta can represent most of the work that was done with semantic networks in AI during the 1960s (*Sowa 1992*). Like Peirce's relational graphs of 1882, those systems took existence and conjunction as primitive, and they connected all the predicates that applied to a single individual with links that were similar to the lines of identity. Most of them also had the same limitations as relational graphs: they could negate a single relation, but they could not show the scope of operators and quantifiers that governed more than one relation. In 1896, Peirce developed a complete version of logic by the simple expedient of combining the relational graphs with the negative contexts. He allowed lines of identity to cross context boundaries: quantification was effectively specified in the outermost context in which a line of identity occurred, and the scope of quantification was defined by the nesting of contexts. Fig. 21.2 shows an existential graph for the Gricean sentence "If a farmer owns a donkey, then he beats it." In Fig. 21.2, the outer oval represents the antecedent or *if*-clause.

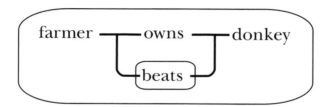

Figure 21.2. Existential graph for "If a farmer owns a donkey, then he beats it."

It contains one line of identity for the farmer and one for the donkey, both of which extend into the *then*-oval. Fig. 21.2 could be translated directly to the following formula in predicate calculus:

$$\sim\exists x\, \exists y (\text{farmer}(x) \wedge \text{donkey}(y) \wedge \text{owns}(x,y) \wedge \sim\text{beats}(x,y)).$$

In prenex normal form with the symbol $\supset$, the formula becomes

$$\forall x\, \forall y ((\text{farmer}(x) \wedge \text{donkey}\ (y) \wedge \text{owns}(x,y)) \supset \text{beats}(x,y)).$$

Again, the formula takes less space, but the graph has fewer symbols. Readability depends largely on prior experience, but people who become familiar with Peirce's graphs find the lines of identity a simpler way of showing connections than variables. Roberts (*1973*) found that Peirce's graphs were more effective than predicate calculus in teaching logic to beginning students, and Ketner (*1991*) wrote an introductory textbook on logic based on them.

Besides defining the notation, Peirce also discovered an elegant set of inference rules for proving theorems expressed in existential graphs. The rules specify the conditions for drawing or erasing a graph in any context, where the conditions depend only on the number of negations in which the graph is nested. A negative context is one that is nested in an odd number of negations; a positive context is nested in an even number of negations. For the Alpha part (propositional logic), there are five rules and one axiom:

- Erasure. Any graph may be erased in a positive context.
- Insertion. Any graph may be inserted in a negative context.
- Iteration. Any graph may be copied in the same context or any context nested inside the context in which it occurs.
- Deiteration. Any graph that could have arisen by iteration may be erased.
- Double negation. In any context, two negations may be drawn around or erased from any set of graphs (including the empty set).
- Axiom. The only axiom is the empty set.

These rules form a complete inference system for propositional logic. For Beta, the same rules apply, but with extensions to specify the conditions for breaking or extending a line of identity within or across contexts. During the 1970s, various approaches to doing logic on semantic networks were defined in artificial intelligence (*Sowa 1987*). Some were as general as Peirce's, but none was as brilliantly simple.

§3. From Existential Graphs to Conceptual Graphs

Although Peirce had a complete graph system of logic, his linguistic analyses were not as fully developed. As an example, Fig. 21.3 shows one of his Gamma graphs for the sentence "You can lead a horse to water, but you can't make him drink" (*Roberts 1973*). The outer oval represents the clause "If there exists a person, a horse, and water." The *then*-oval contains a shaded area to represent possibility, and a shaded area inside a negation to represent impossibility. Altogether, the graph could be read "If there exists a person, a horse, and water, then it is possible for the person to lead the horse to the water and not possible for the person to make the horse drink the water."

Logically, Fig. 21.3 is quite advanced in representing both modality and quantification; but linguistically, it does not represent the details of the sentence structure correctly. The English phrase "to water" is used as an adverb that should be linked to the verb "leads." Peirce's graph, however, shows

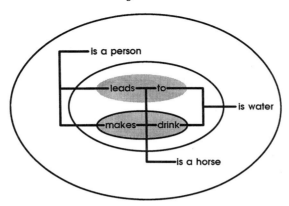

Figure 21.3. Graph for "You can lead a horse to water, but you can't make him drink."

"to" as a relation linking "horse" and "water;" its relationship to the verb is not explicitly shown. The subgraph for the person leading the horse to water would correspond to the following formula in predicate calculus:

$$\exists x\, \exists y\, \exists z (\text{person}(x) \wedge \text{horse}(y) \wedge \text{water}(z) \wedge \text{leads}(x,y) \wedge \text{to}(y,z)).$$

Peirce's subgraph for "make-drink" is also incorrect, since it more closely represents the sentence "A person makes a horse, which drinks water." The person did not make the horse; instead, there is a causal relation between the person's action and the horse's action.

Peirce actually intended leads-to and make-drink as triadic relations, but his notation does not clearly show the valence. The linguist Lucien Tesnière (*1959*) developed his dependency graphs in a way that almost exactly complemented Peirce's. Logically, Tesnière's graphs were no more sophisticated than Peirce's relational graphs of 1882. But he devoted years of careful attention to the details of sentence structure in various languages. The semantic networks in AI have been strongly influenced by Tesnière's graphs as well as other linguistic work on case grammar (*Fillmore 1968*) and thematic relations (*Gruber 1965*). Conceptual graphs are a synthesis of that work with Peirce's logic. Fig. 21.4 shows a conceptual graph where the linguistic relationships are shown by explicit labels: the person is the "agent," the horse is the "patient," and the water is the "destination." The boxes are called *concepts*, and the circles are called "conceptual relations."

In conceptual graphs, the point of quantification is shifted from the lines to the boxes. Each concept box in Fig. 21.4 represents an existential quantifier in a sorted logic. The concept [PERSON] may be mapped to the quantifier ($\exists x$:person), where the *type label* PERSON in the concept box

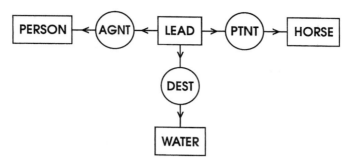

Figure 21.4 Conceptual graph for "A person leads a horse to water."

corresponds to a sort label in a sorted logic. Actions such as *lead* are also quantified: the concept [LEAD] implies that there exists an instance of leading. The circles map to predicates where the arrow pointing to the circle is the first argument and the arrow pointing away is the second argument. The *formula operator* φ, which translates conceptual graphs to predicate calculus, generates the following formula in sorted logic for Fig. 21.4:

$$(\exists x{:}\mathrm{person})(\exists y{:}\mathrm{horse})(\exists z{:}\mathrm{water})(\exists w{:}\mathrm{lead})(\mathrm{agnt}(w,x) \wedge \mathrm{ptnt}(w,y) \wedge \mathrm{dest}(w,z)).$$

To represent the verb "make" in the sense of forcing, the concept type MAKE-FORCE may be used. The action has the horse as patient, since the forcing is done to the horse. But it also has a result, represented by a nested context where the horse is drinking. Fig. 21.5 shows the conceptual graph for the sentence "A person makes a horse drink water."

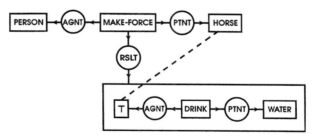

Figure 21.5 Conceptual graph for "A person makes a horse drink water."

In Fig. 21.5, the concept type ⊤ represents the most general type at the top of the type lattice. It commonly occurs on a concept that is coreferent with some other concept whose type is explicitly given. In this case, the dotted line represents a *coreference link* to show that the concept [⊤]

refers to the same individual as the concept [HORSE]. A line of identity in Peirce's sense corresponds to one or more concepts linked by dotted lines. The outermost context containing such a linked line of concepts is the point where the quantification is determined; the more deeply nested concepts are bound by that quantifier. Literally, Fig. 21.5 means that a person performs a make-force action upon a horse with the result that it drinks water. Note that the coreferent concepts like [⊤] typically occur where English and other languages use pronouns; sometimes they correspond to the silent pronouns or *traces* in certain linguistic theories. To construct the conceptual graph that corresponds to Peirce's Fig. 21.3, combine Figs. 21.4 and 21.5 with context boxes for negation and possibility. The result is Fig. 21.6.

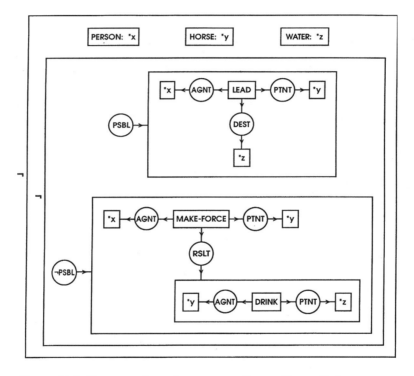

Figure 21.6 Conceptual graph corresponding to Figure 21.3.

Instead of Peirce's shaded and colored areas, conceptual graphs have boxes for all contexts, and the use of a context is shown by the relations attached to it. The relation (PSBL) indicates possibility and (NEG) or its abbreviation ¬ indicates negation. The relation (¬PSBL) is defined by a combination of ¬ and (PSBL); it corresponds to Peirce's shaded area inside an oval. Fig. 21.6 requires seven coreference links, which would crisscross the diagram in a confusing way. As an alternate notation, a concept of the form [*x] represents [⊤] with a dotted line linking it to the outermost concept containing the same symbol *x, in this case [PERSON: *x]. Symbols like *x correspond to variables in predicate calculus, but the term "variable" is actually a misnomer, since the variables do not vary. Logic does not permit anything like the assignment statements in programming languages, where X: = X + 1 causes the old value of X to be replaced. Peirce's lines of identity and the coreference links in conceptual graphs more correctly show that the purpose of a so-called variable is to show that certain arguments of different relations are coreferent.

Although diagrams are a readable way of expressing logical relationships, they take a lot of space on the printed page, and they are not easy to type. For convenience, there is a more compact linear notation that uses square brackets for the concepts and parentheses for the conceptual relations. For a graph like Fig. 21.4 with three arcs attached to the concept [LEAD], there is no way to draw it in a straight line. Instead, a hyphen may be placed after [LEAD] to show that the attached relations are continued on subsequent lines;

> [LEAD] -
>   (AGNT) → [PERSON]
>   (PTNT) → [HORSE]
>   (DEST) → [WATER]

The linear form for Fig. 21.5 requires a symbol *x to show the coreference link:

> [MAKE - FORCE] -
>   (AGNT) → [PERSON]
>   (PTNT) → [HORSE: *x]
>   (RSLT) → [[*x] ← (AGNT) ← [DRINK] → (PTNT) → [WATER]].

Since implication is commonly represented by a nest of two negative contexts, the linear form can be made more readable by writing IF for the opening ¬[and THEN for the nested ¬[. With that convention, the linear form for Fig. 21.6 would be

> IF   [PERSON: *x][HORSE: *y][WATER: *z]
> THEN (PSBL) → [[LEAD] -
>   (AGNT) → [*x]
>   (PTNT) → [*y]
>   (DEST) → [*z]]
>   (¬PSBL) → [[MAKE-FORCE] -
>   (AGNT) → [*x]
>   (PTNT) → [*y]
>   (RSLT) → [[*y] ← (AGNT) ← [DRINK] → (PTNT) → [*z]]].

This graph may be read "If there exists a person *x*, a horse *y*, and water *z*, then it is possible for x to lead y to z and not possible for x to make y drink z." Nested contexts are easier to read in the graph notation, but even the linear notation is more readable than the equivalent in predicate calculus.

§4. Contexts and Discourse Referents

Peirce's form of contexts, which have also been adopted for conceptual graphs, turn out to be important for representing linguistic phenomena that Peirce had not considered in detail: the contexts and referents of connected discourse. Consider the next two sentences:

Sam owns a car. He likes it.

Each of those sentences might be translated to a formula in predicate calculus:

$$\exists x(\mathrm{car}(x) \wedge \mathrm{owns}(\mathrm{Sam},x)). \mathrm{Likes}(y,z).$$

A notation suitable for discourse representation must be able to link the variable y to Sam and z to x. For this example, the second formula could be inserted just before the final parenthesis of the first formula, and the variables y and z could be replaced with Sam and x:

$$\exists x(\mathrm{car}(x) \wedge \mathrm{owns}(\mathrm{Sam},\ x) \wedge \mathrm{likes}(\mathrm{Sam},x)).$$

The technique of inserting new information inside the parentheses of a previous formula is commonly used in many natural language systems. But there are serious limitations: it does not work for universally quantified variables, and there are complex constraints when new contexts are introduced by negations, modalities, or time. To express those constraints, Kamp (*1981a, 1981b*) developed his *discourse representation theory* with a systematic treatment of contexts and discourse referents. Remarkably, Kamp's contexts turn out to be isomorphic to Peirce's. Fig. 21.7 shows Kamp's discourse representation structure (DRS) that corresponds to Fig. 21.2. In Fig. 21.7, the variables x and y represent *discourse referents*, which correspond to existentially quantified variables. The two boxes are contexts, and the arrow represents implication. Kamp's implication does not correspond to the

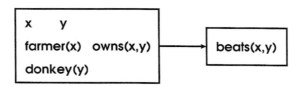

Figure 21.7 DRS for "If a farmer owns a donkey, then he beats it."

symbol ⊃ in predicate calculus, since Kamp makes the additional assumption that variables in the consequent are within the scope of discourse referents in the antecedent. Note that the corresponding predicate calculus does not allow that interpretation:

$$(\exists x \,\exists y \,\text{farmer}(x) \land \text{donkey}(y) \land \text{owns}(x,y)) \supset \text{beats}(x,y).$$

In this formula, the scope of the quantifiers $\exists x$ and $\exists y$ is limited to the antecedent; it does not include beats(x,y). In order to include the whole formula, the quantifiers must be moved to the front and be converted to universals $\forall x$ and $\forall y$.

There is no room in this paper to summarize all of Kamp's examples and arguments, but the important point to note is that his DR structures are isomorphic to Peirce's graphs. Kamp's discourse referents correspond to Peirce's lines of identity; his search rules for resolving anaphora correspond to a search through Peirce's contexts from the inner contexts outward. In fact, Kamp's constraints on anaphora become simpler when stated in terms of Peirce's graphs: Kamp required separate rules for nesting and for implication; but since Peirce nested the consequent of an implication inside the antecedent, only the nesting rule is needed. Some people have tried to reduce the number of primitives in Kamp's form by defining "not-p" as "p implies falsehood." Peirce, however, had a simpler solution: define implication as a nest of two negations.

Peirce's contexts should be distinguished from another graph system with contexts: the *partioned nets* by Hendrix (*1975, 1979*). Instead of letting lines of identity extend across contexts, Hendrix required each concept (discourse referent) to occur only once. But he allowed the contexts to overlap so that the same concept could occur in multiple contexts. Fig. 21.8 shows a partitioned net that corresponds to Fig. 21.2. There are two parti-

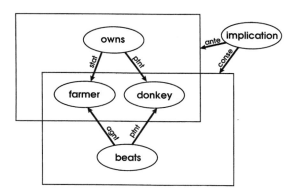

Figure 21.8 Partitioned net for "If a farmer owns a donkey, then he beats it."

tions: one for the antecedent of the implication and one for the consequent. The concepts corresponding to the farmer and the donkey occur in both partitions; the owning occurs only in the antecedent, and the beating occurs only in the consequent. Although Hendrix's graphs also form a complete system of logic, his partitions are not nested in the same way as Kamp's or Peirce's. Therefore, Kamp's discourse representation theory cannot be expressed in terms of Hendrix's graphs. There is also a practical problem in drawing them: if there are more than three overlapping contexts, they cannot be drawn on a plane. Since Peirce's and Kamp's contexts are strictly nested, it is always possible to draw them on a plane.

In conceptual graphs, no concept may occur in more than one context. Fig. 21.9 shows the conceptual graph corresponding to Figs. 21.2 and 21.7.

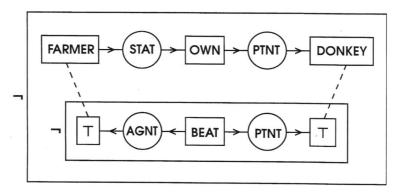

Figure 21.9 Conceptual graph for "If a farmer owns a donkey, then he beats it."

As in Peirce's system, there are two negative contexts, here explicitly marked with the symbol ¬. The concepts [FARMER] and [DONKEY] occur only in the *if*-context. To show references to those concepts in the *then*-context, two concepts of type ⊤ (the top of the type lattice) are shown in that context. They are linked by dotted lines, called *coreference links*, to the concepts [FARMER] and [DONKEY].

Concepts of type ⊤ can be used to represent pronouns in natural language. Since ⊤ is the universal type, it may be coreferent with any concept of any type. Conceptual graphs need these extra concept nodes because they enforce a strict nesting of contexts. By allowing overlapping contexts, partitioned nets eliminate the need for extra nodes. Yet the fact that those extra nodes map directly to English pronouns suggests that there is nothing unnatural about them; on the contrary, it is further evidence in favor of strict nesting. In the linear form, the concept [⊤] with the dotted line becomes [*x], to show the coreference link. Fig. 21.9 then becomes

¬[[FARMER: *x] → (STAT) → [OWN] → (PTNT) → [DONKEY: *y]
    ¬[[*x] ← (AGNT) ← [BEAT] → (PTNT) → [*y]]].

With the symbols "IF" and "THEN" as synonyms for the negative context markers "¬[," the result is the following graph:

IF [FARMER: *x] → (STAT) → [OWN] → (PTNT) → [DONKEY: *y]
    THEN [*x] ← (AGNT) ← [BEAT] → (PTNT) → [*y].

This linear form can be read directly as an English sentence: "If a farmer *x* owns a donkey *y*, then *x* beats *y*." Since the symbols "IF" and "THEN" abbreviate Peirce's contexts, the scopes of the quantifiers are correct. The predicate calculus cannot be read as naturally as an English sentence, since the ⊃ operator does not have the same scoping rules.

Conceptual graph theory also introduces two innovations that go beyond Peirce's and Kamp's treatment of contexts:

> Since every context is a concept, conceptual relations can be linked to them. This feature eliminates the need for Peirce's colored contexts, supports Kamp's 'contexts for points in time, provides a natural way of representing propositional attitudes, and allows metalanguage about language.
>
> The symbol # indicates context-dependent references that have not yet been resolved. These include pseudo-constants like #now or the speech time #s-time. But it is generalized to handle other features that cannot be expressed in predicate calculus, such as definite references, indexicals, and deictics.

Predicate calculus can only represent context dependencies after they have been resolved to a constant or a quantified variable. Concepts containing the symbol # are an intermediate stage with unresolved references. As examples, the phrase "the donkey" becomes [DONKEY: #], and the pronoun "it" becomes [⊤: #]. Indexicals, deictics, and tenses are handled with an *annotation* following the #: [PERSON: #you], [ENTITY: #this], and [TIME: #now]. After the # reference is resolved, it is replaced with a coreference link to another concept in the current context or some containing context. The final form with the # resolved can be translated into predicate calculus. But by providing an intermediate stage with #, conceptual graphs allow the resolution to be postponed until all of the evidence—syntactic, semantic, and pragmatic—can be brought to bear on the analysis. These features are described in more detail by Sowa (*1991*).

§5. Type and Relation Definitions

The first two traditions incorporated in conceptual graphs are Peirce's existential graphs and the semantic networks of AI. The third tradition is the λ-calculus adapted to graphs. The λ-calculus is used to define new types in the hierarchy and determine their place in the partial ordering. They also provide a kind of macro facility that allows the graphs to be expanded and contracted as needed. Besides defining new types that have a permanent place in the hierarchy, λ-expressions can be placed directly in the type field of a concept. These λ-expressions define temporary types that can represent restrictive relative clauses. They have important interactions with quantifiers that will be discussed in Section 6.

A type definition is an assertion that some type label is equivalent to a particular λ-abstraction (*Sowa 1979a, 1984*). Following is the definition of DONKEY-FARMER as a type of farmer who owns a donkey:

DONKEY-FARMER = (λx) [FARMER: *x] → (STAT) → [OWN]
          → (PTNT) → [DONKEY].

The symbol λ introduces x as the variable that marks the formal parameter, which is the concept [FARMER]. The rest of the graph attached to that concept is the *differentia* that distinguishes a DONKEY-FARMER from any other FARMER. Instead of using the symbol λ, the keyword **type** may be used. The following example defines "EMPLOYEE" as a type of PERSON who works for a company for the purpose of earning a salary:

type EMPLOYEE(x) **is**
[WORK] -
  (AGNT) → [PERSON: *x]
  (BENF) → [COMPANY]
  (PURP) → [[*x] ← (BENF) ← [EARN] → (PTNT) → [SALARY]].

This example uses the conceptual relations BENF for beneficiary and PURP for purpose. The purpose is not just a simple concept, but a context that contains a nested graph. The variable *x identifies the concept [PERSON] as the formal parameter; that concept is also coreferent with the beneficiary of [EARN] in the nested context.

Conceptual relations can also be defined by the same mechanism. An *n*-adic relation is defined by a λ-expression with *n* parameters. Following is the definition of the past tense relation:

PAST = (λx) [TIME: #s-time] ← (SUCC) ← [TIME] ← (PTIM) ← [*x].

The defining graph says that the speech time (#s-time) is a successor (SUCC) of some unspecified time, which is the point in time (PTIM) of the situation *x. This definition may be substituted for the PAST relation in the graph for "A cat chased a mouse" (Fig. 21.10).

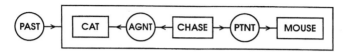

Figure 21.10 Conceptual graph for "A cat chased a mouse."

In the linear notation, the result of substituting the definition for PAST produces the following graph:

[TIME: #s-time] ← (SUCC) ← [TIME] - (PTIM)
    ← [[CAT] ← (AGNT) ← [CHASE] → (PTNT) → [MOUSE]].

This graph can then be translated into predicate calculus to generate the following formula:

$$\exists t(\text{time}(t) \land \text{succ}(t, \textbf{s-time}) \land \text{ptim}(t, \exists x \, \exists y \, \exists z (\text{cat}(x) \land \text{mouse}(y) \land \text{chase}(z) \land \text{agnt}(z,x) \land \text{ptnt}(z,y))))).$$

This formula is identical to the earlier one, but it can be derived by a systematic translation from English to a conceptual graph, then to a conceptual graph with the (PAST) relation expanded, and finally to predicate calculus. If predicate calculus is the assembly language of knowledge representation, conceptual graphs may be considered a higher-level language that simplifies the mapping to and from English. Sowa (*1991*) defines the mapping in two stages: first, a mapping from conceptual graphs into a sorted logic, followed by a second mapping that expands sort expressions into an unsorted predicate calculus.

As these examples illustrate, both type labels and monadic conceptual relations are defined by λ-expressions with one argument. The formula operator φ maps them both into monadic predicates. In any particular example, the primary guideline for choosing a type label or relation depends on the original statement in natural language: content words map into the type labels of concept nodes, and function words and inflections map into conceptual relations. A second guideline depends on use: only concept nodes can be quantified, qualified, or referenced by anaphoric expressions. Some people prefer to represent verbs as relations rather than concepts. That option is possible with conceptual graphs by defining a new relation. Following is a definition of a triadic relation LEAD-TO. (This definition uses the keyword **relation** instead of the symbol λ.)

relation LEAD-TO(x,y,z) **is**
[LEAD]-
   (AGNT) → [ANIMATE: *x]
   (PTNT) → [ANIMATE: *y]
   (DEST) → [PLACE: *z].

Alternatively, one might want to qualify or refer to one of the case relations like AGNT. That is possible by defining AGNT in terms of a concept [AGENT] and the primitive conceptual relation LINK:

relation AGNT(x,y) **is**
[ACT: *x] ← (LINK) ← [AGENT] → (LINK) → [ANIMATE: *y].

When an (AGNT) relation is expanded into the concept [AGENT], further relations or coreference links can be attached to the concept. These transformations allow conceptual graphs to support different views of the same phenomenon. The fundamental guideline is that the mapping from natural language to conceptual graphs should be as simple and direct as possible. After the initial representation in conceptual form has been derived, it can be expanded or contracted as needed by using the definitional mechanisms.

§6. Generalized Quantifiers

In natural languages, quantification has three parts: a *quantifier*, a *restrictor*, and a *scope*. Consider the sentence "Every farmer who owns a donkey beats it." The quantifying word is "every," the restrictor is the phrase "farmer who owns a donkey," and the rest of the sentence is the scope. The standard predicate calculus lacks a restrictor: a quantifier like $\forall x$ allows x to select from anything in the domain of discourse. Sorted logic provides a limited number of sorts that restrict the variable, but natural language allows arbitrarily complex expressions as restrictors.

To accommodate the three parts within a functional notation, some linguists use a three-place form $Q(R,S)$, where Q is the quantifier, R is the restrictor, and S is the scope (*Lewis 1975, van Benthem and ter Meulen 1985, McCord 1987*). Following is the representation for the sentence "Every dog is sleeping" in McCord's LFL notation:

$$\text{every}(\text{dog}(x), \text{sleep}(x)).$$

In this example, the restrictor is $\text{dog}(x)$, which McCord calls the base. The variable x ranges over everything for which the predicate $\text{dog}(x)$ is true. In the scope $\text{sleep}(x)$ (which McCord calls the focalizer), the variable x is bound to occurrences for which $\text{dog}(x)$ is true. For a more complex sentence, LFL may require nested quantifiers in the base, the focalizer, or both.

Following is the LFL representation of "Every farmer who owns a donkey beats it":

every(farmer(x) & exists(donkey(y), owns(x,y)), beats(x,y)).

In this example, the restrictor "farmer(x) & exists(donkey(y), owns(x,y))" contains a nested quantifier with its own restrictor and scope.

In conceptual graphs, the restrictor is placed in the type field of a concept, and the quantifier in the referent field. The scope is the entire context in which the concept occurs. Following is the conceptual graph for "Every dog is sleeping":

$$[DOG: \forall] \leftarrow (AGNT) \leftarrow [SLEEP].$$

This graph actually has two quantifiers: an explicit $\forall$ quantifier on [DOG], and an implicit $\exists$ quantifier on [SLEEP]. By convention, universal quantifiers have precedence over existential quantifiers in the same context. If the restrictor is longer than a single word, the entire phrase, represented as a λ-expression, must go in the type field of the concept. Following is the representation for "Every farmer who owns a donkey beats it":

$$[(\lambda x)\,[FARMER: *x] \rightarrow (STAT) \rightarrow [OWN] \rightarrow [DONKEY]: \forall] -$$
$$(AGNT) \leftarrow [BEAT] \rightarrow (PTNT) \rightarrow [ENTITY: \#].$$

This form is somewhat longer than the LFL form, but it is also more explicit: it shows the thematic roles AGNT, PTNT, and STAT, which are only shown by position in the LFL form. Furthermore, it treats [OWN] and [BEAT] as concepts rather than predicates; it therefore makes them discourse referents in their own right, suitable for further references or qualification. It also represents the word "it" with the concept [ENTITY: #], which contains an unresolved contextual reference. This is an intermediate form that would be linked to [DONKEY] in a separate stage of resolving anaphora. By the rules for expanding the $\forall$ quantifier and the rules for resolving anaphoric references (*Sowa 1984*), this graph can be expanded into exactly the same form as Fig. 21.9, which represents the logically equivalent sentence "If a farmer owns a donkey, then he beats it."

In his original form of discourse representation theory, Kamp did not provide a means of representing the $\forall$ quantifier directly. He had to translate the sentence "Every farmer who owns a donkey beats it" directly into the DRS form in Fig. 21.7. That translation is a complex, non-context-free mapping: it requires the quantifier to be translated to the more primitive DRS form during an early stage of semantic interpretation. With the $\forall$ quantifier and the # symbols in the referent field, conceptual graphs allow the semantic interpreter to proceed with a purely context-free mapping:

Each natural language phrase is mapped into a conceptual graph independent of the way it may be nested inside any other structure. Then the graphs for the separate constituents are joined together to form the interpretation of the sentence as a whole.

Finally, the quantifiers can be expanded, and the discourse referents can be resolved to form the final semantic interpretation.

By postponing the resolution of anaphora, conceptual graphs provide an intermediate stage that allows other processing to be performed. The simple cases are handled in a way that is equivalent to Kamp's, but complex references can be deferred until further background knowledge is considered. After context dependencies and background knowledge have been considered, the final graphs can be mapped into predicate calculus or other knowledge representation languages.

The $\forall$ quantifier can always be expanded into a more primitive form with negation and the $\exists$ quantifier. Other quantifying words in natural language, such as "many," "few," or "most," can only be expanded by a complex paraphrase. For example, "Most of the arrows hit the target" would become "If n is the total number of arrows, then the number of the arrows that hit the target is greater than $n/2$." To handle those quantifiers, symbols of the form @many, @most, or @few may be placed in the referent field of a concept. Instead of defining those quantifiers by expansions into more primitive conceptual graphs, it is easier to leave them in the graphs and give a direct model-theoretic semantics for them. In effect, the evaluation procedure would count the number of instances in the model and apply appropriate criteria for @many, @few, or @most.

§7. Plural Nouns and Sets

Underlying every plural noun phrase are two implicit quantifiers: one quantifier governs a set variable and another governs a variable that ranges over the elements of the set. Consider the sentence "Some dogs are sleeping." The phrase "some dogs" implies that there exists a set s containing at least two dogs. But there is also an implicit universal quantifier: for every x in s, x is sleeping. In conceptual graphs, the *generic set symbol* {★} represents some unspecified set of entities of the type determined by the type label, in this case DOG. The notation {★} @ 2 indicates that the set has at least two elements. Following are the conceptual graphs for the sentences "Two dogs are sleeping" and "Lucky and Macula are sleeping":

[DOG: {★} @ 2] ← (AGNT) ← [SLEEP].
[DOG: {Lucky, Macula}] ← (AGNT) ← [SLEEP].

When the operator ϕ maps conceptual graphs into predicate calculus, it introduces two explicit variables for each set. For the notation $\{\star\} @ 2$, it must introduce a third variable to represent the count of elements in the set. Following are the formulas that correspond to the above graphs:

$$\exists s \, \exists n (\text{set}(s) \wedge \text{count}(x,n) \wedge n \geq 2 \wedge \forall x(x \in s \supset (\text{dog}(x) \wedge \exists y(\text{sleep}(y) \\ \wedge \text{agnt}(y,x))))).$$

$$\exists s (s = \{\text{Lucky, Macula}\} \wedge \forall x(x \in s \supset (\text{dog}(x) \wedge \exists y(\text{sleep}(y) \wedge \text{agnt}(y,x)))))).$$

Since the last formula has a set s with a fixed number of elements, it can be reduced to a form that eliminates s:

$$\text{dog}(\text{Lucky}) \wedge \text{dog}(\text{Macula}) \\ \wedge \exists y_1 (\text{sleep}(y_1) \wedge \text{agnt}(y_1, \text{Lucky})) \\ \wedge \exists y_2 (\text{sleep}(y_2) \wedge \text{agnt}(y_2, \text{Macula})).$$

As these formulas illustrate, plural nouns in English, even when existentially quantified, also introduce implicit universal quantifiers that range over the elements of the set. The universal quantifiers may interact with other quantifiers in the sentence. In these examples, the default reading is that there may be a separate instance of sleeping for each dog. English words like *each* and *all* may override the defaults.

For the sentence "Every trailer truck has eighteen wheels," the universal quantifier for "every trailer truck" has precedence over the implicit quantifiers for "wheels." Following is the conceptual graph:

$$[\text{TRAILER-TRUCK: } \forall] \rightarrow (\text{PART}) \rightarrow [\text{WHEEL: } \{\star\} @18].$$

As before, the symbol $\{\star\}$ represents some unspecified entities of type WHEEL. The notation @18 says that there are 18 of them. The operator ϕ maps this graph into the following formula:

$$\forall x(\text{trailer-truck}(x) \supset \exists s(\text{set}(s) \wedge \text{count}(s, 18) \wedge \forall y(y \in s \supset (\text{wheel}(y) \wedge \\ \text{part}(x,y)))))).$$

The explicit quantifier $\forall$ on the trailer truck has precedence over the implicit quantifiers for the wheels: for each trailer truck x, there exists a set s of wheels.

If a formula contains only existential quantifiers, the order is irrelevant, since they may be freely interchanged without affecting the meaning. But the implicit universal quantifier introduced by $\{\star\}$ may interact with other quantifiers in complex ways[1]:

> Distributive: Dist$\{\star\}$ is the distributive symbol that implies the widest scope for the quantifiers associated with the set.
> Collective: Col$\{\star\}$ is the collective symbol that implies the narrowest scope for the quantifiers associated with the set.

Default: {★} makes the least commitment; it is consistent with either the collective or the distributive interpretations.

To show the differences, consider the sentence "Five blocks are supported by three pyramids." From the background knowledge about the fact that pyramids have pointed tops, a person would assume that the pyramids must be used collectively to support the blocks. Nothing is implied, however, about the number of blocks supported by each group of three pyramids. Following is the conceptual graph for that sentence:

[BLOCK: {★} @5] ← (PTNT) ← [SUPPORT] → (INST)
　　　　　→ [PYRAMID: Col{★} @3].

The operator φ maps this graph into a formula that shows the scope of quantifiers explicitly:

$\exists x(\text{set}(x) \wedge \text{count}(x,5) \wedge \forall u(u \in x \supset (block(u) \wedge$
$\exists y \exists z(\text{set}(y) \wedge \text{count}(y,3) \wedge \text{support}(z) \wedge \text{ptnt}(z,u) \wedge \forall v(v \in y \supset$
$(\text{pyramid}(v) \wedge \text{inst}(z,v)))))))$.

This formula says that there exists a set x of 5 blocks, and for each u in x, there is a set of 3 pyramids and an instance z of support, where the block is the patient of support and each of the three pyramids is the instrument of support. In this formula, the universal quantifier on blocks $\forall u$ has wide scope, including the existential quantifier on the instance of support as well as the quantifiers on the pyramids. The universal quantifier on pyramids $\forall v$ has the narrowest scope. This formula permits, but does not require the instance of supporting and the set of pyramids to be different for each block.

The distributive interpretation requires the instance of supporting and the set of pyramids to be unique for each block. In English, it may be represented by the word "each" in the sentence "Five blocks are each supported by three pyramids." That word is represented by the prefix "Dist" in the conceptual graph:

[BLOCK: Dist{★} @5] ← (PTNT) ← [SUPPORT] → (INST)
　　　　　→ [PYRAMID: Col{★} @3].

The formula for the distributive interpretation has two slight changes from the default: it uses the unique existential quantifiers $\exists!!y$ and $\exists!!z$:

$\exists x(\text{set}(x) \wedge \text{count}(x,5) \wedge \forall u (u \in x \supset (block(u) \wedge$
$\exists!!y \exists!!z(\text{set}(y) \wedge \text{count}(y,3) \wedge \text{support}(z) \wedge \text{ptnt}(z,u) \wedge \forall v(v \in y \supset$
$(\text{pyramid}(v) \wedge \text{inst}(z,v))))))$.

Whereas the default formula allows multiple blocks to share the same set of pyramids and instance of support, the quantifiers $\exists!!y$ and $\exists!!z$ require each block to have its own set of pyramids y and instance of support z. This formula implies that there are exactly 5×3 or 15 pyramids.

The collective interpretation implies that all the blocks collectively are supported by one set of three pyramids, also acting collectively. In English, it is implied by the word "all" in the sentence "Five blocks are all supported by three pyramids." The conceptual graph has the prefix "Col" for both the blocks and the pyramids:

$$[\text{BLOCK: Col}\{\bigstar\} \, @5] \leftarrow (\text{PTNT}) \leftarrow [\text{SUPPORT}] \rightarrow (\text{INST})$$
$$\rightarrow [\text{PYRAMID: Col}\{\bigstar\} \, @3].$$

In the corresponding formula, the quantifier $\forall u$ over blocks and the quantifier $\forall v$ over pyramids both have the narrowest scope:

$$\exists x \, \exists y \, \exists z (\text{set}(x) \wedge \text{count}(x,5) \wedge \text{set}(y) \wedge \text{count}(y,3) \wedge \text{support}(z) \wedge$$
$$\forall u \, \forall v ((u \in x \wedge v \in y) \supset (\text{block}(u) \wedge \text{pyramid}(v) \wedge \text{ptnt}(z,u) \wedge$$
$$\text{inst}(z,v)))).$$

This formula shows a single instance of support z, where all the blocks are the patient of z and all the pyramids are the instrument of z. It implies that there are exactly 3 pyramids.

The distributive interpretation, which implies 15 pyramids, is inconsistent with the collective interpretation, which implies 3 pyramids. Both of them, however, are consistent with the default, which allows 3, 6, 9, 12, or 15 pyramids. In general, the default makes the weakest assumption: it is implied by either the distributive or the collective assumptions. When English sentences are mapped into conceptual graphs, the default symbol may be inserted in the referent field for any plural noun. As the representation is refined, the prefixes "Col" or "Dist" may be added. Sometimes the prefixes are determined by explicit words like "each," "all," or "together." But sometimes background knowledge must be used, such as the knowledge that the pointed top of a pyramid cannot support a block by itself and three of them must be used collectively.

The importance of notation can best be appreciated by comparisons with other systems, such as Montague grammar. Link (*1987*), for example, gave the following formula as a "a Montague-style translation" of the distributive interpretation of "Three men lifted a piano":

$$3 \; \textit{men} \; \text{U} \; \lambda P \, \exists x [(3 \; \textit{men})\text{'}(x) \wedge P(x)]$$
$$^{\text{D}}(\textit{lifted a piano}) \; \text{U} \; ^{\text{D}}\lambda y \, \exists z [\textit{piano'}(z) \wedge \textit{lifted'}(y,z)]$$

This formula contains a great many symbols that are derived from the sentence by a complex process. By contrast, the corresponding conceptual graph has nearly a one-to-one mapping from the original sentence.

$$[\text{MAN: Dist}\{\bigstar\} \, @3] \leftarrow (\text{AGNT}) \leftarrow [\text{LIFT}] \rightarrow (\text{PTNT}) \rightarrow [\text{PIANO}].$$

This conceptual graph captures all the nuances of the Montague notation, and it is just as formally defined. There are many unsolved linguistic problems in representing plurals, and neither notation can solve them by itself. But conceptual graphs provide a more tractable tool for analyzing and representing them, both for linguistic theory and for the practical tasks of knowledge engineering.

Conceptual graphs are the product of thirty years of research on semantic networks in artificial intelligence and computational linguistics. Before that was the work by Lucien Tesnière (*1959*) on dependency graphs in linguistics. But the most important work that makes conceptual graphs into a complete system of logic is the pioneering development of existential graphs by C. S. Peirce. Conceptual graphs are a synthesis of these features:

- Graph structures similar to Peirce's Beta and various semantic networks in AI.
- Contexts derived from Peirce's existential graphs, which are isomorphic to Kamp's discourse representation structures.
- Labels on relations taken from linguistic research on case grammar and thematic roles.
- Definitional mechanisms based on the λ-calculus adapted to graphs.
- Generalized determiners in the referent field for specifying quantifiers, plural nouns, and context-dependent references.
- Operations for building graphs, expanding and contracting definitions, and performing inferences.
- Model theory based on the operations for mapping graphs to graphs.

Peirce's system of existential graphs has provided a sound and resilient framework that has readily accommodated the new extensions. But it is important to remember that Peirce was also the one who invented the linear form of predicate calculus and used it to make major discoveries in logic. Ten years later, he switched to graphs for his major research in logic. Peirce called existential graphs "the logic of the future." There are good reasons to believe that he was right.[2]

NOTES

1. In *Conceptual Structures*, the symbol {★} was used for a collective interpretation and Dist{★} for a distributive interpretation. Since then, the representation has been refined to distinguish the default from both the distributive and the collective interpretations.

2. I would like to thank Charles Bontempo, George Heidorn, and Doug Skuce for making a number of comments and suggestions on an earlier draft of this paper. Their comments have helped me to clarify the issues and strengthen the arguments.

22.

THE INTERCONNECTEDNESS
OF PEIRCE'S DIAGRAMMATIC THOUGHT

Beverley Kent

§1. Introduction

Peirce attributed creative thinking to the mental manipulation of diagrams. He himself thought in visual diagrams—never in words (4.530; MS 619: 8). Indeed, in 1909 he recorded that his lack of facility with linguistic expression and with foreign languages was lamentable (MS 632: 6). Since his voluminous writings often exhibit a fine turn of phrase and his command of differing languages was very impressive, this declaration is surprising.

He attributed his alleged difficulties to an aberrant arrangement of the speech segment of his brain and in this he anticipated modern views concerning the asymmetry of the bicameral brain: It has been found that the tendency for speech to lateralize is correlated with structural differences in the two hemispheres of the brain, discernible even in the foetus. Peirce appended the following footnote to his expression of diffidence:

> I will remark, by the way, that I am led to surmise that this awkwardness is connected with the fact that I am left-handed. For that my left-handedness is not a mere accidental habit, but has some organic cause seems to be evidenced by the fact that when I left the last school where it had attracted attention, I wrote with facility with my right hand, but could not write legibly with my left; and yet when I ceased to make the effort to continue this habit of three years standing, I soon fell back to using my left hand, though I have always used knife, fork, and spoon *at table* just as others do. (*Ibid.*)

The next clause is crossed through but it reveals that Peirce knew that the left hemisphere normally controls speech:

> Now supposing that my cerebral organ of speech is on the left side as in other people. . . . (*Ibid.*)

It is the case that in right-handed persons (which accounts for the majority) language dominance is located in the left hemisphere: the left hemisphere controls speech, writing, mathematical, logical and analytical activities, while the right hemisphere controls spatial relationships, creativity, music and synthesizing activities. In place of the deleted passage, Peirce writes:

> Now since my heart is placed as usual, it would seem that the connections between different parts of my brain must be different from the usual and presumably the best arrangement; and if so, it would necessarily follow that my thinking should be *gauche*. . . . (*Ibid*: 1–2)

Thus Peirce conjectured that the asymmetry in his own case must be atypical.

In the past, left-handedness has been surrounded by superstition to the extent that (as in Peirce's case) left-handers have been encouraged or coerced into writing with the right hand. Even today they repeatedly confront equipment designed for the right-handed. Converting to the right hand has sometimes led to confusion and disorientation among hapless individuals, for very often such retraining involves adapting the spatial hemisphere of the brain to control verbal activities.

That is the negative aspect, but there may also be a positive outcome. The impetus for Peirce's diagrammatic thinking may have stemmed from the extension of his spatially specializing hemisphere to his atypical writing dexterity.

Peirce seems to have suspected that this positive aspect—his facility with diagrammatic thought—was connected with his proclivity to left-handedness, for he expressed frustration at those of us who are right-handed as to the use of our eyes; we *look* with the right eye alone, though we *see* with both, and we are unpractised at using the eye of the mind (MS 298: 11).[1]

In any event, if social pressure occasioned diagrammatic thought, Peirce did not rely on such uncertain schooling: while still a young man he began training himself to think in diagrams, finding it a great advance over algebraic thinking. He traced his own creative initiatives to this systematized diagrammatic thought. Subsequent investigations furnished him with historical confirmation that visual images and muscular imaginations—not words or aural images—provide the best reasonings (NEM 4: 375; 5.363).

Beginning with a suggestion from Berkeley's two volumes on vision (MS 620: 17), Peirce recalled, he conceived the possibility of forming habits from imaginary practice. By exercising the imagination we could visualize the occurrence of a stimulus and mentally rehearse the results of different responses. That which appears most satisfactory, he claimed, will influence actual behaviour as effectively as a habit produced by reiteration in the outside world.[2] Out of this analysis came pragmatism, conceived as a philosophy in which thinking involves manipulating diagrams in order to examine questions.

Later, Peirce developed diagrammatic thinking into a system of logical diagrams—his Existential Graphs. His aim was to have the operation of thinking literally laid open to view; a moving picture of thought. He believed his Existential Graphs would allow logical relationships to be displayed in such an iconic way as to yield solutions to problems which had defied analysis via algebraic logics.

An extension of this method of thinking occurred to Peirce which he expected to be even more potent than that formalized in his Existential Graphs—that of 'stereoscopic moving pictures.' He was prevented from developing the idea, he tells us, because it required expensive apparatus quite beyond his means (NEM 3: 191). Even so, Peirce may have been reaching for just this kind of thought in his tinctured Existential Graphs where he extends his logical diagrams to include modality.

Three-dimensional imagery may also be involved in Peirce's natural classification of the sciences: Comte's hierarchical ordering in terms of decreasing generality becomes, in Peirce's scheme, a series of steps in which the sciences at the top provide principles for those below. This is not a single linear staircase, however. A whole series of ladders are related in a three-dimensional array so as to exhibit the more significant relations of logical dependence among the sciences.

I suggest that those aspects of Peirce's writings which focus on diagrammatic thought—his pragmaticism, his Existential Graphs, and his classification of the sciences—are interrelated in such a way that each sheds light on the others.

Peirce has remarked that his Existential Graphs are similar to his thoughts on any aspect of philosophy (MS 620: 9). And if diagrammatic thought is common to all three, my suggestion should readily be confirmed since it is virtually built in. Nevertheless, I believe the exercise will be illuminating. What, for example, could Peirce have meant by his claim that pragmatism is "a philosophy which should regard thinking as manipulating signs so as to consider questions"? (*ibid.*)

§2. Diagramming Pragmatism

Pragmatism was the outcome of joint discussions in the Metaphysical Club in Cambridge, Massachusetts. It was first given expression by Peirce in 1870, but it is the version in "How to Make Our Ideas Clear," his 1878 article in the *Popular Science Monthly*, which is most widely known.

Although the early statement clearly aligned meaning with concepts and not with action, Peirce did think a cursory reading might lead to that kind of misapprehension. Indeed, he declared that he all but yielded to such an interpretation himself at one point and came close to forsaking pragmatism for that very reason (MS 329: 16; MS 284: 3–4).

On reflection, Peirce thought it was folly to make action, in itself and irrespective of the thought it enacts, the ultimate end of life or of thought, because action itself supposes an end (5.3). He recognized that it was his account of the hardness of an unexperienced diamond which had been most conducive to that error. In his early article "How to Make Our Ideas Clear" Peirce had maintained that calling such a diamond hard or soft was merely a matter of the way in which we use speech. Yet the question for the pragmatic maxim is not what *did* occur, but what action would be appropriate if investigation were pursued sufficiently far.

Disavowing his early account, Peirce maintained that the principle should not be taken in such an individualistic way. The pragmatic end, he said, was the "development of embodied ideas" and could not be attained by single events because individual action is a means rather than the end (5.402, n.2). In "How to Make Our Ideas Clear" Peirce had said the purpose of thinking is to establish a belief, a habit of thought. To go on to consider the hardness of the diamond as simply a verbal question is to regard a habit as being made up of actual events. By attending only to what actually takes place, potentiality is reduced to actuality. Peirce needed to modify his view of possibility for, as long as he did not acknowledge the reality of the unactualized possibility of the diamond's hardness, action would seem to be the be-all and end-all of meaning. In 1878 Peirce had not elucidated the distinction between a law and the set of all actual instances of the law. This is because he had not yet adopted the view that there are real possibilities— possibilities that may never be actualized.

With the integration of quantifiers into the logic of relatives in 1885,[3] Peirce went a step closer to that position. It allowed him to make a distinction between membership and inclusion. And, because he could then indicate his category of Firsts quantitatively by a variable, he reformulated his theory of categories; this time identifying a formal and a material aspect. The logically formal categories were characterized as irreducible relations: Firsts are monadic, Seconds are dyadic, and Thirds are triadic relations; none can be reduced further and all higher relations can be reduced to these three. These formal categories provide a basis for a classification of signs.

The logically material character of those categories was obtained independently by examining experience. Peirce first referred to it as a psychological study, but later he considered it to be a phenomenological study. Firstness is quality understood as undifferentiated, unanalyzable suchness. Secondness is characterized by opposition, action and reaction. Thirdness is representation, mediation, and (ultimately) continuity.

The inclusion of continuity was the upshot of Peirce's reaction to Cantor's set theory. Peirce disagreed with Cantor's definition of continuity, which implied that geometrical continuity has to be thought of as a set of discrete points. Peirce claimed that Cantor's definition could not be used in topology and topology (as it was then conceived) was thought to be presupposed by all of geometry (4.219). He then proposed an alternative definition of a continuum which contains no discrete points. Such a definition is consistent with the theory of real infinitesimals which Peirce wanted to retain.[4] With real infinitesimals, the limit must be contained in the continuum. What this means is that there is no limit to the number of distinct individuals just because it is *not* a multitude of distinct individuals. Peirce says

> there may be a *potential* aggregate of all the possibilities that are consistent with certain general conditions; and this may be such that given any collection of distinct individuals whatsoever, out of that potential aggregate there may be actualized a more multitudinous collection than the given collection. Thus the potential aggregate is, with the strictest exactitude, greater in multitude than any possible multitude of individuals. But being a potential aggregate only, it does not contain any individuals at all. It only contains general conditions which *permit* the determination of individuals. (6.185)

Continuity, then, is quantitatively defined as the collection of all the possibilities which cohere with the general laws whereby they are defined. A qualitative definition is provided by what Peirce refers to as 'Kanticity.' Something is Kantistic, he tells us, if it is continuous such that every part has itself parts of the same kind. An alternative explanation (in terms of points) has it that something is continuous if between every possible pair of points it is possible to insert another point; which is to say the law of the excluded middle does not apply. None of the points are actual because the moment a point is in fact indicated the continuity is disrupted. While it is continuous it has no definite parts and there are no individuals.

Given Peirce's understanding of continuity and his view that laws are, in the final analysis, continuous processes, laws are necessarily general. They are to be understood as holding for all instances including possible but unactualized instances.

By 1897 Peirce recognized that real possibility is an essential consequence of pragmatism. Accordingly, in his later formulation, he noted the significance of real possibility and took every opportunity to redress his own failure to identify this in his 1878 article. It was implicitly expressed there nonetheless.

Moreover, although the wording of the maxim withstood his own critical appraisal for the most part, he did think he might not have been sufficiently aware of the fact that the three grades of clearness are *grades* and not *stages*. He assures us more than once that their differences are qualitative and that they are not meant to supplant one another (MS 649: 3).

Familiarity with a word such that it can be used readily and confidently, and so that others have a fairly good idea of what one has in mind, constitutes the first grade of clearness. It involves recognition as to whether or not a given concept applies to a given image. However, while one may be able to summon an image on demand, one might not be able to spell out the meaning on demand.

A more adequate understanding of a concept is obtainable with the second grade of clearness. This is that abstract and precise definition which results from analyzing a term or concept into its components. Analysis discovers the various broader classes which characterize the sort of thing that is being defined. The two fundamental factors involved here are: 1) generalization, and 2) the single relation of similarity and difference. The larger the number of elementary component qualities found in the analysis, the fewer the individuals which will possess the compound quality. But as the classes become more general our ideas are likely to be somewhat less distinct, for the dividing line between classes becomes less clear-cut. Ultimately, the procedure is confronted with an inexplicable idea—or so it was thought when analysis was confined to a subject/predicate logic as it is in the second grade of clearness.

Now while the value of words is to communicate, a complete exposition of their function, Peirce claimed, would have to await the determination of the end and purpose of humankind. If we assume that that is known in advance of our philosophical investigations, we assume that we already possess the answer to one of the most fundamental questions which lead us to undertake philosophy. Yet, no matter what humanity's purpose might be (whether it be in terms of feeling, of action, or of knowledge), it can be realized only by action. So the necessary and sufficient conditions for action which are enumerated in abstract and general terms in an analytic definition are only applicable when the purpose is known. Consequently, when that is the case, as it is with all artificial things, that thing can be defined. But, Peirce argues, the terms of a definition cannot express the real meaning of a word. And so to the third grade.

The pragmatist realizes that meaning resides in a conditional resolve. It conveys 'pragmatic adequacy' in that the entire intellectual import and value of a word is to be found in the conceivable consequences, for self-controlled conduct, of the acceptance or rejection of that concept; that is to say, the meaning of a concept lies in general habits of action.

According to Peirce,

> the most perfect account of a concept that words can convey will consist in a description of the habit which that concept is calculated to produce.

But how otherwise can a habit be described than by a description of the kind of action to which it gives rise, with the specification of the conditions and of the motive? (5.491)

For this third grade, then, concepts are open-ended. The conceivable consequences may change as the meaning of concepts interact and evolve; elements drop out while others are taken up. Thus, inquiry converges upon the ultimate consensus of the community of investigators—the arbiters of acceptance or rejection of concepts in Peirce's scheme.[5]

In the year in which James popularized pragmatism, Peirce spelled out his maxim in a way which suggests the role of diagrammatic thinking:

> The third grade of clearness consists in such a representation of the idea that fruitful reasoning can be made to turn upon it, and that it can be applied to the resolution of difficult practical problems. (3.457)

Peirce exhorted us to "trace out in the imagination" the conceivable practical consequences of denying or affirming the concept (8.191). This would require manipulating diagrams either in fancy or on paper. And here is where the third grade of clearness abuts on the logic of relations. What the logic of relations shows is that even those simple conceptions thought to be unanalyzable in a subject/predicate logic can be defined effectively just because they imply various kinds of relationship. With the association of contiguity, generalization is no longer moving to a larger class merely, but encompasses an entire system of all those elements linked to one another in a group of connected relations. Within the logic of relations, as we have seen, generality *is* continuity (5.436).

Developing systems for constructing iconic diagrams to aid us in tracing these connections is, Peirce thought, one of the tasks of logic.

§3. Diagramming Reasoning: The Existential Graphs

As early as 1870, Peirce adopted algebraic symbols for the logic of relations. Although algebraic formulae lack perspicuity, they might formalize diagrammatic reasoning nonetheless, for diagrams need not resemble their objects in appearance. It is in respect of the relation of their parts that resemblance is needed (2.282). Thus, the traditional syllogism diagrams the relations between the various terms. In the case of algebraic terms, however, the relations hold by virtue of the meanings ascribed to them and these need bear no resemblance to the relationships they symbolize either. Peirce wanted a more iconic formulation and therefore proceeded to integrate diagrammatic thinking and topology into a system of logical diagrams—his Existential Graphs.

Now the pursuit of topology is an imaging activity. It is described by Peirce as "the study of the continuous connections and defects of continuity of loci which are free to be distorted in any way so long as the integrity of the connections and separations of all their parts is maintained" (4.219). Just as topology focuses on pure hypotheses and ignores the properties of objectively valid space, so a diagram, in Peirce's sense, is an icon of the *form* of the relations as distinct from the actual relations of its objects.

The distinctive and integrating factor linking topology with diagrammatic thinking to produce the Existential Graphs lies in the fact that the integrity of the relevant connections and separations is maintained; i.e., continuities remain continuous and connectives remain mutually conjoined throughout all transitions of even highly plastic distortions and expansions. Similarly, the Existential Graphs diagram the form of the relations in a uniform schema enabling exact experiments to be performed such that throughout all these transformations the premises entrain the consequences.

Peirce's first efforts at constructing logical diagrams resulted in a system which he called 'Entitative Graphs.' These graphs are analogous to the diagrams used by chemists to illustrate the constitution of matter. Even as he was checking the proof sheets of the *Monist* article in which the first system appeared, an improved, more iconic, system occurred to him (MS 280: 21–22). This second system is his Existential Graphs.

These moving pictures, as Peirce describes them, exhibit the action of the mind in thought; they graph the dialogue between various phases of the ego. Thus, each sign has an interpretant which constitutes a new version of thought and that in turn becomes interpreted until finally one comes to the end of the series—to the logically final interpretant which is no longer a sign. This final interpretant is definitive in the sense of attaining the purpose. For intellectual concepts, this is a "conditional determination of the soul as to how it would conduct itself under conceivable circumstances" (MS 298.11).

The Existential Graphs were not intended to convey a perfect, minutely detailed photographic model of reasoning. Nor were they intended to capture every type of reasoning; rather, they were designed to convey what is common to all thought on subjects.

Initially, the Existential Graphs consisted of an Alpha part only (*Roberts 1973: ch. 3*). Alpha provides the foundation for the whole system and is a formulation of the propositional calculus. The Beta part of the Existential Graphs treats of the predicate calculus. And with these two-dimensional diagrams Peirce builds a remarkably iconic system of logic, providing a formal underpinning to his views on diagrammatic thought.

Nevertheless, he was not satisfied. Here is where Peirce incorporates a third dimension into his system. The tinctured Existential Graphs extend his logical diagrams to include modality. This development is the Gamma part of the Existential Graphs.[6] It includes all the elements of Alpha and

Beta and requires no additional symbols. However, the single sheet of assertion is replaced by a book of separate 'phemic sheets.' Peirce is now able to take account of logical possibilities.

With his Existential Graphs, Peirce has provided the groundwork for an extraordinarily iconic system, a system which facilitated his own development of ideas to depths which scholars have only begun to appreciate.[7]

§4. Diagramming Sciences: A Network of Ladders

Three-dimensional imagery, I have suggested, may also be involved in Peirce's natural classification of the sciences.

Peirce thought it necessary to classify the sciences because he believed that such a classification would reveal just how his own ideas of logic and of philosophy differed from accepted views. It would show where philosophy and its subdivisions fit in with other intellectual pursuits and, most importantly, it would indicate what kind of science logic must be. He came to realize that a clear understanding of logic required an examination of its relation to other sciences. A classification scheme would function as a diagram to exhibit those relations most vividly. Logic is the dominant concern of the classification for at least one other reason: Peirce thought that recognition of the relation it holds to other sciences would rescue it from attenuation or (worse) from absorption into some other discipline. He was particularly anxious to quash prevalent tendencies to collapse logic into mathematics, to found it on psychology or on one of the other disciplines which he identified as underlying the logical studies of his contemporaries. In his later writings he rarely embarked on a discussion of logic without first indicating its place within the classification scheme.

Peirce recognized that sciences are not rigidly defined, that borderline studies might require arbitrary distinctions and subsequent revision as new sciences emerge and others become obsolete. A classification which was capable of incorporating these changes would have to be an evolving one.

Aside from a few early attempts at schemes based on the single relation of similarity and dissimilarity,[8] Peirce's classification derived from researches into the logic of relatives and from his method of thinking in diagrams. He explicitly rejected a classification based on similarities and differences in a manuscript entitled "Why I am a Pragmatist," and he undertook to show how he dissents from that approach. "My classification of the sciences," he said,

> purports to be a natural classification such as the taxonomical biologists draw up; and here, less for the sake of the opinion itself than for the illustration it affords of my method of thinking, I wish to express my dissent from a certain form of reasoning that is universal among the taxonomists.

Namely, they are in the habit of taking two forms that appear to be markedly different, and if they can find specimens to bridge the interval between them with a "series" in which successive members show hardly any
perceptible difference, they say that these two forms belong to the same
natural species and that there is but a single natural class to which all the
specimens belong. But I say, on the contrary, that it may be a hard fact of
nature that a whole consists of three or more different divisions, and yet
it may be a pure matter of arbitrary election to which of these three divisions almost any individual member shall be regarded as belonging. (MS
327: 6–7)

It had become clear to Peirce that the unswerving dichotomies, the rigid
chain of values and the linear framework of previous classifications were
not satisfactory. A classification useful to practicing scientists would need to
take a new direction if it were to reveal the intricate and reciprocal relationships among the sciences. The classification which he devised is flexible
and dynamic. Any given science can be seen in terms of the network of
relations to other sciences and that can point the way to its conceivable
effects. This is tantamount to outlining the pragmatic meaning which, in
Peirce's doctrine, would be the most complete way of conveying what is
meant by a given science. It is seen in its third grade of clearness.

In this context, Peirce concluded that a science is best understood in
terms of the activity of those men and women whose lives are animated by
the desire to seek the truth. This conveys the idea of a science as living and
opens up the way to what Peirce called a 'natural classification.'

Early in his investigations Peirce adopted Comte's principle of hierarchical ordering. Since a researcher into positive fact is bound to assume
that certain principles are already established by a logically independent
investigation, he or she must accept, as a provisional assumption, that there
is a science that is antecedent to all the rest (MS 426: 1–2). Peirce maintains
that, with the single exception of mathematics, every science employs without question a principle discovered by some other science, while that latter
science may call upon the narrower one for data, problems, suggestions,
and fields of application.

Providing new facts is the commonest sort of help given to one science
by another. These are treated as direct observations by the receiving science
and melded in with a host of other facts to provide the basis for a generalization. Hence, a given contribution may well be expendable from the point
of view of the generalizing science although it might be necessary for a particular advance in the specific circumstances; but there can be no unique
ordering in terms of data-dependence. Once the generalization is completed the receiving science can reciprocate by providing principles with
which to interpret the observed fact (MS 693b: 374–8). In practice, it is
generally the case that a particular problem cannot be attacked until some

previous problem is more or less solved; and any solution virtually clears the way for the answer to another quite definite problem (MS 601: 10). In one manuscript, the foregoing is given the name *Bateris* (a mounting ladder) (MS 1338: 5). This captures the idea that the

> classification of sciences is a ladder-like scheme where each rung is itself a ladder of rungs, so that the whole is more like a succession of waves each of which carries other waves, and so on, until we should come to single investigations. (MS 328: 20)

Thus, the principle is to be applied to individual sciences, to coordinate groups of sciences, and also to those sciences having divisions with close internal relations (MS 673: 45). Peirce must have envisaged the scheme as a three-dimensional diagram.

The formal differentiating principle which governs the breakdown of the sciences is in terms of Peirce's three universal categories—the ideas of Firstness, Secondness, and Thirdness.

Categorial differentiation comes into operation even prior to the classification of the sciences: since science is regarded as an activity it needs to be considered in relation to other pursuits that occupy people. Thus, Peirce distinguishes as Firsts those who seek enjoyment; these are the most numerous. Seconds comprise those who lead lives of action and who aim at achieving results; included here are the makers of civilizations, the builders of industry, and the wielders of political power. Thirds comprise those whose lives are directed to developing ideas and truth: the men and women of science.

A classification of the sciences is not concerned with the first two divisions, but it is important to recognize that the whole scientific enterprise falls into the category of Third. Science, regarded as the activity of a group of persons in search of truth is one which falls into the category of mind. Were that to be ignored, the allocation of the theoretical sciences to the first of the three divisions of science would be incomprehensible in terms of the categories, although appropriate in terms of principle-dependence.

Sciences, conceived as the activity of the scientist, will not be static. Thus, it would be a mistake to expect Peirce's particular analysis to be definitive, or even that in all divisions sciences should have developed to such a stage that the categories could have universal application. Peirce saw a tendency in the sciences to merge into the more general sciences immediately preceding. Hence, by its very nature, there could be no final scheme[9] unless it be the one in which mathematics was the sole member. It is the principles governing the scheme that must withstand the impact of experience; and it is the principles that dictate the lattice construction.

The positioning of logic is Peirce's major concern. Ultimately, he considers it to be one of three normative sciences, but he adopts several different views on his way to that position. In his early classification schemes, for example, Peirce had considered ethics to be a practical science and separated it from philosophy altogether. Since the aim of scientific activity is critical for pragmaticism, increasingly Peirce recognized that the aim of any activity involves ethics and that in turn involves aesthetics.

The science of logic which emerges from Peirce's investigations is a normative study and does not include an inquiry into formal logic. Peirce held that to be a mathematical pursuit. Mathematics, Peirce claimed, is "the science which draws necessary conclusions" (4.229). The fundamental reason for symbolic logic is, he thought, as a tool for studying necessary or diagrammatic reasoning. The heuretic sciences of aesthetics and ethics have no need of principles investigated in logic. They may proceed unimpeded using deductive logic and a *logica utens* which is the theory one possesses prior to the scientific inquiry. Logic must appeal to aesthetics and ethics for principles and also to mathematics for its study of necessary reasoning.

Peirce reckoned that his first real discovery about mathematical procedure was his recognition of a distinction between two kinds of necessary reasoning: the corollarial and the theorematic (to use Euclidean terminology) (NEM 4: 49). While corollarial reasoning operates on diagrams already constructed, theorematic reasoning requires that the diagram be augmented, as when a line is added to the diagram of a non-trivial basic proposition of Euclidean Geometry, enabling it to be proved. "The construction is by no means implied by the problem or by the postulates of geometry, but it is permitted by them" (*Zeman 1986*: 17).

As J. Jay Zeman remarks, "Non-trivial deductive reasoning is hardly a matter of robotic, linear, left-hemispheric thinking alone. It involves a creative moment, a moment in which the deducer 'constructs,' to use Peirce's term" (*Ibid*: 16). The resulting diagram is a representamen which is chiefly an "icon of relations" (4.418). Zeman believes that these insights into the creative element found in non-trivial reasoning will have an impact on our understanding of the reasoning process in general. Typically, deductive reasoning is viewed as corollarial. Zeman claims that "recognition of the need to construct a diagram or image—an icon of some sort—in non-trivial cases opens the door to a multitude of insights in logic" (*Zeman 1986*: 17). Indeed, this distinction between theorematic and corollarial reasoning has relevance to the operation of abstraction. By 'abstraction' Peirce means "an operation on signs which employs a name for an 'object' which cannot, *in itself*, be said to exist" (*Ibid*: 18). The transformation from concrete to abstract exhibits an important semiotical distinction "that between meaning as *semantic* and meaning as *pragmatics*—the pragmatic dimension of semiotic considers the interpretants, or the effects of sign use; two semantically equivalent statements may have very different effects" (*Ibid*: 20).

Phenomenology occupies the first division of philosophy and is the only other science to which logic might appeal; metaphysics and all subsequent sciences may provide data and problems. Peirce extends the scope of logic to encompass general laws of signs of all kinds and, in this general sense, it is best characterized as semiotic; it is normative semiotic.

Thus viewed in relation to other sciences we are provided with an explanation of normative semiotic in accord with Peirce's third grade of clearness; that is to say, we may see the conceivable effects.

Now the principle virtue of a classification scheme is its simplicity; but this is precisely the character which will prevent it from displaying (but not from signposting) the more intricate relations between the sciences (MS 615: 29). Consequently, even a classification scheme under the aegis of the logic of relations will not yield immediately a complete explanation of any given science.

§5. The After Image

At the turn of the century a number of discrete ideas in Peirce's studies began to coalesce. In 1897, the logic of relations had convinced Peirce that the Philonian or material conditional could not interpret every conditional proposition. Acknowledging the reality of unactualized possibility has several implications: not only does it resolve the diamond dilemma unequivocally, but it shows that the meaning of a concept lies in future possible instances and it provides the link needed to make pragmaticism an intelligible theory (MS 288: 129). Real possibility has critical implications for the relation between logic and ethics also, and might well have precipitated the inclusion of ethics among the divisions of philosophy in the classification of the sciences.

When Peirce had denied that action was the end of thought, he was confronted with the corollary that the pragmatic principle is purpose-oriented. That precipitated the need to inquire into an ultimate purpose which could satisfy a course of action that might be pursued deliberately for an indefinite period (5.135 [1903]). Since the pragmatist has made the end of reasoning the essential element, it becomes necessary to inquire into the logically good. That, as his classification reveals, involves the morally good and ultimately depends on the nature of the *summum bonum* found out by aesthetics. For Peirce, the general ideal for pragmaticism is expressed as "the growth of concrete reasonableness" or "the continual embodiment of idea potentiality." It is an evolving end and receives its impetus from all three normative sciences and from the interrelations between them.

The particular end for the science of thought would be reasonableness. Whether or not there is such reasonableness is not for logic to determine.

However, any assurance vouchsafed by metaphysics has itself to be founded on logic. Thinking about it will not alter the real object, in Peirce's view, since that is just what he means by 'real'; the real is that which is such as it is independently of what you or I think it to be. In the final analysis, then, what guarantees that the conclusion of any reasoning is true? Peirce points out that such assurances must needs be unreasoned, so it can only have the character of faith. But unless we do assume that the universe is characterized by a reasonableness on which the ultimately destined opinion might converge, there is no possibility of knowledge. If it should turn out to be destined to a limited fulfillment there would, at least, be an approximate reality.

While the end of reasoning indicates the secondary role of ethics and the primary role of aesthetics with regard to pragmaticism, reasoning itself testifies to the direct connection with ethics. Reasoning is controlled thought and the inquiry into the operation of self-control, Peirce had argued, is the business of ethics. Given the ethical principle that it is futile to criticize what you cannot control and that future facts are all that we can (in a measure) control, pragmaticism, which asserts that meaning refers to deliberate or controlled conduct, must maintain that the conclusion of a piece of reasoning refers to the future. Thus, the future-directedness of pragmaticism is borne out by ethics and endorses the conclusion derived from the recognition of real possibilities.

Pragmaticism, Peirce said, is "a theory in regard to the common nature of the meanings of all concepts" (MS 298: 1); and that which it makes common to every concept is shown in an influence on possible conduct. Since the Existential Graphs provide a moving picture of the mind's action during thought (or, more specifically, during what is common to thought) it should reveal what is essential to all significations of concepts (4.534 n.1). Thus, the Existential Graphs are capable of an analysis of thought which is appropriate to patterning pragmaticism.

Peirce did not claim that the Existential Graphs are unique in providing such an analysis. What he claimed was that it "provides a singular and signal facilitation of that achievement" (MS 298: 5).

The pragmatic principle is a special application of the scientific method and is, therefore, a principle resulting from the investigations of methodeutic, the third division of logic; and the Existential Graphs form a system of logical diagrams. Thus, any illumination cast on logic by relating it to other sciences might also reflect on pragmaticism and on the Existential Graphs. On the other hand, by exhibiting the logical relations between the sciences, Peirce's classification scheme is expected to convey the pragmatic meaning of any given science. Thus, the classification of the sciences provides a perspective which makes these various interconnections explicit.

N O T E S

1. Our eyes also involve cerebral specialization, but Peirce was speaking metaphorically here.

2. Peirce related an incident from his youth when, with the whole family "at table," a lady's dress caught fire. Charles' younger brother extinguished it with singular alacrity. When asked to explain the swiftness of his reflexes, the young Herbert reported that he had mentally rehearsed appropriate action after Mrs. Longfellow had died from just such an accident (5.487 n.1).

3. Developed by Peirce and one of his Johns Hopkins students, O. H. Mitchell (3.359–403). While this was an independent discovery, it came six years after Frege published his *Begriffschrift*.

4. Peirce was not one to hesitate to adopt or disclaim a theory merely because he was alone in doing so. Infinitesimals were unfashionable then and only recently have been reconsidered.

5. That consensus follows from the reality of objects. It is not a claim that inquiry will go on indefinitely.

6. An earlier version of Gamma (not completed) is described by Roberts (*1973*: ch. 5).

7. Peirce, himself, has remarked that his Existential Graphs are similar to his thoughts on any aspect of philosophy (MS 620: 9).

8. A classification in terms of genus and species is based on the single relation of similarity and its converse. A science in such a scheme would be exhibited in its second grade of clearness.

9. This reflects the open-endedness already noted in the third grade of clearness.

23.

WHAT IS DEDUCTION?

E. James Crombie

Peirce's division of inferences into deduction, induction, and abduction is well-known, controversial, and basic to an understanding of his thought. As early as 1865 and as late as 1911, Peirce claimed this classification as his own invention (Harvard Lecture VIII, W1: 267). In 1865, he claimed that every argument "may be resolved into à priori [i.e., deductive], à posteriori [i.e., abductive] and inductive elements" and "that these classes do not run into each other" (W1: 286). In 1911, he said that he had "proved" that deduction, induction, and retroduction (which is the term he was then using for abduction) constitute "three absolutely disparate ways of reasoning" (NEM 3: 177) and that he had "very strong *probable* reasons for believing there is no fourth kind" (NEM 3. 177–178). In the cognition series of 1868, Peirce further suggested that even among the unconscious and uncontrolled operations of mind there is none which does not "conform to the formula of valid inference" (5.262).

Restricting ourselves for the moment to inference properly so called, that is to say to modes of thought subject to what Peirce calls logical self-control, and inspiring ourselves more from the later than from the earlier Peirce, we will say that abduction is that mode of reasoning which generates hypotheses in order to explain what purport to be the known facts. With the same restriction, we will say that induction, in the Peircean sense of the term, is that mode of reasoning by which we attempt to determine the degree of match between a given hypothesis and facts as yet unobserved (cf. 2.96 [1902], 8.209 [c.1905]). Deduction, finally, is that mode of reasoning whereby we determine what must necessarily (or probably[1]) be the case if a given hypothesis is true.

The definition of "deduction" just given seems clear enough at first glance. We note, however, that one problem with this definition is that it applies only to *successful* deductions, while the definitions given for induction and abduction are not restrictive in this way. Thus a bad or unsuccessful attempt to produce an explanation is still, for all that, an abduction, whereas an unsuccessful attempt to determine what necessarily must be so, given the truth of a set of premises, does not, under the proposed definition, count as a *poor* or *unsuccessful* deduction, because it does not count as a deduction at all.[2] One way around this problem is, of course, to make deductive validity not the *actual* but the *desired* result of arguments counting as deductive. Thus Copi, when explaining the difference between deductive and non-deductive arguments, writes: "Only a *deductive* argument involves the claim that its premises provide *absolutely conclusive* evidence [for the truth of its conclusion]" (*Copi 1967*: 4).

Peirce in 1893, uses the verb "pretends" to the same effect: "*Demonstrative reasoning* [writes Peirce], pretends to be such that it is logically impossible for the premises to be true while the conclusion is false" (2.447). Both Copi's definition and this one of Peirce's make the definition of deduction turn on what an argument *claims* or *purports* to be or to do, rather than on what the argument really is or does.

So far, so good. Now we can have unsuccessful deductions in much the same way as we can have unpromising abductions and sloppy inductions. This may not seem to be much of a cause for celebration, but such is the price of symmetry.

There is, however, a further difficulty with both our original definition of deduction and the proposed emendations. The problem lies in a kind of circularity which results from the presence in the *definiens* of expressions like "must necessarily," "logically impossible," and "absolutely conclusive evidence." One wonders immediately—or at least one *would* wonder, if only we had not become so jaded—what the "cash value" of these words might be. The concept of necessity has a long and muddy history but clearly in *this* context the proof of necessity or of what is logically possible or impossible involves *deduction*. Absolutely conclusive evidence, for its part, requires nothing less. So that, in effect, we have defined a deductive argument either as being or purporting to be a deductive argument.

But this unpromising result need not disappoint us since, as Peirce wrote in his 1878 article which announced pragmatism and the pragmatic maxim to the world, "nothing new can ever be learned by analyzing definitions" (5.392), their function being that of setting our existing beliefs in order.

Now it is perhaps worthwhile at this point to make a brief digression concerning what the difference between deductive and non-deductive modes of reasoning is *not*. "Deducible," for example, does not mean "certain." The contrast between deductive and non-deductive reasoning is thus *not* between reasoning whose conclusions are certain and reasoning whose conclusions are not. It is, for example, far more certain (as far as the author of these lines is concerned, in any case) that the sun will come up tomorrow than that there is a largest finite mathematical group. And this is so in spite of the fact that the former is the result of non-deductive inference, while the latter is a conclusion which has been reached by a long labor of deduction.

This confusion is inherent in attempts, common in the history of philosophy, to restructure our world view in such a way that it would rest solely upon and be deductively derivable from what is seen to be indubitably or certainly the case. In opposition to this tendency, philosophers like Thomas Reid in the 18th century, C. S. Peirce in the 19th and early 20th century and, closer to us, Wittgenstein[3] and Anthony Quinton,[4] among others, have been at pains to point out some of the confusions inherent in such projects. Peirce's fallibilism can, for its part, be seen as a denial of the identity which has been supposed to exist by some philosophers between logical necessity and certainty. Fallibilism asserts that the probability of making a mistake in any kind of judgment or inference is never exactly equal to zero. There is, for example, according to Peirce, a very small probability approaching but not equal to zero that we have been in error all these years in expecting that whenever we have two of something, and then acquire two more, we must then necessarily have four of whatever it is. Peirce also mentions how certain propositions which for centuries had been taken to be necessary and universal truths were subsequently recognized either to be false or to have a less universal application than had been believed. A notable example of this is the dictum that the part is greater than the whole. We now realize that this does not apply, for example, to the relative multitude of even numbers and integers. Every even number is an integer, while not every integer is an even number; so that we can say that the even numbers form a proper part or subset of the integers. Yet it would be false to say that there are "more" integers than even numbers, as can be easily shown. Another argument in favour of fallibilism with respect to deductive reasoning is that mathematicians are continually finding mistakes in each other's proofs,[5] so that deductive reasoning cannot be said to afford certainly in the sense that there is any kind of incorrigibility or immunity from error (Cf. 1.248 [1902]).[6]

Deduction, we may conclude in summary—and this is one of the results of Peirce's fallibilism—cannot be distinguished from non-deductive reasoning by means of the certainty (in the sense of absolute incorrigibility or indubitability) which has been erroneously attributed to the former but not to the latter. Even relative certainty is not a completely reliable criterion, since there exist inductive conclusions whose certainty is greater, for most of us at least, than the certainty we attach to some purported deductions.

Having terminated our little digression on what the difference between deductive and non-deductive modes of reasoning is *not*, in virtue of our initial attempts to define deduction in terms of logical necessity and possibility, we now bring the pragmatic maxim to bear on the matter.

Verbal definitions like the ones just considered bring us to what both Peirce and "the [logic] books" (5.292) recognize as the second "step toward clearness," the first degree of clearness being a certain kind of familiarity with the notion to be defined. In our case, the first step towards clearness concerning deduction would be to "have a feeling" for what deduction is, and to be able to use the term correctly in sentences. The *error* of what Peirce calls "the books" lies in their omission of "all mention of any higher perspicuity of thought" (5.292). Such perspicuity is attained, according to Peirce, by reflecting on the nature of belief and inquiry and by considering that "the root of every real distinction . . . consists] in . . . a possible difference of practice" (5.400).

So far, however, the definitions of deduction we have considered have had very little to say about practice. Deduction being a kind of reasoning, and reasoning a kind of *doing*, we thus need a bit *more* than a description of what the result would be like if we happened to be so fortunate as to succeed in such an endeavor. What we need is some indication of *how* to engage in the activity in question, and perhaps also of the circumstances under which such activity is appropriate. Peirce's definition of deduction as involving the production of *diagrams* representing ideal states of affairs is thus a natural outgrowth of his pragmatism (later, pragmaticism).

Now this idea did not spring fully formed from Peirce's forehead. There is, for example, no entry for the terms "diagram" or "diagrammatic" in the index of any of the first three volumes of the new edition of the *Writings of Charles S. Peirce*, which take us to 1878. Don Roberts, however, dates Peirce's interest in diagrammatic thinking from 1870 onwards. Roberts identifies Peirce's letter to O. H. Mitchell of 21 December 1882, as containing "perhaps the first attempt by anyone to apply diagrams to the logic of relatives in general" mentioning however that Leonard Euler "as early as 1761" and John Venn in 1880 had used circle diagrams "to treat the logic of classes"— and that Peirce himself acknowledged a filiation between his entitative graphs and the chemical diagrams of William K. Clifford (1877–1878 onwards) (*Roberts 1973*: 17–18. See also *Ketner 1986* and *1987*).

By the mid-1890s, we thus find Peirce defining deduction as:

> that mode of reasoning which examines the state of things asserted in the premises, forms a diagram of that state of things, perceives in the parts of that diagram relations not explicitly mentioned in the premises, satisfies itself by mental experiments upon the diagram that these relations would always subsist, or at least would do so in a certain proportion of cases, and concludes their necessary, or probable, truth. For example, let the premiss be that there are four marked points upon a line which has neither extremity nor furcation. Then, by means of a diagram,

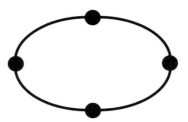

we may conclude that there are two pairs of points such that in passing
along the line in any way from one to the other point of either pair, one
point of the second pair will be passed an odd number of times and the
other point an even (or zero) number of times. This is *deduction.* (1.66)

The point about deduction involving "mental experiments" is crucial.
In the passage just quoted, these "mental experiments" do not yet seem
to involve any actual modification of the material diagram. The passage
quoted may also be found somewhat troubling because of the extension of
the notion of deduction to cover not only reasonings yielding necessary
truth concerning certain relations which would "always subsist" but also
those yielding "probable truth" concerning relations which "would do so
in a certain proportion of cases." Dealing with this extension of deductive
inference to include certain kinds of reasoning involving probability goes
somewhat beyond the scope of the present essay.[8] The passage quoted below, however, casts some further light on this question, as well as on one
which is of crucial interest to us here, namely the matter of experiments
upon the diagram resulting in a modification of the latter. In 1903, Peirce
wrote:

> A *Deduction* is an argument whose Interpretant represents that it belongs
> to a general class of possible arguments precisely analogous which are such
> that in the long run of experience the greater part of those whose pre-
> misses are true will have true conclusions. Deductions are either *Necessary*
> or *Probable.* Necessary Deductions are those which have nothing to do with
> any ratio of frequency, but profess (or their interpretants profess for
> them) that from true premisses they must invariably produce true conclu-
> sions. A Necessary Deduction is a method of producing Dicent symbols by
> the study of a diagram. It is either *Corollarial* or *Theorematic.* A corollarial
> Deduction is one which represents the conditions [i.e. premisses] of the
> conclusion in a diagram and finds from the observation of this diagram,
> as it is, the truth of the conclusion. A Theorematic Deduction is one
> which, having represented the conditions of the conclusion in a diagram,
> performs an ingenious experiment upon the diagram, and by observation
> of the diagram, so modified, ascertains the truth of the conclusion. (2.267.
> See also 5.162, 6.417, and 4.572.)

In this definition, merely "mental" experiments upon the diagram have become assimilated to "observation," while the possibility of actually modifying the diagram has been explicitly allowed for. This advance in clarity has its price, however, as the tidiness of the tripartite division of arguments may now be threatened since it now appears that deduction itself includes both abductive and inductive elements.

The devising of an "ingenious experiment" one might perform on a diagram, first of all, is clearly abductive; a "bright idea" which comes to one in such a context is clearly a *hypothesis* concerning manner of proof. Such a hypothesis may or may not pan out, as is the case in the so-called empirical sciences. (Note that the problem of defining deduction is turning into a problem concerning the demarcation between empirical and non-empirical or formal sciences.)

The tripartite division is also threatened in another way, since, on the other hand, the working out of the idea involved in the ingenious experiment is a process the nature of which is obviously *inductive*. Thus, in the kind of deduction which Peirce calls "theorematic," abduction and induction turn out to form part of the very structure of the argument. This is embarrassing in that deduction, in turn, is involved in the definitions which Peirce usually gives of induction. This is defining A in terms of B and C, then turning around and defining B in terms of A. These objections do not of course refute the thesis that deduction, abduction, and induction are "three absolutely disparate ways of reasoning" (NEM 3: 177/Letter to Kehler [1911]), but the fact that one or more of them enter into the definition of the others does make it difficult to advance the stronger claim that these three ways of reasoning can be considered as primitive notions in a (formalizable) general theory of reasoning.

There are numerous passages in which the essential role played by abduction and induction in deductive inquiry is confirmed. Sense perception also has its place in getting inquiry going in the first place. In an article published in 1908, Peirce wrote:

> Every inquiry whatsoever takes its rise in the observation, in one or another of the three Universes, of some surprising phenomenon, some experience which either disappoints an expectation, or breaks in upon some habit of expectation of the *inquisiturus*; and each apparent exception to this rule only confirms it. (6.469)

In this passage, the expression "every inquiry whatsoever," is sufficiently broad to cover deductive inquiry. There would furthermore seem to be no reason to assume that deductive inquiry is not coextensive with deductive reasoning, in which case the latter clearly consists of experiential, abductive, inductive and deductive elements, involving us in some kind of a circle or regress. Alternatively, we might replace the *tripartite* division of modes of reasoning with a *bipartite* one consisting of abduction and induction alone, with deduction dropping out altogether as a separate form of reasoning. What had previously been construed as a difference between modes of reasoning would now reappear as the distinction between the different "Universes" in which inquiry may take place, mathematics belonging to one universe and natural science to another.

In any case, if we stick with the tripartite division of modes of reasoning we are struck with the above-noted untidiness. But if we attempt to transform the deductive/non-deductive distinction into one between the "Universes" in which inquiry is pursued, we run up against the problem that Peirce proposed not two but, as was his wont, *three* of them.

Earlier in the article we have just quoted from, Peirce had defined these "three Universes of experience familiar to us all" (6.455) as consisting in the first case of "mere Ideas," in the second of "the Brute Actuality of things and facts," and in the third of "everything whose being consists in active power to establish connections between different objects" (*ibid.*). Peirce went on to explain, on the subject of the elements of this third Universe, that "such is everything which is essentially a Sign—not merely the body of the Sign, which is not essentially such, but, so to speak, the Sign's Soul, which has its Being in its power of serving as intermediary between its Object and a Mind" (*ibid.*).

Note that Peirce spoke of these universes of inquiry also as universes of *experience*. Before pursuing our line of argument, it is thus perhaps useful to remind ourselves what meaning Peirce attached to the word "Experience" (with a capital E). In the paragraph immediately preceding the one where Peirce introduces the three Universes of Experience, Peirce defined an "Experience" as "a brutally produced conscious effect that contributes to a habit, self-controlled, yet so satisfying, on deliberation, as to be destructible by no positive exercise of internal vigour" (6.454).

We may now ask ourselves in which "Universe of Experience" what is ordinarily called deduction may be said to take place, according to Peirce. We may *ask* ourselves this question, but the present author has no definite answer to give. Here are some suggestions however: The diagram being manipulated embodies and represents an idea, belonging to the *first* Universe; it is thus a sign, belonging to the *third* Universe; but it is perhaps merely the *body* of the diagram which is manipulated, according to mechanical rules, so that we may be dealing with the *second* Universe of things and facts; the rules themselves, however, obviously belong to the third Universe. The iconicity required of diagrams suggests Firstness and the first Universe, however.

This line of inquiry might, if pursued further, lead us to a firmer basis for the demarcation between deductive and non-deductive inquiry. Pursuing this line, however, would lead us beyond the intended scope of the present chapter.

For present purposes it is sufficient to examine a little further the implications of Peirce's theory of inquiry for his thesis concerning the distinctness and disparateness of deduction. We have already quoted the first sentence of Peirce's paragraph beginning "Every inquiry whatsoever." We now go back to this paragraph beginning with the third sentence:

> The inquiry begins with pondering these phenomena in all their aspects, in the search of some point of view whence the wonder shall be resolved. At length a conjecture arises that furnishes a possible Explanation, by which I mean a syllogism exhibiting the surprising fact as necessarily consequent upon the circumstances of its occurrence together with the truth of the credible conjecture as premises. On account of this Explanation, the inquirer is led to regard his conjecture, or hypothesis, with favor. . . . The whole series of mental performances [described here] . . . I reckon as comprising the First Stage of Inquiry. Its characteristic formula of reasoning I term Retroduction. (6.469)

"Retroduction" is a term sometimes used by Peirce instead of "Abduction."

Peirce goes on, in subsequent paragraphs, to explain how the hypothesis is tested, through deductive "examination of the hypothesis" in the "Second Stage of Inquiry" (6.470) and the third, inductive stage, which is itself divided into three parts, the third of which "passes final judgment on the whole result" (6.472).

In a draft of Lecture II of the 1903 Lectures on Pragmatism, Peirce confesses that he has

> sometimes been tempted to think that mathematics differed from an ordinary inductive science hardly at all except for the circumstance that experimentation which in the positive sciences is so costly in money, time, and energy, is in mathematics performed with such facility that the highest inductive certainty is attained almost in the twinkling of an eye. (NEM 4: 158)

This confirms that the idea which constitutes the central pivot of this paper had also occurred to Peirce. That it worried him, at least to some extent, is confirmed by the very next sentence in the same manuscript. "But it is rash," writes Peirce, "to go so far as this" (*ibid.*). This is an unusual sentence for Peirce, whose normal reflex was to allow himself to be carried as far as an idea would take him. Rashness and prudence belong to the world of "vital concerns," not to that of theoretical inquiry (cf., e.g., 1.635–637 [1898]).[9] By way of explaining *why* it would be rash to push the parallel between inductive science and mathematics "so far as this," Peirce writes: "The mathematician, unless he . . . deludes himself . . . , reaches conclusions which are . . . general and yet, *but for the possibility of mere blunders*, are absolutely infallible" (*ibid.*, emphasis added). This is about as close as we are ever likely to come to finding Peirce grasping at straws. Infallibility, as we have seen, albeit sketchily, and as Peirce well knew, is a Cartesian will-o'-the-wisp, especially as a demarcation criterion between deductive and non-deductive inquiry. It is, rather, instinct, not mathematical reasoning, which is for practical purposes infallible.

Smelling blood, then, let us return to Peirce's definition of deduction as diagrammatic reasoning and survey the extent of the damage. Clearly the troublesome part of the definition concerns what Peirce calls *theorematic* deduction. Perhaps theorematic deduction could be construed as corollarial deduction using a new diagram, or using new diagrammatic elements. This is not very elegant. But let us suppose, for the moment at least, that this reduction can be performed. Next, let us consider what might be so distinctive about *corollarial* deduction, as compared with non-deductive modes of thought. What we seem to be left with are two elements: production of the diagram, and observation of the diagram.

Now one is entitled to wonder just *where* the deduction is here. We may begin, I think, by provisionally eliminating the *observation* of the diagram as the candidate for the locus of what is distinctive about deduction, since such observation is a species of perceptual judgments and perceptual judgments enter into the structure of both abduction and induction, so that the hope is slight that we will find in these perceptual judgments the *disparateness* we are looking for, especially in the light of the further fact that Peirce thought of perceptual judgments as being analogous to and shading off into genuine abductions, not deductions (cf., e.g., 5.183). Before going on, however, we will make a mental note to the effect that one possible point of difference here is that observation in deduction is always observation of an artifact, namely the *diagram*, whereas this is not always the case with the observation involved in induction and abduction.

Having thus provisionally eliminated observation of the diagram as the locus of the allegedly distinctive character of deduction, we now consider the *production* of the diagram. Are we to conclude that the disparateness of deduction from the other modes of reasoning revolves around *drawing a diagram*? Is the mere production of the diagram (whether by drawing or by some other means) already distinctively deductive, the other aspects dropping away as shared with other modes or inessential? Perhaps so, but three objections come rapidly to mind.

First, this definition of deduction is certainly not one which is intuitively obvious. This is not a knock-down objection, of course, but some sort of reply is required.

A second objection is that this definition of deduction seems to suggest that what is distinctive about deduction is that it involves the production of a sign, namely the diagram. But deduction is not alone in involving the production of a sign. Abduction, for example, also involves the production of an explanatory hypothesis, which is also a sign (of course)—and which ought to be such as to allow deductions to be performed upon it, so that it would seem that it is *abduction* here which produces the diagram and deduction which only transforms it. We might add that perception itself—which, as we have just mentioned, Peirce thought of as being analogous to and shading off into abduction—also involves the production or quasi-production of a sign, not in the sense that the perceiver actually devises it but in the sense that it wells up, so to speak, from the uncontrolled regions of mind and intrudes, often brusquely, upon consciousness.

Thirdly, what are we to make of certain other garden varieties of sign production, such as speech and writing? It might be thought in fact that the definition of deduction as the production of a diagram strains the bounds of paradox by requiring us to consider speaking and writing as particular cases of deduction. Speech and writing are, after all, capable of being made sufficiently diagrammatic to allow much deductive reasoning to be carried on by their means, so that any kind of verbal or graphic production might be thought of as the production of a diagram. The uncharitable reaction here is to conclude that we are faced with what amounts to a *reductio ad absurdum* of the whole idea.

We will attempt to answer these objections shortly. In the meantime, and in preparation for these answers, we will consider briefly another reason to think that we may have been barking up the wrong tree by putting so much emphasis on *arguments*—and hence in the deductive case on diagrams—while giving relatively little attention to the *process* of thought which arguments are thought of as representing. After all, did not Peirce himself call attention to the fact that the correspondence between these processes and the arguments which purport to represent them is far from perfect. For one thing, an argument is a static representation of a process which is not static. As Peirce wrote in his "Minute Logic" of 1902:

> A man goes through a process of thought. Who shall say what the nature of the process was? He cannot, for during the process he was occupied with the object about which he was thinking, not with himself nor with his motions. Had he been thinking of these things his current of thought would have been broken up, and altogether modified; for he must then have alternated from one subject of thought to another. Shall he endeavor, after the course of thought is done, to recover it by repeating it, on this occasion interrupting it, and noting what he had last in mind? Then it will be extremely likely that he will be unable to interrupt it at times when the movement of thought is considerable; he will most likely be able to do so only at times when that movement was so slowed down that, in endeavor-

ing to tell himself what he had in mind, he loses sight of that movement altogether; especially with language at hand to represent attitudes of thought, but not movements of thought. Practically, when a man endeavors to state what the process of his thought has been, after the process has come to an end, he first asks himself to what conclusion he has come. That result he formulates in an assertion, which, we will assume, has some sort of likeness—I am inclined to only a conventional one—with the attitude of his thought at the cessation of the motion. That having been ascertained, he next asks himself how he is justified in being so confident of it; and he proceeds to cast about for a sentence expressed in words which shall strike him as resembling some previous attitude of his thought, and which at the same time shall be logically related to the sentence representing his conclusion, in such a way that if the premiss-proposition be true, the conclusion-proposition necessarily or naturally would be true. That argument is a representation of the *last part* of his thought, so far as the logic goes, that is, that the conclusion would be true supposing the premiss is so. But the self-observer has absolutely no warrant whatever for assuming that that premiss represented an attitude in which thought remained stock-still, even for an instant. (2.27 [1902])

The argument is thus to be considered more as a *defence* of an inference than as a faithful image of it. It is only the "self-defence" of a process of thought which "is clearly broken up into arguments"—for "there is no fact in our possession to forbid our supposing that the thinking process was one continuous (though undoubtedly varied) process" (*ibid.*).

Perhaps then (but only perhaps), if we want to know what deduction is and what the difference between deductive and non-deductive modes of thought might be, we should be looking at the thought processes themselves, rather than at the arguments used to defend them. This could turn out to be difficult if it is true, as Peirce argued in detail in 1868 and as the passage just quoted from 1902 reminds us, that we have no introspective power enabling us to scrutinize the thought processes themselves while they are underway. With this warning in mind, however, let us examine Peirce's systematic classification of all mental action, both conscious and unconscious, including both those modes of thought subject to what Peirce called logical self-control and those which are not. According to this classification, abduction, deduction, and induction *properly so called* are merely the self-controlled manifestations of three fundamental modes of thought which, in their uncontrolled and unconscious occurrences we might call (although Peirce does not) proto-abduction, proto-deduction, and proto-induction. This is what Peirce means when he says that all mental action "conforms to the formula of valid reasoning" (5.282). In 1868, Peirce argues in detail to rule out any kind of mental action which would not belong to one of the three basic modes, including, for example, any kind of direct determination of the mind by its object, which is what Peirce calls "intuition" and what

Russell would call "acquaintance." Every cognition or "sign," according to Peirce, is derived from or "determined by" previous signs. Proto-abduction, for its part, involves the sudden but possibly evanescent emergence of a pattern or connectedness among previous thought signs—as for example in the "filling-in" of the blank spot in the retina, or in the "seeing" of a moving object as the result of the extremely complicated pattern of stimuli provided by an image moving over the receptors of the retina. Proto-*induc*-tion results in the maintenance and reinforcement of such a pattern, as occurs in practising the piano, or in its dissolution, as occurs when the expectation generated by a belief is not fulfilled. Proto-deduction, finally, consists in acting in conformity to a pattern or general rule when the appropriate occasion presents itself; it is the reaction of the mind under the dominion of a habit or idea to a stimulus "recognized" as involving that idea, a sort of unconscious analog of *modus ponens*, we might say. Examples of proto-deduction would include the salivation of Pavlov's dog and other cases of stimulus-response both learned and biologically "wired-in."

We might here (briefly) entertain the hope that we can shed light on the difference between deductive and non-deductive modes of thought by studying the differences between proto-deduction and the other modes of proto-reasoning and in general we might hope to better understand all three modes of reasoning by considering their uncontrolled analogs.

There are certainly abundant cases of processes which show strong analogy to logically self-controlled thought. Even beyond the limits of the individual human person, Peirce saw analogs of inference at work in, for example, the evolution of human institutions, weights and measures, and animal species. Induction is identified with a Lamarckian-style evolution or *agapasm*—where change is the result of a striving after or "love" of the result. Abduction, in turn, is identified with the spontaneous variations required by Darwinian evolution or *tychasm*. (The selection component of the Darwinian process would have to be considered as inductive, here—in spite of not being very lovely or agapastic.) And deduction, finally, is associated with catastrophine evolution (associated with the names of Cuvier, Clarence King, and Weismann) or *anancasm*—which is change brought about by some kind of *necessity*.[10]

The brief presentation just given here of Peirce's general theory of mental action, both controlled and uncontrolled, is of course somewhat sparse and schematic, so that the reader may be pardoned for not seeing in it the kind of promise which the author and perhaps Peirce himself once saw in it. The point to be made here, however, is that whatever promise we may see in this general theory of mental action of providing a better account of the difference between deductive and non-deductive inferences than that obtained by examining argument-forms is largely an illusion. This is because the basis for the classification of non-self-controlled modes of

thought is itself derived from the normative study of self-controlled inferences. Furthermore Peirce clearly says (in 1902—cf. 2.27) that to use an
argument to represent an uncontrolled inference is a defective and misleading representation of it, since recourse to an argument, being essentially a defence and justification of an inference, presupposes that the inference is subject to self-control and thus in a special sense of "*could*," that
we *could* have reasoned otherwise—or at least that in the future we could
reason otherwise. Thus logic casts but an imperfect light on psychology,
although as Peirce suggests in his 1868 articles we perhaps have no other
light to cast upon it—other than that shed by *physiology*. Psychology has, in
other words, only reflected light to cast back on our mastery of logical form.
So much then for our study of mental process *qua* process.

It is now time to look again at our battered and somewhat tattered definition of deductive reasoning as involving the production (and observation—since this aspect has only been provisionally "put on hold") of a diagram, with or without intervening manipulations and alterations. We also
formulated a series of three objections to the close association of deduction
with the production of a not-so-distinctive sort of sign. We are now in a
position to try to answer them.

We deal first of all with the third of these objections which raised questions concerning whether the proposed definition of deduction does not
have the bizarre result of making it hard to distinguish deduction from
simply talking and writing.

Our answer to this objection will be in two parts. In the first part, we
admit that the production of diagrams for whatever purpose is a species
within the genus of sign-production in general. We further admit, thus taking by the horns the attempt to saddle us with a *reductio*, that speech and
writing can be viewed as proto-deductive activities, since they are governed
by usage, which can at least potentially be described by rules. Where the
consciousness of the rule applied is slight or absent, we will say that we are
dealing with an uncontrolled analog of deduction. Where the application
of the rule is conscious, particularly where there is hesitation or deliberation, we clearly have a case of deduction properly so called. On this view,
then, there is no particular problem or mystery involved in allowing the
that the production of diagrams belongs to the same class of activity as
these.[11] The shading off of deductive sign production into proto-deductive
spontaneous speech or writing is analogous to the shading off of abduction
properly so called into proto-abductive perceptual judgments noted by
Peirce in his Lectures on Pragmatism of 1903 (cf. 5.183).

As a *second* reply to our third objection, we point out that diagrams
produced for the more deliberate variety of deductions are typically and,
we might add, preferably, much more *iconic* in nature than the usual run
of sign-production. This iconicity may be mediated, in fact it must be me-

diated, by various conventions, as is the case even in the Existential Graphs. But for Peirce, the more iconic the representation the better—with the caveat that it is better that the diagrammer should, as far as possible, refrain from representing details which are not *relevant* to the matter at hand. In other words, the diagram ought to be both as iconic as possible and as vague as possible. The result is the greatest *perspicuity*.[12]

In reply to the second of our three objections to defining deduction in terms of the production of a diagram, it would now seem possible to differentiate between the *abductive* production of a sign, on the one hand, and the process of "examin[ing] the state of things asserted in the premises, [and] form[ing] a diagram of that state of things" (1.66) involved in deduction, by simply pointing out that abduction produces a sign which adds an element of unity and connectedness which the premises of the objects referred to in the premises would not have had otherwise. This is quite contrary to what is aimed at in producing a diagram for purposes of deductive inquiry.

Are we ready now, finally, to say what deduction *is* and to differentiate it from non-deductive reasoning? Perhaps not entirely, for much remains to be said, but I will give it a try since we need a provisional hypothesis as a guide. My proposal may also go some distance towards a reply to the *first* of our three objections, concerning the counter-intuitiveness of what we seemed to have discovered about what Peirce's definition of deduction has to be. Briefly, our proposed answer is largely based on Peirce's later insights on *self-control*: deliberate reasoning is a species of doing. This is why logic is a province of ethics and perhaps also of what is today called "action theory." Now, contrary to what is commonly taken to be the case in the description of merely physical events over which we have no control and for which we assume no responsibility, the description of an *act* or an *action* requires some mention of its *purpose*, and some mention of the *means* envisaged. This idea is at least as old as Kant's ethics. The distinctive purpose of abduction is the perception of possible interconnectedness among thoughts. Induction, on the other hand, is engaged in for the purpose of acting upon the degree to which we will *act* in conformity to a thought or a sign. Deduction, finally, has the role and purpose of rendering our ideas clearer in the sense of being more distinct and explicit with a view to injecting order into whatever inductive testing we decide to undertake. Thus deduction is an important tool in the self-control of induction and even abduction.

The diagrams produced (and used) by deductive reasoning are *means* towards ends. Practice (reference to which is required by the third degree of clarity) revolves around *means*, certainly, but some reference to *purpose* is required if the description/definition is to be complete, or (completeness being a somewhat presumptuous ideal) if it is to at least *approach* completeness. Purpose, we might add, in the case of diagrams used for deduction, is "in the eye of the beholder"—hence the emphasis in 2.267 (quoted earlier) on what the *Interpretant* of the diagram produced must do in order for what is going on to count as deduction. This is the "observation" of the diagram which we had provisionally eliminated as the locus of what was distinctively deductive. We now see that *both* of the elements of corollarial deduction earlier distinguished are essential. Deduction is defined by the unique combination of means and ends, of diagram and interpretant. The crucial point turns upon how the argument or diagram is "studied."

I began this chapter by suggesting that definitions of deduction in terms of the barest notion of deductive validity (e.g., like the one which defines "demonstrative reasoning" as "pretend[ing] to be such that it is logically impossible for the premises to be true while the conclusion is false" (2.447)) achieve at most the *second* level of clarity. I went on to suggest that the third level of clarity might be obtained by a definition referring to the use of *diagrams* as a *means*. My study of this possibility, however, indicates that, in addition, some reference is required to the *purpose* for which the "practice" of deduction is engaged in. Obtaining order and arrangement among the various elements of what we know is certainly part of the story, although this is a purpose which deduction shares with the practice of *definition*. It is suggested that in deduction this purpose of order and arrangement is perhaps subordinate to the concern to maximize *perspicuity*—although this word occurs with surprising infrequency in the Peircean corpus. The principal insight here has to do with the importance of perspicuity for logical self-control and the concern with the "defence" of an inference inherent in the deductive use of diagrams. A less obvious but no less important point has to do with the use of deduction in the self-control of induction and abduction.

NOTES

1. Cf. Note 8 below.
2. There is a similar problem with regard to the language of perception: during the performance by the magician, did I "see" a headless man? Or was what I "really saw" a man with his head in a black bag against a black background in circumstances where the lighting was purposely made deceptive? Cf. J. L. Austin, *Sense and Sensibilia* (1962).
3. Cf. *Ludwig Wittgenstein 1972*. This posthumous manuscript is mainly a reply to G. E. Moore's famous articles "Proof of the External World" and "Defence of Common Sense"—and contains a number of paragraphs which, on the surface, contradict what Peirce says about the possibility of being mistaken about "2 + 2 =

4" (in Wittgenstein's case "12 × 12 = 144"); this is the case with *some* of the following: §§ 38, 42, 43, 155, 193ff., 303f., 426, 447f., 650–655.

4. Anthony Quinton, *The Nature of Things 1973*. See especially chapter 6 ("Certainty"), pp. 143–171. Before tackling the question, "Are basic statements certain?", Quinton distinguishes five senses of the word "certain" "as it appears in philosophical discussions" (pp. 143f): (1) characterized by *psychological* indubitability (i.e., by subjective assurance or the incapacity to believe otherwise; but what is "certain" in this sense is perfectly capable of being false); (2) logically necessary (but necessary truths are capable of being doubted); (3) self-authenticating (as in the case of statements such as "I think," "I am writing in English," etc.—which constitute a different kind of puzzle than Descartes thought); (4) incorrigible (which is the philosophically interesting sense, perhaps; in any case fallibilism maintains there *are* no incorrigible beliefs—cf. pp. 150f); (5) beyond all reasonable doubt (which is the "ordinary" sense of the word). Many philosophical discussions are vitiated by confusions between these different senses of "certain."

5. To give just one example, a "faulty argument" by H. Lebesque, in *1905*, was "discovered in 1917 by Suzlin who then proceeded, together with Luzin to create the theory of A-sets" (*Kurepa and Schön 1974*: 465; cf. also p. 465n.33, n.34).

6. It might seem that we have suddenly switched here from talk of deductive arguments to talk about necessary truths. This is so, but in order to avoid quibbles here, it should be mentioned that by an operation known as "conditionalization" it is always possible to turn a valid deductive argument into a theorem of logic or necessary truth of the form "If *P*, then *C*," where *P* is the conjunction of all the premises and *C* is the conclusion. Thus what was merely an *argument* becomes a *statement*. If the argument is deductively valid, then the statement is necessarily true. But this is simply a reflection of the fact that the *consequent* of the statement can be *deduced* from its *antecedent*—which is what the deductive reasoning represented by the argument established in the first place. Thus, in spite of the existence of such things as theorems of mathematics and logic which seem to be affirmed both categorically and as necessarily true, a reference to logical necessity always involves some reference to what can be deduced from one or more premises in a deductive argument.

7. Since this quotation has been somewhat gerrymandered, here is a more complete version of the passage in question: "[T]he whole function of thought is to produce habits of action. . . . Thus we come down to what is tangible and conceivably practical, as the root of every real distinction of thought, no matter how subtle it may be; and there is no distinction of meaning so fine as to consist in anything but a possible difference of practice" (5.400).

8. This extension of deduction to inferences involving probability could be the occasion for wondering whether what Peirce calls "deduction" has not invaded the domains of abduction and induction, completely undermining his tripartite distinction between the modes of reasoning. The answer to such an objection is that so-called "probable deductions" obviously belong to what we would call the *mathematical* theory of probability and possibly also to the *applications* of such a developed theory to concrete cases (cf., e.g., 5.145, 6.595). Thus the *a priori* and diagrammatic character of such inferences remains a property distinguishing them from induction and abduction proper, as does the circumstance that the premises diagrammed would represent an "ideal" state of affairs which need not necessarily be supposed to obtain in reality.

9. The reference is to Peirce's first lecture on "Detached Ideas on Vitally Important Topics," bearing the title "Philosophy and the Conduct of Life" (cf.

1.616n.). I did not find the words "rashness" and "prudence" in the text—although Peirce does address himself to the question of what is "sensible" (cf. 1.677).

10. Cf. 6.12–17 [1891], 1.104, 6.302–307 [1892], 1.105–109 [c.1896]. Cf. Also my *1970*: 10–17, and 22ff., and my *1970*: 3–34. Peirce seems to have coined the term "catastrophine."

11. Deduction thus requires a diagram, which in turn requires a deduction, which in turn requires a diagram. . . . There is no first cognition.

12. In his *Logic Notebook* of 1865 (*MS* 117), Peirce writes (entry of Dec. 14): "Hegel makes a great boast of the fact that his Logic develops [*sic*] its own method. Mine pursues a rational method of which the logic itself is but the deduction and proof. Moreover I am not forced to make my book unintelligible in order to follow mine, but on the contrary it is the very procedure which *perspicuity* demands" (W1: 340).

24.

PEIRCE AND THE STRUCTURE
OF ABDUCTIVE INFERENCE

Tomis Kapitan

§1. Introduction

According to Peirce, abduction is "the process of forming an explanatory hypothesis" (5.172) which "must cover all the operations by which theories and conceptions are engendered" (5.590) including not only the *invention* of hypotheses but *selection* of them for further consideration:

> The first starting of a hypothesis and the entertaining of it, whether as a simple interrogation or with any degree of confidence, is an inferential step which I propose to call *abduction*. This will include a preference for any one hypothesis over others which would equally explain the facts. (6.525)

Peirce also described abduction as the *only* logical mechanism for originating new ideas and hypotheses (5.171), the most important kind of reasoning (L231: 59), the only hope of arriving at truth (2.786), as resting upon an instinctive insight into Thirdness (5.173), and as distinct from either induction or deduction (5.146). Any of these descriptions is grist for an independent study, but let us focus upon certain theses characteristic of abduction in its scientific setting and prominent in his post-1900 writings on the topic.[1]

Inferential Thesis. Abduction is, or includes, an inferential process or processes (5.188–189, 7.202).

Thesis of Purpose. The purpose of "scientific" abduction is both (i) to generate new hypotheses and (ii) to select hypotheses for further examination (6.525); hence, a central aim of scientific abduction is to "recommend a course of action" (MS 637: 5).[2]

Comprehension Thesis. Scientific abduction includes *all* the operations whereby theories are engendered (5.590).

Autonomy Thesis. Abduction is, or embodies, reasoning that is distinct from, and irreducible to, either deduction or induction (5.146).[3]

The last of these, stemming from Peirce's threefold distinction among the elementary kinds of reasoning which he once called the "Key of Logic" (2.98), is among the most fascinating yet perplexing of his claims. What is its rationale, and for the tripartite division itself? Regarding *what* factor(s) are the kinds of reasoning distinguished in a way that does not permit a fourth? The *explicative/ampliative* and *necessary/non-necessary* contrasts are inadequate for the task, and while one might think that Peirce's very definitions of deduction, induction, and abduction settle the matter straight-away, they alone do not preclude subsuming abduction under one of the other kinds.[4] Furthermore, demarcations in terms of methods, purposes, forms of argumentation, and conclusion-types, sanction a multifold classification (see Section 6). Let us try to resolve the matter with a closer look at the structure of abductive reasoning.[5]

§2. Inference

To understand abduction it is necessary to become clear about Peirce's view of inference. Contrary to many, he did not use "inference" to refer to an ordering of propositions, rather, to an act of reasoning described as "the conscious and controlled adoption of a belief as a consequence of other knowledge" (2.442, 2.144, 5.109), which "consists in the thought that the inferred conclusion is true because *in any analogous case* an analogous conclusion *would be* true" (5.130). Its aim "is to find out, from the consideration of matters and things already known, something else that we had not known before" (MS 628: 4). The terms "inferring" and "reasoning" are usually used synonymously by Peirce, though at one point he indicated that inference is just a single step in the reasoning process. Despite this statement, inference and reasoning are characterized almost identically there (*Baldwin 1901*: I, 542; II, 426) and elsewhere.

Besides suggesting a premiss-conclusion connection, phrases like "as a consequence of" and "because" in these passages have a *causal* significance in that inference "produces" or "creates" a belief in the mind of the reasoner (2.148). Peirce described a *belief* as "holding for true," or "any kind of holding for true or acceptance of a representation" (NEM 4: 39–40), and "to say we really believe in the truth of any proposition is no more than to say we have a controlling disposition to behave as if it were true" (MS 652: 15).[6] However, this characterization poses a problem, for Peirce also contended that abduction is not a matter for belief (5.589), that belief is out of place in science (5.60, 1.635), and that non-truth-valued items can be inferred (MS 293: 37). Conveniently, he also spoke of inference in terms of *acceptance*, which seemed to be a broader type of assent or "favorable attitude towards" a proposition with which positive belief was sometimes contrasted (MS 873: 23). As such, while an abductive inference might not be productive of a full-fledged belief, it can still result in an acceptance. The choice of terms is important if inference is to be defined so as to achieve requisite generality and, accordingly, "acceptance" will be given the nod.

Three steps are essential to any inference: colligation, observation, and judgment (MS 595: 35). Inference begins with *colligation*, the conjoining of distinct propositions and asserting the whole (2.442–443, 5.579).[7] One then deliberately contemplates the colligated data (MS 595: 30/7.555) which, in turn, results in an *observation* that the conclusion *C* will be true if the premiss *P* is. Next follows the *judgment* which includes an acceptance that what is observed in the premises yields, by following a rule, that conclusion and, so, an acceptance of the conclusion (7.459, 2.444). Control is exercised over the judgment, colligating and contemplating, but not over the observing (7.555), and it is control that subjects inferences to norms. An inference is valid or not according as it follows a method (norm) which is conducive to satisfying the aim of reasoning, viz., the acquisition of truth.

Let us summarize the more central features of inference in the following theses.

(1) Inference is a conscious, voluntary act over which the reasoner exercises control (5.109, 2.144).[8]

(2) The aim of inference is to discover (acquire, attain) new knowledge from a consideration that which is already known (MS 628: 4).

(3) One who infers a conclusion *C* from premiss *P* accepts *C as a result* of both accepting *P* and approving a general *method* of reasoning according to which if any *P*-like proposition is true so is the correlated *C*-like proposition (7.536, 2.444, 5.130, 2.773, 4.53–55, 7.459, L232: 56).

(4) An inference can be either *valid* or *invalid* depending on whether it follows a method of reasoning it professes to and that method is conductive to satisfying the aim of reasoning, namely, the acquisition of truth (2.153, 2.780, 7.444, MS 692: 5).

Every inference involves an acceptance of a conclusion caused by an acceptance of the premises together with the reasoner's *application* of a general method of inference pattern. The latter reflects a habit encoded in a *leading principle* of the inference (2.588–589) and included in the agent's *logica utens* (2.186; 5.108, 130). Logic as "critic" studies the correctness of methods for establishing beliefs (3.429), and the inference patterns it proposes are normative in indicating what one is *warranted in accepting*. Increasingly,

Peirce felt that logicians should explore the conditions under which non-necessary reasoning is valid (MS 660: 3).

§3. Abductive Form

Peirce's 1878 model of reasoning by "hypothesis" from result and rule to case (2.623) is familiar:

(F1)   All *A*s which are *B* are *C.*   (Rule)
         This *A* is *C.*   (Result)
         Therefore, the *A* is *B.*   (Case)

Abduction differs from deduction which moves from rule and case to result and from induction which goes from case and result to rule, and it is this tidy contrast in terms of syllogistic permutations that very likely led Peirce to trichotomize inference in the first place.[9]

The contrast, and with it, (F1), are superseded in his subsequent work (2.102; MS 441: 30, 475: 13). (F1) restricts the types of statement that qualify as premises and conclusions, and unlike valid deductive forms, does not guarantee a good abductive inference. In refining it, Peirce stressed that the novel hypothesis is envisioned in the course of explaining some surprising phenomenon (the "result") and that what one is entitled to infer is something less than the hypothesis itself. In his 1903 lectures on Pragmatism, he tentatively presented a different pattern to which abductions ought to conform:

(F2)   (1)   The surprising fact, *C* is observed;
       (2)   But if *H* were true, *C* would be a matter of course, hence,
       (3)   there is reason to suspect that *H* is true.[10]

In depicting inferences, terms like "therefore" and "hence" must be understood as having both a causal and a normative force, indicating not only a transition from some acceptings (beliefs) to another, but also what would be a permissible transition. In depicting inference *forms*, however, these terms serve only the latter function. (See *Harman 1986*: 1–10 for more on the distinction between reasoning and argument.)

At one level of abstraction, the premises of (F2) accurately portray scientific procedure; research typically begins with a problem (puzzling or surprising phenonomenon) and aims at a solution in terms of an explanatory hypothesis. Yet, the form is problematic. For one thing, it does not seem valid; there are any number of "wild" hypotheses that would explain why you are now reading this paper were they true, but that alone does not provide reason to think that they are true (*Achinstein 1970*: 92). More is required to legitimately conclude, select, or make plausible a hypothesis than what (F2) reveals and, if anything, the "logic" of abduction should bring this out. Peirce did offer more. Keenly aware that there are numerous hypotheses which would explain any given fact were they true (5.591), he viewed it as a "serious problem" whether a given hypothesis should be entertained at all (6.524). Discussing this second task of abduction—*viz.*, hypothesis selection or determining a hypothesis' pre-testing *merit*—under the labels of "economy" and "economy of research" (L75: 284; 7.218–223), he claimed that to abductively infer *H* to be reasonable is justifiable only if *H* is economical, or, to be precise, more economical than competing hypotheses. The result is that an accurate rendition of abductive form must include a further premiss that *H* is more economical than its (envisioned) competitors.

   (F2) is also misleading in the form of its conclusion. Already by 1901, Peirce had written, "By its very definition abduction leads to a hypothesis which is entirely foreign to the data. To assert the truth of its conclusion ever so dubiously would be too much" (MS 692: 26). He offered a number of weaker alternative conclusion-types, and after 1908, a claim concerning the "plausibility" of the hypothesis was among the more prominent candidates, where a hypothesis is *plausible* if it is explanatory, testable (though not yet tested), and "of such a character as to recommend it for further examination" (2.662). In more exact terms, to match the comparative nature of judgments of economy and capture the idea that plausibility comes in degrees (8.223; 6.469, 480–488; 2.662), the proper abductive conclusion is a comparative assessment of plausibility. The upshot is that (F2) yields to,

(F3)   (1) Some surprising fact *C* is observed.
       (2) If *H* were true then *C* would be a matter of course.
       (3) *H* is more economical than its envisioned competitors.
       Hence
       (4) *H* is more plausible than its envisioned competitors.

For the moment, let us consider (F3) as a more accurate picture of what Peirce had in mind when he spoke of a normative format according to which the selection of hypotheses for further examination should take place.[11]

§4. The Generation of the Premisses

If we take Peirce's Comprehension Thesis seriously and include all that is relevant to the generation of hypotheses under the heading of "abduction," then (F3) represents only the last and final phase of abductive reasoning. Before one can so "select" a hypothesis from among its competitors, considerable thinking must go into the generation of the premisses themselves, that is, into the discovery of surprising data, the derivation of explanatory conditionals containing novel hypotheses, and the comparative assessment of rival explanations. Peirce had a good deal to say here which a complete account of his theory of abductive reasoning must incorporate. By examining the reasoning behind each premiss, perhaps we can determine whether the Autonomy Thesis might be anchored somewhere in these earlier stages of the abductive process.

Premiss (1). Considerable thinking goes into the delineation of problems, requiring the researcher to draw upon a body of background assumptions as well as relevant perceptual data. Abductory thinking begins with juxtaposing the unfamiliar (the surprising fact) with the familiar (the background) (L75: 286–287; 7.188). The claim that C is *surprising* is intelligible *only* against a body of background expectations, illustrating that the premiss cannot be understood without an implicit reference to that background. Indeed, an isolated fact, not contrary to what is expected, calls for no explanation at all (7.192–201).

The judgment that a phenomenon is surprising follows upon the observation that it is contrary or improbable given what is expected from previous information (L75: 177; 2.776, 7.188–200), hence, in need of explanation. This judgment, in other words, is inferred, and knowledge begins with this "discovery" (7.188).[12] What *kind* of reasoning is involved? Colligation *qua* conjoining, is deductive. Recognition of what is *contrary* to what is expected or *improbable* given what is expected, is determined by focusing upon the *relations* between premisses and conclusion. To "see" that P is contrary to Q is to realize that Q implies $\sim P$, in which case the observation underlying a judgment of contareity is exactly the sort underlying any necessary deduction. If P is improbable given Q, then the observation is one which gives rise to a probable deduction. In either case, the inference that the phenomenon *is* surprising, in *want* of explanation, has every appearance of being deductive.

Premiss (2). A breach of expectation stimulates a demand for explanation 7.191) which, if successful, results in acceptance of an explanatory conditional. Peirce suggested that our acceptance that C would as a matter of course be explained by H, is itself the product of inference, e.g., in discussing his paradigm example of abduction, Kepler's discovery of elliptical orbits (2.96, and see also 1.72–74). There is some question, however, whether the initial conceiving of H qua antecedent arises as a *result* of inference. Here he wrote that the abductive suggestion comes to us "like a flash," of its being an act of "instinctive insight" tending to make us guess correctly nature's laws (5.604, 181), or abduction itself as being "neither more nor less than guessing" (MS 692: 24), and of guessing as being an instinctive power (6.491, 7.48). At the same time, he emphasized that one guesses *on the basis of* other information (MSS 692: 27–36, 595: 37); we distinguish good or "reasonable" guesses from poor ones (MS 873: 11), and given that there are "trillions" of possible hypotheses to explain given facts, we expect the operation of rational constraints warding us away from idle and fruitless guesses (5.172, 591; 7.38). Hypotheses are not generated fortuitously (MS 475: 20; 7.48, 6.476). Such remarks point to the inferential nature of guessing, implying that it is under our control.[13]

Peirce attempted to resolve this apparent clash of intuitions regarding instinct and inference in his 1903 Pragmatism lectures, suggesting that the hypothesis initially emerges in the *observation* of the colligated whole through an uncontrolled insight into the world of ideas, i.e., into Thirdness as given in perception (5.150, 160, 173, 209–212; 7.198). The self-control typical of inference becomes relevant when one decides to "adopt" the inference, thereby judging the conclusion to be true since it is a consequence of the information one began with. In itself, however, self-control of any kind is purely "inhibitory" and "originates nothing" (5.194). Thus, while the novel conceiving of premiss (2) is *caused* by prior cognitions, and its content "suggested" by the facts, not everything suggested is inferred from cognition of the facts:

> There are, as I am prepared to maintain, operations of the mind which are logically exactly analogous to inferences excepting only that they are unconscious and therefore uncontrollable and therefore not subject to criticism. (5.108)

Such operations occur when one cognition is determined by another without our being aware of it, and should properly be called "associational suggestions of belief," not "inferences" (5.441). The observation so engendered is the creative and instinctive insight that given what one already knows, *H* will explain *C.* But the *guess,* the "deliberate acceptance" consequent upon examining the suggested connection (MS 451: 18), is the reasoned *adoption on probation* (MS 692: 36) of the explanatory conditional. So, while the initial conception of the conditional premiss is not a result of inference, acceptance of it is.[14]

The guessing that typifies *hypothesis generation* may well be deductive. The very phrase "as a matter of course" in premiss (2) indicates a degree of intuitiveness, a point underscored by the fact that an *explanatory* conditional conveys a connection of necessity or high probability (8.231, 7.36). Peirce emphasized that once a hypothesis has been identified, it is a matter of *deduction* to establish *further conditionals* in which the hypothesis is the antecedent and predictions are the consequents (7.115n.; MS 473: 9–10). Ideally, among these predictions will be those whose very "incredibility" makes their consequents similar to the "surprising" fact *C.* It then appears that the inferential acceptance of the explanatory conditional—the *judgment*, not the creative *observation*—results by a deductive transition from the background assumptions. Finally, it is striking that Peirce described *theoric* deduction as a "power of looking at facts from a novel point of view" (MS 318: 42), which is "very plainly allied to Retroduction" differing only in being "indisputable" (MS 754: 8).[15]

Premiss (3). Several considerations must be taken into account before a hypothesis may be "chosen" for further examination (7.219). Obviously, the hypothesis must be explanatory if true, and, similarly, testable (7.220), but under the heading of "economy" Peirce also included (a) the cost (in time, money, and effort) of testing the hypothesis (6.533, 7.230); (b) the intrinsic value of the hypothesis in terms of its "naturalness" and "likelihood" (7.223); (c) the fact that the hypothesis can be readily broken down into its elements and studied (MS 692: 33); (d) the hypothesis' simplicity (i.e., that it is more readily apprehended, more facile, more natural or instinctive) (L75: 286; 6.532, 477); (e) the breadth of the hypothesis or the scope of its predictions (L75: 241; MS 457: 37); (f) the ease with which the hypothesis can be falsified (L75: 285); (g) the testability of the hypothesis by means of severe tests based on "incredible" predictions; and (h) the hypothesis' analogy with familiar knowledge (MS 873: 16).

The evaluations that emerge are *comparative*, often phrased in terms of *preference*, in which case a judgment of economy is itself comparative (7.220–231; see also NEM 4: 37–38, L75: 285–286, 2.786, and MS 475: 37). The underlying inferences are best scrutinized by means of the standard logic of preference or decision-making in which deductive considerations abound, especially in eliminating hypotheses by means of disjunctive syllogisms (7.37). But the reasoning to the premisses of any such deduction are likely to be inductive, in the mode of qualitative inductions. How do we know, for instance, if H is more cheaply testable than H'? We consider the kind of hypothesis that H is, and based on our familiarity with past testing procedures, we reason inductively that H is the kind of hypothesis that could be examined more cheaply, more quickly, etc. Similar inductive considerations would permeate decisions based upon the other criteria as well.

§5. Plausibility and Abductive Form

If grounds for the tripartite division of reasoning and the uniqueness of abduction cannot be located in the inferences that generate the premisses of (F3), attention must shift back to the final phase of abductive thinking, hypothesis selection, whether of form (F3) or otherwise. Is there any reason to think that a unique kind of reasoning is manifested at this juncture?

A more rigorous scrutiny of (F3) is needed. According to the Thesis of Purpose, the aim of every abduction is to "recommend a course of action." How does (F3) do this? Judging from what Peirce says, one might suppose that its conclusion implies,

> (a) There is reason to suppose that *H* is worth pursuing (examining further).

Yet this statement falls short of sanctioning a *recommendation*; if a plausible theory has a character to "recommend it for further examination" (2.662), what is needed is something stronger than a claim that there is "some" reason, or "*prima facie*" grounds for so doing (*Curd 1980*: 214, and *Achinstein 1987*: 433). To capture the sense in which the hypothesis "merits" inductive examination (2.786, 8.223), an (apparently) implied normative is more forceful:

> (b) One ought, insofar as one desires an explanation of *C*, examine *H* further.

But although a normative is often used to *support* a practical directive like a recommendation, it is not *identical* to a directive. To make explicit the sense in which an abduction "recommends," (b) must sanction (imply) something like,

> (c) *H* is recommended for further examination.

Or, if reference to an agent is required,

> (d) It is recommended, for one who desires an explanation of *C*, to further examine *H*.

Again, a recommendation can be phrased interrogatively, and since Peirce often claimed that the hypothesis should be asserted only as an "interrogation," implying by this that the hypothesis merits testing (MS 692: 26), the abductive conclusion might equally be,

> (e) Why not pursue *H*?

addressed to oneself or others. Or, phrasing the hypothesis itself interrogatively (say, through the appropriate "wh-movements" of the grammarians), practical import can also be conveyed by,

> (f) *H*?

together with appropriate stress. Each of (c)–(f) are suitable forms of practical directives, though for convenience, the schematic (d) will be treated as canonical.

A strict literal reading of the Thesis of Purpose would suggest that (F3) be replaced by an explicitly "practical" inferential pattern formed by adding a second practically-oriented conclusion as follows:

(F4) (1) Some surprising fact *C* is observed.
      (2) If *H* were true then *C* would be a matter of course.
      (3) *H* is more economical than its envisioned competitors.
      Hence,
      (4) *H* is more plausible than its envisioned competitors.
      Hence,
      (5) It is recommended, for one who desires an explanation of *C*, to further examine *H*.

Any of the other suggested directives would equally suffice, and the actual chain of reasoning might also include the normative (c) between steps (4) and (5). Perhaps Peirce had something close to (F4) in mind when he wrote that abduction "commits us to nothing. It merely causes a hypothesis to be set down upon our docket of cases to be tried" (5.602). The phrases "to be set down" and "to be tried" reflect a *decision* concerning future action, and provide one interpretation of the "acceptance" in which a practical inference culminates.

However, (F4) also faces difficulties. For one thing, though the concluded recommendation of form (5) in (F4) might be appropriate given the information in the premises, it is not truth-valued, in which case there is no obvious format for viewing (F4) as valid within Peirce's construal of validity in terms of truth-producing virtues. For another, (F4) does not square with all the texts. Peirce constantly spoke of a hypothesis *itself* being "adopted" as a result of an abduction, albeit "problematically" (2.777) or "on probation" (MS 873: 22). These passages suggest that a special *mode of acceptance* is operative in correct abductive reasoning, a "probationary" mode which, falling short of belief in the hypothesis (5.60, 589) is still a favorable attitude towards it. With the hypothesis itself as a conclusion, the correct pattern of inference is something like,

(F5) (1) Some surprising fact *C* is observed.
      (2) If *H* were true the *C* would be matter of course.
      (3) *H* is more economical than the envisioned competitors.
      Hence,
      (4) *H* is more plausible than its envisioned competitors
      Hence, probationally,
      (5) *H*

Here, the qualifier "probationally" indicates that what is warranted is mere probational adoption, not the sort of *belief* appropriate to deduction and induction (5.170, MS 754: 5).

What sort of "acceptance" is probational adoption? Peirce was reluctant to say that abduction results in a full-fledged *belief*, yet minimally, probational adoption would seem to show some degree of favor towards the hypothesis. Perhaps the attitude stands to plausibility as belief does to truth, i.e., it concerns holding a hypothesis to have, not truth, but a comparatively high degree of plausibility (2.776). Given the normative character of plausibility, accordingly, accepting a hypothesis as plausible is not to believe *it*, rather, to accept that it ought to be pursued. However, were probational adoption of H *identical* to accepting that H is more plausible than its competitors—thus, that it ought to be pursued—then the transition from (4) to (5) in (F5) would be redundant, since it would duplicate what one is already warranted in doing in moving to step (4). The result would bring us back to (F3).

Keeping probational adoption of H distinct from either belief in H or acceptance that H ought to be pursued, leaves the possibility that it is a practical attitude, for instance, a willingness to submit H to further test. What one then accepts, strictly speaking, is a course of action, not a theoretical hypothesis, and the relevant mode of acceptance is more akin to a recommending, an intending, or to what Peirce himself called a *resolve* (5.538, 1.592). He explicitly allowed that one might entertain a proposition in ways other than belief:

> One and the same proposition may be affirmed, denied, judged, doubted, inwardly inquired into, put as a question, wished, asked for, effectively commanded, taught, or merely expressed, and does not thereby become a different proposition. (NEM 4: 248)

A willingness to examine H, then, is a favorable attitude towards H, and consequently, a readiness to act in certain ways, perhaps not the "controlled and contented" disposition to act that *belief* is (NEM 4: 249); only a readiness to submit H to testing. But if this is all the attitude amounts to, it is difficult to distinguish a *probational adoption of H* from a *willingness to further examine H*, in which case (F5) would not differ from (F4), since the favorable attitude towards H warranted by (F5) would reduce to the warrant for adopting a recommendation.

The most likely means of forcing a distinction between (F5) and (F4) is to insist that probational adoption is a low degree of belief (MSS 475: 42, 652: 11–16), and that when Peirce said that abduction is not a matter for belief he meant a relatively high degree of "positive" belief (MS 873: 23; see also 2.662 where a *high* degree of plausibility to a hypothesis justifies us in "seriously inclining towards belief in it"). The matter is difficult to adjudicate by appeal to the texts, and it is doubtful that Peirce had a clear and consistent picture of what is to be inferred in an abduction, or, for that matter, a clear distinction between belief and resolve (intention).

Whatever the final assessment, the very fact that the two approaches represented by (F4) and (F5) are intertwined within Peirce's writings gives some grounds for speculating on their connection. If we insist on their distinctness, one way to combine the probational acceptance of a hypothesis with the desired practical force of abduction is by supporting the following equivalence:

> A probational acceptance of *H* is warranted iff the recommendation to further examine *H* is appropriate

in which case (F5) is a *valid* argument iff (F4) is.

§6. The Tripartite Division of Reasoning

Returning to our main concern, how can the three kinds of reasoning be distinguished so as to anchor the Autonomy Thesis and exclude a fourth? Since the answer to the question cannot be located in the reasoning that generates the abductive premises (Section 4), we must reexamine hypothesis selection. Before doing so, however, notice that each of Peirce's attempts to distinguish reasoning in terms of distinct stages of inquiry (5.171, 6.475), purposes (2.713, 6.472), and forms (2.619–644) fails to secure the tripartite division. There are, for instance, at least four distinct "stages" of inquiry within abduction alone, and the division of the three kinds of induction (2.755f.) can appeal to both forms and purposes, as can the probable/necessary and corollarial/theorematic contrasts within deduction. Thus, none of these modes of demarcation establishes the division in a way that would preclude a fourth, fifth, etc.

More promising alternatives concern, first, the different ways in which each kind of reasoning can be "justified," and second, the different sorts of "logical goodness" (validity, soundness, excellence of argument) belonging to each. *Justification* of a kind of reasoning concerns whether the procedures it employs actually promotes attainment of the goals for which it exists. Hence, what is to be justified is a particular manner of inferentially accepting one claim on the basis of others. But this depends upon the existence of an *aim* to serve as a basis for justification, since the question of "ought" has no meaning except relatively to an *end* (5.594). For Peirce, the end of reasoning *per se* is the gaining of truth or knowledge (2.153), especially, knowledge of something not previously known (MS 628: 4). Moreover, we "are always justified in presuming, for the purposes of conduct, that our sole end may be reached," otherwise we would be immobilized (L75: 271; compare MS 634: 9–10). Intentional action presupposes a hope, if not a likelihood, or success, apart from which justification is meaningless. The reasoner must assume that knowledge is attainable; it is the logician's task to determine the general conditions for such attainment (NEM 4: 196) and, accordingly, to distinguish among the different sorts of logical goodness, i.e., "truth-producing" virtues.

Peirce provides different justifications for the three kinds of inference. A deductive inference is justified (when performed through valid patterns) because it never leads away from truth; its justification is based upon the necessity with which the conclusion follows from the premises (2.778). Induction is justified if performed correctly (e.g., by conforming to rules of sampling) (2.725–740), because its conclusion "is reached by a method which, steadily persisted in, must lead to true knowledge in the long run of cases of its application" (7.207). This, in turn, is an expression of the "self-correcting" power of induction, whereby it would eventually correct any single inference to an erroneous conclusion (5.145, 2.769).[16]

Abductive inference can be justified in neither of these ways; it provides no guarantee or high probability of generating truth from truth, and its method might never lead to the truth in the long run. How can it be justified in terms of the ultimate end of reason? Peirce offered at least three distinct yet related justifications of abduction: (a) *The Evolutionary Justification*: The human mind, having evolved under the influence of natural laws, has a "natural tendency" (or instinct) to think as nature is (MS 876: 5). Man's mind is attuned to the truth of things in order to discover what he has discovered (6.476). (b) *The Success Justification:* By sampling many abductions, we see that the results of reasoning abductively are beneficial (MS 637: 6–9; 2.270, 786), for humans would not have survived without having knowledge and this requires abductive thinking (5.603, NEM 4: 320). (c) *The Desperation Justification*: Abduction is the only hope of attaining a rational explanation (2.777, 5.145), of regulating future conduct rationally (2.270), of attaining our purposes of reaching truth (2.786) or of comprehending the universe (L75: 272). Unless we reason abductively we cannot know anything of positive fact (MS 475: 43; 5.603, 171; 7.219).

These differ; (a) and (b) assume that abduction has been successful, inviting not only an explanation why it is successful but encouraging confidence on inductive grounds. Of course, (a) is suspiciously circular and (b) gives a debatable description of the evidence insofar as it assumes that we *have*, on occasion, generated scientific *knowledge* through abductive means. If we generalize to all abduction, on the other hand, (b) begins to look more impressive (L477: 57–8; 2.786). (c), the weakest mode, makes no success assumption, relying solely on the empirical claim that our only (or, at least, our best) hope of attaining knowledge is through abductive reasoning, though even (c) must be supplemented with the "hypothesis" that the facts *do* admit of rationalization (7.219–220).

The distinction in terms of *logical goodness* (excellence of argument) or *validity* (also called "soundness": 5.161) is closely related to the issue of justification, for a particular method of reasoning (inferential pattern) cannot be expected to have a truth-producing virtue unless it is a kind of reasoning that can be justified in terms of reason's end. Now an inference is valid if it follows (pursues) a method that (i) it professes to follow, and (ii) has the truth-producing virtue it is supposed to have (2.780), viz., *qua* being an inferential method of one of the three elementary kinds of reasoning.[17] Each kind has its unique brand of validity (2.781, 5.161):

Deduction: A deductive argument is valid insofar as it invariably (necessary deduction), or very probably (probable deduction), yields truth from truth.

Induction: An inductive argument is valid insofar as it approximates to truth in the long run.

Abduction: An abductive argument is valid insofar as it leads to truth if truth can be attained at all.

Differences in validity-type can also be put in terms of differences in the *assurance* that a given method is truth-producing insofar as an inference is valid iff it provides some assurance of being truth-producing. In distinguishing abduction from induction in 1911, Peirce wrote,

> the distinction that is undoubtedly of highest importance between Reasonings (and I call anything a "Reasoning" where one belief or tendency to believe causes another) is that which consists in the *nature of the assurance* being different. (L231: 56)

The distinction among types of assurance can be put simply in terms of *security* and *uberty*. Security is a measure of the confidence that we are avoiding falsehood in the inferential transition, while uberty reflects the "productiveness" in knowledge gained, i.e., that we are increasing our grasp of truth. While we lose in security in moving from deduction to induction and, again, in going to abduction, we gain in uberty at each step (8.384–388).[18]

§7. The Autonomy Thesis Again

We have uncovered two ways in which Peirce attempted to secure his tripartite division, one appealing to justification, the other to validity. Does either succeed? The question does not concern the adequacy of his account of justification or validity, nor whether there *are* the three sorts of justification and validity that he says. Nor is the question whether there are inferences, inferential patterns, or methods of reasoning which are abductively valid or which can be justified in just the way that a good abductive inference is justified. Perhaps *all* "logically good" inferences or inference methods are so valid or justifiable, since abductive goodness is but the weakest assurance of having reached truth, i.e., good deductive and inductive arguments also qualify as abductively valid and justifiable on Peirce's definitions. The question is whether there *are* any modes of valid reasoning which can be justified *only* abductively, for unless there are, Peirce's claim to have unearthed a third mode of reasoning irreducible to deduction or induction cannot be defended. In short, Peirce's definitions secure neither his tripartite division nor the Autonomy Thesis without an additional existence proof.

Are there any grounds for thinking that the final phase of abduction, hypothesis selection, whether depicted in terms of (F4) or (F5), is neither deductive nor inductive? Consider (F4). How are its conclusions related to the steps upon which they are established? Since Peirce defined validity in terms of *truth* (5.191; MS 628: 4), recommendations, imperatives, or interrogatives are not, strictly speaking, validly inferable. On the other hand, if *propositions* are properly questioned, commanded, or recommended, as Peirce suggested (NEM 4: 248), one might construe the validity in terms of the truth-value of the propositions correlated to the inferred questions or recommendations. But there is no obvious way of doing this, since the conclusions of *valid* ampliative inferences can be false even if their premises are true. In particular, a recommendation to do an action A can be appropriate even though the corresponding performance proposition is false. Because of this, *either* Peirce must have had in mind a broadened sense of validity that does not make essential reference to truth-productiveness *or* (F4) cannot be what he took to be the pattern of valid inference underlying hypothesis selection.

Some of Peirce's words can be interpreted to a favor a broadened concept of validity, e.g., his simple formula that an inference is valid "if it possesses the sort of strength that it professes and tends toward the establishment of the conclusion in the way in which it pretends to do this" (5.192). Perhaps a method can also be valid if it has the "virtue" of establishing conceptions having a semantic value other than truth. A question or recommendation, for example, might be deemed *appropriate* given the information in the premises. As far as inferences to directive go, validity might be determined by,

(A) A recommendation to further examine H is appropriate for one who desires an explanation of C iff one is permitted (or, one ought), so far as an explanation of C is desired, to examine H further.

In other words, it is the truth of the normative that fixes the appropriateness of a corresponding directive. It must be emphasized that given the standard logics of normatives (*Castañeda 1975*), considerations about which recommendation is appropriate *all things being considered* could yield a different judgment if overriding normatives sanction a different recommendation. Examining *H* might be appropriate given certain ends but not other, even higher, ends. Also, the workability of (A) requires that there be some means of determining when a normative is true. Here, the most Peirce offered was a rough teleological basis, "ends determine oughts," a formula anchoring truth-values and allowing some normatives to "override" or cancel others.[19]

The distinction among the three sorts of validity survives this broadening. As before, deductive validity guarantees that one passes from truth to a "preferred" semantic value, whether truth or appropriateness; inductive validity yields the preferred value in the long run; while abductive validity secures the preferred value if anything does. Returning, now, to (A) and treating it either as explicative of appropriateness or as a necessary "synthetic" truth, the move from step (4) to (5) in (F4) is straightforwardly *deductive*, since (4) *entails* the normative given Peirce's construal of plausibility (see Section 5). Consequently, the inference from (4) to (5) is deductive.

But if the second conclusion of (F4) is established deductively, what about the first? One of Peirce's strategies for determining the presence of deductive validity is to ask whether the relevant leading principle is necessarily true, in this case,

> (B) If *H* is the most economical explanation of surprising phenomenon *C*, then *H* is more plausible than any of its competitors.

The answer is that (B) is a necessary truth assuming Peirce's account of economy as the correct theoretical basis for deciding plausibility. In that case, the first conclusion of (F4) is also deducible, and the judgment is irresistible that insofar as (F4) is a valid pattern at all, it is deductively valid in the broadened sense of deductive validity. Consequently, if (F4) is the favored form, the Autonomy Thesis collapses.

What about (F5)? It is distinct from (F4) only if the probationary adoption of *H* differs from the attitude of endorsing a recommendation and, as indicated, is an attitude favoring *H* itself, say, a low degree of belief. Its validity, unlike that of (F4), would then involve the reasoner's transition to a truth-valued claim, and there would be no need to broaden the notion of validity to accommodate other semantic values. What we would discover is that the move from step (4) to (5) is certainly not deductively valid since instances of the conditional,

> (C) If *H* is more plausible than its envisioned competitors then *H*

not only lack necessity but are often false. No conditional linking the premises of (F5) to the bare assertion of *H* could fare any better. Moreover, not only would the truth of the conclusion not be guaranteed or be made very probable, but (F5) provides no assurance of approximating truth in the long run, in which case it is not inductively valid either. Hence, *insofar as* (F5) is "valid" at all, its validity must be of some third sort, perhaps what Peirce described as "abductive" validity. Only in this way can the Autonomy Thesis be established.

Have we the desired existence proof? Everything depends upon distinguishing probational adoption from both (positive) belief and the practical endorsement of a recommendation. I have argued that it is not evident that Peirce succeeded in doing this, and for this reason, his overt discussion of abduction does not secure a difference between (F5) and (F4). An explicit existence proof is, therefore, wanting—thought perhaps it can be unearthed in the development of the Peircean system.[20]

NOTES

1. By 1898, Peirce acknowledged that he had previously confused abduction (also called "hypothesis," "retroduction," and "presumption") with a species of probable inference (NEM 4: 183; see also 2.102), and in 1910 wrote:

> the division of the elementary kinds of reasoning into three heads was made by me in my first lectures and was published in 1969 in Harris's *Journal of Speculative Philosophy.* I still consider that it had a sound basis. Only in almost everything I printed before the beginning of this century I more or less mixed up Hypothesis and Induction. (8.227)

In his later writings, abductive thinking extends beyond the confines of scientific discovery, with "scientific retroduction" being but one species of a generic type of reasoning (MS 637: 5–6). The insight underlying this extension is that we are constantly engaged in creative thinking at some rudimentary level insofar as we seek solutions to problems, e.g., whenever we are thinking of what to do (*Lieb 1988*), what to say, or how to interpret or describe what we observe (MS 692: 27–28). We are always searching for "action-guiding" hypotheses in a manner similar to the way a scientist looks for explanatory mechanisms to guide research, though under differing time constraints (MS 637: 5).

2. The passage from MS 637: 5 was written in 1909, and Peirce indicated that a recommendation is the aim of both "practical and scientific retroduction." Emphasis on the practical aspect of abduction, at least as concerns the conduct of inquiry, permeated his later discussion of the topic, for example, in MS 318: 188 (1907). See also, 8.322 (1906) where he writes: "By 'practical' I mean apt to affect conduct; and by conduct, voluntary action that is self-controlled, i.e., controlled by adequate deliberation." Despite the obvious connection of abduction to practical thinking Peirce did not, to my knowledge, offer a sustained discussion of practical reasoning as such.

3. One might wish to rename what I am here calling the "Autonomy Thesis" the "Irreducibility Thesis," in order to distinguish it from the claim that the logic

of discovery is autonomous from the logic of evaluation, in the sense that each domain incorporates sharply different sorts of considerations. This is how "autonomy" is used in *Thagard 1981*, and the claim is one version of what is labelled "Divorce Thesis" in *Nickles 1985*: 183.

4. His syllogistic contrast faded in his later discussion of abductive form, and while the division of reasoning into explicative (analytic) and ampliative (synthetic) reasonings was retained (2.664), it is unable to demarcate abduction from induction. Both points hold for the distinction between necessary and probable inference which Peirce also used. Increasingly, he refused to characterize inductive validity in terms of probability (2.781, 5.170; MSS 293: 20, 652:12), writing that induction "lends no definite probability to its conclusion" (2.780), and "does not render its conclusion any more *probable* than it was before" (MS 475: 8). Nor does abductive validity have anything to do with probability (2.102). By 1910 he labelled induction as "verisimilar" or "likely" reasoning while abduction was dubbed "plausible reasoning" (MS 652: 13–16). More dramatically, induction is not a matter of any relation between premises and conclusion, since "it is not justified by any relation between the facts stated in the premises and the fact stated in the conclusion; and it does not infer that the latter fact is either necessary or objectively probable" (7.207). This is overstated, for induction and abduction require some such relation if the hypothesis is to explain the surprising phenomenon and imply various predictions (though this is only a *necessary* condition for validity in either case). Also, he characterized the business of reason as finding "connections between facts" (7.198), and claimed that it depends on a belief in a relation between what the premises and conclusion assert (MS 668: 2). The premises are a *sign* of the conclusion, and given that *illation*, the "fundamental logical relation" (3.440) is also the "primary and paramount semiotic relation" (2.444n.), perhaps the tripartite division can be drawn in terms of species of illation. But Peirce never appealed to *different* relations between premises and conclusion in making the tripartite division.

5. Peirce continually denied that there was any need to recognize a fourth category of reasoning. In 1911 he wrote,

> I have constantly been on the alert for a fourth kind of reasoning, and have yet never found the least vestige of any. . . . I think myself entitled to presume, for the present, that there is no such fourth form. (MS 856: 6–9)

He stressed that argument by analogy is a mixture of the three elementary kinds (2.733, 7.98) while other familiar types of argumentation, e.g., statistical deduction or qualitive induction, are species of deduction and induction respectively.

6. For Peirce, a belief is a habit according to which one would act in certain ways in given circumstances, specifically, according to the logical consequences of the proposition believe (MS 873: 24). Moreover, a belief is described as a habit one is aware of, satisfied with, does not struggle against, and which can be acquired merely by imagining situations and the behavior they call for.

7. Colligation is also described as a classificatory procedure, in which one brings phenomena under an abstract determination by noting certain common and relevant features in them, as indicated in Peirce's entry under "colligation" in *Whitney 1889*, 2: 1104. *Tursman 1987*: 19–22 contains a nice discussion of this aspect of colligation.

8. In places, Peirce spoke of unconscious and uncontrolled "inference," e.g., 7.444 [1893], but this possibility ceased to qualify later characterizations. Having an irresistible idea "suggested" to the mind may constitute observation, but is not

the inference itself (1.606). Not every case of belief causing belief is inference (4.53), e.g., the inference-like "associational suggestions of belief" (5.441, 108) that Peirce spoke of in discussing associational psychology (MS 318: 38).

9. Peirce suggested as much in his 1903 Lowell Lecture on Abduction, where, commenting on his earlier contrast between deduction and induction, he writes: "With this hint as to the nature of induction, I at once remarked that if this be so there ought to be a form of inference which infers the Minor premiss from the major and the conclusion" (MS 475: 10).

10. See 5.189–191. (F2) is the focus of concern in *Hanson 1958*: 85–90, and has dominated the discussion of Peircean abduction ever since (and, following Hanson, I have taken the liberty of using *H* in place of Peirce's *A*.) It is significant, however, that by 1911, Peirce wrote: "I do not, at present, feel quite convinced that any logical form can be assigned that will cover all 'Retroductions.' For what I mean by a Retroduction is simply a conjecture which arises in the mind" (L231: 55).

11. *Achinstein 1970*: 93–94 and *Curd 1980*: 214 present analogous forms. Peirce's notion of plausibility has been variously interpreted. According to *Thagard 1981*, Thomas Goudge felt that the conclusion of an abduction is a statement of the form "it is plausible to entertain *H*" with plausibility being a mode of acceptance (8.222). Thagard, on the other hand, finds plausibility to be a property of propositions, a reading more solidly grounded in the texts (8.223, 2.662, 6.469–476 and MS 652: 16).

12. The observation that causes one to be surprised is not inferred, but a direct encounter with Secondness. However, a judgment that a certain phenomenon is surprising and, thus, merits examination, is typically made after careful thought. That problem-generation is creative is emphasized by many, e.g., *Heinz Pagels 1982*: 304). See also *Root-Bernstein 1988*: 34 which stresses that rational planning is involved not only in generating novel theories, but also in the discovery of problems: "the best scientists know how to surprise themselves purposely."

13. Doubts about the ability of Peirce's account of abduction to explain how hypotheses are originated are expressed in *Frankfurt 1958, Alexander 1965*, and *Brown 1988*. More positive assessments are given in *Hanson 1958, Achinstein 1970, Anderson 1986*, and *Roth 1988*. I have discussed the issue more fully in *Kapitan 1990*.

14. Concerning suggestion and association, see also 7.202, 2.776, and 5.171. In discussing the associational psychologists in 1907 Peirce wrote that "The action by which, an association having once been established, that act by which in accordance with it, one idea calls up another they called *suggestion*. I shall use this terminology . . . " (MS 318: 38). Suggestion is treated as a genus of which inference is only a species (7.443, 1.606). In MS 293: 7–8, he allowed that one could follow a rule of inference in accepting a conclusion without actually reasoning, since the latter requires seeing that the conclusion is justified. It is important to realize that Peirce defines "abduction" as the first *adoption* of a novel hypothesis (7.202, 6.525), not as the first *conceiving* of it. While every inference is tacitly rule-governed, the agent's initial "irresistible" acceptance of it may be tempered by subsequent evaluation of it by the norms of reasoning one accepts (1.606). Thus, it is important to distinguish among (i) initially conceiving that one might reason in a certain manner (the creative observation), (ii) actually inferring in that manner (guessing), and (iii) evaluating the reconstructed inference. Similar distinctions are made in *Achinstein 1980*: 121 and *Curd 1980*: 203.

15. Carolyn Eisele has noted a parallel between the type of ingenuity it invokes and the *il lume naturale* which Peirce said was central to the process of abductive guessing (*Eisele 1982*: 337). *Langley et al 1987*: 14, classifies the heuristic procedures

of hypothesis generation as "inductive" since they search for general theories from finite data, though it points out that Newton's derivation of the inverse square law of universal gravitation was a deductive process that utilized an algebraic heuristic (pp. 54–7). *Zahar 1982* champions the deductivist approach to hypothesis discovery, viewing the theoretical innovations of Maxwell and Einstein as well as that of Newton as embodying deductive processes. *Pera 1980, 1987,* on the other hand, finds the invention of hypotheses to be inductive, though, equating inductive with ampliative, he does not view discovery in terms of establishing an explanatory conditional.

16. Peirce attempted a deductive justification of induction when he interpreted the phrase "in the long run" in terms of an endless series, and argued that an endless series exists only by conforming to a law or pattern which determines succession. So, if all finite subsequences exhibit a regularity (pattern), this regularity must hold throughout (5.170).

17. In places, Peirce distinguished validity from *strength* (5.192, 2.780) allowing that one argument can be stronger than another though both are valid, e.g. an induction based on more instances, a deduction with a more probable conclusion, or an abduction whose hypothesis has fewer competitors. In 5.192, he defined validity in terms of strength by writing that an argument is valid "if it possesses the sort of strength it professes and tends toward the establishment of the conclusion in the way in which it pretends to do this." However, he did not provide an explicit definition of strength, and in the later discussions of abduction it ceased to be of importance.

18. Distinguishing kinds of reasoning in terms of differing grades of assurance is echoed by Peirce's 1909 distinction citing different *modes of acceptance* (MS 638: 5) correlated to the different assurances provided by the three kinds of reasoning, i.e., probationary adoption is warranted in abduction, provisional acceptance in induction (2.731; 5.591), and "positively asserting" in deduction (MS 473: 17–18).

19. A thorough discussion of practical reasoning and of the rationale and grounds for assigning semantic values to imperatives, intentions and the like, are offered in *Castañeda 1975.* His strategy is more complicated than what (A) conveys, however, for the normative claim must have an implicit index reflecting a particular ground ("ideal harmonization of ends") upon which it is based, and the strategy for determining the appropriateness or "legitimacy" of directives (his "practitions") is more complicated than here suggested. Despite the strength and comprehensiveness of the account, I have expressed doubts about his explanation of legitimacy (in *Kapitan 1984*).

20. I am indebted to Nathan Houser, Christian Kloesel, and the staff at the Peirce Edition Project for advice and assistance given me during the development of this paper, and to the National Endowment for the Humanities, the Southern Regional Education Board, and East Carolina University for financial support.

25.

LOGIC, LEARNING, AND CREATIVITY IN EVOLUTION

Arthur W. Burks

§1. Introduction

Charles Sanders Peirce was the first to suggest that evolution is a learning and a creative process, akin to inductive logic and to the logic of discovery (1.103ff.). He did this one hundred years ago, strongly influenced by Charles Darwin's theory of biological evolution, but unaware of Gregor Mendel's profound discovery.

Peirce used the term "logic" broadly, so broadly as to conceive of evolution as a logical process! By the end of this paper it will be clear why and how he did this. To some extent he was following Immanuel Kant. Kant had used the term "logic" broadly—the central part of the *Critique of Pure Reason* (*Kant 1781*) is devoted to "transcendental logic." Peirce was a student of the *Critique*, and a lifelong student of logic.

Here are some of Peirce's statements about logic.

The very first of distinctions which logic supposes is between doubt and belief, a question and a proposition. Doubt and belief are two states of mind which feel different, so that we can distinguish them by immediate sensation. (7.313 [The Logic of 1873])

The Darwinian controversy is, in large part, a question of logic. (5.364 ["The Fixation of Belief"])

. . . the logic of the universe . . . (6.189 ["The Logic of Continuity"])

. . . the logic of evolution and of life . . . (6.218 ["The Logic of Events"])

Peirce's logic treated abduction and induction, as well as deduction (*Burks 1946*). It covered icons and indices, as well as symbols (1.191, 559). It included the theory of inquiry and the methods of science, as well as grammar and the analysis of arguments.

For Peirce, logic was a normative science (*Burks 1943*). He was committed to this view early, and he stated it explicitly later on.

> He who would not sacrifice his own soul to save the whole world, is illogical in all his inferences, collectively. So the social principle is rooted intrinsically in logic. (5.355 [1869]; 5.355 Cf. 2.654 [1877])

> The genius of a man's logical method should be loved and reverenced as his bride, whom he has chosen from all the world. (5.387 [1877])

> [L]ogic is a *normative* science . . . it not only lays down rules which ought to be, but need not be followed; but it is an analysis of the conditions of attainment of something of which purpose is an essential ingredient.
>   For that which renders logic and ethics peculiarly normative is that nothing can be either logically true or morally good without a purpose to be so. (1.575 [1902])

> [T]he logical norms . . . correspond to moral laws. (1.609 [1902])

I think that Peirce made doing logic (in this broad and normative sense) his main mission in life.

I'll begin with a presentation of Peirce's theory of evolution. Then I'll discuss the theory in terms of current knowledge of biological evolution and the logical theory of automata, and from the perspective of my own philosophy of logical mechanism (*Burks 1990*).

§2. Peirce's Theory of Evolution

Peirce's logical theory of evolution was actually a theory of the origin, history, and future of the cosmos. He published it as a series of articles in *The Monist* between 1891 and 1893:

"The Architecture of Theories" (6.7–34)
"The Doctrine of Necessity Examined" (6.35–65)
"The Law of Mind" (6.102–163)
"Man's Glassy Essence" (6.238–271)
"Evolutionary Love" (6.287–317)

This theory marked a high point in his philosophic thought, encompassing much of what he had done before and providing the basis for later developments in his philosophy.

Peirce's evolutionism was a triad of doctrines or principles: *tychism* (absolute chance), *synechism* (continuous growth), and *agapism* (evolutionary love). These three principles do not sound like logical principles, and Peirce rarely mentions logic in the five papers, though he does connect synechism to his logic of relations and to deduction, induction, and hypothesis (6.113, 144). But Peirce was interdisciplinary, and he treated the evolution of the universe from several perspectives: as a psychological process of learning and discovery, as a logical process of induction and abduction, as a physical and biological process, and as a creative and progressive goal-directed process.

Peirce's theory of evolution covers the history of the universe from its beginning in a primeval chaos to an ultimate state of regularity and perfection.

> [I]n the beginning—infinitely remote—there was a chaos of unpersonalized feeling, . . . without connection or regularity. . . . This feeling, sporting here and there in pure arbitrariness, would have started the germ of a generalizing tendency. . . .
>
> Thus, the tendency to habit would be started; and from this, with the other principles of evolution, all the regularities of the universe would be evolved. At any time, however, an element of pure chance survives and will remain until the world becomes an absolutely perfect, rational, and symmetrical system, in which mind is at least crystallized in the infinitely distant future. (6.33)

> [T]he evolution of the world . . . proceeds from one state of things in the infinite past, to a different state of things in the infinite future. The state of things in the infinite past is chaos, tohu bohu, the nothingness of which consists in the total absence of regularity. The state of things in the infinite future is death, the nothingness of which consists in the complete triumph of law and the absence of all spontaneity. Between these, we have on *our* side a state of things in which there is some absolute spontaneity counter to all law, and some degree of conformity of law, which is constantly on the increase owing to the growth of *habit*. The tendency to form habits or tendency to generalize, is something which grows by its own action, by the habit of taking habits itself growing. Its first germs arose from pure chance. There were slight tendencies to obey rules that had been followed, and these tendencies were rules which were more and more obeyed by their own action. (8.317)

Tychism (chance), synechism (continuity), and agapism (evolutionary love) name different aspects of this evolutionary process.

Tychism is the doctrine that the basic laws of the universe are probabilistic (chance) connections among elementary feelings (psychological Firsts), and that the primeval chaos consisted of random connections among such feelings. The "slight tendencies to obey rules" in the last sentence of the quotation above are such probabilistic connections. Peirce was the first to suggest that the ultimate laws of nature are probabilistic.

Synechism is the doctrine that the laws and systems of the universe evolve gradually in the strict mathematical sense of continuity. This evolution is a growing, developmental, learning process. It is governed by what Peirce calls variously "the law of habit", "the law of mind", "the law of association", and "the tendency to generalize." Peirce speaks also of "habit taking" and "spreading."

Peirce's law of mind is a generalization from human principles of learning, reasoning, and sign usage to the cosmos, and is thus a cosmic law of learning, reasoning, and sign usage. In accordance with this law, chance connections between elementary panpsychic feelings have evolved by a continuous tychistic process into two kinds of systems, the matter of physics and chemistry, on the one hand, and living systems on the other. We humans are the most advanced living systems, and we are capable of learning, reasoning, and using language consciously and intentionally.

Agapism is the doctrine that the synechistic evolution of tychistic laws is guided by evolutionary love, which "makes development go through certain phases, having its inevitable ebbs and flows, yet tending on the whole to a foreordained perfection" (6.305). Peirce's concept of evolutionary love is not easy to understand. It is reminiscent of the attractive aspect of Plato's Idea of the Good, and it was influenced by Aristotle's teleology.

Indeed, Peirce later expressed agapism in terms of final causes.

> There is efficient causation and there is final, or ideal, causation. If either of them is to be set down as a metaphor, it is rather the former. Pragmatism is the correct doctrine only in so far as it is recognized that material action is the mere husk of ideas. (8.272)

> To say that the future does not influence the present is untenable doctrine. It is as much as to say that there are no final causes, or ends. The organic world is full of refutations of that position. Such action constitutes evolution. . . . (2.86)

> The evolutionary process is . . . not a mere evolution of the *existing universe*, but rather a process by which the very Platonic forms themselves have become or are becoming developed. (6.194)

The laws governing physical systems and those governing living systems differ critically with respect to the issue of determinism and probabilism. Peirce emphasizes that probability and statistics are widely applicable to the physical sciences. But there are some laws and systems of the universe that are deterministic and reversible, or nearly so. Classical dynamics is the paradigm example, though Peirce insists that the claim of determinism can never be verified empirically because of the ubiquity of measurement errors.

The standard scientific view of deterministic laws and systems such as Newtonian mechanics is that they are invariant over time. Peirce held, in contrast, that these laws and systems are the product of cosmic evolution. The primeval chaos was unlawful, and deterministic laws and systems gradually evolved from it in accordance with the principles of tychism, synechism, and agapism.

The one intelligible theory of the universe is that of objective idealism, that matter is effete mind, inveterate habits becoming physical laws. (6.25)

[W]hat we call matter is not completely dead, but is merely mind deadened by the development of habit. (6.158)

Thus the deterministic systems of today are evolutionary products of chaos.

In contrast to deterministic systems, living systems are characterized by spontaneity and statistical variation.

The law of habit exhibits a striking contrast to all physical laws in the character of its commands. A physical law is absolute. What it requires is an exact relation. Thus, a physical force introduces into a motion a component motion to be combined with the rest by the parallelogram of forces; but the component motion must actually take place exactly as required by the law of force. On the other hand, no exact conformity is required by the mental law. Nay, exact conformity would be in downright conflict with the law; since it would instantly crystallize thought and prevent all further formation of habit. The law of mind only makes a given feeling *more likely* to arise. It thus resembles the "non-conservative" forces of physics, such as viscosity and the like, which are due to statistical uniformities in the chance encounters of trillions of molecules. (6.23)

Peirce believed that human freedom is incompatible with determinism, and that the spontaneous aspect of human habits is the basis of free choice.

Peirce's cosmic theory of evolution was an extension to the whole cosmos of his earlier cognitive evolutionism and pragmatism. His three papers on intuitive knowledge in the *Journal of Speculative Philosophy* (1868–1869; 5.213–357) expound a cognitive evolutionism:

[E]very thought must be interpreted in another, [for] all thought is in signs. (5.253)

[T]he mind is a sign developing according to the laws of inference. (5.313)

[L]ife is a train of thought. (5.314)

[The] existence of thought now depends on what is to be hereafter. . . . (5.316)

His pragmatism of the 1870's was a development of this cognitive evolutionism into an evolutionary account of common sense and scientific inquiry. Inquiry is an intellectual process of adapting to the environment via the sequence: belief, doubt, investigation, and revised belief.

Good inquiry uses all three modes of inference: hypothesis (analogy, abduction), induction, and deduction. Tychism can be thought of as a cosmic generalization of abduction, and synechism as a cosmic generalization of induction. These parallels need explanation.

Abduction is the logic of discovery, involving guessing, which is probabilistic. Tychism is the principle of cosmic discovery, also involving probabilistic sampling. The human ability to guess right about nature is a manifestation of tychism in the human stage of evolution. Peirce said that we are born with an innate capacity to guess plausible hypotheses from among a set of infinite possibilities, and that otherwise science would not be possible.

Induction is a form of human learning, involving the adaptation of lawlike generalizations (beliefs, habits) to information from the environment. Synechism is cosmic learning, the evolution of laws of nature. Synechism is also a generalization of the psychological law of association.

Deduction does not correlate as well with agapism as induction correlates with synechism and abduction with tychism. Nevertheless, Peirce makes some connections between deduction and agapism.

> The mind works by final causation, and final causation is logical causation. Note, for example, the intimate bearing of logic upon grammatical syntax. Moreover, everything in the psychical sciences is inferential. (1.250)

And, as we saw at the beginning of this paper, Peircean logic is a normative science, and "logical norms . . . correspond to moral laws" (1.609). This includes deduction as well as induction and abduction.

The preceding remarks show how Peirce's theory of cosmic evolution grew out of his earlier philosophical views. These views were influenced by Darwinian evolution, and Peirce's philosophical theory of evolution was also directly influenced by Darwin's theory. But in his theory, Peirce extended the temporal span of evolution to the infinite past and the infinite future. In content, the chance variations of biological evolution became the absolute chance of tychism, and the gradualness of change between neighboring space-time points became the strict mathematical continuity of synechism.

Agapism (final causality) is the third doctrine of Peirce's theory. Final causality was not a factor in Darwinian evolution, which emphasized the role of competition and survival of the fittest. Peirce saw that evolution is creative in many respects, and he believed that evolution progresses toward some ideal state. He thought that competition and the survival of the fittest could not explain these advances, but that agapism could.

Evolution is a growth process, resulting in increased variety and increased complexity. Peirce emphasized this positive aspect of evolution and down-graded the negative side.

> Everywhere the main fact is growth and increasing complexity. Death and corruption are mere accidents or secondary phenomena. . . . [T]here is probably in nature some agency by which the complexity and diversity of things can be increased. . . . (6.58)

The "some agency" to which Peirce refers here is agapism.

Evolution has produced complex holistic systems, and final causality played an essential role in this.

> Efficient causation is that kind of causation whereby the parts compose the whole; final causation is that kind of causation whereby the whole calls out its parts. Final causation without efficient causation is helpless; mere calling for parts is what a Hotspur, or any man, may do; but they will not come without efficient causation. Efficient causation without final causation, however, is worse than helpless, by far; it is mere chaos; and chaos is not even so much as chaos, without final causation; it is blank nothing. (1.220)

The human person—rational, intentional, and consciously goal-directed— is the prime example of a complex holistic system.

Men and women are the most complex organisms on the earth. We have powers of learning, discovery, and creation far beyond those of other biological species. Peirce took this fact to show that evolution is progressive, and put forth agapism to account for this fact. For example, in "Evolutionary Love" he says:

> [E]volution[ary] . . . development go[es] through certain phases, having its inevitable ebbs and flows, yet tending on the whole to a foreordained perfection. Bare existence by this its destiny betrays an intrinsic affinity for the good. (6.305)

Compare the first quotation in this section (6.33).

To summarize: by the doctrine of tychism, chance plays an exploratory or searching role in evolution; by the doctrine of synechism, long range improvement is a gradual developmental process. But tychism and synechism together are not sufficient to account for evolutionary progress; that is, for growth, increased variety and complexity, and advance toward perfection. Chance and continuity need guidance, otherwise nature might just evolve from one kind of chaos to another.

Because Peirce aimed at a complete philosophical theory of evolution, he needed to explain progress. Given the information available to him at the time and his faith in evolutionary progress, it was reasonable for him to postulate agapism as the source of progress. He was basically putting the traditional doctrine of final causes in a new metaphysical context.

Peirce's theory of evolution was cosmic in scope, extending from the initial chaos of panpsychic feelings through the present time and on to the infinite future. He thought that the evolution of mind and matter proceeded in parallel. Where he says above that matter is "effete mind" or "mind deadened by the development of habit," he really means that matter consists of panpsychic feelings connected by probalistic habits that have evolved to a probability value close to unity. In contrast, a human mind consists of panpsychic feelings interrelated by complicated probabilistic habits, the whole system being conscious and intentionally goal directed, capable of making free choices.

Contemporary science holds that the universe began with a big bang about 10 billion years ago, and that biological evolution begain with self-copying chemicals about 4 billion years ago. I have now summarized Peirce's philosophical theory of the *whole* evolutionary process. In the remainder of this paper I will focus on the *biological part* of evolution, both as Peirce conceived it and as we understand it today. After writing this paper I wrote a much longer paper on Peirce's philosophical theory of the whole evolutionary process, "Peirce's Evolutionary Pragmatic Idealism" (*Burks 1996*).

§3. The Problem of Evolutionary Creativity

Peirce used agapism (evolutionary love, final causality) to explain evolutionary creativity and progress. He merged these two concepts, whereas we will distinguish them, making *creativity* factual and *progress* evaluative.

But first we need to define the concept of fitness. Peirce said that evolution is characterized by both increased variety and increased complexity. Each of these involves creativity, the creation of new types of organisms, some of which are selected as "fit" or adapted to the environment.

"Fitness" is an engineering-like term. It should be defined independently of survival, so as to allow for adaptive explanations of the degrees to which organisms and their relatives survive. The fitness of an organism is its ability to grow, prosper, and produce fit offspring. It does this in an environment that contains other organisms, including some of its own species. Those aspects of the whole environment that are relevant to its evolution constitute its adaptive *niche*. The niches of different species of organisms often overlap.

Thus an organism is *fit* to the degree that its physical, physiological, and computational abilities and goals enable it to survive, procreate, and successfully raise *fit* kin. This definition is recursive, as by the generational nature of evolution it must be. Since nature and evolution are very complex and non-linear, the concept of fitness is highly context dependent (*Burks 1984*: 45).

Consider now Peirce's observation that evolution produces increased variety in nature. Evolution creates new kinds of organism, and sometimes these survive and develop into new species, adding to the variety of nature. The niche of a new species will generally overlap the niches of several existing species, and so the new species will compete with the old, though it may also cooperate with some of them.

Sometimes the niche of a new species will have resources not used by any other species. For example, there might be a type of grain in the environment not eaten by any previous species but useful as food to the new species. Such a resource gives the new species an evolutionary advantage. Thus the evolutionary increase in variety that Peirce observed is due in part to the availability of previously unused resources. These resources are often the organisms of other species.

Sometimes environmental possibilities can only be exploited by new organisms that are more complex than previous ones. The role of human intentionality and manual skills in our evolution is an example. Humans developed complicated forms of agriculture by cultivating grains and herding mammals as livestock, and by building large dams and irrigation systems. These achievements were possible because we had planning and executing capabilities superior to those of prior species.

Environmental opportunities that can only be exploited by more complex organisms account for much of the increase in organic complexity that evolution has produced.

One can assess the complexity of organisms very roughly by estimating the informational content of their genetic programs, or, in the case of animals, by estimating the computational complexity of their brains and nervous systems. By either criterion, evolution on earth has produced more and more complex organisms, as Peirce noted. (See, for example, the two graphs in *Sagan 1977*: 26.) Rough estimates of the variety of living things on earth over the course of evolution would similarly validate Peirce's claim that evolution has generally increased organic variety.

Thus evolution is creative in the sense of producing new kinds of organisms, some of which are more complex than their predecessors. I will call this phenomenon *evolutionary creativity*. Evolutionary creativity is a factual, non-evaluative concept, even though it is not precisely defined.

Peirce saw that evolution is creative in this sense, though he mixed this factual belief with his belief in progress. For he held that evolutionary increases in variety and complexity are also increases in perfection. Moreover, his belief that evolution is progressive was basic to his philosophy of religion. I will call the central concept of his stronger claim *evolutionary value progress*, to distinguish it from the weaker, factual concept of evolutionary creativity just defined.

Peirce held that Darwinian evolution only partly explained evolutionary creativity, and I think he was correct on this point. Competition for survival selects those members of a new generation that are best adapted to the environment. But this does not explain how the characteristics of the offspring can be improved enough over the characteristics of the parents to allow natural selection to produce a new generation superior to the old.

Darwinism did not have a satisfactory account of how characteristics evolve. There were two main theories: the blending theory of inheritance, and Lamarck's theory of the inheritance of acquired characteristics.

The blending theory was that each feature of the offspring results from a blending or mixing of the corresponding features of the parents. But blending did not account for such phenomena as "reversion," in which a trait reappears in a family after being absent for several generations. Moreover, no satisfactory mechanism for blending had been proposed.

Jean Baptiste Lamarck had suggested that acquired characteristics could be inherited, but he had no idea of how this might take place. Moreover, many essential characteristics of organisms cannot be acquired by individual effort, and the existence of these characteristics could not be explained by Lamarckianism.

Thus Peirce saw that evolution is creative in producing a sequence of ever more complicated organisms and societies of these organisms, and that Darwinism did not fully explain this creativity. Peirce also believed that this evolutionary sequence tended statistically toward perfection. I accept the first thesis but am very doubtful of the second. Not only is it much more speculative than the first, but its main concept of value progress is much more problematic than the idea of increasing complexity. For this reason, I will not consider the question of value progress further in this paper.

There has been a tremendous increase in our knowledge of biological evolution since Peirce postulated agapism to explain evolutionary creativity. Present knowledge constitutes a very strong case for a naturalistic or logico-mechanical explanation of evolutionary creativity.

The key that unlocked the door to understanding the mechanisms of evolution was discovered by Gregor Mendel in the 1860s. In his justly famous experiments with peas, he showed that there are discrete pairs of factors controlling an inherited trait category, such as tall versus short. Moreover, one factor of each pair is dominant and the other recessive, with the consequence that some of the inherited characteristics of an organism are hidden.

Mendel's experiments were not known to the Darwinian community of scholars in the nineteenth century, and hence not to Peirce. Also, Mendel had no mechanism to account for the operation of dominant and recessive inheritance. His results were achieved independently by others, however, in the early 1900s, and his work was then brought to light. Genes were soon discovered, so there was a mechanism to explain how the characteristics of an offspring were derived from those of its parents. The naturalistic account of evolution has advanced steadily since then.

Mutations were discovered, and the fact that they can be induced by heat and x-rays. Crossing over is the exchange of genetic material between homologous chromosomes, that is, chromosomes with the same linear sequence of genes. By World War I, the positions of genes on a chromosome could be mapped by the frequency with which they crossed over. Rates of human gene mutations were estimated by the 1930s. Concurrently with the growth of genetics, there were many discoveries in physiology and evolutionary biology that advanced our knowledge of organisms, how they function, and how they evolve from one to another.

R. A. Fisher launched the mathematical foundations of evolutionary genetics in the 1920s. He gave a statistical proof that the blending theory of inheritance could not account for the variance in properties between succeeding generations (*Fisher 1929*: 37, 50–51). The blending theory of inheritance was one source of Peirce's synechism, his infinitesimal theory of continuity being the other source. But by the time of Fisher's proof, Peirce's requirement of strict continuity in evolution had already been refuted by the success of Mendelian discreteness.

Although Fisher's theorem about trait variance and inheritance undermined the strict continuity aspect of Peirce's synechism, it did exemplify Peirce's methodology. In "The Fixation of Belief" Peirce said that "Mr. Darwin proposed to apply the statistical method to biology" (5.364). Fisher's theorem was the first deep probabilistic theorem about evolution. (Mendel's result was combinatorial, though of course it yielded statistical distributions.)

Thus it has been proven that evolution is only approximately continuous, not mathematically continuous as Peirce thought. Evolution proceeds in small discrete steps, like a digital computer rather than an analog computer. Still, evolutionary creativity does take place gradually.

Even this brief survey shows that by World War II there was considerable understanding of the discrete mechanisms of heredity. After the war, the rate of progress was greatly accelerated, and it is now proceeding at a revolutionary pace. The Watson and Crick double-helix model of DNA was a key step. Much has been learned since about the genetic code and how the information in the genome is used to direct the construction of an organism. Segments of DNA can be replicated and modified. There is a start at decoding the three billion base pairs of the human genome.

Much has also been learned about other aspects of biology and evolution; and these subjects have in turn joined the physical sciences (biochemistry, biophysics) and engineering (bioengineering). Biology has also developed strong connections with computer science and automata theory. This connection began at about the time that James Watson and Francis Crick discovered the double helix arrangement of DNA. John von Neumann suggested a cellular automaton model of self-reproduction that has been worked out in many ways. He also suggested a robot model, which helps show how engineering considerations enter into adaptation and fitness. We'll return to this subject later (Sec. 8, Computer Models of Evolution).

What is the bearing of these results of Peirce's doctrine of agapism (final causality)? Evolutionary biology has found no use for final causes, and this is a very strong argument against agapism. My position is that present knowledge of evolution combined with philosophical arguments involving automata establishes a naturalistic or mechanistic account of biological evolution. Logic and language play basic roles in this account, as Peirce saw, and so it is a logico-mechanical account. Biological evolution can be viewed in terms of a mathematical-statistical theorem of evolution calculable on a computer, a recursive law of evolution, and computer and robot models of evolution. This doctrine is part of my philosophy of logical mechanism (*Burks 1972, 1977, 1979, 1984, 1986a, 1986b, 1988a, 1988b, 1990*).

Because one of Peirce's motives for agapism was to account for the creativity of evolution, let us see how creativity can be explained in logico-mechanical terms. Although evolution is a gradual process, it has many successive stages or levels, with new types of entities and systems emerging at each one. So far, evolution has progressed through self-replicating chemicals, bacteria, cells, communities of cells, organisms, sexual reproduction, and societies of organisms, adding each to the earth's ecology.

Evolutionary creativity takes different forms in different stages of evolution. We will focus on the important case of diploid sexual reproduction, in which each organism has a matched pair of genomes (strings of chromosomes), and each parent contributes a genome to the offspring. It is reasonable to call the genetic material of each organism a *genetic program*, because it consists of sequences of instructions (genes) that directed the construction of the organism.

To simplify our presentation, we borrow an idea from the economists and assume a single species with non-overlapping generations. The production of a new generation is an *evolutionary cycle* or recursive step with two phases:

1. (*Generation phase*) A male and a female
   a) pair for reproduction (sexual selection)
   b) and produce a fertilized egg with a new genetic program that is compounded from their genetic programs (genetic generation).
2. (*Environmental selection phase*)
   Under the direction of this genetic program, the fertilized egg develops into an adult organism. The offspring competes for the resources of the environment with other organisms of the same and different species. It is often helped by its parents.

Some of the new organism survive and go on to produce the next generation of the species. This evolutionary cycle is presented here as a biological cycle, but it can also be thought of as a cycle in robot or automaton evolution, and hence is fundamentally logico-mechanical (*Burks 1984*).

Repeated over a succession of generations in an appropriate context of environment and other organisms, this evolutionary cycle has produced new kinds of organisms and new species. Thus it is clearly creative.

Both phases are essential for this creativity. If the generation phase produced no offspring that were better adapted than their parents, then the selection phase could not yield a succession of improved generations. When the generation phase does produce some superior offspring, the selection phase is needed to choose them over the poorer offspring.

The generation phase has two subphases, sexual selection and genetic generation. Darwin recognized the importance of sexual selection. But the key to evolutionary creativity is in genetic generation, which neither he nor Peirce knew about. We will give a logical analysis of this in the next section.

§4. The Logic of Evolutionary Creativity

Peirce always had a broad sense of logic, a sense that included inquiry and semiotics. After he developed his cosmic theory of evolution, his sense of logic became even broader. Thus he concluded his fourth Harvard lecture on pragmatism (1903) by saying:

> Therefore, if you ask me what part Qualities can play in the economy of the universe, I shall reply that the universe is a vast representamen, a great symbol of God's purpose, working out its conclusions in living realities. Now every symbol must have, organically attached to it, its Indices of Reactions and its Icons of Qualities; and such part as these reactions and these qualities play in an argument [they also] play in the universe—that Universe being precisely an argument. . . .
>
> Now as to [the function of Qualities] in the economy of the Universe. The Universe as an argument is necessarily a great work of art, a great poem—for every fine argument is a poem and a symphony—just as every true poem is a sound argument. (5.119; cf. 5.107)

Peirce's thesis that cosmic evolution is a kind of argument presupposes his extremely broad use of "logic," as does his inclusion of synechism, agapism, and all of semiotics within logic. I prefer to construe "logic" more narrowly, though broadly enough to include logical grammar, the logic of discovery (Peirce's abduction), inductive logic, the theory of modelling, and automata theory (both deterministic and probabilistic). Even on this narrower concept of logic, there is much that is logical about "bio-logical" evolution. There is a sense in which biological evolution is an iterated, highly parallel, and coherent system of inductive arguments.

The historical story starts with Gregor Mendel's discovery of cases in which some properties of the offspring were the result of discrete combinations of the properties of the parents, some properties being dominant and others recessive. His experiments had the genius of simplicity. He bred many generations of a strain of pea plants whose properties came in pairs:

tall versus dwarf, yellow or green, etc. We'll describe his results with respect to tall/dwarf.

First, he repeatedly self-pollinated each kind, and found that the two kinds behaved differently. The dwarf peas bred true, that is, the offspring of dwarf peas were always dwarf. In contrast, the offspring of tall peas were mixed, some breeding true but others being hybrids that produced mixtures of true breeders and hybrids. Second, he cross-pollinated true-breeding tall peas with dwarf peas, and got all tall peas. He then repeatedly self-pollinated these and got a mixture of 25% true dwarfs, 25% true tall peas, and 50% hybrids!

He explained these results by the simple theory that the height of each pea plant was determined by the interaction of two underlying factors, TALL and SHORT, and that TALL was "dominant" and SHORT was "recessive" in this strain of peas. This was indeed a stroke of genius. Mendel's verified theory not only refuted the blending theory of inheritance, it also said something very basic about how the traits of parents influence the traits of their offspring. What came to be known as genes are the instructions of genetic programs, and paired instructions such as those for the properties tall and short are called "alleles."

It is relevant to our analysis that Mendel's experiments embodied the logical idea of a recursion. First, Mendel's very handling of properties was logical in the sense that he treated them as discrete (present or absent), rather than as continuous or a matter of degree. Moreover, since recessive properties are hidden by dominant properties, it was necessary for Mendel to *iterate* reproductions over several generations to uncover recessive properties and to determine the ratios with which they were manifested.

With this background we return to the evolutionary cycle introduced in the last section. During the genetic generation subphase of this cycle, the genetic program of an offspring is derived from those of the parents by a statistical process of recombination. We shall show that recombination is a natural probabilistic logic of discovery.

Each diploid genetic program is a matched or homologous pair of genomes, each genome is a string of chromosomes, and each chromosome is a string of genes. Paired genes determine the properties of the developed organism, sometimes in the mode of dominant-recessive, as Mendel discovered, but usually in more complicated ways. During recombination half of the material of each parent is copied and the resultant genomes are combined to make a genetic program for the offspring. Thus the offspring receives half of its genes from each parent, and the pairing of homologous genes is different so that the dominant-recessive relations are changed. Recombination explains how the properties of the offspring are derived from those of its parents.

The selection of half of the genes from each parent is a matter of chance, involving *mutation* and *crossover*. A mutated gene is a new building block, added to the gene pool of a population. Many mutations are lethal, while some are beneficial. The viability of a mutation is tested by the survival of the mutated organism.

The process whereby a parent generates a gamete (egg or sperm) containing half of its genetic material is called "meiosis" or "reduction division." In reduction division the genetic material for a gamete is produced by "crossing over" the two genomes. Crossover is most simply explained with a diagram or icon. Let the two genomes of a parent be of the forms

(G) _____x_____
(H) . . . . x . . . . .

where x marks the crossover spot. The two new gametes would then contain the genomes

(G′) _____ . . . . . .
(H′) . . . . _____

The probabilistic operator of crossover greatly increases the number of new genetic programs that can result from sexual generation. To see this, abstract from the variations introduced by mutation. Without crossover during meiosis, the parent GH would contribute one of two genomes to the offspring (G or H), and its mate would contribute one of two, yielding only four possible offspring. With crossover, there are twice as many possible genomes from each parent as there are possible points of crossover.

Peirce was greatly impressed by the variety and diversity in the universe, and invoked tychism (objective chance) as a partial explanation of it (6.57–59). The probabilistic operator of crossover is an important contributor to biological variety and diversity. The number of possible offspring that can result from a mating is the product of the number of possible gametes from the mother and the number of possible gametes from the father. These numbers increase as the number of crossover points increase, and the latter numbers increase as evolution produces longer genomes. Humans have lengthy genetic programs, and crossover results in extensive reshuffling of their genes.

With this background we can explain the creativity of an evolutionary cycle. New genetic programs are produced by recombination, which involves mutation and crossover. These new programs constitute competing hypotheses as to what kinds of organisms are viable in the given environment or ecology. The environmental selection phase then chooses the next generation from among these.

The foregoing constitutes a logico-mechanical explanation of evolutionary creativity for the important case of organism with diploid (double) genetic programs reproducing sexually. Somewhat similar accounts can be given for simpler forms of genetic reproduction. Consider now the implications of these results for Peirce's tychism-synechism-agapism.

The probabilistic nature of mutation and crossover confirms Peirce's tychism insofar as biological evolution is concerned. His synechism is partly confirmed. Evolution is not continuous in the mathematical sense, but it is gradual, and it is essential to evolutionary creativity that it operate in small steps. Moreover, this gradualness produces large steps. Successive levels or organisms emerge in evolution, and become stable contributors to later levels. Cells, organisms, organs, and societies are examples.

The situation is more complex with respect to agapism (final causality). Peirce was correct on a number of important points. He understood Darwinian evolution very well, and saw that it did not fully explain creativity. He saw that the traditional mechanical view of causality did not seem to account for several basic features of biological evolution: evolution is a learning (inductive) and discovery (abductive) process, it involves sign usage in his general sense, it deals with teleological or goal-directed systems, and it is holistic in nature (cf. the current concept of ecology). Peirce realized that all of these features needed explanation. But history has proved Peirce to be wrong in thinking that an irreducible kind of final causality is required to explain evolutionary creativity. For modern genetics provides a logico-mechanical explanation.

It *is* prima facie surprising that goal-directed systems should result from the operation of classical mechanical causality, in which the effect occurs no earlier than the cause. I think the explanation is logical: biological evolution is carrying out an inductive reasoning process that involves an indirect use of mechanical causality.

To see how evolutionary induction works, consider the history of a typical sexual species over a succession of evolutionary cycles. Any such species is part of an "ecology" that includes the encompassing environment and other organisms. Now, the genetic programs produced at each cycle are in effect hypotheses as to the kind of species that is well fitted to survive and prosper in this ecology. Each such *genetic program hypothesis* is tested by the ongoing success of its product or outcome, namely the organism it creates or attempts to create.

This evolutionary process has several features that justify our calling it inductive. A genetic hypothesis is not tested directly by the environment, but indirectly, by the success of its consequence, the organism it creates. Moreover, verification is comparative, because there are other organisms of the same species that compete for the resources of the same niche. Also, this verification procedure is probabilistic, for many chance factors effect whether the organism succeeds or fails.

Now these features of evolution are the defining features of Bayesian induction, which is an application of Bayes' theorem of inverse probabilities. There is a set of competing hypotheses with probabilistic consequences. The hypotheses are not directly confirmable, but their consequences are, so that confirmation is indirect (cf. *Burks 1977*, secs. 2.5, 10.6). Hence the evolution of repeated cycles of diploid sexual reproduction is a natural Bayesian inductive process.

Similar analyses can be given for other stages of evolution (e.g., asexual reproduction) and for other levels (e.g., the origin of species, the success of individual genes). Our earlier analysis was in terms of a genetic program and the organism it creates or attempts to create. But a gene is a lower level hypothesis, for it plays a role in the survival of each organism that has it. Even if the organism does not survive, some of its genes probably will. The biologist's concept of "inclusive fitness" takes account of the fraction of genes an organism shares with each of its "fit" relatives.

We have given a logico-mechanical explanation of the creativity of an important stage of biological evolution. It constitutes a strong argument against Peirce's thesis that agapism or final causality is needed to explain evolutionary creativity, while at the same time confirming his view that biological evolution is an inductive process, a grand "argument." We go next to discuss further aspects of reasoning in biology.

§5. Reasoning and Semiotics in Evolution

We saw in Section 2 that Peirce's law of mind is a generalization from human principles of learning, reasoning, and sign usage, extended back through time. He generalized the psychological law of the association of ideas, the reasoning forms of logic (abduction, induction, and deduction), and human sign usage (icons, indices, symbols). Though he extended these to the beginning of the universe, we are focusing here on their application to biological evolution.

Peirce called regularities in the biological domain "habits," so that associations of ideas, operative rules of inference, and semiotic practices are habits, and learning is a matter of acquiring and changing habits. Restricted to the biological domain, his law of mind is that learning, reasoning, and sign usage developed gradually from the beginning of biological evolution all the way up to conscious human learning, reasoning, and language usage. Peirce gave some examples to substantiate his thesis, and what we now know establishes his thesis pretty conclusively.

He classified inferences into three categories: deduction, (probabilistic) induction, and hypothesis (abduction). In his 1883 paper, "A Theory of Probable Inference," he gave an example of a deductive argument occurring in the nerves of a frog's leg.

Neural systems communicate with electrical signals produced by chemical means. Neural networks can transform these signals logically. [They can also store them in reverberating cycles.] In 1883 Peirce argued that neurons can almost ("virtually") reason.

> [A] syllogism in *Barbara* virtually takes place when we irritate the foot of a decapitated frog. The connection between the afferent and efferent nerve, whatever it may be, constitutes a nervous habit, a rule of action, which is the physiological analogue of the major premiss. The disturbance of the ganglionic equilibrium, owing to the irritation, is the physiological form of that which, psychologically considered, is a sensation; and, logically considered, is the occurrence of a case [minor premiss]. The explosion through the efferent nerve is the physiological form of that which psychologically is a volition, and logically the inference of a result [conclusion]. (2.711)

The following syllogism of form Barbara illustrates what Peirce had in mind:

> Every stimulus of type *A* produces a response of type *B*
> This is a stimulus of type *A*
> Therefore, this will produce a response of type *B*.

In "The Law of Mind" (1892), Peirce transformed deduction, induction, and abduction (hypothesis) into aspects of his cosmic law of mind, and stated without qualification that the frog's leg reasons.

> The three main classes of logical inference are Deduction, Induction, and Hypothesis. These correspond to three chief modes of action of the human soul. In deduction the mind is under the dominion of a habit or association by virtue of which a general idea suggests in each case a corresponding reaction. But a certain sensation is seen to involve that idea. Consequently, that sensation is followed by that reaction. That is the way the hind legs of a frog, separated from the rest of the body, reason, when you pinch them. It is the lowest form of psychical manifestation.
>
> By induction, a habit becomes established. Certain sensations, all involving one general idea, are followed each by the same reaction; and an association becomes established, whereby that general idea gets to be followed uniformly by that reaction.
>
> Habit is that specialization of the law of mind whereby a general idea gains the power of exciting reactions. But in order that the general idea should attain all its functionality, it is necessary, also, that it should become suggestible by sensations. This is accomplished by a psychical process having the form of hypothetic inference. . . .
>
> Thus, by induction, a number of sensations followed by one reaction become united under one general idea followed by the same reaction; while, by the hypothetic process, a number of reactions called for by one occasion get united in a general idea which is called out by the same oc-

casion. By deduction, the habit fulfills its function of calling out certain reactions on certain occasions. (6.144–146)

Thus Peirce realized that a neural system carries out deductive inferences. He also realized that a digital calculating machine does that. In 1886, he had the idea of using electromagnetic relays to represent "and" and "or," and he suggested that with this basis one could build a relay version of Babbage's analytical engine. Moreover, he recognized that such a machine could be used for the calculations that Leibniz envisaged for settling arguments posed in his universal language (*Burks and Burks 1988*: 333–348).

We show next that signs in this broad sense play an essential role in biological evolution, thus confirming his belief that human language is the evolutionary outcome of a gradual semiotic process. Peirce's classification of signs into icon, index, and symbol is useful here (*Burks 1949*).

A sign is something that represents or signifies an object to an interpretant. An icon represents its object by being similar to it. A picture, the scale drawing of a machine, and a wiring diagram of a computer are examples of icons. A photocopy of an original document is also an example, and in this case the similarity can be near-perfect.

An index represents its object by being in existential relation to it. Examples of indices are acts of pointing, "right" and "left" as determined by the speaker's orientation, and "this hat" accompanied by a gesture. A symbol represents its object by virtue of a conventional rule. The color word "red," "triangle" for a three-sided figure, and "classical physics" are examples of symbols.

Signs can, of course, belong to more than one category, as a labelled graph and the examples of indices illustrate. On my theory, a proper name is an index (*Burks 1951*). Peirce defined the concepts of icon, index, and symbol for human languages, but by his law of mind they apply to earlier stages of evolution in reduced form. Generally speaking, icons appeared first in evolution, then indices, and then symbols.

Self-replication involves iconicity, the copy being a near perfect icon of the original. Self-replication is ubiquitous in evolution, ranging from self-replicating chemicals, through the generation of genetic programs, and on to cultural forms such as repeated story-telling, copying manuscripts by hand, printing, and electrostatic copying.

Evolution has produced many important variations of iconic reproduction. Reproduction is sometimes positive (the copy being identical or close to the original) and sometimes complementary (as in complementary strands of DNA and photographic negatives). Reproduction usually produces minor variations, such as genetic mutations. These statistical variations play an essential trial-and-error or searching role in evolution, and thus illustrate Peirce's tychism.

Evolved indices of fitness play an important role in evolution. In a sexual species, a male's internal strength and absence of disease may be signified by some external sign that the female employs when choosing a mate. Examples of such indices are: the length of a peacock's tail, the loudness of a male frog's song, and the brightness of a wild cock's wattle. Biologists have established that females who choose by these criteria tend to have stronger offspring.

Such an index of male health and strength may co-evolve with its use by females in breeding; that is, there may be a co-evolution of the sign and its use. Suppose the health and strength of males is manifested by some property that can be observed by females. For example, a peacock that survives despite the disadvantage of a tail longer than needed is likely to be stronger than one that survives with a standard length tail. Now compare a female that has an accidental tendency to use such an external manifestation in selecting a mate with a female that does not have this tendency. The former female's offspring will probably be stronger than the latter female's offspring, and will generally out compete the latter.

Iteration of this step by natural selection will gradually strengthen the surviving females' habit of using that external index in selecting a mate. It may also bring about an increase in the strength of the external sign; for example, longer peacock tails may co-evolve with their use by females as criteria for mate selection. Thus indices come to play an important role in phase (1a) of the evolutionary cycle (sexual selection) described in the previous section. This phenomenon illustrates inductive learning in evolution, Peirce's law of mind, and what he called "spreading," as well as his concept of an index.

Animal mimicry illustrates the deceptive use of indices. The monarch butterfly is poisonous to predators, and they have learned to avoid it. A non-poisonous butterfly has evolved an appearance that closely resembles or mimics the appearance of a monarch, and so predators avoid it also! Hence the human practice of lying and deceit has its evolutionary antecedents in the animal kingdom.

Icons and indices may also work together in evolution. For example, part of a molecule may serve as a tag (index) to latch onto the corresponding (iconic) part of another molecule. This is an important mechanism in the immune system, which has been designed by evolution to protect the body from foreign substances. Early in the life of an organism its immune system learns to distinguish non-body from body, foreign substances from internal substances. In the human body there are very many billions of different kinds of antibodies, each with a tag corresponding to a possible antigen.

When an antibody defender meets an antigen invader, the antibody compares its tag with that of the antigen, and if the tags correspond (complement one another) the antibody directs the destruction of the antigen. This is a finely tuned mechanism, for foreign carbohydrates, nucleic acids, fatty acids, and proteins are not very different from internal ones. The system fails in the case of an autoimmune disease, such as rheumatoid arthritis, multiple sclerosis, and perhaps cancer.

Our analysis of genetic creativity in the two previous sections shows that learning, reasoning, and sign usage operate in genetics. Genetics also has a natural, primitive grammar. DNA molecules of adenine, cytosine, guanine, and thymine are coded as triplets. Strings of triplets represent properties; compare the "one gene, one enzyme" hypothesis. Chromosomes are strings of genes, and a genetic program is composed of chromosomes.

Mutation and crossover are stochastic grammatical operators. Mutation changes a single unit of a genetic program. Crossover produces a new genome by taking the first part of one genome and attaching the last part of the homologous genome. Looked at conceptually, a mutation results in a new "concept" to be tested by the environment, whereas a crossing over results in a new arrangement of old "concepts" that needs to be tested. Crossover makes larger changes in genetic programs than mutation, and thus contributes to faster evolutionary change in a species. As we will see in the next section, the longer the life span of a species, the stronger the role of crossover in its evolution.

Let us turn now from communication within an organism and give a few examples of communication between and among organisms.

Karl von Frisch discovered that honey bees use a dance language to communicate the location of food to one another. After returning to the hive from a source of nectar, a successful bee does a dance. Different features of the dance indicate the direction and distance of the source.

Ants have well-developed societies, with queens, workers, fighters, farmers, hunters, and slaves. One type of ant colony uses many chemical pheromones for communication. These are secreted from glands onto the ground, and then recognized and responded to by other ants of the colony.

According to Peirce's principle of synechism, human languages evolved gradually from pre-human languages. The first human languages were vocal, with gestures. Copying took the form of repeated story telling. Spoken languages were augmented later by written forms, and later still by the technical languages of mathematics and science.

Automatic language processing and communication began with clocks and music boxes. It proceeded through automatic looms and mechanical calculating devices to electromagnetic forms: telegraph, telephone, and punched-paper devices. Beginning fifty years ago, the computer revolution has produced rapidly growing powers of calculation, display, and interaction between human and machine (*Burks and Burks 1988*).

The human use of language is consciously and intentionally goal-directed. It employs models of actual, imaginary, and desired situations. These models range from the analogic (iconic) and verbal use of human memory and imagination to digital computer simulations and displays. They function both in prediction and in planning.

The great evolutionary advantage of signs lies in the fact that they are much easier to manipulate than their objects. This is true for sign systems that function inside an organism as well as for sign systems that operate between organisms.

We have concentrated on the semiotic aspect of evolutionary creativity because Peirce emphasized it. A language presupposes organisms that use it, and as languages have become more complex, so have the other aspects of these organisms: their structures, energy processors, input and output systems, and overall organizations. Evolution has been creative in all of these respects.

§6. Competition, Cooperation, and Complexity

Two sections ago we showed how the two-phase cycle of genetic reproduction is creative, the production of new genetic programs from old amounting to the production of new hypotheses of adaptability. Using this result we argued that Peirce's doctrine of agapism is not needed to account for evolutionary creativity. Then we gave examples of sign usage and reasoning from several stages of evolution, to demonstrate that Peirce was correct in thinking that these played essential roles in evolution.

Let us look next at the creative results of iterating the two-phase evolutionary cycle, and in particular, at the interaction of competition and cooperation in this process.

Darwin emphasized competition in his "survival of the fittest." Peirce, in opposition, emphasized cooperation (evolutionary love). We now know that this is too simple an opposition, a false dichotomy. For competition and cooperation both operate in evolution, and they interact in complicated ways. The environmental selection phase of the evolutionary cycle involves competition with other organisms for resources in the environment, but it also involves cooperation between parents and their offspring, and sometimes with other related or unrelated organisms (*Axelrod 1984; Dawkins 1976; Maynard Smith 1978, 1982*).

Repeated evolutionary cooperation led gradually to the construction of compounds of various degrees of complexity. Cells, colonies of cells, organisms with organs, societies of organisms, and societies of societies—all originated in this way. In such construction processes, some organic building blocks are used repeatedly; the four-chambered heart is an example.

A very interesting example of how the interaction of competition and cooperation can be creative is to be found in the host-parasite relationship.

We observed in the previous section that for a sexual species an outward indexical sign of male health and strength may co-evolve with the use of that sign by females in choosing mates. Now it turns out that internal asexual parasites, one of the main sources of ill health and weakness, play a deep role in the evolution of sexual species.

Because sexual reproduction requires the cooperation of two parents, it is less efficient than asexual reproduction. William D. Hamilton has suggested an advantage of sexual reproduction that compensates for this inefficiency. His hypothesis has been somewhat confirmed by field observations and by computer simulations (*Hamilton 1982; Hamilton and Zuck 1982; Hamilton, Axelrod, Tanese 1990;* cf. *Burks 1990*: 492–495).

That hypothesis concerns the co-evolution of metazoan hosts and their internal parasites. Sexual organisms are usually or periodically hosts to much smaller asexual parasitic organisms. A host species and its parasite species are in evolutionary competition, with the rate at which each species evolves being crucial to its success. Since the host organism is much more complicated than any of its parasites, its regeneration cycle is much longer than that of its parasites. Other things being equal, a parasite species would out-evolve its host species (though in doing so it might be destroying its own environment!).

However, the sexuality of the host species tends to compensate for this difference in regeneration span. Sexual generation combines information from two genomes, involves dominant-recessive relations, and employs crossover. As we noted in the last section, crossover operates at a higher grammatical level than mutation. Whereas mutation changes single letters, crossover rearranges chunks of genetic code, and thus may lead to larger changes in the offspring than mutation.

As a consequence of the genetic difference between asexual and sexual reproduction, the host species can achieve much more genetic change per generation than its asexual parasite species. This involves cooperation between sexes. In other words, to compete with its parasites the host employs sexual cooperation to compensate for the fact that its life span is longer than that of its parasites. Notice that the sexuality of the hosts helps their parasites by providing a long-term environment for them.

Our host-parasite example illustrates the complexity of competitive and cooperative interactions in evolution and helps explain how they can be creative. It is of philosophical interest to note a similarity to the interaction of competition and cooperation in human culture, in long-range strategies of mixed honesty and cheating, for example.

The above analysis of cooperation and competition shows that Peirce was much too naive in the preachy aspect of his claim that logic is a normative science (*Burks 1943*). For example,

I do aver, and will prove beyond dispute, that in order to reason well, except in a mere mathematical way, it is absolutely necessary to possess, not merely such virtues as intellectual honesty and sincerity and a real love of truth, but the higher moral conceptions. (2.82)

[A] man must prefer the truth to his own interest and well-being and not merely to his bread and butter, and to his own vanity, too, if he is to do much in science. (1.576; cf. 7.87)

There are plenty of historical examples that disprove these generalizations. Moreover, Peirce's naive view about competition and cooperation runs counter to what we just saw of how the evolution of parasites has made an important contribution to the evolution of humans.

We have now described a few of the many important ways in which the evolution of organic systems is creative. The basis of this evolutionary creativity is physical and chemical, but, as Peirce foresaw, pre-conscious semiotic communication and processing emerged in evolution and played an essential role. In my terminology, these creative evolutionary processes are logico-mechanical. We will discuss the logical aspects of evolution in the next two sections, and its philosophical implications in the last.

Evolution has yielded a mutli-branched continuum of biological forms. As evolution proceeded these forms usually increased in complexity, both in number of parts and in organization. Homo sapiens is the most complicated species to appear so far on the earth, although there may be more complicated species elsewhere in the universe.

Let us look at ourselves from both the external and the internal points of view. We are complex, non-linear, holistic, dynamic, hierarchical feedback systems. Our genomes have perhaps 4 billion base pairs, our nervous systems perhaps 10^{11} neurons and 10^{14} synapses. We can make artifacts, manipulate the environment, communicate with each other, and play games in much more developed ways than any other existing species. Evolution has produced in us long-term interests and long-term episodic memories. These are manifested in our cultures and our social organizations. We even bury time capsules and send messages to outer space.

We humans can do these things in part because we are conscious and can direct our attention to objects that do not exist. We can create imaginary universes, models of possible situations, and plans of action. We develop goals, both short-term and long-term, and work consciously to achieve them. A goal may be hierarchical and sequential, and a plan for achieving it may be dynamic. Intentional goal-seeking is an iterated cycle of planning, data collection, information processing, and action.

In connection with his pragmatism, Peirce defined a concept as a conscious habit of action.

The elements of every concept enter into logical thought at the gate of perception and make their exit at the gate of purposive action; and whatever cannot show its passports at both these two gates is to be arrested as unauthorized by reason. (5.212)

His emphasis on logic and reasoning implies that there may be considerable information processing between sensory input and action output. Thus we can say more generally that a concept is a complex relational structure. Conceptual models of all kinds, ranging from the visual to the symbolic, are examples. Homo sapiens is unique in its ability to employ such general concepts in adapting to its environment.

While consciousness emerged gradually in evolution, human consciousness is distinctly more complicated and more advanced than the consciousness of any other vertebrate. Consciousness is organizational in nature. It is a relatively efficient real-time control system, capable of both short-term and long-term goal-seeking. The conscious subsystem of a person is an operative and directive model of that person in, and interacting with, the environment.

Immediate experiences—such as sense data, feelings of pain, and memories—are signs in this internal control system that tell it what is going on outside it. Self-consciousness consists in focusing one's attention on one's own role in a real or imagined model or plan; one experiences a sign of oneself that is a part of oneself. The semiotic analog is a sentence that refers to itself (*Burks 1986a, 1990*).

§7. Economics and Evolution

One form of human competition is economic, and in his discovery of the theory of evolution, Darwin was influenced by the existing theories of human competition for resources. Peirce pointed this out when advising his Johns Hopkins logic students on how to do original science.

[T]he higher places in science in the coming years are for those who succeed in adapting the methods of one science to the investigation of another. That is what the greatest progress of the passing generation has consisted in. Darwin adapted to biology the methods of Malthus and the economists [Adam Smith, David Ricardo]; Maxwell adapted to the theory of gases the methods of the doctrine of chances, and to electricity the methods of hydrodynamics. (7.66)

My colleague John Holland has adapted ideas from economics and evolution to develop a class of computer systems for learning and discovery, called "classifier systems" (*Burks 1986c; Holland 1975, 1985, 1986, 1989, 1990a, 1990b; Holland and Burks 1987, 1989; Holland et al. 1986; Riolo 1989a,*

1989b). The processes carried out by these systems can be viewed in Peircean terms (*Burks 1990*: 449–455).

Think of a problem-solving robot whose internal computer is programmed with a classifier system. The data, called "messages," take three forms, as in Peirce's pragmatism. There are *input messages* reporting sensory stimuli and *output messages* controlling actions. And there are *internal messages* that store the results of intermediate computations. (Peirce treats these last implicitly via his emphasis on logic and reasoning.)

The instructions of a classifier system are hypothetical ("if . . . then _____") rules similar to Peirce's practical conditionals. A *classifier* responds to each message that satisfies its antecedent, and transforms that message into a new message in accordance with the content of its consequent. A classifier is so-called because it responds to the *class* of messages satisfying its antecedent.

There are also two-condition classifiers, in which one of the two conditions can be negative. We are describing a classifier system as a computer program, but a classifier system or any part of it could be expressed in computer hardware.

Classifier systems have been used for successful trial-and-error learning. How they do this is best explained by treating a classifier system or program as a three-level hierarchy of

(L_1) basic performance system
(L_1L_2) L_1 plus a market economy learning system
($L_1L_2L_3$) L_1L_2 plus a genetic discovery system

$$(L_1)$$

Computation is organized into a succession of *major cycles* or transformations of a single *message list*. At the beginning of a major cycle new input messages from the robot's sensors are added to the message list. Then every classifier is applied to every message to generate a new message list. Output messages are sent to the robot's effectors, and the remainder of the new messages become the message list for the next major cycle to process. (Note that a given class of messages can be remembered for more than one major cycle by means of a copying classifier that performs this temporary memory function.)

A basic performance program can be written to direct the robot to perform a particular task. Consider next a classifier system that enables the robot to improve its performance over a succession of trial runs.

$$(L_1L_2)$$

An artificial market economy is added to the basic performance system L_1. The message list is limited in size, and whether a message is carried over to the next major cycle depends on the results of a competition. Each classifier is given an initial strength or capital to use in a market auction.

When a classifier C generates a message m, it stakes some of its strength on m by bidding in the auction to have m carried over. If C's bid is sufficient, m is carried over, and C then pays its supplier(s) for the input message(s) it used to generate m. Similarly, if a classifier C* uses m to make a message $m*$ that is carried over to the following major cycle, C* pays C for the use of m. A successful message producer will receive more than it pays out and will then be able to outbid its competitors in the auction more often.

This market economy is fueled from outside by rewards of strength given to the classifier program whenever the robot succeeds in its task. The reward is divided among those classifiers active at the time of success. Over repeated problem-solving runs, these external rewards are distributed by the auction system to classifiers that contribute to successful performance earlier in the trial runs. This pay-off distribution system is called the "bucket brigade algorithm" after the ancient and medieval method of passing buckets of water to be thrown on a fire.

In economic terms, this learning system is a market economy. Each classifier is an information producer, buying messages, transforming them, and competing in auctions to sell them to other information producers. A classifier (computer instruction) that generally produces useful messages prospers relatively to competing classifiers that are less successful in getting their messages used. A classifier system is a paradigm of an information-based economy.

After a classifier system has learned to solve a problem, its successful classifiers are reacting to useful messages and producing useful messages. Similarly, its cooperating groups of classifiers are transforming useful sequences of messages into extended series-parallel conditionals of thoughts and responses. In Peirce's terminology, these successful classifiers and groups of classifiers have learned the natural classes, relational structures, and laws of the environment; they are accordingly pragmatically useful concepts, theories, and habits of action. (Peirce said that a concept is a kind of habit, a conscious habit of action.)

$(L_1L_2L_3)$

A genetic procedure, called "the genetic algorithm," is added to the classifier system L_1L_2. Periodically, some of the strongest classifiers are selected and used to produce new classifiers by the genetic operations of mutation and crossover. The offspring are substituted for some of the poorest classifiers. As the classifier system continues to run in the same environment, these new classifiers are tested by the market economy. Thus they function as hypotheses for the classifier system to use in solving the given problem.

As we noted in Section 5 (Reasoning and Semiotics in Evolution), mutation changes single characters and crossover rearranges sequences of characters. Each classifier is a construct of these building blocks of letters and sequences of letters. When the robot is actually learning to solve a problem, there is a statistical correlation between the success of a classifier and the contributions of its symbolic building blocks. Thus the market economy of a classifier system identifies statistically successful building blocks, the genetic procedure constructs new hypotheses out of these, and the market economy then tests these new hypotheses. This procedure is continuously iterated.

All of these operations—the market-economy auction, the choice of classifiers for the genetic procedure, and the genetic operations on classifiers—are stochastic. This use of chance insures that the set of possible outcomes is adequately sampled. It, like biological evolution, confirms that chance is essential for creativity and learning; compare Peirce's tychism.

With this sketch of the classifier version of the logic of evolution, let us analyze the respects in which it is Peircean and those in which it is not. The basic computer instructions are conditional rules that convert input messages and internal messages into output messages and new internal messages. These instructions are modern computer versions of Peirce's practical conditionals.

Peirce's theory of learning was based on a generalization of the psychological law of association. He and his student Joseph Jastrow had shown experimentally that association is probabilistic (7.21–48). Peirce called his law of learning by various names: the law of mind, the general law of mental action, the tendency to generalize, the law of habit, and the law of habit-taking. He also spoke of the continuous spreading of ideas (synechism). His basic principle was that repeated successful uses of a habit tended to strengthen it, and that useful habits spread to cover related situations. He applied this principle to concept learning as well as to statement learning, for he recognized that adaptation to a problematic situation requires both appropriate or adaptive concepts as well as knowledge of adaptive or true connections among these.

Here is another place where competition plays a constructive role that Peirce did not recognize. He viewed induction as a process in which the inquirer tracks a relative frequency. It is better to view induction as a Bayesian procedure of testing competing hypotheses (*Burks 1977*: chs. 2, 3, 8). We used this inductive logic earlier to explain the creativity of diploid sexual reproduction (sec. 4). Similarly, in a classifier system, the genetic algorithm generates new classifiers (practical conditionals) and the market economy tests them over repeated problem-solving runs.

Peirce proposed a logic of abduction (discovery) and he wrote about several cases of intellectual discovery, including that of Darwin. He saw that concepts as well as theories needed to be tested inductively (pragmatically).

But he did not realize that competition among concepts and among theories could improve their adaptive utility, and of course he had no idea that there was a genetic language operating indirectly to increase adaptation. Instead, he invoked the final causality of agapism to explain biological adaptation and scientific progress.

Agapism is relevant to competition in another way. The concept of a free market is used in classifier auctions (L_1L_2) in a completely ethical and hence artificial way, since there is perfect honesty in transactions. Every classifier pays its debts! But any actual market has some cheating, and if the cheating becomes excessive the market disintegrates. Hence an actual market depends on morality, and Peirce thought that morality depended on agapism. We mention this to illustrate that evolutionary interactions of competition and cooperation are exceedingly complex and hierarchical.

§8. Computer Models of Evolution

We turn now to another respect in which biological evolution is logical. An automaton representation of a computer is a logical system, and, conversely, a computer is a functioning logical system. John von Neumann's automaton models of self-reproduction were the first application of logic to evolution (*von Neumann 1966* and Part IV of *1986*; *Burks 1970, ed. 1970, 1984; Laing 1975, 1977*).

Von Neumann worked on two successive models of self-reproducing automata. The first was robotic and kinematic. A self-reproducing robot was conceived as a three-dimensional organization of several kinds of parts: sensing devices, structural components, computing components (switches and memories), communication components, and effectors. When the robot is placed in an environment of such parts, it selects the needed parts and assembles them into a duplicate of itself.

Von Neumann's second model of self-reproduction was based on a two dimensional cellular automaton framework or "space," like an infinite checkerboard. Each square cell of the space contains the same small automaton, and each such automaton is connected to its four immediate neighbors. Included in the set of states of the automaton is a blank state, and the system can be started by setting a finite number of the cells to non-blank states.

One can choose the starting states of connected areas of this system so that these areas constitute various kinds of logical switches, storage elements, and communication channels. Finite automata can be constructed from those building blocks, and also a storage tape that is indefinitely expandable. A universal Turing machine U can then be made by connecting such a storage tape to a universal finite automaton.

Von Neumann chose the basic cell states (29 in number) so that a universal constructing automaton *C* can also be constructed in a finite area of cells. When *U* and *C* are connected to form *U* + *C*, the result is a self-repro-ducing system. For if a description of automaton *U* + *C* is placed on its tape and this system is started, the system will construct a duplicate of itself in a blank area of the cellular space. This duplicate will include a copy of the original tape. Hence the automaton *U* + *C* with a description of itself on its tape is a self-reproducing automaton. This formal construction of an au-tomaton from its description should be compared with the construction of an organism from its genetic program (Secs. 3 and 4).

Self-reproduction is an essential element of evolution. The next step in viewing evolution logically is to think of the whole process of biological evolution in automaton terms. Evolution takes place in small but discrete steps, sometimes slowly and sometimes rapidly. There has been a gradual advance in complexity of goal-directed organisms from the start of evolu-tion on the earth to the present time. With reference to Peirce's synechism, I have called this "the teleological continuum" (*Burks 1984, 1988a, 1988b*).

Although the teleological continuum is biological, it can also be inter-preted in terms of automata or robot-like machines. This interpretation can be done at various levels: behavior, functioning, and the entire evolutionary process. Corresponding to each chemical, cell, organism, or society of the teleological continuum, there is a finite automaton that can behave equiva-lently to that entity. Even more strongly, there is a finite automaton with a similar internal organization that can model the internal operation of that entity. The teleological continuum may be viewed as an evolutionary suc-cession of such automata.

Thus evolution may be usefully studied by computer simulations, in which not only the evolving entities but their environment would be repre-sented. Such simulations would trace how organisms adapt to their envi-ronment and to each other, taking into account the interactive changes of both. It would simulate the drives, desires, and goals of organisms, as well as their experiences, beliefs, and actions.

Computer simulations can show how successive adaptations of organ-isms interact with the evolution of their genetic programs. Simulations could also be used to explain the structures and characteristics of cells, organisms, and organizations of these, and to explain the origin of species and the nature of their innate abilities (*Burks 1979* sec. 4, 1990; *Langton 1989*).

Simulations of evolution and evolutionary explanations are a special class of probabilistic simulations and explanations. Since probability is in-volved, these must deal with statistical spectra of alternatives. This point can be illustrated with the claim made by some philosophers that the theory of evolution cannot be predictive, but only explanatory. Their argument is that evolution theory cannot predict, as physicalist theories can; rather, after some fact occurs that is in accord with the theory of evolution, that accordance constitutes an explanation.

This argument overlooks the spectral nature of probabilistic explanations and predictions. Each state of a probabilistic system has many different consequences, with a probabilistic distribution over them. Hence in probabilistic cases neither explanations nor predictions are all-or-none as in the deterministic case, but only matters of degree. Insofar as a probabilistic theory can explain an observed result, that theory can predict that result, qualified by the same probability.

There are many interesting evolutionary issues that might be explored by computer simulation. Peirce observed that evolution has produced increased variety and complexity. What are the circumstances under which variety or complexity increase? We know only of evolution on the earth, but evolution in radically different environments can be explored by computer simulations. Heraclitus said that one can never step in the same stream twice. If that were literally true—i.e., the environment was in complete flux—then no organic evolution of stable systems could occur. Also, humans could not compete with simple organisms in a very simple environment, and hence could not develop by evolution in such an environment.

In Section 6 we mentioned an evolutionary hypothesis that has been investigated by computer simulation and observation: Hamilton's hypothesis that diploid sexual reproduction plays a key role in the evolutionary race of metazoa with their internal parasites. There are two competing factors in this evolutionary race: the time span between generations, and the amount of genetic change per generation. A metazoan host is much larger and more complex than its internal parasites, and so takes much longer to mature and reproduce than they do. On the other hand, the host species is diploid and sexual, and hence changes much more per generation than its parasites, which are haploid and reproduce asexually.

We argued in Section 4 that evolutionary creativity can be understood as a form of genetic hypothesis construction. Put in these terms, Hamilton's hypothesis about the evolutionary race between metazoa and their internal parasites can be expressed by saying that, per generation, sexual reproduction is more creative than asexual.

Because of its complexity, evolution cannot be simulated in detail. One should study typical cases and general principles and tendencies, and use simulation interactively with observation and experimentation. This limitation on the applicability of computer simulation is not unique to evolution, or to the biological sciences, but holds for the physical sciences as well. Moreover, it is not just a limitation on probabilistic systems, but holds for deterministic systems also. For it is characteristic of complex *non-linear systems* generally.

Organisms, even simple ones, are adaptive (teleological) non-linear systems. Biological evolution is a process for developing or "creating" such systems; in a certain sense, it is itself such a system. It follows that biological evolution is often chaotic in the sense in which that term is used in modern mathematical physics and computer science.

This chaos is not the chaos of Peirce's tychistic primeval chaos, for a non-linear chaotic system can be deterministic. It would be better to call this kind of chaos *computational chaos,* because the future states of the system are not generally predictable from the present state by finite computational means. (See *Gleick 1987* for a good popular exposition of the nature of chaos.)

Since the chaotic character of biological evolution restricts its "simulability" by computer, this topic deserves further discussion. The key issues can be illustrated by comparing two deterministic dynamic systems that impinge on our everyday life: our solar system and our weather.

The differential equations of Newtonian mechanics describe the behavior of the solar system. Starting from any state of the system, one can compute any other state, past or future. Pierre Laplace's classic definition of determinism was based on this feature of Newtonian physics.

> We ought then to regard the present state of the universe as the effect of its anterior state and as the cause of the one which is to follow. Given for one instant an intelligence which could comprehend all the forces by which nature is animated and the respective situation of the beings who compose it—an intelligence sufficiently vast to submit these data to analysis—it would embrace in the same formula the movements of the greatest bodies of the universe and those of the lightest atom; for it, nothing would be uncertain and the future, as the past, would be present to its eyes. (*Laplace 1814*: 4; cf. *Burks 1975*)

Two aspects of Laplace's statement need to be distinguished; determinism and infinite accuracy. He is assuming that the postulated "intelligence" can both observe and compute with perfect accuracy, whereas all actual observations and computations are finite and approximate. As Peirce emphasized in arguing for his tychism, a measured state of any natural system is subject to observational errors. Actual computers work with numbers of finite length, and hence their results have round-off and truncation errors.

However, in the case of the solar system, and of linear systems generally, it is fairly easy to limit the effects of inaccuracies of measurement and computation. Observational errors can be reduced with better instrumentation, and computational errors generally have a relatively small effect on the results and can be reduced by employing longer numbers. Modern electronic computers have been used to make extensive computations of the past and future motions of our solar system. These calculations go back to the eclipses observed by the Babylonians, and far into the future.

Thus in simulating the solar system one can limit the effects of observational and computational errors, and make long-range predictions. Such is not the case for the weather. The difference between the two kinds of systems is a fundamental difference between their differential equations: those for the solar system are *linear*, while those for the weather are *non-linear*. In non-linear systems small differences can amplify into tremendous differences, and this makes it more difficult to study these systems.

A linear system is so-called because the sum of any two solutions is also a solution. Suppose input i_1 to a linear system produces output o_1 and input i_2 produces output o_2. It follows from the linearity that input $i_1 + i_2$ will produce output $o_1 + o_2$. Addition is a linear operator, as contrasted with such operators as multiplication and logical conjunction. The term, "linear differential equations," then, is appropriate for differential equations that have this additive property.

We have discussed the differential equations of Newtonian mechanics that govern the solar system so that we could contrast them to the differential equations of the weather. But for digital calculation in each case the differential equations are transformed into difference equations, and these are expressed in some computer language as a computational algorithm. It is therefore simplest to attribute linearity and non-linearity to the systems themselves, the solar system being linear and the weather being non-linear.

The additivity property of the solutions to a linear system does not in general apply to non-linear systems. Consequently, a very small change in the state of a non-linear system may be amplified so as to produce a tremendous difference in effect. In other words, two states that are only slightly different can lead to very different histories of the system. Since such small differences are intrinsic to approximate calculations, the computed results are chaotic in the sense of computational chaos.

Because logic is also non-linear, the essence of non-linearity can be illustrated by a simple computer switching mechanism. Consider an input switch that is part of a complicated computer system for controlling the nuclear arsenal of some world power. This computer system controls when and how the world power will launch a nuclear attack.

The input switch mechanism controls the start of an attack. For security reasons, there are two toggle inputs e and s to the switch. One military officer controls the enable toggle e. A different officer, in another location, controls the actual start toggle s. The two toggles feed into an "and" gate, the output of which causes the computer control system to launch the nuclear missiles.

Let 0 mean "off" and 1 mean "on." Then neither input 01 nor input 10 to the input switch will cause the control system to launch and direct a nuclear attack. But the *sum* 11 of these two inputs will cause such an attack, and quite possibly a nuclear holocaust. Thus, when this non-linear control system is in certain states, *a very small change in the input (a single bit change) is amplified into a very large change in the output.* Moreover, this small change would have a devasting effect on the exceedingly complex world system of which the control is a very small part.

The relevance of this example for the issue of chaos and predictability can be illustrated by considering a visit to the earth by a group of humanoids or robots from outer space. The visitors carefully observe everything on the earth, including the nuclear launch system, and make statistical summaries of their results. With their powerful computers they then simulate the future of humankind. But they would very likely miss the key roles played by the two nuclear control officers. Consequently, their simulated predictions of what would happen on earth would be chaotic.

It is important that the non-linear weather system is deterministic, because this shows that chaos is a phenomenon of deterministic as well as non-deterministic systems. There is a type of chance underlying chaos, but it is the chance inherent in computation, not the chance of indeterminism. If one had a complete (and so infinite) description of the starting state of a deterministic dynamical system and a computer that could process real numbers (e.g., an infinitely precise analog computer), one could make perfect predictions, as Laplace said in his formulation of determinism.

Thus chaos limits the power of a finite computing system, human or robotic, to make predictions. We humans have long known that there are limits to our powers of prediction, but the limit imposed by chaos in non-linear systems was discovered only recently, when powerful electronic computers were used in the attempt to predict the behavior of non-linear dynamic systems such as weather.

This completes my remarks on the use of computers to simulate evolution. Biological evolution is a very complex inductive logical process, and computers are the most powerful mathematical tool we have for studying its general principles. Still, evolution cannot be simulated as a whole and in detail because of its complexity and non-linearity.

I will conclude this section by discussing the bearing of computational chaos on Peirce's tychism and on his studies of great men.

Consider tychism. The fact that chaos can occur in a deterministic system undermines Peirce's arguments for tychism and free will. Since a deterministic system can be chaotic, observations that a particular system is chaotic do not show that that system is non-deterministic (*Burks 1990*: 443–437).

The phenomenon of computational chaos is relevant to Peirce's study of "great men" (7.256–266) and to the "great man" issue in history: Can one person make a very large difference in the course of history? There are cases where the actions of a single man or woman do seem to alter that course. The historical context must be ripe for this to happen, however, and the relevant context is far more complicated than the particular person's actions. So how can it be shown that the supposedly great man actually made a great difference in history?

History is a non-linear system. The phenomenon of chaos in non-linear systems suggests that indeed the actions of a single person could be causally effective and amplify into a tremendous difference in history. But the phenomenon of chaos also suggests that such a causal effect could not be distinguished by computer simulation from the amplification of random effects. Hence the great man theory of history must remain a speculation.

§9. Conclusion

Peirce's cosmic theory of evolution was a panpsychism: the universe began as a chaos of primitive feelings and gradually evolved the material and mental laws and systems of today. This process is governed by three principles:

1) tychism—the laws and systems of nature are probabilistic
2) synechism—a cosmic law of gradual habit-taking and generalization governs the growth of laws and systems
3) agapism—evolution is guided by final causes

I will make a few summary comments about these doctrines, and then discuss some of the issues they raise from my own perspective.

Using Darwin's evolutionism, Peirce generalized his earlier cognitive evolutionism and adaptive pragmatism to produce this novel theory of evolution. The speculation of an original chaos is ingenious, and Peirce showed deep insight into the role of probability in the physical sciences. But it is an exaggeration to call astronomical evolution a learning process, so that Peirce's theory of evolution is most plausible for biological organisms and their organizations. As we have shown in this paper, biological evolution is an inductive learning and discovery process, and it has many other logical and semiotic aspects.

Starting from the materials of physics and chemistry about 4 billion years ago, evolution has gradually produced an increasing variety of more and more complex species. It has created human beings with their abilities to use language, to reason and create mental models, to dream and imagine, and to carry out long-range plans.

Peirce took these facts to show that evolution is creative and progressive, and he crafted his theory so as to explain these and other aspects of evolution. Tychism-synechism-agapism was a plausible explanation relative to what was known at the time. In my opinion, Peirce's theory was the best of the nineteenth century philosophical theories of evolution. His theory was highly original and was based on a deep understanding of what was then known about evolution. I also think it is the best non-reductive philosophical theory of evolution to this date.

Tychistic chance in the form of sampling is essential to evolution, which is a natural logic of discovery. The actual mechanisms are semiotic in a broad sense, and at a certain level of complexity they involve Mendelian genetics. These facts were not known in Peirce's time, but Peirce was correct in his emphasis on chance and semiotics.

Consider next Peirce's synechism, the claim that the relational structure of the mathematical continuum is basic to evolutionary learning. Mendelian discreteness proves that this is not literally the case. Nevertheless, Peirce showed great originality in searching for forms of standard mathematics and logic appropriate to evolution. We now know that the mathematics of evolution is non-linear, but non-linear mathematics was not studied much in Peirce's time. Indeed, it cannot be studied very effectively without the use of powerful computers. The phenomenon of chaos, discussed in the last section, is an example. Mathematical logic is really a branch of non-linear mathematics, and Peirce showed vision in seeing that logic is relevant to evolution.

Most important, Peirce was correct in holding that there is no sharp line between humans and the simple organic systems from which humans have evolved, nor between these simple organic systems and the physical and chemical starting systems from which they evolved. This deep insight reduces the problem of explaining the origin of humans to the problem of understanding the gradual process of evolution and its starting point.

Agapism was the last doctrine of Peirce's evolutionary triad. He thought that, in addition to chance and the law of habit-taking, evolutionary love or final causation was needed to explain evolutionary creativity and value progress. Because tychism and synechism are not sufficient to explain these, he invoked agapism (evolutionary love) to do so. This was a natural philosophical step, since agapism is a form of the traditional doctrine of final causes. The *élan vital* of Henri Bergson (*1907*) and the *entelechy* of Hans Driesch (*1908*) are similar doctrines.

Peirce's tychism-synechism-agapism is hospitable to religion; indeed, for him it resolved the conflict between science and religion. The theory's panpsychism makes subjective feelings basic to both matter and mind, and its evolutionism gives an explanation of how material and mental laws and systems were created gradually. Peirce thought that tychism provided a metaphysical basis for a libertarian theory of free will and responsible choice.

Moreover, as we just noted, Peirce was fundamentally optimistic about progress and the ultimate domination of good over evil. He found competition less important than cooperation, and love stronger than hate. These beliefs and attitudes are features of many religions.

I turn now to my own philosophical theory of evolution, which is part of my philosophy of logical mechanism (*Burks 1972, 1977, 1979, 1984, 1986a, 1986b, 1988a, 1988b, 1990*). This philosophy augments traditional atom-compound materialism with ideas from logic, automata theory, computer modelling, evolutionary genetics, and other subjects that treat complex non-linear systems. It holds that logico-mechanical accounts can be given of biological evolution and the systems it has produced.

The philosophy of logical mechanism defines reduction in terms of the idea of an embedded subsystem. In this sense, biology is reducible in principle to physics and psychology to biology. A free human is an embedded subsystem of a deterministic system, and consciousness is an embedded subsystem of a physiological system (*Burks 1975, 1977, 1979, 1986a, 1990*).

My views contrast sharply with Peirce's, and yet they have been strongly influenced by his. I agree with him that there is no sharp line between humans and the simpler living systems from which humans have evolved, nor between these simpler living systems and the starting systems of evolution. I also agree that logic and semiotics permeate the process of biological evolution.

On the other hand, I disagree with Peirce about the metaphysical character of the beginning of biological evolution, on the role that values play in directing evolution, and hence on the relation of mind to matter. Peirce held that evolution started from groups of elementary or subjective feelings, weakly associated, and that agapism (final causality) plays an essential role in evolution. In contrast, I think that evolution started from non-sentient physical systems, and that evolution is a logico-mechanical process that can be completely understood in terms of principles of logic, automata, and the physical sciences.

This last claim was illustrated in Section 4 (The Logic of Evolutionary Creativity) with an analysis of the two-phase evolutionary cycle. The generation phase of the cycle produces new genetic programs from old by probabilistic logical operations on them. The environmental phase then selects the best adapted of these; this phase is covered by Darwin's formula "survival of the fittest." Genetic programs are natural hypotheses, and the production of new genetic programs from old is a natural logic of discovery. Since an organism is a product or consequence of its genetic program interacting with the environment, these genetic hypotheses are being selected indirectly according to their consequences in the environment. Evolution is thus a natural Bayesian inductive procedure of indirect evaluation and selection.

Peirce emphasized cooperation (agapism, evolutionary love) in opposition to competition as the main driving force of evolution. While many have viewed evolution in terms of this simple dichotomy, it overlooks the creativity of the step-by-step interaction of competition and cooperation. We gave several examples: mating, organic systems aggregating to become larger organic systems, the evolutionary race of sexual and asexual reproduction in host and parasite, and the efficiency of an honest market economy.

We saw a moment ago that Peirce's agapism was designed to explain evolutionary creativity. But in this paper we have shown how to explain evolutionary creativity in terms of logical and semiotic principles, the materials of the physical and biological sciences, and the laws of these subjects. Thus there is a logico-mechanical explanation of evolutionary creativity.

Because evolution is logico-mechanical, it is a natural design and construction process. Now, analysis and reduction are the inverses of design and construction. Biological evolution began with non-sentient physicochemical systems. Mental systems evolved from this starting point by a logico-mechanical process, and so they are reducible to them. A detailed understanding of evolution would constitute a proof of reduction. The human mind has many aspects: immediate experiences or subjective feelings, functional consciousness, intentional goal-directedness, and self-consciousness. I conclude that mind in all these aspects is reducible to the materials and laws of the physical sciences (*Burks 1972, 1984, 1986a, 1986b, 1988a, 1988b, 1990* sec. 3).

My theory is the metaphysical opposite of Peirce's panpsychism. Nevertheless, it incorporates Peirce's thesis that evolution is logical and semiotic in character. Biological evolution is an abductive discovery process and an inductive learning process. Evolutionary replication is iconic, tags on indices play essential roles in the development and operation of organic systems, and genetic programs are natural symbolic expressions with grammatical structures. Nature learns and reasons pre-consciously to produce humans that do so consciously.

Thus Peirce's foresight on many important aspects of evolution has been substantiated by modern genetics and the theory of logical automata.

ACKNOWLEDGEMENTS

I am grateful to the National Science Foundation for research support on Grants IRI *86*10225 and *89*10225. My wife Alice, Edward Jenner, Rick Riolo, and Don Roberts have contributed many helpful comments.

I have learned much about evolution and how to computerize it from a faculty seminar with Robert Axelrod, Michael Cohen, John Holland, Rick Riolo, Carl Simon, and occasionally William Hamilton.

The reader may benefit from my definition of "computer theory" in *The Cambridge Dictionary of Philosophy*, edited by Robert Audi (*Audi ed. 1995*).

26.

A NONMONOTONIC APPROACH TO TYCHIST LOGIC

Ana H. Maróstica

§1. Introduction

Tychist logic—a sort of universal evolutionary logic—is composed of the set of laws ruling our mind and the universe. The hallmark of those laws is that they evolve in such a way that everything is continuously permeated by those laws. This Tychist logic at the human level is related, according to my view, to Peirce's logic of the process of scientific inquiry. Correspondingly, the logic of the scientific reasoning has three parts. They are the ones underlying the abductive, deductive, and inductive inferences involved in the scientific inquiry itself.

I think that Peirce's scientific reasoning is nonmonotonic. Which is to say: scientists often draw conclusions on the basis of partial information that they later retract when they obtain more accurate or complete information. My main claim will be that human Tychist logic uses nonmonotonic notions, principles, and rules. I plan to show which of those apply for each part of this logic of scientific inquiry. The main nonmonotonic tools I will use are the following: in the abductive part, J. McCarthy's circumscription technique, and some restrictions on grammar given in relevance logic by A. R. Anderson and N. D. Belnap Jr. (*1975*).[1] In the deductive part, I will use N. Rescher's and R. Brandom's logic of inconsistency in its semantic part, as well as relevance logic in grammar and in proof theory. Finally, in the inductive part, my main logical tool will be R. Reiter's theory of default reasoning.

I plan to give a logical scenario for the three parts of Tychist logic at the human level and to show the unifying connection among them. I will defend the hypothesis that a unifying element is provided by Artificial Intelligence's analysis of the logic underlying nonmonotonic reasoning.

Instead of the classical division of metalanguage into syntax (formation and transformation rules), semantics, and pragmatics *à la* Carnap, I propose the following division (where pragmaticist meanings involve informal correspondence rules between symbols and interpretants; for example the meaning of "$P \rightarrow_R Q$" is "if-then relevantly"—this is related to the *pragmatic maxim* [8.33]):

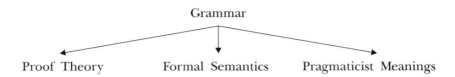

Since the examples will be presented in natural language with some formalizations, I will need a brief description of a formal language L (a first order-language) whose symbols allow the construction of expressions, terms, and formulas needed in the description of the abductive, deductive, and inductive parts of scientific inquiry.[2]

§2. Abduction

Peirce's theory of scientific inquiry conceives the three basic types of reasoning—abduction, deduction, and induction—as the main components of the scientific method. Abduction is that process of making an hypothesis, in fact, an explanatory hypothesis conceived as a mere conjecture, or a fair guess.

The natural form of that inference is: "The surprising fact C, is observed; But, If A were true, C would be a matter of course; hence, there is a reason to suspect that A is true" (5.189).

The hypothesis A ought to fulfill several conditions to be good. It must explain the surprising fact. It must be capable of verification (it must be subject to the test of experiment), and it must be a simple one: "Before you try a complicated hypothesis, you should make quite sure that no simplification of it will explain the facts equally well" (5.60).

Here, we have already all the elements we need in order to give a brief characterization of abduction using the concepts and logic of Artificial Intelligence.

The logical characterization of abduction requires the grammar described in note 2 and the pragmaticist meanings. I agree with N. Belnap and T. Steel in the following: it would be wrong to present abduction (Logic of Questions and Answers) in the form of a deductive system (i.e., to add to its grammar a derivative proof theory). My reason for this is slightly different from the ones given by them. Abduction, as well as induction, are types of reasonings different from deduction. In both abduction and induction, the premises give to their conclusions only partial support, and the form (contrary to deduction, in which the good reasonings have a different form from the bad ones) is exactly the same in the good arguments as in the bad ones. In deduction, we can derive the good arguments using natural rules. In abduction, we cannot do that. Here we need criteria for determining which are the good and the bad arguments according to their content (most of them having the same form).

Since, according to Peirce, abduction is the process of making or adopting an explanatory hypothesis due to the observation of a surprising fact, we can conceive that process in its genesis as a logical mechanism of question and answer:[4]

Figure 26.1

The first phase analyzes the input-question into its constitutive parts; the second phase uses the information from the first phase to construct a representation of the surprising fact as a statement with meaning, and the third phase uses the information from the second phase in order to answer the question.

Following Belnap and Steel (*1976:* 84–87, with some modifications) we can consider "inquiry-questions" as single-example questions (why-question type) about some surprising fact requiring an explanatory hypothesis as an answer. Their form is something like "Why$_i$ does c have P?" or "Why$_i$ is c a P?." For these types of questions, we can use the following pattern:

$$?\text{Why}_i \ (x \parallel Px, c)$$

in which "Why$_i$" means a special type of why-question, the inquiry question (i). That inquiry-question is the "*request.*" Moreover, "$x \parallel Px, c$" is the *subject.* Inside the subject there is a variable x (the *queriable* in Belnap's terminology), which is that variable in the matrix which required an answer, the explanatory hypothesis. The answer to that question has the following form: "Because, A," where "A" in its simplest form could be "c is Q" or "c has Q." For inquiry questions there is always an answer, although there is not a unique answer because there could be an infinite number of answers to that question. The interesting answers in everyday life and in science depend on the inquirer's mind, not on the logical mechanism. For example, Kepler and Einstein were able to provide more interesting answers to the surprising facts they faced than the ones given by other people.

To sum up: Inquiry-questions are questions that can be put in English (or in any other language), in the form of an interrogative sentence of which the following is true: (i) the sentence begins with the word *why* (the request), (ii) the remainder of the sentence has the structure of an interrogative sentence (the subject) designed to ask a question which always has at least one answer (correct or not), (iii) the subject of the sentence asks for facts, not for opinions (especially in science), and (iv) the presupposition of the sentence must be a surprising fact. An inquiry question will be said to be in *normal* form if it satisfies (i)–(iv).

Since there is only one type of inquiry-question (a subtype of why-questions), I will ignore why-questions whose normal forms are not in the indicative mood, whose presuppositions refer to human acts or intentions or mental states, and I will not consider inquiry-questions whose answers cannot be put in the form "because A," where A indicates a position reserved for declarative sentences. The retriever, after receiving the semantic interpretation will give that answer. *Example*: I will use in a simplified way a famous example of abduction given by Peirce (1.71–74) as an example of the inquiry-question-answer relationship. It is related to the work of Kepler (1571–1630) as it is presented in his *De Motibus Stellae Martis*. The inquiry-question was: Why does Mars have problems with calculations of positions related to a circular orbit (i.e., Mars has positions $P_1 \ldots P_n$, and $P_1 \ldots P_n$ are not positions related to a circular orbit)? Using the patter for this type of question, we have:

(I)   ?Why ($x \parallel x$ has problems with calculations of position related to a circular orbit, Mars)

The *parser* analyzes the two parts of the question, and separates the *subject* from the *request*. The subject will provide the *presupposition* of the inquiry-question. In this case, the presupposition of (I) is the following declarative sentence:

(II)   Mars has problems with calculations of positions related to a circular orbit.

(II) is what Peirce calls the surprising fact. The answer to question (I), according to Kepler, is:

(III)   Because Mars has an elliptic orbit.

Now we have all the elements of the natural argument. The premiss is the surprising fact, and the conclusion is the explanatory hypothesis.

(IV)   Mars has problems with calculations of positions related to a circular orbit (i.e., Mars has positions $P_1 \ldots P_n$, and $P_1 \ldots P_n$ are not positions related to a circular orbit (c)).

Mars has an elliptic orbit. (A)

Using Peirce's notation, (c) is the surprising fact, and (A) is the explanatory hypothesis. What Peirce calls the leading principle, that general missing premiss which makes the connection between the premiss and the conclusion, is in the Mars case: All planets with elliptic orbits have problems with calculations of positions related to circular orbits.

Now we can reconstruct (in a very simplified way) Kepler's abduction:

(V)   All planets with elliptic orbits have problems with calculations of positions related to circular orbits (Leading Principle).[5] Mars has problems with calculation of positions related to a circular orbit (c).

Mars has an elliptical orbit. (A)

The general logical form of this type of reasoning is, according to Peirce (2.511–514 [1893]), the following:

Any M is, for instance, P', P'', P''', etc.
S is P', P'', P''', etc.
Therefore, S is M.

Translating in a simplified way Peirce's formalization to the symbolization of our first-order language L, (V) becomes, (VI).

(VI) $(\forall x)(Qx \rightarrow_R Rx)$
Rc

Qc

I have said before that Peirce gave some criteria for abduction: the hypothesis arrived at must be an explanatory hypothesis, it must be able to be verified in experiment, and it should be a simple one. In this paper, I will be concerned only with simplicity.

In order to cope with the problem of simplicity as a criterion, I will use John McCarthy's circumscription method of nonmonotonic reasoning. McCarthy's circumscription is a *rule* of conjecture that can be used by a person (or a program) for jumping to certain conclusions.[7] Namely, the objects that can be shown to have a certain property (P) by reasoning from certain facts (C) are all the objects that satisfy P. More generally, circumscription can be used to conjecture that the tuples $<x_1, \ldots, x_n>$ are all the tuples satisfying this relation. Thus, McCarthy circumscribes the set of relevant tuples. The result of applying circumscription to a collection (C) of facts is a *sentence schema* that asserts that the only tuples satisfying a predicate $P(x_1, \ldots, x_n)$ are those doing so according to the sentences of the fact C. Conclusions obtained from circumscription conjecture that C includes all the relevant facts, and that the objects whose existence are obtained from C are all the relevant objects. Usually humans do not think in myriads of ways to understand a certain problem and easily capture the essence of the problem. We immediately associate a package of additional knowledge with such description as "rivers normally are much broader than a boat" in the famous problem of missionaries and cannibals.[8] However, this extension is performed in a minimal way, *i.e.*, no objects or properties are assumed that are not normally associated with a scenario like the one under consideration. Circumscription offers a technique to simulate such a behavior in a mechanical way.

Applying the former discussion to abduction, we can claim that McCarthy's circumscription rule is able to capture the idea of simplicity of a hypothesis inherent in Ockham's razor. Only those objects should be assumed to exist which are minimally required by the explanatory hypothesis.

McCarthy (*1980*) defines the circumscription formula $A(P)$ in first order language in which we circumscribe a certain predicate $P(x_1, \ldots, x_n)$ in that formula. Given a predicate P and a formula $A(P)$ containing P, the circumscription of P by $A(P)$ can be thought of as saying that the *P*-things are those needed to satisfy $A(P)$ and no more.

The circumscription of P in $A(P)$ is the following scheme which describes from the metalanguage a set of well-formed formulas in the object language:

(VII) $A(P)/P = \mathrm{df}\ A(\Phi) \wedge (\forall x_1) \ldots (\forall x_n)[\Phi(x_1, \ldots, x_n)$
$\rightarrow {}_R P(x_1, \ldots, x_n)] \rightarrow {}_R \cdot (\forall x_1) \ldots (\forall x_m)[P(x_1, \ldots, x_m)$
$\rightarrow {}_R \Phi(x_1, \ldots, x_m)]^9$

In (VII), A is a sentence of first-order language containing a predicate $P(x_1, \ldots, x_m)$; Φ is a predicate parameter for which we may substitute an arbitrary expression. $A(\Phi)$ is the result of replacing all occurrences of P in A by the predicate expression Φ. In this definition, the first conjunct, $A(\Phi)$, expresses the assumption that Φ satisfies the conditions satisfied by P. The second conjunct expresses the assumption that the entities satisfying Φ are a subset of those that satisfy P. The second part of the formula, the consequent, is the converse of the first entailment. This is the reason that, in that case, Φ and P must coincide. *Example*: Let us consider an application of the circumscription rule to the conclusion of Kepler's natural abduction (IV).

(VIII) Mars has problems with calculations of positions related to a circular orbit.

$[A(P)/P]$ Mars has an elliptical orbit

In this case, the predicate "P" is related to the elliptic motion, namely, "to be an elliptic orbit." There are essential elements in Kepler's calculation of Mars' orbit, for instance (i) the type of velocity of Mars in its orbit (inversely proportional to its distance from the Sun), (ii) that the force which moves a planet is located in the Sun, and (iii) the position of the Sun in the orbit of Mars, etc.

Among the irrelevant elements that Kepler did not consider when he was calculating Mars' orbit, we have, for instance, (i) the interplanetary forces (that were actually left aside), (ii) the nature of natural spheres and the way in which they move (following Tycho Brahe's ideas rather than Copernicus' view on the issue).

This was the natural circumscription performed by Kepler. Now, in order to express this circumscription in relation to the conclusion of (VIII), I will use the circumscription scheme (VII). There, we must circumscribe the predicate "to be an elliptic orbit." In a very simplified way, the scheme (VII) will do the job.

$$(IX) \quad \Phi(\text{Mars}) \wedge (\forall x)[\Phi(x) \rightarrow_R \text{Elliptic orbit } (x)] \rightarrow_R (\forall x) \\ [\text{Elliptical orbit } (x) \rightarrow_R \Phi(x)]$$

(IX) means that the predicate parameter "Φ" contains only those essential elements included in the concept of "being an elliptical orbit" (like the type of Mars' velocity, etc.). In that formula, Mars is one of those entities (planets) satisfying the restrictions of Φ. In the second part of the scheme, we have that the essential elements are the only elements considered in the calculation of elliptical orbits.

To conclude: Circumscription is a rule of conjecture, because in the circumscription of a predicate, we assume that certain entities satisfy a given predicate only if they have to, on the basis of a collection of facts given by experience. Moreover, circumscription is nonmonotonic because, if we add new elements in the circumscription of a predicate, we can apply that predicate to more entities. Finally, circumscription (as a rule) functions, in the abductive part, as a criterion of simplicity for explanatory hypotheses.

§3. Deduction

The main focus of the present section is the predictive character that Peirce ascribed to deduction, *i.e.*, to "the prediction of Effects, [or observable consequences] which is accomplished by deduction" ("A Theory of Probable Inference," in *Peirce ed. 1883*). Of course, Peirce thinks that "a hypothesis can only be received upon the ground of its having been verified by successful prediction" (*ibid.*).

Furthermore, Peirce conceived deduction as "the only necessary reasoning. . . . It starts from a hypothesis, the truth or falsity of which has nothing to do with the reasoning, and of course its conclusions are equally ideal" (5.145). *Example:* Even though Peirce did not continue with the process of scientific inquiry for Kepler's example, let me go ahead with its reconstruction. One prediction could be:

> All planets with elliptical orbits have problems with calculation of positions related to circular orbits. Jupiter has an elliptical orbit.
> _____
> Jupiter will have problems with calculations of positions related to a circular orbit.

In considering the deductive character of this type of prediction, it is convenient to supply the deductive part not only with the grammar and formal semantics provided in note 2, but also with the appropriate *derivation rules*. One standard way of providing these rules is by the *natural deduction rules*. Using these rules, it is possible to give fairly natural and well-structured proofs, and express the forms of arguments which arise in logical practice. The main idea of natural deduction is to allow the making of temporary hypotheses, with some device usually being provided to facilitate the bookkeeping concerning which of these hypotheses have been used, and when they have been discharged.[10]

For the rules of introduction and elimination of connectives and quantifiers, I will use the device given by M. Dunn (*1986:* 139–145) of an effective subscripting of each hypothesis made with distinct numerals (α, β, etc.), and then passing this numeral along with each application of a rule. In this way, we can keep track of which hypotheses are used. In this way, we can allow more than one formula to occur on a line. We need the rules of introduction (I), and elimination (E) of commas, connectives, and quantifiers illustrated in Fig. 26.2.

I have made some restrictions on grammar in which "$P \supset Q$" $\neq$ "$P \rightarrow>_R Q$," because the conditional statement in classical logic means only "$\neg P \vee Q$." In science, a central goal is to unify nature by uncovering connections. Those connections are expressed in an "if-then relevantly" type of statement. And those statements indicate that the antecedent is relevant to the consequent. Besides, in the deductive part, we have to use another restriction provided by relevance logic. We need to restrict the rules of inference too, in order to prevent, at the syntactic level (proof theory), the presence of irrelevant theorems and conclusions, as well as of syntactical inconsistencies. Thus, in relevance logic, Ackermann's Rule Υ (or *Modus ponens* for the material conditional "$\neg P \vee Q$") is not allowed without restrictions.

$$\frac{P, \neg P \vee Q}{Q} \; MP \qquad \frac{\neg P, P \vee Q}{Q} \; DS$$

This rule can be considered as a version of the disjunctive syllogism rule. If we do not put restrictions on this rule, we can transform the deductive

COMMA-I	COMMA-E	∧-I	∧-E	∨-I
Γ α Δ α ——— Γ, Δ α	$\dfrac{\Gamma, \Delta}{\Gamma}$	$\dfrac{\Gamma, P, Q \;\; \alpha}{\Gamma, P \wedge Q \;\; \alpha}$	$\dfrac{\Gamma, P \wedge Q \;\; \alpha}{\Gamma, P, Q \;\; \alpha}$	$\dfrac{\Gamma, P, Q \;\; \alpha}{\Gamma, P \vee Q \;\; \alpha}$

∨-E	→$_R$-I	→$_R$-E	¬-I	¬-E
$\Gamma, P \vee Q \;\; \alpha$ $\Gamma, P \quad \{\theta\}$ $\Delta \quad\quad \beta \cup \{\theta\}$ $\Gamma, Q \quad \{1\}$ $\Delta \quad\quad \beta \cup \{1\}$ ——— $\Delta \quad\quad \alpha \cup \beta$	$\Gamma, P \;\; \{\theta\}$ $\Gamma, Q \;\; \alpha$ ——— $\Gamma, P \to_R Q \;\; \alpha\text{-}\{\theta\}$ (provided $\theta \in \alpha$)	$\Gamma, P \to_R Q \;\; \alpha$ $\Gamma, P \quad\quad \beta$ ——— $\Gamma, P \quad\quad \alpha \cup \beta$	$\Gamma, \neg P \;\; \{\theta\}$ $\Delta \wedge \neg\Delta \;\; \alpha$ ——— $\Gamma, P \;\; \alpha\text{-}\{\theta\}$	$\Gamma, P \;\; \{\theta\}$ $\Delta \wedge \neg\Delta \;\; \alpha$ ——— $\Gamma, \neg P \;\; \alpha\text{-}\{\theta\}$

∀-I	∀-E	∃-I	∃-E
$\dfrac{P \;\; \alpha}{(\forall x)P \;\; \alpha}$	$\dfrac{(\forall x)P \;\; \alpha}{P(t/x) \;\; \alpha}$ where x is a variable, and t is a term and is free for x in P.	$\dfrac{P(t/x) \;\; \alpha}{(\exists x)P \;\; \alpha}$ where x is a variable, and t is a term and is free for x in P.	$(\exists x)P \;\; \alpha$ $[P] \;\; \{\theta\}$ $\gamma \;\; \beta \cup \{\theta\}$ ——— $Q \;\; \alpha \cup \beta$

Figure 26.2

system in an inconsistent system.[11] For example, in *Dunn 1986*: 186, we could have:

(1)	$P \wedge \neg P$	A
(2)	P	1, ∧ –E
(3)	$\neg P$	1, ∧ –E
(4)	$\neg P \vee Q$	3, ∨ –I
(5)	Q	2, 4 MP

If we accept MP without restrictions, we have to accept the above proof: from contradictory premises we can obtain any sentence (Q, for example). If we use the restrictions of relevance logic, at the grammar and proof theory level, we cannot derive conclusions (in "if-then" form) with an irrelevant antecedent. This means that, when we arrive at the last phase of the scientific inquiry, the inductive part, we cannot use irrelevant evidence in order to confirm a given hypothesis. Relevance logic deals with the notion of implication (or "if-then relevantly") in which the antecedent suffices relevantly for the consequent. Moreover, in this way the paradoxes of confirmation can be avoided.[12] Of course, relevance logic is also a type of nonmonotonic logic because, to prevent the collapse of deductive logic, it creates restrictions to the rules of derivation at the syntactic level.

Another restriction that is very useful in the deductive part of Peirce's logic of inquiry is the one given by N. Rescher and R. Brandom in *The Logic of Inconsistency (1979)*. In order to avoid semantic inconsistencies, Rescher and Brandom modify the semantic part instead of modifying proof theory.[13]

According to Rescher and Brandom, for tolerating inconsistencies without producing a logical anarchy we cannot extend beyond the so-called weaker inconsistency. This type of inconsistency is permitted if we accept the prospect that for some genuinely possible world $w \in K$:

(1) $t_w(P)$ and $t_w(\neg P)$, for some P,
      where $t_w(P) = \text{df } [P]_w = +$,
      and "$[P]_w = +$" means in Rescher-Brandom terminology
      "P obtains, or its ontological status is 'on'."

The other cases of inconsistencies could lead us to a logical chaos:

(2) To accept the prospect that for some genuinely possible world
      $w \in K$:
      $t_w(P \wedge \neg P)$, for some P (Strong inconsistency)
(3) To accept the prospect that for some genuinely possible world
      $w \in K$:
      $t_w(P \wedge \neg P)$, for all P (Hyperinconsistency)
(4) (Logical Chaos) To accept the prospect that for some genuinely
      possible world $w \in K$:
      $t_w(P)$, for all P,
      (and accordingly $t_w(P)$ and $t_w(\neg P)$, for all P)

In other words, Rescher and Brandom hold a *minimal* consistency—the exclusion of any but the weakest form of inconsistency. This is not a *constitutive* feature of the world but a *regulative* feature (a Kantian perspective), an indispensable aspect of our very concept of the world. It is not the world itself that is self-contradictory (consistently with Peirce's view of it according to which the world is logically structured without contradictions), but the assumptions we are being asked to make about it.

Rescher-Brandom explored a theory of schematic and inconsistent worlds. Their semantics can accommodate a Meinongian theory of objects.[14] Such a theory of objects admits a more tolerant perspective on inconsistency. Accordingly, they try to treat inconsistencies as localizable anomalies rather than as a logical chaos.

By reducing the demand for consistency to the minimal consistency, *i.e.*, by rejecting every inconsistency beyond weakest inconsistency, we can grasp one aspect of our understanding of the world. This aspect is very common in scientific practice. There are many scientific statements that scientists obtain as the conclusion of predictions and they "merely entertain" those statements. They are used for their pragmatic qualities other than truth. All members of a set of mutually inconsistent statements cannot be rationally assented to, but each member of such a set can acquire confirming evidence in the inductive phase. When we have evidence for the truth of each of two incompatible claims, it is quite rational to entertain both. However, the fact that they are inconsistent means that we must mentally flag them to guard against indiscriminate future use. One or the other is false. At best, both can be "approximately true," or "partially true." All this is consistent with Peirce's view of approximately true hypotheses, for the time being, which cannot be asserted as definitely true according to the evidence at hand.

We can say, for example, that the statement "light is a longitudinal wave" and the claim that "light exhibits asymmetry with respect to its direction of propagation" are two incompatible but individually confirmable hypotheses. The ultimate judgment of nineteenth century physics was that there is an element of truth in both. Deciding what role two such confirmed but incompatible statements might play in continued inquiry cannot be purely a formal matter. Following Rescher's proviso, in those cases, it is advisable to isolate those claims and to be cautious until the checking of induction in the long run allows the scientist to give a definite answer. Local inconsistencies can be viewed as disequilibrium examples which work like Peircean doubts to force people to achieve equilibrium (belief). I will come back to this concept of equilibrium in Section 5.

The logic of inconsistency is another type of nonmonotonic logic because we can deduce predictions, but, if among those predictions we find inconsistencies, either we have to give up those inconsistencies if they are beyond type 1 or we should be very cautious and isolate them. Sometimes the reason for inconsistent conclusions is that the premises themselves are inconsistent. It is obvious that here the restrictions are not in theorems or derivability, but in semantic matters. Peirce's logic of scientific inquiry has to deal with scientific statements having a conditional form, like the universal generalizations. Peirce was a scientist; so, we would agree with the restriction given to conditional forms in relevance logic, because, in scientific practice, the antecedent must be relevant to the consequent. And this restriction holds for every part of the process in which there is a conditional sentence. For example, in abduction, the leading principle has that form. Moreover, we must remember that the leading principle reappears in deduction, and in induction as the conclusion of inductive generalizations. In the inductive part, the "if then relevantly" form is very important because induction is related to truth in the long run.

In addition, as I have stressed above, one of the tasks of science, according to Peirce, is to obtain deductive predictions (observable consequences) related to the hypotheses arrived at by abduction. This means that all the rules of deductive derivations have to be in accordance with the restriction of the "if-then relevantly." In that part, Peirce would not accept, in the logic of the scientific inquiry, a material *modus ponens*, because being a variation of the disjunctive syllogism rule (DS), it would allow the inference of any proposition whatsoever from contradictory premises. This would be against scientific practice. So, it is necessary to restrict *modus ponens* as well. If the premises are statements accepted by the theory, only if they are relevant to the conclusion can we detach the latter from the former.

The restrictions stated by Rescher-Brandom within the logic of inconsistency are in accordance with Peirce's view of that issue. Thus, Peirce claimed (4.79) that "Logic teaches us to expect some residue of dreaminess in the world, and even self-inconsistency." Here, he is talking about the logic of scientific inquiry that has to deal with an evolving universe. Sometimes this universe could present to us contradictory facts as novelties, and it might be the case that the given evidence cannot help us to decide which is the correct alternative. The evolving logic of scientific inquiry plus the resources of the logic of inconsistency allow us to cope with these local inconsistencies and treat them as local anomalies. In fact, the logic of inconsistency teaches us the way to isolate those conflicting alternatives and to stop any prediction from those alternatives until evidence can tell us which one is correct.

§4. Induction

In this section, I shall attempt to give a simplified description of Peirce's conception of quantitative induction. When Peirce talked about induction, he was considering mainly quantitative induction. Its main characteristic is that it allows us to determine a ratio, a probability-value, according to the requirements of statistics, moving from sample to population. With quantitative induction, the whole process of the scientific inquiry comes to an end.

Already in 1878 Peirce said that in this type of induction "we generalize from a number of cases of which something is true and infer that the same thing is true of the same proportion of the whole class" (W3: 326).[15]

In 1883, in "A Theory of Probable Inference" (*op. cit.*), Peirce gave more details about induction, especially about its logical form. There, he distinguished between two forms of quantitative induction:

Form 1. "S', S'', S''', etc., form a numerous set taken at random from among the M's.
S', S'', S''', etc., are found to be—the proportion ρ of them—P's;
Hence, *probably* and *approximately* the same proportion, ρ, of the M's are P's."

This is the form of a statistical generalization: given the proportion ρ of the sample being *P*'s, probably there is about the same proportion in the whole lot (population) of *P*'s.

Form 2. Sometimes Peirce says that this is an inference from "a sample to the whole lot." Using the same symbolism as in Form 1, we could write:

S', *S''*, *S'''*, etc., are taken at random from among the *M*'s.

S', *S''*, *S'''*, etc., are found to be *P*'s;

Hence, probably all *M*'s are *P*'s.

Example: Having an elliptical orbit, Jupiter has problems with calculations of positions related to a circular orbit.

Having . . . Venus . . .

Having . . . Saturn . . .

Hence, probably all planets with elliptic orbits have problems with calculations of positions related to circular orbits.

The key question is now, What are the *criteria* according to which an inductive inference, in general, and an inductive generalization, in particular, are considered legitimate? In other words, how do we know that from what has been true of hitherto examined members of a class we can infer something about the whole population? What are the conditions that make an inductive argument a good one? Can we avoid incorrect inductions (inductive fallacies) and try to obtain correct ones?

In inductive arguments, the form does not help to make distinctions between good and bad inductions. This explains the need for criteria related to the information provided by such types of arguments. And this is primarily a semantic matter. In inductive generalizations (statistical or not), criteria related to the number of elements in the sample set, or the variation of those elements, allows us to get a *random* sample. Increasing sample size obviously tends to decrease the probability of an erroneous conclusion, but, in some cases, even a sample which includes a high percentage of the population may, nevertheless, be biased. However, we can increase our confidence in estimates made from relatively small samples if we take into account the kind of sample being used.

There are several means by which random samples (those samples in which, in the election of their elements, no human preference intervenes) can be obtained. The method employed varies according to the subject matter under investigation.

If, among the n observed instances of A's, taken at random, m have been found to be instances of B, we expect that m/n of A's are B's. However, in the scientific practice, we continue to search for further instances of A and constantly modify the estimated ratio (m/n) as new data accumulate.

This way of proceeding in scientific practice guarantees that the sample will be as random as possible. The whole procedure (which, by the same token, can be used as a criterion for randomness) is captured by the *default rule* given by R. Reiter (with some modifications) in "A Logic for Default Reasoning" (*Reiter 1980*).

The nonmonotonic logic for default reasoning allows us to make default assumptions from incompletely specified situations. They will be corroborated, modified, or rejected by subsequent observations. It is this property which leads us to the nonmonotonicity of a logic of defaults.

What we believe about the world is approximately true, because almost always we find some exceptions. This is very common in scientific practice. For instance, using Reiter's example, if we say that the birds that we know all fly and we find exceptional cases like penguins and ostriches, for example, we cannot conclude that "all birds fly," but a better conclusion would be "most birds fly."

In symbols, the first and second arguments would be

(1) $B(a)$ and $F(a)$
       $B(b)$ and $F(b)$
       . . .
       . . .
       $B(n)$ and $F(n)$
       ───────────────
       All B's are F's

(2) $B(a)$ and $F(a)$
       $B(b)$ and $F(b)$
       . . .
       . . .
       $B(i)$ and $\neg F(i)$
       ───────────────
       Most B's are F's

where among the $B(i)$ could be penguins, for example.

In "Most B's are F's," we have to give up some percentage in the conclusion. The reason is that we found exceptional cases. The way to reduce the percentage of the conclusion, if exceptional cases are found, is captured and expressed by Reiter's *Default Rule* (DR). This rule has approximately the following form:

$$\text{Rule (DR)} \quad \frac{P : MQ1, \ldots, Mqn}{R}$$

where P, $Mq_1, \ldots$, Mq_n, are well-formed sentences of first order language. P is called the *prerequisite* of the default, and R is its *consequent*. The type of default used in this paper is a closed normal default. A default is *closed* if none of P, Q_1, ..., Q_n, R contain a free variable. A *normal* default is that default in which $n=1$ and $Q_1=R$ in the above rule.

(DR) is a type of rule of inductive support. It has to do with the content of the premises and conclusion. Now, we can rewrite (2) as follows:

$$\frac{B(n) : \quad MF(n)}{F(x), \text{ for most } x\text{'s}}$$

where "most" has to be translated into some concrete percentage in the case of birds.

Moreover, we can rewrite Kepler's example, in which the conclusion is a simple generalization in the following way:

$$\frac{\text{Elliptical orbit}(n) : M \text{ Problems}(n)}{\text{Problems}(x), \text{ for all } x\text{'s}}$$

where "Problems" is an abbreviation of "problems with calculations of positions related to circular orbits."

Using defaults, we can say: If nothing is known to the contrary, we assume that with planets with elliptical orbits we will have problems with calculations of positions related to circular orbits.

As a final and systematic summary of this section, a flow-diagram for the inductive phase of scientific inquiry follows below, in which the real inquirer is, according to Peirce, the scientific community. We can say that this diagram is the expression of the sub-process of inductive generalization and takes into account the restrictions and rules discussed above.

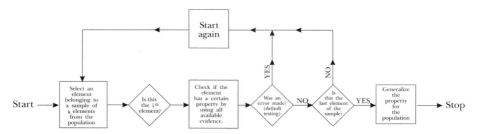

Figure 26.3

§5. Heuristic Principles

In 1908 (2.759) Peirce started thinking of *qualitative induction*[17] as a type of *linking principle* (heuristic in nature) connecting the three main types of reasoning involved in scientific inquiry.[18] There, he says that

Qualitative induction consists in the investigator's first deducing from the

retroductive [abductive] hypothesis as great an evidential weight of genuine conditional predictions as he can conveniently undertake to make and
to bring to the test. . . .

He called it "qualitative induction" because its function is similar to the
one performed by induction: a type of checking which hypotheses obtained
by abduction might work. But, its goal is not to determine any probability
value for the hypotheses according to the evidence at hand. However, "it is
of more general utility than either of the others [types of induction], while
it is intermediate between them, alike in respect to security and to the scientific value of the conclusion" (2.759).

To clarify the role of qualitative induction and its relationship to the
whole process of scientific inquiry, it is necessary to deal with Peirce's three
grades of approach to certainty, namely, plausibility, verisimilitude, and
probability.[19] It is my view that only the first two grades, plausibility and
verisimilitude, belong to qualitative induction. Probability is related to
quantitative induction.

By *plausibility*, Peirce means the degree to which a statement or hypothesis "recommends itself to our belief," independently of any factual evidence. We tend to accept a hypothesis because of a kind of instinct to regard
it favorably. The plausibility of a hypothesis is what recommends it for further inductive examination. Then, its plausibility makes us continue working with a hypothesis, previous to any checking.

Verisimilitude (or likelihood) is something that applies further to qualitative induction if the hypothesis has passed the plausibility test. It might
happen that, given a certain statement or hypothesis, (a) we do not have
enough evidence to consider the hypothesis proved, and (b) we would consider the statement proved if that evidence which is not yet examined continues to be of the same virtue as that already at hand, or if the evidence
that never will be examined should be like that which is at hand. Then,
plausibility and verisimilitude function as regulative principles or rules su-
pervising the process of scientific inquiry.

It would be "too easy" to use the concepts of "epistemic utility" that
Hempel uses in *Aspects of Scientific Explanation* (*1970*: 76–77) when he tries
to explain the concept of "pursuit of truth" (that is involved in plausibility).
Similarly, it could be thought that, in order to explain verisimilitude, one
could use the same concept as characterized by Popper in, for example,
Objective Knowledge (*1975*: 47–60).

However, I think that if we do that, we will betray Peirce's view. He
stressed many times that plausibility and verisimilitude do not involve probabilities or quantities. Hempel's "epistemic utility," in turn, involves a "content measure" function that assigns to every formula in a certain language

L, a number in the interval from 0 to 1. On the other hand, in Popper's verisimilitude, the concept of "content of a sentence" in a given language plays a crucial role. But that concept involves the epistemic probability of that science.

Since Peirce said that qualitative induction's characters are estimated according to their significance not using quantitative notions, I will present qualitative induction as a part of a semantical network comprising the process of scientific inquiry (see Fig. 26.4).

Artificial Intelligence again provides a solution with its heuristic methods.[20] In that network, the principles involved in qualitative induction, *i.e.*, plausibility and verisimilitude, can be considered as informal heuristic principles or rules. For example, plausibility performs a heuristic search among

THE SEMANTIC ARCHITECTURE OF THE PROCESS OF SCIENTIFIC INQUIRY

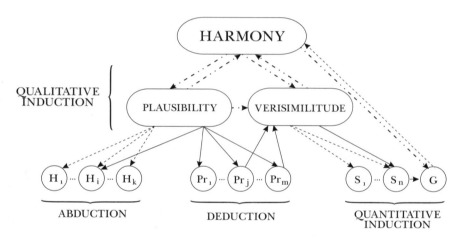

Figure 26.4

the explanatory hypotheses arrived at by abduction. Using that rational instinct that qualitative induction shares, in some way, with abduction (that capability of guessing about the unexplored portion of the search space which will be the most promising hypothesis) plausibility can explore, evaluate, and select the most promising hypothesis.

However, to do that requires having a way to evaluate such an estimate. The crux of the issue is determining how to assess promise. I claim that "the pursuit of truth" of plausibility is connected with a universal principle called "*equilibrium principle*" (in Peirce's words, "harmony principle") that is at the top of the process of scientific inquiry regulating it. A system (or a process) is in equilibrium if the system (or process) does not tend to undergo any further change of its own accord.[21] With this help, plausibility can select the hypothesis showing the maximum self-equilibrium, and the heuristic function selects that hypothesis.

We can define, in an informal way, which is the most plausible (or more promising) hypothesis by using some equilibrium graphs given by A. Müller *(1991)*.

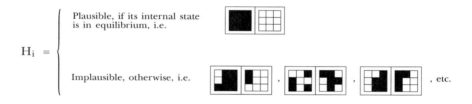

$$H_i = \begin{cases} \text{Plausible, if its internal state is in equilibrium, i.e.} \\ \\ \text{Implausible, otherwise, i.e.} \end{cases}$$

where H_i is a sentence of a given language. There is a heuristic function (h) choosing the most plausible hypothesis. For example, if we have a series of hypotheses,

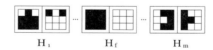

$$H_1 \qquad H_f \qquad H_m$$

in a first search, $h(H) = H_f$, because it is the most promising hypothesis; the one showing the maximum self-equilibrium.

I have to emphasize that the work of plausibility is performed before any checking of quantitative induction. When the most promising hypothesis has been selected, qualitative induction activates deduction in order to predict observable consequences related to that hypothesis. After that, it activates quantitative induction for checking the most promising observable predictions. If the evaluation of the checking of all the elements of the sample set is considered successful, then qualitative induction allows quantitative induction to generalize for the population set. The last part of this work is performed by another type of heuristic principle, verisimilitude, or likelihood. It supervises, by using another heuristic function, the partial checking of the verisimilar elements of the sample set related to that promising hypothesis.

$$S_i = \begin{cases} \text{Likely, if } E_R^+(S_i) > E_R^-(S_i) \\ \text{Unlikely, otherwise.} \end{cases}$$

This means that one statement belonging to the sample set is verisimilar (or likely) if the relevant positive evidence is greater (or "more important") than the relevant negative evidence.

To sum up, the heuristic search (or "things to do") performed by qualitative induction can be sketched in the following way:

(1) Select the most promising explanatory hypothesis according to its internal equilibrium (Plausibility).

(2) Activate deduction in order to predict observable consequences related to that hypothesis. If it is not possible to get observable predictions, the search ends unsuccessfully.

(3) Select the next most promising hypothesis (*i.e.*, the one closest to its internal equilibrium). If several hypotheses qualify, select one randomly.

(4) Activate deduction in order to predict observable consequences related to that hypothesis. If this is done, activate quantitative induction for checking those observable predictions. If the evaluation of the checking of all the elements of the sample is considered successful (applying verisimilitude), the process will end with the generalization for the population set, performed by quantitative induction.

(5) Otherwise, come back to step (3) and select the next promising hypothesis.

§6. Conclusion

I have tried to emphasize several points in the present chapter. First that the task of presenting induction and abduction as formal deductive systems creates several problems. One of the main ones is that formulating the rules of abduction and induction appears to be a more difficult enterprise than doing the same for deductive logic. Deductive logic is a matter of "yes" or "no." A deductive argument is valid or not, and this issue depends only on the form of the arguments. In abduction and in induction, the form is not enough for deciding if some given argument is correct or not. There the strength of the arguments has to be measured according to the content (and not only of the form) of the arguments. Perhaps the answer to this type of problem could be to introduce at least a syntactic-semantical system for the whole logic of inquiry (not only for one of its parts).[22] The basic idea would be to present syntactical and semantical systems for all the parts plus some connections between the parts as linking principles.

The second point is that Artificial Intelligence is able to provide some tools (like nonmonotonic logics and heuristic techniques that are, in some way, nonmonotonic too) or making rigorous, and updating the three parts of Peirce's logic of inquiry. However, Peirce's logic of inquiry can, in turn, allow Artificial Intelligence's nonomonotonic logics to have a broader scope: (i) to apply circumscription techniques not only to deduction but to abduction as well, (ii) to use default principles in the inductive part, and (iii) to apply relevance logic to the whole process of scientific inquiry.

Some final remarks on the equilibrium principle are due. Peirce did not use the word "equilibrium" but "harmony." He strongly believed that there is a general harmony between the structures of the universe and our mind (7.438). That harmony makes man capable of becoming tuned with the universe through the use of the scientific method: " Successful research . . . is conversation with nature. The macrocosmic reason, the equality occult microcosmic law, must act together or alternately till the mind is in tune with nature" (6.568). This harmony or equilibrium between the parts of the system of the universe (properly mind and nature) is reproduced among all the parts composing the mind and nature, and so on. This could involve a regress deep down to the very atoms and so forth. In each level, all the components must have that equilibrium.

This harmony is not static, but dynamic. In fact, according to Peirce, all those systems evolve toward perfection. In other words, everything evolves in the direction of an increasing equilibrium. Thus, when doing science, the passage from doubt to belief (as in knowledge, in general) is a process tending to restore the equilibrium. Any doubt constitutes a state of disequilibrium. When doubt is overcome (state of belief), equilibrium is restored, and the new equilibrium is more stable than the previous one.

This equilibrium or harmony principle can be considered as an expression of consistency of the universe and our mind. This consistency is a dynamic one that can tolerate those local inconsistencies mentioned in the deductive part (when using Rescher's and Brandom's logic of inconsistency). And, we can say that the property of consistency (deductive system) can be considered as a manifestation of this more encompassing self-consistency involved in the harmony or equilibrium principle.[23]

N O T E S

1. The most important scholars who contributed to *Anderson and Belnap 1975* are J. M. Dunn and R. K. Meyer. From now on, following M. Dunn (*1986*), I will mention the authors as Anderson, Belnap and Co.
2. The *grammar* of L is supplied with:

    (1) A set of symbols; this set contains
        a. The parentheses: (,), [,], {, }
        b. The connector symbols: $\neg$, $\wedge$, $\vee$, $\to_R$
        c. The quantifier symbols: $\forall$, $\exists$
        d. Identity symbol: =
        e. Function symbols: a, b, c, . . . , f, g, h,. . . . To each function symbol is attached an integer n, its rank. The function symbols of rank 0 are the constants: a,b,c, . . .
        f. Variables: w, x, y, z
        g. Predicate symbols, P, Q, R,. . . . To each predicate symbol is attached an integer n, its rank. The predicate symbols of rank 0 are the *statements*. The parentheses are used as usual, and terms and formulas are defined recursively in the usual way.
    (2) *Restriction*
      In every first order language, it is usual to accept the following definitions:

$$P \supset Q = \mathrm{df} \, \neg \, P \vee Q \text{ and}$$
$$P \equiv Q = \mathrm{df} \, (P \supset Q) \wedge (Q \supset P).$$

Let me call "$\supset$" and "$\equiv$," "material conditional" and "material biconditional," respectively. Following Anderson, Belnap and Co. (*1975*: 473), "$P \to_R Q$" is not equivalent to "$P \supset Q$," because in the relevant implication (or the "if-then relevantly"), the antecedent must be relevant to the consequent drawn. this is an important restriction, because in the semantic analyses of scientific interpretations which are related to interpretations of the world, to use the relevant implication avoids interpretations of "$P \to_R Q$," as "Irrelevant evidence implies (or confirms) the hypothesis." Moreover

    (3) (a) *Expressions* are finite sequences of symbols $S_1, \ldots, S_n$. The empty sequence is an expression. An expression is closed if it does not contain variables.
         (b) The set of *terms* is the smallest subset of expressions such that:
            • every variable is a term
            • every constant is a term
            • if f is a function symbol of rank n and $t_1, \ldots, t_n$ are terms, then $f(t_1, \ldots, t_n)$ is a term. In what follows, the notation $f(t_1, \ldots, t_n)$ include the case where f is a constant ($n = 0$).

   - That's all
(c) The *atomic formulas* are:
   - The statements
   - The expressions of the type $P(t_1, \ldots, t_n)$ where P is a predicate symbol of rank n and $t_1, \ldots, t_n$ are terms. The notation $P(t_1, \ldots, t_n)$ includes the case where P is a statement ($n = 0$)
(d) The set of *formulas* is the smallest subset of expressions such that:
   - Every atomic formula is a formula
   - If P and Q are formulas and if x is a variable, then $(P \vee Q)$, $(P \wedge 30Q)$, $\neg P$, $(P \rightarrow_R Q)$, $\exists x \, P$, $\forall x \, P$, are formulas. An occurrence of a variable x in a formula P is bound, if it appears in a sub-formula of P having the form $\forall x \, P$ or $\exists x \, P$. If not, it is free. A closed formula (or a statement) is a formula which contains no *free* occurrence of any variable. If not, it is an open formula.
   - That's all.

Turning now to the semantic part of L (formal semantics), I characterize an interpretation "*I*" of L (Peirce would have said "a denotation of L") with a domain D (where D is a non-empty set), as a function such that it takes the variables and constants as arguments, and yields some values in the domain D. Then, in *Formal Semantics*, an interpretation I of L, with a non-empty domain D, is a function such that:

(1) For every function symbol f of rank n, it assigns a function $I[f]$ from D^n to D.
(2) For every predicate symbol P of rank n, it assigns a function $I[P]$ from D^n to $\{T,F\}$.
(3) For every term t, it assigns $I[t] \in D$.
   (a) $I[f(t_1 \ldots, t_n)] = I[f](I[t_1], \ldots, I[t_n])$.
   (b) $I[P(t_1, \ldots, t_n)] = I[P](I[t_1], \ldots, I[t_n])$.
   (c) $I[F] = 0$ and $I[T] = 1$. Besides, for every formula P and Q:
        $I[\neg P] = 1 - I[P]$,
        $I[P \wedge Q] = \min \{I[P], I[Q]\}$;
        $I[P \vee Q] = \max \{I[P], I[Q]\}$; and
        $I[P \rightarrow_R Q] = I[P] \rightarrow_R I[Q]$.
   (d) If x is a variable and P a formula, then $E[I,x]$ is defined as the set of interpretations with the same domain as I, which agree everywhere with I, except possibly on x.
        Then, we have:
        $I[\forall x P] = 1$ iff $I'[P] = 1$ for all I' in $E[I,x]$,
        $I[\exists x P] = 1$ iff $I'[P] = 1$ for some I' in $E[I,x]$.

3. Mainly, I plan to use the *Semantics For a Question-Answering System* by William A Woods (*1967*). I will use, as well *The Logic of Questions and Answers* by Nuel D. Belnap, Jr. and Thomas Steel (*1976*). Finally I also use, with some modifications, "Why-Questions" by Sylvain Bromberger in *Mind and Cosmos* (*1960*), and John McCarthy's "Circumscription, A Form of Nonmonotonic Reasoning," in *Artificial Intelligence* (*1980*: 27–39).

4. I am modifying the diagram presented by W. A. Woods (*1967*: 2–2), in order to explain this logical mechanism or process.

5. According to Peirce, the content of A (the explanatory hypothesis) is already present in the leading principle, "If A were true, C would be a matter of

course." In our case, the leading principle is "All planets with elliptical orbits have problems with calculations of positions related to circular orbits."

6. However, the form of this argument (Kepler's argument) is exactly the same as the form of the following bad abduction:

> Any "subtle matter" serves as a medium of transportation
> (Leading Principle)
> Ether is a medium of transportation (C)
> _____
> Ether is a "subtle-matter." (A)

The problem is that some arguments in which the premises give only partial support to their conclusions (I am using Hempel's double line to indicate that fact) have the same form.

7. John McCarthy (*1980*) presents circumscription as a rule. (See note 3.) There McCarthy defined "domain circumscription" and "predicate circumscription." *Domain circumscription* conjectures that the known entities are all there are. *Predicate circumscription* assumes that entities satisfy a given predicate only if they have to, on the basis of a collection of facts. In "Application of Circumscription to Formalizing Common-Sense Knowledge" (*1986*: 89–116) he presents a new version of circumscription which is called by him *formula circumscription A(P)*, where P is a predicate-variable, and it is defined in second-order logic. I will use the 1980 version of circumscription: circumscription as a rule. There is some discrepancy between McCarthy's use of circumscription and my own. He treats circumscription in the common framework that includes first-order logic. My approach is related more closely to the logic namely abductive logic, which is different from deductive logic.

8. In his 1980 paper, McCarthy gave some applications of the circumscription rule for formalizing common-sense knowledge. An example is the missionaries-cannibals puzzle. Three missionaries and three cannibals are to cross a river with a boat that carries exactly two persons. They have to do so in a way that at no time the cannibals outnumber the missionaries on either side of the river. In this problem, by using common sense, we circumscribe the possibilities, eliminating those ways that could misunderstand the story due to its lack of precision. For example, we eliminate ways like the following: to use the boat as a bridge, or to conjecture that there might not be a solution because the oars might be broken, etc.

9. We can get a more general schema if we circumscribe several predicates jointly. For instance, the circumscription of $P_1, \ldots, P_n$ could be expressed in the following formula:

$$A(\Phi_1, \ldots, \Phi_n) \wedge \{(\forall \bar{x}_1)[\Phi_1(\bar{x}_1) \rightarrow_R P_1(\bar{x}_n)] \wedge \ldots \wedge$$
$$(\forall \bar{x}_n)[\Phi_n(\bar{x}_n) \rightarrow_R P_n(\bar{x}n)] \cdot (\forall \bar{x}_1)[P_1(x_1) \rightarrow_R \Phi_1(\bar{x}_1)] \wedge \ldots \wedge$$
$$(\forall \bar{x}_n)[P_n(\bar{x}_n) \rightarrow_R \Phi_n(\bar{x}_n)]\}$$

where $(x_1), \ldots, (x_n)$, are tuples satisfying P_i and Φ_i.

10. I am using Michael Dunn's ideas and terminology as given in "Relevance Logic and Entailment," in *1986*: 117–224. In the rules provided, the formulas $P \wedge Q$, $P \vee Q$, $\neg P$, are zero-degree formulas. They could be either formulas of relevance logic or classical logic. $P \rightarrow_R Q$, where P and Q are zero-degree formulas, is considered first-degree implication. It could be considered either a formula of relevance logic or a metalinguistic logical relation between two formulas P and Q. An entailment $P_1 \vee \ldots \vee P_n \rightarrow_R Q_1 \wedge Q_2 \wedge \ldots \wedge Q_n$ is called a normal tautological entailment.

11. See M. Dunn (*1986*) for details of the Meyer-Dunn proofs of the admissi-

bility of Υ in relevance logic. Those proofs involve semantic notions like "metavaluation of formulas." See R. K. Meyer and M. Dunn "E, R, and Υ" *(1969)*, and M. Dunn *(1986)*. In this system of relevance logic, material modus ponens is admissible iff $\vdash_R P$ and $\vdash_R P \supset Q$, then $\vdash_R Q$.

12. This is one of the claims made by C. Kenneth Water in *1987*.

13. In order to deal with the problem of semantic inconsistencies, I will introduce, in an informal way, a set $K=\{w_1, w_2, \ldots, w_n\}$ of possible worlds (or cases) to the elements given in the interpretation (see note 2).

14. The possible-world semantics presented in the Rescher-Brandom book is composed by two types of non-standard worlds. Those anomalous worlds are: (1) *Schematic worlds*: these worlds are indeterminate, incomplete with respect to properties, laws, etc. Each world is ontologically indecisive in points of P vs. $\neg P$. The schematic Meinongian objects (those objects that are totally indeterminate like "Hamlet's hat size") are in these types of worlds; (2) *Inconsistent worlds*: these worlds are ontologically overdeterminate. There are some states of affairs such that both P and $\neg P$ obtain in them. They accept overdeterminate objects, *i.e.*, objects possessing every property (contradictory objects). Peirce (5.448) considers these two types of worlds, but he used the terms *vagueness* and *generality*. Thus, he said that "everything is general insofar as the principle of excluded middle does not apply to it." According to Rescher, the law of noncontradiction, if it fails, gives inconsistent worlds.

15. The underlined expression shows that Peirce had started to relate, even before 1880, first truth with the inductive part of the process of inquiry, and, second, probability with induction. It is not my purpose to discuss Peirce's conception of truth in this chapter. However, let me say that he conceived human truth as the agreement between the final opinion (or *representamen*) and reality (or the real object). See, for example, W3: 273, 2.652, and MS 383.

16. More appropriately, instead of "all *B*'s are *F*'s," we have to use a probability statement, in the sense of "approximation": "Most *B*'s are *F*'s."

17. I must acknowledge that Peirce is sometimes obscure when talking about qualitative induction. In 1909 (MS 842), he says that qualitative induction makes something crucially basic for the later work of quantitative induction. Given a certain hypothesis, it provides an evidential weight of genuine predictions.

18. I could have characterized Peirce's qualitative induction as consisting of a set of *bridge principles*. However, I do not want to relate these types of principles with the ones mentioned by Carl Hempel (see, for example, *Philosophy of Natural Science* (*1966*: 72–75), since Hempel's bridge principles "indicate how the process envisaged by the theory are related to empirical phenomena with which we are already acquainted, and which the theory may then explain, predict or retrodict." They connect part of the theory to empirical phenomena. Peirce's principles, in turn, connect different parts of the scientific process itself.

19. In two letters written in 1910, one to P. Carus (8.233ff.) and the other to S. Barnet, Peirce mentioned plausibility, likelihood (or verisimilitude), and probability as the three grades of approach to certainty.

20. A *heuristic* (it means "serving a discovery") is any criterion, method, or principle for deciding which among several alternative courses of action promises to be the most effective in order to achieve some goal. It is like those hints or rules of thumb that are used to guide one's actions. But, in opposition to algorithms, heuristic plans do not offer an *a priori* guarantee that a solution can be found, if one exists. Heuristic methods look promising, and work in most of the cases, but they are not infallible. All heuristics share these features: (a) they apply to specific

situations; (b) their advice is certainly worth considering. Many of their rules have the form: "If a certain situation (a pattern) occurs, then do the following." The "thing to do" is the heuristic to be applied when the corresponding pattern is found. In all the cases in which there is a heuristic search, we need a *search space* which is associated with problems requiring intelligence. Using heuristic search techniques, we can reduce that domain to a manageable size. In all those problems in which heuristics are required to solve them, we have to follow more or less the same steps: (1) any example presents us with repetitive choices among a multitude of exploration paths that might be taken to reach a solution. (2) We have exhibited a *heuristic* function that serves to indicate which among all those cases seem the most promising (*i.e.*, seems to lead to the solution). (3) The heuristics are simple to calculate (formal heuristics) or evaluate (informal heuristics) relative to the complexity of finding a solution and, although they do not necessarily always guide the search in the correct direction, they quite often do. (4) Moreover, whenever they do not, we can resort to some recovery scheme to resume the search from a different point. The key is that these heuristic techniques save time by avoiding many futile explorations and checking.

21. For example, a physical system is in equilibrium when all the transformations are compensated: when for each possible transformation there is a corresponding possible transformation of equal value, oriented in the opposite direction.

22. For example, Juham Pietarin in *1972*: 9–10, claims that Carnap's system of quantitative inductive probability fails to accomplish the task of constructing probability measures that adequately represent rational degrees of belief in inductive generalizations and express how these beliefs are affected by the instances of generalizations. He adds that the "distinction between lawlike and accidental generalizations cannot be taken into account if they are formalized in standard logical calculus (quantification theory), and so all inductive measures of Carnap-type must necessarily fail in representing rational degrees of belief. . . . " Reichenbach also criticized Carnap's approach. In *The Theory of Probability* (*1949*: 457) Reichenbach says that "theories like those of Carnap and Helmer-Hempel-Oppenheim may be said to develop a deductive conception of probability, because they derive the values of probabilities from observational material by deductive methods alone. I should like to summarize my criticism of all such methods by comparing them with the inductive conception of probability, expressed in the frequency interpretation, for which the derivation of probability value presupposes the use of inductive inferences." However, it is well known that Reichenbach's conception is also susceptible to criticism.

23. For more logical and mathematical details about the structure of harmony theory, see P. Smolensky's *1988*: 194–281. Smolensky presents a very interesting relationship between probability and harmony: the higher the harmony, the greater the probability.

27.

THE DEVELOPMENT OF PEIRCE'S THEORIES OF PROPER NAMES

Jarrett Brock

Peirce's earliest public remarks about proper names appear in a series of lectures on logic delivered at Harvard in 1865. They are provided along with the earliest version of the icon, index, symbol trichotomy.[1] Instructively, these terms and/or their provisional equivalents are defined in terms of denotation or extension and connotation or comprehension. Thus icons (copies, likenesses) are said to connote without denoting, indices (signs, mere signs, conventional signs) are said to denote without connoting, and symbols or general terms are said to denote by virtue of connoting.[2]

Peirce uses this trichotomy to define "logic" as follows:

> A first approximation to a definition, then, will be that logic is the science of representations in general, whether mental or material. This definition coincides with Locke's. It is however too wide for logic does not treat of all kinds of representations. The resemblance of a portrait to its object, for example, is not logical truth. It is necessary, therefore, to divide the genus representation according to the different ways in which it may accord with its object. . . .
>
> The second kind of . . . [accordance with an object] is the denotation of a sign, according to a previous convention. A child's name, for example, *by a convention made at baptism, denotes that person. Signs [of this type] may be plural but they cannot have genuine generality because each of the objects to which they refer must have been fixed upon by convention.* . . . Signs, therefore, in this narrow sense are not treated of in logic, because logic deals only with general terms. (W1: 169–170 [1865] *emphasis added)

Unlike the general terms which logic requires, signs such as proper names represent their objects "without any real and essential correspondence" (W1: 323 [1865]). In other words "they represent no character of their objects" (W1: 468 [1866]).

Let us call the view that proper names are signs whose references are fixed upon by baptismal convention and which accordingly denote without connoting, the indexical view of proper names. This view of proper names receives no further mention until after Mitchell and Peirce discover quantification and Peirce redefines his categories in terms of his discovery of irreducible triadic, dyadic, and monadic relations. These changes lead to the provision of a legitimate role for indexical symbols in logic and a redefinition of the icon, index, symbol trichotomy. As a result, the indexical view of proper names becomes Peirce's official view and is developed in interesting and credible ways.

In the meantime and in the very same series of lectures in which indexical proper names are declared to be beyond the scope of logic, Peirce puts forward a quite different view about proper names. This view is introduced by Peirce's denial that there is a "logical distinction between universal judgments such as *all men are mortal* and singular judgments such as *George Washington was a great man*" (W1: 252 [1865]). In making this remark, Peirce's principal aim is to deny that universal and singular judgments differ in extension.[3]

> This kind of quantity [extension] has reference to the sub-classes a class can be broken up into. Thus All men are animals may be broken up into . . . All good men are animals and All bad men are animals. . . . Now let us take a singular proposition. George Washington is mortal may be broken up into Young George Washington is mortal and Old George Washington is mortal and therefore stands in the same predicament as a universal [i.e., being subject to "break up"]. (1: 252–253 [1865])

This passage reveals Peirce to hold that the extension or denotation of "George Washington" is (somehow) greater than and different from the extension of both "Young George Washington" and "Old George Washington," and that these latter also differ in extension or denotation. Apropos of proper names, the suggestion is that the reference of "George Washington" varies in sentences containing these phrases. Apropos of objects, this kind of talk implies and Peirce asserts that George Washington is a class and that "George Washington" is the name of that class.[4] It is also implied that Young George Washington is either a sub-class and/or a member of George Washington, etc. It is further implied and Peirce asserts that proper names have intensions (comprehensions, connotations) as well as extensions. How so? One possible answer is provided by the following text:

Being	*Nothing*
All breadth	All depth
No depth	No breadth

We can conceive of terms so narrow that they are next to nothing, that is have an absolutely individual sphere. Such terms would be innumerable in number. We can also conceive of terms so high that they are next to *being*, that is have an entirely simple content. Such terms would also be innumerable.

Simple terms	Individual terms

But such terms though conceivable in one sense—that is intelligible in their conditions—are yet impossible. You never can narrow down to an individual. Do you say Daniel Webster is an individual? He is so in common parlance, but in logical strictness he is not. We think of certain images in our memory—a platform and a noble form uttering convincing and patriotic words—a statue—and the man whom that statue was taken for and the writer of this speech—that which these are in common is Daniel Webster. Thus, even the proper name of a man is a general term or the name of a class, for it names a class of sensations and thoughts. The true individual term the absolutely singular *this* and *that* cannot be reached. Whatever has comprehension must be general. (W1: 460–461 [1866])

This text suggests that the intension of "Daniel Webster" is the set of features common to all of our images, experiences, and thoughts of Daniel Webster. This is an interesting suggestion regarding the intensions of what Peirce is calling proper names here, but it appears to have been dropped. In any case, we need a vocabulary to characterize the kind of view or theory of proper names to which such suggestions belong.

Let us call the view that proper names are general terms denoting classes and accordingly have intensions as well as extensions, the collective or descriptive view of proper names. This view, which is the first to be developed by Peirce, is obviously inconsistent with the indexical view. (A given proper name cannot both have and lack a connotation, for example.) It is also obvious, though perhaps not to Peirce himself, that the collective view was devised to provide what Peirce took to be logically acceptable substitutes for purely indicative names. As such, especially if one accepts Peirce's final opinion regarding proper names, the collective theory isn't a theory about proper names at all. It is a theory about how to replace proper names by predicates or descriptions and individuals by collections or sets. The consideration of Peirce's position from this replacement point of view, though unavoidably touched upon in this chapter, is not our main concern. This chapter is devoted to the statement and evaluation of the collective theory as advertised, i.e., as a theory of proper names. It will trace its development up to the point at which Peirce appears to abandon it in favor of an indexical theory. This is a relatively manageable task, since the only significant further development of the collective view of names is provided in Peirce's 1870 "Description of a Notation for the Logic of Relatives."

Peirce's paper on the logic of relatives contains the most important appearance and perhaps the clearest statement of his collective view of proper names together with what he later calls a proof "that no sign can be absolutely and completely determinate" (5.506 [1905]). We are interested in this proof insofar as it is an attempt to show that proper names can be indeterminate.

An important condition for both understanding and criticizing this "proof" is the De Morgan-Peirce notion of a universe of discourse.

> I propose to use the term "universe" to denote that class of individuals *about* which alone the whole discourse is understood to run. The universe, therefore in this sense, as in Mr. De Morgan's, is different on different occasions. In this sense, moreover, discourse may run upon something which is not a subjective part of [or individual in] the universe; for instance, upon the qualities or collections of the individuals it contains. (W2: 366 [1870])

So every universe of discourse contains and must contain individuals. These are the referents of singular terms and/or proper names. But what does "individual" mean? Here is Peirce's multiply surprising answer.

> In reference to the doctrine of individuals, two distinctions should be borne in mind. The logical atom, or term not capable of logical division, must be one of which every predicate may be universally affirmed or denied. For, let A be such a term. Then, if it is neither true that all A is X nor that no A is X, it must be true that some A is X and some A is not X; and therefore A may be divided into A that is X and A that is not X, which is contrary to its nature as a logical atom. Such a term can be realized neither in thought nor in sense. Not in sense, because our organs of sense are special. . . . When I see a thing, I do not see that it is not sweet, nor do I see that it is sweet; and therefore what I see is capable of logical division into the sweet and the not sweet. . . . In thought, an absolutely determinate term cannot be realized, because, not being given by sense, such a concept would have to be formed by synthesis, and there would be no end to the synthesis because there is no limit to the number of possible predicates. A logical atom, then, like a point in space, would involve for its precise determination an endless process. We can only say, in a general way, that a term, however determinate, may be made more determinate still, but not that it can be made absolutely determinate. Such a term as "the second Philip of Macedon" is still capable of logical division—into Philip drunk and Philip sober, for example; but we call it individual because that which is denoted by it is in only one place at one time. It is a term not *absolutely*

indivisible, but indivisible as long as we neglect differences of time and the differences which accompany them. Such differences we habitually disregard in the logical division [classification] of substances. In the division of relations, etc., we do not, of course, disregard these differences, but we disregard some others. There is nothing to prevent almost any sort of difference from being conventionally neglected in some discourse and if I be a term which in consequence of such neglect becomes indivisible in that discourse, we have in that discourse $[I] = 1$. (W2: 389–390 [1870])

This frequently cited passage introduces an interesting and forward looking notion of semiotic and ontological discourse relativity but it fails to make a convincing use of it with respect to names and individuals. This can be shown by global consideration of Peirce's alleged proof, by a consideration of the criterion of determinacy/indeterminacy which is supposed to be involved in this proof, and with respect to examples which are supposed to show the indeterminacy of proper names.

It should first of all be noted that Peirce's argument or proof is simultaneously and hence equivocally about signs (words, concepts), objects (people), and sign-object relations (extension, denotation, reference). As such, it is as ambiguous as anything one could find in Aristotle or Kant. If one factors in possible and actual ambiguities regarding Peirce's use of "proper name" and logical and semantical difficulties relating to his failure to distinguish classes or sets from individuals, the resulting problem of interpretation is considerable.[5] To make matters worse, the overall design of the argument is obscure and none of its important assumptions are either explained or justified. For example, Peirce gives nothing by way of explanation or justification for the crucial assumption that individuals must be absolutely determinate, which he also seems to have an interest in denying, or for the equally crucial assumption that "there is no limit to the number of possible predicates." Yet the pairing of these assumptions is necessary to get to the conclusion that we cannot form or construct a concept of a completely determinate and/or atomic object. Peirce also assumes, without argument, that the only way one could construct a concept of a logically atomic object would be to produce an infinitely complete description of it. For Peirce, this means that the only way to identify an individual or to achieve a uniqueness of reference to an individual by means of a description is to produce a complete individual concept.

If we think of our would-be individual concept—the would-be intension of a collective or descriptive proper name—as a list of predicates and negations of predicates we can see why Peirce might have embraced this notion. An unused or unlisted predicate or negation might effect a division of any so-far determinate individual into two—one possessing and one lacking the appropriate property. Immunity to division requires, per impossible, that all of the predicates be used up. Peirce must also be assuming that no two individuals could satisfy the same infinite descriptive list, otherwise his insistence of completeness of description would be unmotivated.[6]

In case of division, our would-be atomic term (word or concept) would not achieve the minimal extension of one, but would have an extension of two or more. Hence the would-be descriptive proper name of the hitherto undivided individual would become applicable to many, i.e., general or ambiguous. Thus, as "triangle" is applicable to isoceles triangles and non-isoceles triangles, "Philip" becomes applicable to Philip Drunk and Philip Sober. As Peirce puts it considerably later "A sign must, from the nature of it, be applicable to different objects, supposing there happen to exist any such objects. This [is] true even of a proper name. Philip of Macedon may stand for Philip drunk, or for Philip sober, or for the collective Philip" (MS 9: Sec. 2 [1903?]).

Given what has been said thus far, it seems obvious that the individuals Peirce is discussing are constituted by lists or sets of properties and that, consequently, any changes in their lists would entail "losses of identity" and the creation of new identities in place of the old. Given this notion of identity, the only securely self-identical or unitary objects are changeless objects—be they eternal objects or their merely instantaneous cousins. Thus, in a footnote to the text we are considering, Peirce suggests and elsewhere asserts "That no object is individual but that the things the most concrete have still a certain amount of indeterminacy. Take Philip of Macedon for example. This object is logical[ly] divisible into Philip drunk & Philip sober; and so on; and you do not get down to anything completely determinate till you specify an indivisible instant of time, which is an ideal limit not attained in thought or in *re*" (MS 1104: 2).[7] Eventually, Peirce settles into the view that the only real or true individuals are instantaneous events or the instantaneous states of spatiotemporal singulars such as Philip of Macedon.

Peirce's Parmenidean or Leibnizian notion of identity is dominant in the universe of discourse in which "differences of time and the differences which accompany them" *are not* neglected. In this discourse, which is simply assumed to have privileged status with respect to what is absolutely or really so, Philip drunk is necessarily a different individual from Philip sober. Furthermore, Philip amusingly drunk would have to be a different individual from Philip disgustingly drunk and so on. As Peirce suggests, an infinite vocabulary would be needed to keep up with this situation.

Now let us consider how individuality is defined relative to a discourse in which "differences of time and the differences that accompany them" *are* habitually or conventionally ignored or *neglected*. The text suggests that such neglect is sufficient to take us from the privileged discourse to this habitual or ordinary discourse. Thus, if we take "Philip" as understood in the privileged universe and disallowed divisions of its object to be effected by temporal differences, we will end up with "Philip" referring to something which is individual in the sense that it occupies "only one place at one time." The same disallowance will yield a universe in which questions about the denotation of proper names do not normally or perhaps cannot come up. Hence, in this universe, proper names can serve as the individual terms that logic requires.

This is an attractive picture, but it is misleading and incomplete. To be told to neglect (all?) temporal differences is not to be told that what we call an individual in the resulting discourse counts as such because "it occupies only one place at one time."[8] This is a new and different conception of individuality.

In addition, to be told to neglect certain differences is not to be told what differences or predicates will yield identifying descriptions in the resulting discourse. (We still need to know the intension of "Philip" in order to determine its extension.) But on this question, which must be answered if there is to be a collective or descriptive substitute for purely denotative or indexical names, Peirce has nothing to say. This is disappointing with respect to the completion of a collective theory of names, but insignificant with respect to the viability of Peirce's claim that proper names can be indeterminate. For let us suppose that Peirce had supplied an identifying description to serve as the intension of "Philip" in the habitual universe. Then "Philip" would denote an atomic or indivisible set. In the privileged universe, "Philip" would have a different intension and a different extension, namely a set with sub-sets. In neither universe would the referent of "Philip" be a person, though Peirce's identification of persons and sets undoubtedly encouraged him to believe otherwise. This semantical situation entirely undercuts Peirce's attempt to show that "Philip" can be indeterminate, since that would require "Philip" to be the same name in all cases. All that is really common to the different cases is "Philip" qua vocable. Let us turn, then, to Peirce's criterion for determinacy and indeterminacy.

Peirce's stated criterion for indeterminacy and/or divisibility is explicitly quantificational and is easily seen to be applicable to indisputably general terms. For example, it is neither true that all triangles are isoceles nor that no triangles are isoceles and so it is true that some triangles are isoceles and some triangles are not isoceles. Hence the term "triangle" is indeterminate and the class of triangles is divided into the sub-class of triangles that are isoceles and the sub-class of triangles that are not isoceles. That Peirce is obliged to apply the same criterion to proper names is obvious and is at least partially conceded in remarks such as the following: "Every term . . . must have some generality. For it is not universally true that Philip is drunk nor [universally true] that Philip is sober" (MS 288B: 213 [c1905]).

Following this line of reasoning, one would expect Peirce to provide inter-pretations for sentences containing subjects such as "All Philip" or "All Philips" and of sentences such as "All Philip is drunk" or "All Philips are drunk," etc. Such peculiar sentences would be permissible on a collective view of names, but Peirce never produces them. In fact none of the exam-ples Peirce produces are quantified, and those that are implicitly quantified do not involve putting the quantifiers in such an odd position.[9]

What Peirce probably had in mind, given his examples, are quantificat-ions over times and states. Thus "Philip is drunk" could be tortured into "Philip is always (at all times) drunk" and so on. Given this rather implau-sible quantificational scheme, "Philip is neither drunk nor sober" would be rendered as "Philip is neither always drunk nor never drunk (always sober)" from which it would follow that "Philip is sometimes drunk and Philip is sometimes sober."[10] (Unquantified variations might have it that Philip is not essentially drunk, etc.) The point to note with respect to such cases is that—apart from Peirce's conceptions of identity and reference—they provide no reason for worries about either the "division" of Philip or about variations in the reference of his name. The same is true of the somewhat more com-plicated cases involving quantification over Philip's states.

By quantifying over states we might render "Philip is neither drunk nor sober" as "It is neither the case that all Philip states are states of a drunken man nor that no Philip states are states of a drunken man" from which it would follow that some Philip states are drunken and some are not. Here we would get a division of the set of Philip states, but not a division of Philip unless we assume, as Peirce might, that Philip is the set of Philip states. The discourse appropriate for statements such as "All Philip states are states of a drunken man" would appear to contain singular terms standing for Philip states and a predicate of the form "_____is a Philip state." This predicate is a general term which determines the set of Philip states, not Philip's proper name. Whether or not there is a nominative occurrence of "Philip" in such a statement depends on how "_____ is a Philip state" is defined. But in any case, no worries about Philip's identity will turn up.

This concludes our discussion of Peirce's failure to apply his stated cri-terion to his cases. It remains to discuss a case of his favorite sort with-out the benefit of the usual quasi-formal background. For this purpose I have chosen a pragmaticized example of Peirce's indeterminacy thesis with respect to proper names. By considering this example we may discover an a posteriori linguistic version of the a priori thesis we have been consider-ing.

Every sign has a single object, though this single object may be a single set or a single continuum of objects. No general description can identify an object. But the common sense of the interpreter of the sign will assure him that the object must be one of a limited collection of objects. Suppose, for example, two Englishmen to meet in a continental railway carriage. The total number of subjects of which there is any appreciable probability that one will speak to the other perhaps does not exceed a million; and each will have perhaps half that million not far below the surface of consciousness, so that each unit of it is ready to suggest itself. If one mentions Charles the Second, the other need not consider what possible Charles the Second is meant. It is no doubt the English Charles the Second. Charles the Second of England was quite a different man on different days; and it might be said that without further specification the subject is not identified. But the two Englishmen have no purpose of splitting hairs in their talk; and the latitude of interpretation which constitutes the indeterminacy of a sign must be understood as a latitude which might affect the achievement of a purpose. (5.448 n. 1 [1905])

There are actually two examples here. The first concerns a possible question as to whether "Charles the Second" as it occurs in some sentence context, say "Charles the Second was a foolish King," is to be interpreted as referring to the English Charles the Second or the French Charles the Second. This is a perfectly straightforward question about reference. There actually are two different persons called Charles the Second and therefore, in a suitable context, there could be a question about which person is being referred to.

The second example is a familiar but peculiar one having to do with whether or not the fact that Charles the Second of England was "quite a different man on different days" renders "Charles the Second of England" indeterminate in reference. Peirce thinks that this is the same kind of case as the first. It just fails to come up because the Englishmen, unlike logicians or semioticians, have no purpose of splitting hairs. But the cases are different.

If Charles the Second of England were literally a different man on different days, and this is difficult if not impossible to imagine,[11] one might be able to insert a question about reference here. What is meant, however, is that Charles's behavior was quite different on different days or perhaps even that Charles's personality was "multiple" in a clinical sense. However, such changes as these do not affect the reference of a person's name.

Nevertheless Peirce thinks that there are different objects of some sort—immediate objects, intentional objects, whatever—generated by such situations. Here is a possible scenario as to how ordinary language may have suggested this idea to him.

Let us suppose that Charles the Second would occasionally make stupid or outrageous decisions under the influence of alcohol but was sly and astute when not so influenced. Let us suppose further that one of our Englishmen asserts that "Charles the Second was a foolish King." Without doing any violence to English, the second Englishmen could ask the following Peircean questions:

Do you mean Charles Drunk or Charles Sober or Charles on the whole?

Are you referring to Charles Drunk, Charles Sober, or Charles on the whole?

Are you talking about Charles Drunk, Charles Sober, or Charles on the whole?

Peirce understands such questions as requests for an identification of the object or referent of "Charles the Second." But this is to misunderstand such questions. The reference of "Charles the Second" is in no way problematic. The assessment of the claim that Charles was a foolish King is what is at stake. We can imagine our second Englishman saying, "That was certainly true of Charles when he was drunk, but not when he was sober." For Peirce, such talk might clinch the case for a plurality of Charleses and hence for the referential indeterminacy of Charles's name. He would definitely be encouraged to take this position by his attempt to replace proper names by descriptions true of their bearers. But our Englishman is only responding to the original claim—perhaps suggesting that it needs qualification if it is to be accepted.

Suppose our first Englishman, the utterer of "Charles the Second was a foolish King," said, in response to the listed question, "I meant Charles on the (as a) whole." There are two things to be observed regarding such a case. First, that this new statement may simply mean that Charles, throughout his life, was more often foolish than not. It certainly does not mean, "I meant Charles as the collection of his states." For, contra Peirce, a man is not a collection. Second, it should be observed that "Charles the Second was a foolish King" has been replaced by "On the whole, Charles the Second was a foolish King." The latter statement does not have the same meaning as the former, though it may be more secure with respect to truth value. It is, perhaps, what the speaker "intended to say," but didn't. Similarly, if the utterer says "I meant Charles drunk," he is not providing a "further determination" of the reference of "Charles the Second" in the form of an odd object. He is saying that, in the interest of truth, it would have been better to say "Charles the Second was a foolish King when drunk."

Peirce's way of thinking about reference (extension, denotation, breadth, etc.) systematically confounds questions about what a statement or its subject denotes with questions about whether a statement is true or not. His collective theory of names reflects this confusion. Peirce rids himself of at least part of his collective theory of names. But he never gets untangled with respect to reference and truth.

Though Peirce does not explicitly notice or announce any changes of view, he rejects both the collective theory and the principal assumption behind it in 1885. These rejections appear most forcefully in an unpublished review of Royce's *The Religious Aspect of Philosophy*. In this review, Peirce responds to Royce's polemic against his conception of reality, as follows:

> Dr. Royce's main argument in support of his own opinion, to the confusion of Thrasymachus [Peirce], is drawn from the existence of error. Namely, the subject of an erroneous proposition could not be identified with the subject of the corresponding true proposition, except by being completely known, and in that knowledge no error would be possible. The truth must, therefore, be present to the actual consciousness of a living being. This is an argument drawn from Formal logic, for Formal logic it is which inquires how different propositions are made to refer to the same subject, and the like. . . . [Unfortunately] Dr. Royce's argument from formal logic overlooks one of the most important discoveries that have lately resulted from the study of that exact branch of philosophy. He seems to think that the real subject of a proposition can be denoted by a general term of the proposition; that is, that precisely what it is that you are talking about can be distinguished from other things by giving a general description of it. Kant already showed, in a celebrated passage of his cataclysmic work, that this is not so; and recent studies in formal logic have put it in a clearer light. We now find that, besides general terms, two other kinds of signs are perfectly indispensable in all reasoning. One of these kinds is the *index*, which like a pointing finger exercises a real physiological *force* over the attention, like the power of a mesmerizer, and directs it to a particular object of sense. One such index at least must enter into every proposition, its function being to designate the subject of discourse. Now observe that Dr. Royce does not merely say that there are no means by which an erroneous proposition can be produced; what he says is that the conception of an erroneous proposition (without an actual including consciousness) is absurd. If the subject of discourse had to be distinguished from other things, if at all, by a general term, that is, by its peculiar characters, it would be quite true that its complete segregation would require a full knowledge of its characters and would preclude ignorance. But the index, which in point of fact alone can designate the subject of a proposition, designates it without implying any characters at all. (8.41 [1885])

After finally opting for a modified theory of indexical subjects, Peirce consistently denies that proper names have connotations of senses. But, as has been shown, he subsequently and frequently asserts that proper names are or can be indeterminate. This practice raises questions of consistency which ought to be addressed in a more complete study of Peirce's theories of names and individuals.[12]

NOTES

1. By "earliest" I mean earliest published in W1 (1857–1866).

2. This way of characterizing the icon, index, symbol trichotomy is also applied to arguments. An induction, for example, "informs us as to extension but not as to comprehension, that is it represents a representation which has extension without comprehension, which is an Index" (W1: 485 [1866]). For further references regarding terms see W1: 174 [1865], 323 [1865], 467–468 [1866].

3. Peirce also denies that there is any essential difference in "number" between universal and singular judgments. He does so by pointing out that "All the blackboards in this room coincide in point of number with this one" (W1: 252 [1865]. His point is that a universal judgment may yield knowledge of only one thing rather than many, but his language suggests the sort of substitution of unit classes for individuals which one can sometimes find recommended in text presentations of syllogistic logic. For syllogistic purposes, we are told, we may replace subjects such as "George Washington" with expressions of the form "All members of the class containing only George Washington." It is worth noting at the outset that Peirce exhibits no apparent interest in this device. But see notes 5 and 8 for further discussion.

4. Here I paraphrase Peirce's claim that "Daniel Webster, for example, is a class embracing Daniel Webster under 50 years of age & Daniel Webster over 50 years of age" (W3: 9 [1872]).

5. On the question of sets and individuals, Richard Martin claimed that when Peirce "speaks of a 'logical atom, or term not capable of logical division' . . . the context makes clear that the logical atom is a unit class" (*Martin 1976b*: 236). Yet Martin also points out that "Peirce apparently nowhere [between 1867–1885?] distinguishes between an individual (or singular) and the class whose only member is that individual" (*Martin 1976b*: 238). My contention is that since Peirce failed to make this essential distinction, the assertion that "the logical atom is a unit class" is anachronistic. It is also inconsistent with Peirce's explicit assertions that individuals are classes. Martin notwithstanding, the concept of an individual put forward in the Logic of Relatives paper, and at 3.216 [1885], is a concept of an indivisible class.

6. Peirce later relieves his notion of individuality of its dependence on the principle of the identity of indiscernibles via the doctrine of hecceity. See 3.611 [1901].

7. In the footnote to W1: 290–291 Peirce claims that "whatever lasts for any time, however short, is capable of logical division, because in that time it will undergo some change in its relations."

With respect to strict individuals and slightly less than strictly individual singulars, the following texts are interesting.

> The individual . . . is that which is in every respect determinate. It is, therefore, the instantaneous state of an existent. The *singular* . . . is that which

has a continuity of existence in time and at each instant is absolutely determinate (MS 280: 27 [1905]).

> Philip is not, on the whole, either drunk or sober. The only strict individual is the living deed, the actual event. . . . (MS 515: 23^P in a variant sequence)

8. According to Richard Martin, "We are not to construe this passage as stating that individuality is to be determined by occupying 'only one place at one time,' for even there further determination is possible—for example, Philip kind, Philip brave. Some further discourse might be available in which Philip kind is one singular, Philip brave another, and both at the same place and time" (*Martin 1976*: 237). The interpretative principle which seems to be operative here is that Peirce cannot mean what he says, because it is mistaken. This is literally being charitable to a fault! But Martin's counterexample, which aims to show that two "individuals" may occupy the same place at the same instant, though difficult to evaluate, is to the point. My own inclination would be to argue that such problems depend, in part, upon the false or at least unnecessary assumption that individuals must be completely determinate. I have no doubt, for example, that the actual and perfectly individual Philip was neither drunk nor sober through many intervals of time and, if you like, at *many an instant*, e.g., when he was in borderline states.

9. If Peirce actually had the notion of a unit set, as Martin suggests, he could use a plural form to interpret "Philip is drunk," etc. The form would be "All members of the set containing only Philip are drunk," etc. But Peirce would then have to explain the occurrence of "Philip" in this formula and would face a dilemma in doing so. For either "Philip" occurs indexically (non-collectively) or collectively. If "Philip" occurs indexically, then the "collective subject" is parasitic upon the non-collective or indexical subject. If "Philip" occurs collectively, as a unit set, then our collective subject would read "All members of the set which contains only (all members of?) the set which contains only Philip" or something of this sort. But then the question about the occurrence of "Philip" will recur.

10. That Peirce may have had some such interpretation in mind is suggested by many texts already considered and by the following:

> Time is that diversity of existence whereby that which is existentially a subject is enabled to receive contrary determinations in existence. Philip is drunk and Philip is sober would be absurd, did not time make the Philip of this morning another Philip than the Philip of last night (1.494 [1896]).

> Philip is not, on the whole, either drunk or sober. (MS 515: 23^P in a variant sequence)

That the quantifiers probably belong on times is implicitly recognized in Manley Thompson's remarks concerning Philip's identity (internal quotations from 3.93).

> The second Philip of Macedon is a singular from the point of view of a human lifetime. A singular statement such as "The second Philip of Macedon is mortal" is true because the predicate applies to the subject during every moment of its existence. But "The second Philip of Macedon is drunk" is not true of this same subject. It is true only of a certain temporal part of this subject, and is therefore true of a different singular. Accurately expressed the statement has the form "The second Philip of Macedon at time t_1 is drunk." But then two statements, one affirming that Philip at t_1

is drunk and the other affirming that Philip at t_2 is sober, are about different singulars. Yet we say that they are about the same subject because "in the logical division of substances" we "habitually disregard differences of time and the differences that accompany them" (*Thompson 1981*: 138).

These exegetical remarks capture Peirce's drift remarkably well. The unargued assumption which they articulate is roughly this: Statements and/or predicates which "apply to" or are "true of" a subject at different times (always, sometimes, t_1, t_2) are or must be "about" (refer to) different subjects or different "temporal parts" of the same subject. Though Peirce never explicitly states his position exactly like this, he must have held something similar.

11. Would we have to contemplate the possibility that Charles was Henry the Eighth yesterday, the King of Siam the day before, and is now Charles again?

12. For an excellent account of what I have called Peirce's indexical theory of proper names see *Hilpinen 1982* and *Hilpinen 1989*. Both of Hilpinen's papers profitably compare Peirce's theory to contemporary theories. The latter relates Peirce's theory to many other concerns in Peirce's philosophy of language and in contemporary philosophy of language as well.

28.

CHARLES PEIRCE'S THEORY OF PROPER NAMES

Jeffrey R. DiLeo

§1. Methodological Considerations

Charles Peirce's thoughts on the semantic and pragmatic roles of proper names are distributed throughout his writings. From as early as the Harvard Lectures of 1865 where he wrote that a proper name denotes "by a convention made at baptism" (W1: 170) through the first decade of this century Peirce time and again concerned himself with proper names. He recognized them as an essential feature of language, writing that "[e]very language must have proper names" (2.328). Although the depth and intensity of his studies and interest concerning proper names varied over the years, they nevertheless still showed a clear anticipation of contemporary semantic theories of proper names, particularly those which emphasize the indexical dimension of singular reference by proper names (for example *Castañeda 1989* and *Burge 1983*).

Any attempt to consider Peirce's theory of proper names is presented with a problem at the outset; although Peirce developed doctrines that may be considered as components of a formal theory of proper names, and commented on proper names per se with some frequency, he never systematically developed a complete theory of proper names. That is to say, his work on proper names was never expounded in any systematic way, nor was it ever the main topic of any single comprehensive work. His discussions of proper names are to be found in many different contexts and no paper contains more than a cursory treatment of *aspects* of a theory of proper names. Such being the case, it might seem presumptuous to assume that Peirce ever even had something like a general theory of proper names. However, it is possible to extrapolate a general theory from the rich corpus of writings that Peirce has left us on proper names without distorting his views on the subject. Our discussion will be restricted to the years 1891–1914, which have been called by Max Fisch (*1967*: 192) Peirce's "*Monist* Period."

As such, this essay may be regarded in at least two ways: first, as a construction of a Peircean theory of proper names that is consistent with his *Monist* period work on proper names; and second, one might regard it as a *reconstruction* of his late theory of proper names. It is hard to believe that even though Peirce wrote extensively on proper names, that he nevertheless did not have one complete program of analysis in mind. Therefore, I regard this paper as a *prolegomenon* to a reconstruction of Peirce's complete theory of proper names. As such, it may be regarded as an introduction to Peirce's analysis of proper names. Its aim is to present some of Peirce's main theses concerning proper names—theses that represent the core of his general theory.

Our discussion will begin with a brief survey of Peirce's doctrine of signs for, as we shall see, his whole *Monist* period analysis of proper names is grounded in this doctrine. Section 2 will provide those unfamiliar with Peirce's semiotic with a general account sufficient to set a context for the subsequent discussion of his analysis of proper names.

§2. Logic as Semeiotic: Peirce's Theory of Signs

Peirce's analysis of proper names was made through the conceptual framework generated by his semeiotic (or semiotic). As such, given that he believed that "[l]ogic, in its general sense, is . . . only another name for *semiotic*, the quasi-necessary, or formal, doctrine of signs" (2.227), his analysis of proper names was not only semiotical, but also logical.

For Peirce (2.93), logic was "the science of the general necessary laws of Signs" and had three branches: *Critical Logic* or "logic in the narrow sense"—"the theory of the general conditions of the reference of Symbols and other Signs to their professed Objects, that is, it is the theory of the conditions of truth"; *Speculative Grammar*—"the doctrine of the general conditions of symbols and other signs having the significant character," and *Speculative Rhetoric*—"the doctrine of the general conditions of the reference of Symbols and other Signs to the Interpretants which they aim to determine." This division of logic or semeiotic is reflected today in the *semantics, syntactics, pragmatics* distinction primarily attributable to the work of Charles Morris. In a division reminiscent of Peirce, the positivist Morris (*1937*: 4ff.) divided semiotic into semantics, syntactics, and pragmatics. Morris' division of semiotic has been widely accepted and followed, e.g., by Carnap and Tarski.[1]

"A sign," wrote Peirce around 1897,

is something which stands to somebody for something in some respect or capacity. It addresses somebody, that is, creates in the mind of that person an equivalent sign, or perhaps a more developed sign. That sign which it creates I call the *interpretant* of the first sign. The sign stands for something, its *object.* (2.228)

That is to say, a sign is the subject of a triadic relation, involving the *sign* itself, an *object* or what the sign stands for, and an *interpretant* or the equivalent sign which the first sign creates in the mind of the person apprehending it. Of this definition of a sign, Charles Morris (*1970:* 19) correctly observed the following significant features: "it does not limit signs to language signs; it does not introduce the term 'meaning'; the interpretant is said to be 'in the mind'; the term 'sign' is not completely clarified since the interpretant of a sign is itself said to be a sign; there is no reference to action or behavior."

Peirce embedded his semiotic in the metaphysics of his categories. He held that all phenomena or experience whatsoever possess three modes of being or aspects, specifiable under the categories of *firstness,* or the mode of being of that which is such as it is, positively and without reference to anything else; *secondness,* or the mode of being of fact, struggle or "hereness and nowness" whose very essence is its "thisness"; and *thirdness,* or the mode of being of that which is such as it is in bringing firstness and secondness into relation with each other. His definition of the sign from around 1902 better reflects the relationship between semiotic and the categories, although it is more obscure to those who are unfamiliar with the categories. "A *Sign,* or *Representamen,*" wrote Peirce, "is a First which stands in such a genuine triadic relation to a Second, called its *Object,* as to be capable of determining a Third, called its *Interpretant,* to assume the same triadic relation to its Object in which it stands itself to the same Object" (2.274). One virtue of this definition of the sign as opposed to the earlier definition is that the interpretant is not necessarily "in the mind." It should be noted that Peirce thought that in order

> to get more distinct notions of what the Object of a Sign in general is, and what the Interpretant in general is, it is needful to distinguish two senses of "Object" and three of "Interpretant." It would be better to carry the division further; but these two divisions are enough to occupy my remaining years. (NEM 3: 842)

It would have been interesting to have seen how Peirce would have further developed the sense of interpretant and object.

For Peirce, the three senses of interpretant were called the *immediate interpretant, dynamical interpretant,* and *final interpretant.* He described the immediate interpretant as "the interpretant as it is revealed in the right understanding of the Sign itself, and is ordinarily called the *meaning* of the sign." The dynamical interpretant "is the actual effect which the Sign, as a Sign, really determines" (4.536); "it is the volitional element of Interpretation" (NEM 3: 844). The final interpretant is that "which refers to the manner in which the Sign tends to represent itself to be related to its Object" (4.536); it is "that which *would* finally be decided to be the true interpretation if consideration of the matter were carried so far that an ultimate opinion were reached" (NEM 3: 844).

On the other hand, Peirce labelled the two senses of object as the *immediate object* and *dynamic object*. The former is "the Object as the Sign itself represents it, and whose Being is thus dependent upon the Representation of it in the Sign" (4: 536); it is "the Object as cognized in the Sign and therefore an Idea" (NEM 3: 842). While the latter "is the Reality which by some means contrives to determine the Sign to its Representation" (4.536); "the Object as it is regardless of any particular aspect of it, the Object in such relations as unlimited and final study would show it to be" (NEM 3: 842).

In discussing Peirce's distinction between the object and the interpretant of a sign it is hard not to recall Gottlob Frege's distinction between the *Sinn* (sense) and *Bedeutung* (denotation or reference) of a sign. It is also interesting to note here that Frege was working out these ideas around the same time as Peirce for his "Über Sinn und Bedeutung" was first published in 1892. Yet although Frege's distinction between the *Sinn* and *Bedeutung* of a sign (word, sign combination, expression) is similar in some respects to Peirce's distinction between the interpretant and the object of a sign, the two doctrines are quite different. The most obvious difference is that Peirce's three senses of interpretant and two senses of object do not directly correspond with Frege's *Sinn* and *Bedeutung* distinction. This results in the possibility of parallels on a number of different levels. For example, it might be argued that Peirce's immediate/dynamic object distinction parallels Frege's *Sinn* and *Bedeutung*, or that one sense of Peirce's interpretant and one sense of his object parallel Frege's distinction. The consequence of this is that there is no clear parallel.

The triadic character of the sign relation led Peirce to distinguish three main divisions of signs: (1) the sign in itself, (2) the sign as related to its object, and (3) the sign as interpreted to represent its object. In turn, (1), (2), and (3) are phenomena which each possess three aspects of being and as such are specifiable under the categories of Firstness, Secondness, and Thirdness. As a result, all of the three divisions become trichotomies.

The first division (trichotomy) specifies three types of representamens. It is subdivided into the *qualisign*, a quality which is a sign; the *sinsign*, an actual existent thing or event which is a sign; and the *legisign*, a law which is a sign. The second division demarcates three ways the sign may be related to its object. This division is subdivided into the *icon*, a sign which refers to the object that it denotes merely by virtue of characters of its own, and which it possesses just the same whether the object exists or not; the *index*, a sign which refers to the object that it denotes by virtue of being really affected by that object; and the *symbol*, a sign which refers to the object that it denotes by virtue of a law. Finally, the third division distinguishes three ways in which interpretants may represent signs as related to their objects. It is subdivided into the *rheme*, a sign of qualitative possibility for its interpretant; *dicisign*, a sign of actual existence for its interpretant; and *argument*, a sign of law for its interpretant (2.233–272).

Consequently, the application of Peirce's principle (2.235–237)—which claims that a first can determine only a first; a second can determine only a second or a first; and a third can determine a third, a second, or a first— to the three trichotomies yields Peirce's ten classes of signs: (1) (rhematic iconic) qualisign; (2) (rhematic) iconic sinsign; (3) rhematic indexical sinsign; (4) dicent (indexical) sinsign; (5) (rhematic) iconic legisign; (6) rhematic indexical legisign; (7) dicent indexical legisign; (8) rhematic symbolic (legisign); (9) dicent symbolic (legisign); and (10) argument (symbolic legisign) (2.254–264).[2]

§3. Our Starting Point: The Semeiotic Classification of Proper Names

In his letter to Lady Welby of 12 October 1904, Peirce (SS: 35) wrote that proper names, as one of the ten principal classes of signs, were *rhematic indexical legisigns*. Around 1903, he had defined a rhematic indexical legisign as "any general type or law, however established, which requires each instance of it to be really affected by its Object in such a manner as merely to draw attention to that Object" (2.259). Although Peirce did not explain why he classified proper names as rhematic indexical legisigns in his letter to Lady Welby, he did briefly describe the rhematic, indexical, and legisignic aspects of proper names to her.

Peirce told Lady Welby that a proper name, as a legisign, "has a definite identity," though usually admits of "a great variety of appearances" (SS: 32). He described the indexical character of proper names as that of being determined by their *dynamic object* "by virtue of being in a real relation to it" (SS: 33). Finally, he said that as rhemes proper names are "not true nor false, like almost any single word except 'yes' and 'no,' which are almost peculiar to modern languages," and also are defined as signs which represent their "signified interpretant as *if it were* a character or mark (or as being so)" (SS: 34).

Peirce's remarks to Lady Welby on 12 October, 1904 and, in particular, his classification of proper names as rhematic indexical legisigns are crucial to an understanding of his views on proper names. This classification presents us with a gateway leading into Peirce's views on proper names. It is one of the few places where Peirce *explicitly* classifies proper names within his tenfold classification of signs. Also, it is significant that he does not classify proper names among any of the other nine classes. Was this meant to be an exclusive classification? I think so, and have presented a defence of this view elsewhere (*DiLeo 1989*).

This classification is significant for, as we have seen from our brief look into Peirce's semeiotic, it reveals at least (a) what a proper name in itself is as a representamen (namely, a legisign), (b) the way in which a proper name as a representamen is related to its (dynamic) object (namely, indexically), and (c) the way in which its interpretant is related to its object (namely, rhematically). Each of these characteristic features of proper names is developed more closely by Peirce in various other writings, thus corroborating the validity and soundness of his classification of proper names as rhematic indexical legisigns. Through a closer examination of these three semiotic characteristics of proper names, one may gain a solid preliminary understanding of Peirce's general theory of proper names. Let us begin with a look at some of Peirce's remarks on the rhematic character of proper names.

§4. Rhemes, Indices, and Propositions: On the Rhematic Character of Proper Names

Peirce's rheme belongs to his trichotomy of signs, which distinguishes the ways in which interpretants may represent signs as related to their objects. This triad is divided into rhemes (or senses), dicents (or dicisigns), and arguments; a division which corresponds to the traditional distinction between term, proposition, and argument.

Around 1903, Peirce (2.250) defined a rheme as "a sign which, for its Interpretant, is a Sign of qualitative Possibility, that is, is understood as representing such and such a kind of possible object," adding that "[a]ny rheme, perhaps, will afford some information; but it is not interpreted as doing so." In other words, a rheme is a sign which is interpreted by its final interpretant as representing some quality which *might* be found in a possibly existing object. He continued by stating that

> If parts of a proposition be erased so as to leave *blanks* in their places, and if these blanks are of such a nature that if each of them be filled by a proper name the result will be a proposition, then the blank form of proposition which was first produced by the erasures is termed a *rheme*. According as the number of blanks in a rheme is 0, 1, 2, 3, etc., it may be termed a *medad* (from μηδέν, nothing), *monad, dyad, triad,* etc., rheme. (2.272)

The description of a rheme as a "blank form of proposition" was used frequently by Peirce in his *Monist* period. The above definition of a rheme stresses that it may have any number of "blanks," yet in other definitions the emphasis is different. For example, in his 1906 paper for the *Monist* entitled "Prolegomena to an Apology for Pragmaticism" the terms rheme and predicate were pointed out by Peirce as synonymous and were defined as a "blank form of proposition which might have resulted by striking out certain parts of a proposition, and leaving a *blank* in the place of each, the parts stricken out being such that if each blank were filled with a proper name, a proposition (however nonsensical) would thereby be recomposed" (4.560). Notice that this suggests that any proper name(s) may be used to fill the blank(s) of any rheme. For example, it would be just as appropriate to fill the blank of the rheme "_____ is a woman" with "Plato" as it would be to fill it with "Sappho." The only difference being that one would be regarded as nonsense, whereas the other would be perfectly sensible. To understand this better, let us now look at Peirce's proposition a bit more closely.

As to the general structure of the proposition, Peirce held that it consisted in a *predicate* and a *subject* (or subjects) such that a "proper name or term filling the blank of a rheme, is a *subject* of the proposition in which it occurs" and the "*rheme*, considered as part of a proposition, is its *predicate*" (MS 280: 103). Yet a proposition is a sign, and as such must have an interpretant and object. Peirce wrote that the "interpretant of a proposition is its predicate; its object is the things denoted by its subject or subjects" (5.473). Thus, the proposition "Plato is a woman" would be described by Peirce as a propositional sign (dicisign) with an iconic interpretant that misrepresents its object. In other words, Peirce's notion of the proposition can account for "false predications."

Part of the function of the proposition is to indicate objects. Around 1904 Peirce wrote that a "proposition is a sign which separately, or independently, indicates its object" (NEM 4: 242). The subject or subjects of the proposition each "denotes an individual object irrationally by virtue of being actually connected with that object" (MS 280: 97). A proper name then is always found in the subject position of a proposition, and its interpretant is always a rheme. It may be noted here that a proper name, because of its relationship with the rheme, cannot be a proposition *per se*, nor can it express a proposition. That is to say, a proper name needs an interpretant in order to be or express a proposition.

Finally it should be noted that Peirce's proposition is different from what he called an *assertion*. He wrote to Lady Welby (SS: 34) that a "*proposition* as I use that term, is a dicent symbol" where a "dicent is not an assertion, but is a sign *capable* of being asserted."

In effect, what Peirce's rheme does is widen the traditional notion of a term in order to accommodate his semeiotic concepts of the object (subject) and interpretant (predicate) of the propositional sign. Peirce wrote (NEM 4: 246) that

> On the whole, it appears to me that the only difference between my rhema and the "term" of other logicians is that the latter contains no explicit recognition of its own fragmentary nature. But this is as much as to say that logically their meaning is the same; and it is for that reason that I venture to use the old, familiar word "term" to denote the rhema.

Thus, the subject and the predicate of a proposition may be called terms. For Peirce, the logical term, which is a class name, is a rheme (4.538). It should be noted that the terms of a proposition can sometimes be very complex. For example, he wrote that "the Subject of the proposition, 'Whatever Spaniard there may be adores some woman' may best be regarded as, 'Take any individual, A, in the universe, and then there will be some individual, B, in the universe, such that A and B in this order form a dyad of which what follows is true,' the Predicate being '_____ is either not a Spaniard or else adores a woman that is _____' " (2.330). The subject of this proposition would be considered a term by Peirce.

As a point of reference to possibly more familiar work, one might note that Peirce's rheme is analogous to what Bertrand Russell called a "propositional function." Russell (*1903*: 19) explains (but does not define) the notion of a propositional function as follows: "Φx is a propositional function if, for every value of x, Φx is a proposition, determinate when x is given." Similarly, Peirce's rheme might be explained as follows: " '_____ is x' is a rheme if, for every proper name that fills the blank, '_____ is x' is a proposition, determinate when the blank is filled." For Peirce, the x in "_____ is x" is some property or quality, such that it might be embodied in an object, possibly either existing or fictitious, and the rheme is monadic. It must be noted here that x cannot be another proper name, for if it is, then the rheme is actually dyadic, rather than monadic. For example, Peirce writes that "in the proposition 'Boz is Charles Dickens,' the subjects are Boz and Charles Dickens and the predicate is "*identical with*" (2.341). The resulting dyadic rheme is "_____ identical with _____." This serves to show that predicates are not always monadic and that identity statements are treated as dyadic rhemes.

One point of difference between Peirce and Russell concerned the truth values of their respective rhemes and propositional functions. Whereas Peirce (SS: 34) held that "a rheme is any sign that is not true nor false" while a proposition is either true or false (2.310), Russell (*1903*: 20) stated that a "propositional function in general will be true for some values of the variable and false for others." Finally, let us look a bit closer at the relationship between the proper name and the rheme.

Peirce viewed the proper name and the rheme as joined together in the proposition by an index. He wrote that it "may be asked what is the nature of the sign which joins 'Socrates' to '_____ is wise,' so as to make the proposition 'Socrates is wise,' " and replied that "it is an index" (NEM 4: 246). Thus, it was Peirce's view that part of the function of an index was to provide "Socrates" and "_____ is wise" with "something to indicate that they are to be taken as signs of the same object" (NEM 4. 246). He stated that it was the *replicas* of the two signs "Socrates" and "_____ is wise" that were connected in the proposition, not the two signs "Socrates" and "_____ is wise."

> We do not say that "_____ is wise," as a general sign, is connected specially with Socrates, but only that it is so as here used. The two replicas of the words "Socrates" and "wise" are *hic et nunc*, and their conjunction is a part of their occurrence *hic et nunc*. They form a pair of reacting things which the index of connection denotes in their present reaction, and not in a general way; although it is possible to generalize the mode of this reaction like any other. There will be no objection to a generalization which shall call the mark of junction a *copula*, provided it be recognized that, in itself, it is not general, but is an *index*. (NEM 4: 247)

Let us now turn our attention to the intriguing indexical character of proper names.

§5. Indices and Haecceity: On the Indexical Character of Proper Names

One of the central features of Peirce's analysis of proper names was his treatment of them as *indices* or *indexical* signs. In general, Peirce classified indices under the trichotomy of signs which divided signs according to their relation to their objects. He considered each indexical sign as a sign of its object by virtue of a real relation to it. In 1901, Peirce (2.305) wrote that an index is a "sign, or representation which refers to its object not so much because of any similarity or analogy with it, nor because it is associated with general characters which that object happens to possess, as because it is in dynamical (including spatial) connection both with the individual object, on the one hand, and with the senses or memory of the person for whom it serves as a sign, on the other hand." Peirce (2.306) then attributed three characteristic marks or functions to indices which, in turn, distinguished and differentiated them from other signs:

(1) Indices direct the attention to their objects by blind compulsion;
(2) Indices bear no significant resemblance to their objects; and
(3) Indices refer to individuals, single units, single collections of units, or single continua.

As we shall see, all of these three characteristic functions of an index are in turn characteristic functions of proper names.

Section 4 showed us that proper names occur in propositions as subjects. In addition, Peirce (2.357) maintained that "[e]very subject partakes of the nature of an index, in that its function is the characteristic function of an index, that of forcing the attention upon its object." Therefore, since the subject or subjects of every proposition must be some type of indexical sign, every proper name must be regarded as some type of indexical sign. But of what type of indexical sign are proper names?

Three types of indexical signs were described as candidates for the subject of a proposition by Peirce in his "Syllabus," c. 1902. He wrote that

> [e]very subject of a proposition, unless it is either an Index (like the environment of the interlocutors, or something attracting attention in that environment, as the pointing finger of the speaker) or a Subindex (like a proper name, personal pronoun, or demonstrative) must be a *Precept*, or Symbol, not only describing to the Interpreter what is to be done, by him or others or both, in order to obtain an Index of an individual (whether a unit or a single set of units) of which the proposition is represented as meant to be true, but also assigning a designation to that individual, or, if it is a set, to each single unit of the set. (2.330)

Yet, although indices, subindices, and precepts, are considered as different types of indexical signs which share the common feature of designating an object or objects to their interpreters, proper names take on only two of these three indexical relations. That is to say, the range of possible indexical functions of proper names is not as broad as the general indexical character of the subject(s) of a proposition.

Peirce viewed the indexical character of proper names as *twofold*. In 1902 he wrote that

> A proper name, when one meets with it for the first time, is existentially connected with some percept or other equivalent individual knowledge of the individual it names. It is *then*, and then only, a genuine Index. The next time one meets with it, one regards it as an Icon of that Index. (2.329)

In other words, a proper name may function as an indexical sign in a proposition or an assertion as either a *genuine index* or as a *subindex* (also called a Hyposeme). Hence, for Peirce, proper names have both a *genuine indexical character* or function and a *subindexical character* or function.

Furthermore, although the twofold indexical character of proper names is in accordance with Peirce's claim that every subject of a proposition must be either an index, a subindex, or a precept, proper names do not take on all of the indexical characteristics of the subject(s) of a proposition. That is to say, proper names were not regarded as precepts for Peirce, which were taken to be *definite descriptions*. As precepts, definite descriptions describe to their interpreter what is to be done in order to obtain an index of an individual of which the proposition is represented as meant to be true. So, definite descriptions and proper names differ in at least one respect according to Peirce; they hold different indexical relationships to their objects. Let us now turn to a further consideration of the indexical and subindexical characteristic functions of proper names.

The *genuine indexical character of proper names* suggests that proper names may function as pure indices. According to Peirce, a proper name is only a genuine or pure index—"like something attracting attention in that environment, as the pointing finger of the speaker" (2.330)—if one has not previously been acquainted with the proper name, or if one has met with the proper name before, but in meeting with it again, one does not remember that one has met with it before, or some analogous scenario. Our first meeting with proper names either in speech or in writing "only says 'There!' It takes hold of our eyes, as it were, and forcibly directs them to a particular object and there it stops" (3.361). Yet proper names can only be genuine indices on our first meeting them. After this they become icons of that index, and take on a different indexical character and *serve* a different function.

For Peirce, the *subindexical character of proper names* attributes a type of *degenerate* indexical function to proper names. While the secondness of the *genuine indexical character of proper names* is an "existential relation" (2.283), existing as a matter of brute fact, "known *a-posteriori*, and not uniquely identifiable through general description alone" (*Savan 1976: 28*), the secondness of the *subindexical character of proper names* is a "reference" (2.283) which "directly refer[s], and need only refer, to the images in the mind which previous words created" even though it "may accidentally and indirectly, refer to existing things" (2.305).

Subindices, or hyposemes, are "signs which are rendered such principally by an actual connection with their objects" (2.284), yet differ from genuine indices in that they are not singular individuals. "Thus," wrote Peirce "a proper name, personal demonstrative, or relative pronoun or the letter attached to a diagram denotes what it does owning to a real connection with its object, but none of these is a [genuine] Index, since it is not an individual" (2.284). That is to say, a "genuine index and its Object must be existent individuals (whether things or facts), and its interpretant must be of the same character" (2.283), whereas a subindex calls "upon the hearer to use his powers of observation, and so establish a real connection between his mind and the object" (2.287).

The function of the *subindexical character of proper names* is to order and aid in the retrieval of information concerning some real or fictitious object denoted by the proper name. In his 1906 paper from *The Monist* entitled "Prolegomena to an Apology for Pragmaticism" Peirce explained that the first time one hears a Proper Name pronounced, it is but a name, predicated, as one usually gathers, of an existent, or at least historically existent, individual object, of which, or of whom, one almost always gathers some additional information. The next time one hears the name, it is by so much the more definite; and almost every time one hears the name, one gains in familiarity with the object. (4.568)

This passage emphasizes the collective function and role of the *subindexical character of proper names*. As one gathers additional information concerning the proper name, the belief that one is becoming more familiar with the object increases. This point was further elucidated in an 1897 writing entitled "Recreations in Reasoning," in which Peirce wrote that

What we call a Thing is a cluster or habit of reactions, or, to use a more familiar phrase, is a centre of forces. . . . [W]hen man comes to form a language, he makes words of two classes, words which denominate things, which things he identifies by the clustering of their reactions, and such words are proper names, and words which signify, or mean, qualities, which are composite photographs of ideas of feelings, and such words are verbs or portions of verbs, such as are adjectives, common nouns, etc. (4.157)

Again, the collective role of proper names is emphasized by Peirce's use of the term "cluster."

In the passage above (4.157), it is interesting to note that whereas a *proper name* (or proper noun—Peirce uses these terms interchangeably) is a word which "denominates things," which things we then identify by the clustering of their reactions, a *common noun* is a word which "signifies," or means, qualities which are composite photographs of ideas of feeling. This view is similar to J. S. Mill's approach to proper nouns and common nouns. Clearly for Peirce, proper nouns do not have *senses* or meaning, for "there is no verb wrapped up in a proper name" (2.328).

There is no reason for saying that *I, thou, that, this,* stand in place of nouns; they indicate things in the directest possible way. It is impossible to express what an assertion refers to except by means of an index. A pronoun is an index. A noun, on the other hand, does not *indicate* the object it denotes; and when a noun is used to show what one is talking about, the experience of the hearer is relied upon to make up for the incapacity of the noun for doing what the pronoun does at once. Thus, a noun is an imperfect substitute for a pronoun. . . . Allen and Greenough say, "pronouns indicate some person or thing without either naming or describing." This is correct—refreshingly correct; only it seems better to say what they *do*, and not merely what they don't. (2.287n.)

This passage helps us to better understand the different roles of the twofold indexical character of proper names. The closest that proper names come to being pure demonstratives is when they function as genuine indices. Nonetheless, what demonstratives *do* is *indicate*, and indicating is neither describing nor naming. Thus, for Peirce, naming is an indirect form of indicating or demonstrative reference; a fact that is reinforced and supported by the subindexical character of proper names. According to Peirce, this indirect indication has to be mediated by the experience of the hearer, and this experience may be gained through two processes.

On 12 November 1908 Peirce wrote that by "a Proper Name, I mean a name of anything considered as a single thing; and this thing which the Proper Name denominates must have been one with which the Interpreter was already acquainted by direct or indirect experience" (MS 612: 31). Peirce then goes on to describe two processes which either independently or through some mixture of the two enable us to gain this experience of anything considered as a single thing. They are as follows:

> [*The Direct Process:*] The interpreter should first by his own personal experience become sufficiently acquainted with that to which the Proper Name applies, and subsequently with that Proper Name as denominating that thing. (MS 612: 32)

> [*The Indirect Process:*] By overhearing various assertions into which the Proper Name enters, he should discover, first, that it is a Proper Name, and then of what sort of object it is a Name, and finally that he should gain a greater or less mass of information about the history of the single object, sufficient to make it stand out in his indirect experience as contrasting with other things of the same sort. (MS 612: 32)

It is interesting to note that these two processes resemble Russell's distinction between "knowledge by acquaintance" and "knowledge by description." Let us now consider why Peirce regarded proper names as indices rather than symbols.

In a letter to Christine Ladd Franklin, Peirce addressed this question, writing that a symbol "is the sign of its object only and *merely* because it will be interpreted as such, not because of this as a matter of fact."

> Thus, a proper name is not a symbol. The first time you hear it, it is an Index. Afterward habit makes it a legisign but it always remains an Index.[3]

Let us reflect on this for a moment.

For Peirce, a symbol is a sign which is determined by its object "only in the sense that it will be so interpreted" and "depends either upon a convention, a habit, or a natural disposition of its interpretant" (SS: 33).

> But a symbol, in itself, is a mere dream; it does not show what it is talking about. It needs to be connected with its object. For that purpose, an *index* is indispensable. (4.56)

It must be remembered that for Peirce proper names have an inherent indexical connection with their objects, whereas symbols "denote the objects that they do by virtue only of there being a habit that associates their signification with them" (4.544). A symbol may be considered in relation to any object depending on the relation of the symbol to its object, whereas a proper name has either some real connection or an existential relation with its object, which is known only through experience and is not uniquely identifiable only through a general description. Peirce (4.544) explained that although proper names might be regarded as symbols, because "[a]ll general, or definable Words, whether in the sense of Types or of Tokens, are certainly Symbols. . . . They should probably be regarded as Indices, since the actual connection (as we listen to talk), of Instances of the same typical words with the same Objects, alone causes them to be interpreted as denoting those Objects."

One of the reasons Peirce thought that the subject of subjects of a proposition must be indexical signs was illustrated by him through an analysis of the subjects of the proposition "Ezekiel loveth Huldah." He wrote that the subjects of this proposition,

> Ezekiel and Huldah must, then, be or contain indices; for without indices it is impossible to designate what one is talking about. Any mere description would leave it uncertain whether they were not mere characters in a ballad; but whether they be so or not, indices can designate them. (2.295)

Once again, in Peirce's view the object of the proposition "is the things denoted by its subject or subjects (including its grammatical objects, direct and indirect, etc.)" (5.473). Indexical signs play a crucial role in propositions, for not only do they bring together the subject(s) and rheme to form the proposition (as described in Section 4) and designate the object(s) of the proposition, they also enable us to determine the world (fictional, real, etc.) of the object(s) of the proposition. This is an important feature of indices for the "real world cannot be distinguished from a fictitious world by any description" (2.337).

Yet Peirce claimed that it is more than just simply an index that identifies the world of the object(s) of the proposition for "no language (so far as I know) has any particular form of speech to show that the real world is spoken of" (2.337). Peirce (2.357) wrote that

> if somebody rushes into the room and says, "There is a great fire!" we know he is talking about the neighbourhood and not about the world of the *Arabian Nights' Entertainments*. It is the circumstances under which the proposition is uttered or written which indicate that environment as that which is referred to. But they do so not simply as index of the environment, but as evidence of an intentional relation of the speech to its object, which relation it could not have if it were not intended for a sign.

In other words, the *context* of the proposition plays a major role in the determination of both the object of the proposition and its ontological status. In the case of spoken propositions (or assertions), "tones and looks act dynamically upon the listener, and cause him to attend to realities" (2.337), that is to say, help him determine the nature of the world that has been spoken of to him. Let us now look at the semantic function of proper names a bit closer.

As an indexical sign, a proper name directs our attention by blind compulsion. It forces our attention to its object. It is said by Peirce to "compel" us to have an experience, to look around us, and/or share an experience with a speaker. A proper name is said by Peirce to put us into real active connection with what is being talked about. What then is the nature of the object of a proper name? What is it that our attention is being forced to or directed toward by a proper name?

Peirce provided an answer to these questions in his 1897 article for *The Monist* entitled "The Logic of Relatives" (3.456–552) wherein he wrote that an

> indexical word, such as a proper noun . . . has force to draw the attention of the listener to some hecceity common to the experience of speaker and listener. By a hecceity, I mean, some element of existence which, not merely by the likeness between its different apparitions, but by an inward force of identity, manifesting itself in the continuity of its apparition throughout time and space, is distinct from everything else, and is thus fit (as it can in no other way be) to receive a proper name or to be indicated as *this* or *that*. Contrast this with the signification of the verb, which is sometimes in my thought, sometimes in yours, and which has no other identity than the agreement between its several manifestations. That is what we call an abstraction or idea. (3.460)

Now, the use of *haecceity* in this passage was the result of the profound influence that scholastic philosophy had on Peirce, and of the importance which he attached to realism, in the scholastic sense, as against nominalism. An understanding of the nature of haecceity is necessary for an understanding of Peirce's analysis of proper names in general, and his notion of the object of a proper name in particular. Although a full account of Peirce's *haecceitism* is out of the question at this time, a few words are in order.[4]

Haecceity was a term used by Scotus to denote the formal property of an object or person in virtue of which it is uniquely individuated as just *this* or *that* object or person. The word "haecceity" functioned like "uniqueness" according to Scotus, and did not add anything to the character of the object. Peirce, in turn, used a notion of haecceity similar to that of Scotus. Peirce wrote that

A sign which denotes a thing by forcing it upon the attention is called an *index*. An index does not describe the qualities of its object. An object, in so far as it is denoted by an index, having *thisness*, and distinguishing itself from other things by its continuous identity and forcefulness, but not by any distinguishing characters, may be called a *hecceity*. (3.434)

Thus, it is the haecceity, "thisness," or "hereness and nowness" (1.405; 8.266) of objects, which is denoted by a proper name, and it is haecceity which enables the objects to be unambiguously distinguishable. The indexical character of proper names serves a purely denotative function, although it does not denote any properties or qualities of objects or bare particulars (haecceity for Peirce is *not* a bare particular).

In general, haecceity can be understood as that which renders existence and individuality upon objects: whenever there is haecceity, there is individuality, existence, and thinghood. The presence of haecceity is not an affirmation of the existence of the haecceity as *per se* any thing, but is rather the affirmation of non-ego, otherness, reaction, and struggle—in other words, secondness. For Peirce, a proper name *denotes* some haecceity of an individual, such that when the proper name is used it has the force of drawing the attention of the listener to some haecceity common to the experience of speaker and listener. Once again, haecceity for Peirce is an inward force of identity, manifesting itself in the continuity of its apparition throughout time and space that is distinct from anything else. It is this self-same haecceity that Peirce considers "fit" to receive a proper name. It should be noted here that rhemes or predicates do not have the same relationship with haecceity as proper names, "[f]or a predicate is of an ideal nature, and as such cannot be a mere haecceity" (2.341).

Finally, although it is possible to regard Peirce and Russell as having held similar views as to the indexical character of proper names, this is wrong. Unlike Russell, Peirce did not contend that demonstrative reference was the only provider of genuine indices, and that in all other cases proper names were merely disguised descriptions. Peirce had a different notion of indexicality in mind which allowed him to distinguish between two different types of indexical functions for proper names, and to differentiate proper names from definite descriptions by their differing characteristic indexical functions.

§6. Legisigns and the Meaning of Proper Names

As noted earlier, Peirce told Lady Welby that proper names are legisigns, and as such, (a) they have a definite identity, and (b) they are usually admitted through an appearance. What did he mean by this?

A legisign, wrote Peirce, "is not a single object, but a general type which

... signifies through an instance of its application, which may be called a *Replica* of it" (2.246). The replicas of rhematic indexical legisigns are rhematic indexical sinsigns (2.259). Thus, for Peirce, a proper name *appears* to us in a proposition or assertion as a sinsign of a legisign. In other words, a proper name is a type which is embodied in a token. Such a token of a type Peirce called an "instance" of the type (4.537).

On the other hand, the interpretant of a rhematic indexical legisign represents it as an iconic legisign (2.259). Peirce defined an iconic legisign as "any general law or type, in so far as it requires each instance of it to embody a definite quality which renders it fit to call up in the mind the idea of a like object" (2.258). Iconic legisigns exist only through tokens which are governed by legisigns. They can aid their interpreters in the identification of the objects of proper names, if their interpreters have associated certain icons of definite qualities with certain proper names.

> Yet it does not follow and could only very rarely be true that the name *signifies* certain defining marks, so as to be applicable to anything that should possess those marks, and to nothing else. For not to speak of the fact that the interpreter only uses the marks as aids in guessing at his acquaintance's identity, and may possibly be mistaken, however extraordinary they may be, there will be no one *definite* set of marks which the name signifies rather than another set of equally conclusive marks. If there were any mark which a proper name could be said *essentially* to signify, it would be the continuity of the history of its object. (MS 283: 317–318)

In other words, even though a set of definite qualities, defining qualities and/or icons may be associated by the interpreter with a proper name and may aid the interpreter in identifying the object of the proper name, they are not part of the *meaning* of the proper name.

Proper names are a type of indexical sign, and as such do not have meanings. Peirce wrote that

> *meaning* is the associations of a word with images, its dream exciting power. An index has nothing to do with meanings; it has to bring the hearer to share the experience of the speaker by *showing* what he is talking about. (4.56)

So, even though (a) a proper name may be introduced to us through the process of indirect experience by means of or with the help of descriptive signs, and (b) the referent of a proper name may be secured with the help of a description or descriptions, these descriptive signs *do not* give the proper name meaning, nor are they part of the meaning of the proper name. For Peirce, a proper name is different from a description, and the meaning of a proper name is not a set of definite qualities, defining qualities and/or icons that may aid the interpreter in securing the reference of the proper name. Proper names and definite descriptions function differently in experience, and are not the same.

Furthermore, no description can describe the referent of a proper name, for by its very nature, the denotation of a proper name, that is a haecceity, is beyond description and undescribable. A proper name does not signify, rather it denotes or shows "the continuity of the history of its object"; in other words, some haecceity. For Peirce, the object of a proper name is clearly distinguished from its meaning. In "Pragmatism" (5.6 [c.1905]), Peirce wrote that

> The object of a sign is one thing; its meaning another. Its object is the thing or occasion, however indefinite, to which it is to be applied. Its meaning is the idea which it attaches to that object, whether by way or mere supposition, or as a command, or as an assertion.

Nevertheless, one might have the impression that in some respect, proper names can be said to have at least an *essential* signification. Let us look into this a bit more closely.

The question of the meaning of proper names is directly addressed by Peirce in his *Monist* paper of 1905 entitled "What Pragmatism Is" (5.411). Therein, Peirce poses the question "What do you make to be the meaning of 'George Washington'?" (5.429) and responds with the following claims:

(1) Every proposition professes to be true of a certain real individual object, often the environing universe.
(2) Pragmaticism fails to furnish any translation or meaning of a proper name, or other designation of an individual object.
(3) Pragmatistic meaning is undoubtedly general, and the general is of the nature of a word or sign.
(4) Individuals alone exist.
(5) The meaning of a word or significant object ought to be the very essence of reality of what it signifies.

From this, one may conclude that the meaning of "George Washington" *ought to be* the very essence of reality of what George Washington signifies, wherein this very essence of reality, as described earlier, is a haecceity. This seems to be in accordance with Peirce's general views that a proper name "has a certain denotative function peculiar, in each case, to that name and its equivalents," and that "every assertion contains such a denotative or pointing-out function." In other words, that proper names have denotation.

Yet, Peirce then went on to say that the pragmaticist *excludes* this individual, or particular meaning from the rational purport of the assertion because pragmatistic meaning is undoubtedly general, and the general is of the nature of a word or sign (3). Nevertheless, for Peirce, the *like* of this individual meaning, "being common to all assertions, and so being general and not individual" may enter into pragmatistic meaning (4.429). The *generality* that may enter into pragmatistic meaning is of two types: *objective* and *subjective.*

According to Peirce, *objective generality* is characteristic of all common nouns, whereas *subjective generality* is characteristic of both proper names and common nouns. He explained this notion through an example:

> A statue of a soldier on some village monument, in his overcoat and with his musket, is for each of a hundred families the image of its uncle, its sacrifice to the Union. That statue, then, though it is itself single, represents any one man of whom a certain predicate may be true. It is *objectively* general. The word "soldier," whether spoken or written, is general in the same way; while the name "George Washington," is not so. (5.429)

Thus, the objective generality of common nouns, such as "soldier" is different from the generality of proper nouns, such as "George Washington." Nevertheless, Peirce continues by stating that

> each of these two terms remains one and the same noun, whether it be spoken or written, and whenever and wherever it be spoken or written. This noun is not an existent thing: it is a type, or form, to which objects, both those that are externally existent and those which are imagined, may *conform*, but which none of them can exactly be. This is subjective generality. (5.429)

That is to say, not only are common nouns and proper names subjectively general, rather all words are subjectively general, but not objectively general. Hence, the pragmaticistic meaning of proper names is subjectively general.

§7. Concluding Remarks

Peirce's theory of proper names was in the spirit of J. S. Mill's famous theory of proper names as presented in his *System of Logic* (*1843*); a work with which Peirce was quite familiar and critical of as early as his Harvard Lectures of 1865. Mill's basic thesis was that proper names have denotation, but not connotation. That is, for Mill, whereas a common noun like "soldier" has both a connotation and a denotation such that it connotes those properties which would be specified in a definition of the word "soldier" and it denotes all soldiers, a proper name, on the other hand, only denotes its bearer, that is to say, it refers to a specific object without the mediation of a property or attribute. Today there are a variety of variations of Mill's direct reference approach to proper names advanced by Keith Donnellan, Hilary Putnam, Saul Kripke, and others. One of the most celebrated and debated of these "new theories of reference" is that of Saul Kripke as expounded in *Kripke 1972.*

Kripke, arguing against views such as Bertrand Russell's and Gottlob

Frege's, defended the view that names are genuine indices by arguing that causally based reference preserving links can sustain names as genuine indices. Kripke's move was significant because it sidestepped the mistaken Russellian view that demonstrative reference is the only provider of genuine indices, which in turn led him to hold that in all other cases, names are really just disguised descriptions. As such, one can see some of the affinities between Peirce's and Kripke's views. At some point it might be useful to parse out those affinities, as well as Peirce's relation to numerous other approaches to proper names.

Hector-Neri Castañeda (*1989*: 46) claimed that any "useful" theory of proper names must provide an account of:

(1) The semantic roles of proper names as parts of a language system;
(2) The main structure of the phenomenon of singular thinking reference carried out in a speech act or a mere thinking episode by means of a sentence containing a proper name;
(3) The main structure of the communication of singular reference from the speaker of a sentence containing a proper name to the hearer of such a sentence.

Does Peirce's theory of proper names stand up to these criteria? Absolutely, even though we have only scratched the surface with respect to the full exposition of these accounts. Nevertheless, at least by Castañeda's standards, Peirce's theory is a useful account of proper names. Whether it is a sufficient account is another question.

Peirce viewed proper names as semiotic phenomena, and sought to articulate a theory of reference by names rooted in semiotic concepts. It is interesting to note that the semantic and pragmatic roles of proper names are not necessarily independent for Peirce, and as such Morris' semantics-syntactics-pragmatics distinction is incompatible with Peirce's view of language in general, and in particular, with the view of the role of proper names in language. For Peirce, semantics is often determined by pragmatics in that the relations between expressions and objects are in many cases determined by the activities of the speaker and the interpreter. In this respect, Peirce's semantics has a pragmatic character such that the semantic and pragmatic roles of proper names become interrelated.[5]

N O T E S

1. For example, Rudolf Carnap (*1960:* 288–289) defined "semiotic" as "a general theory of signs and their applications, especially in language; developed and systematized within Scientific Empiricism." According to Carnap, semiotic has

three branches: Pragmatics—"theory of the relations between signs and those who produce or receive and understand them"; Semantics—"theory of the relations between signs and what they refer to"; and Syntactis—"theory of the formal relations among signs."

2. According to Peirce, designations among the ten classes in parentheses are superfluous (2.264). It should also be noted that Peirce further divides signs into sixty-six and 59,049 classes. These further classifications will not be pursued in this paper.

3. This undated letter fragment is from the Columbia University Rare Book and Manuscript Library special collection of Christine Ladd Franklin material. It is significant for its explicit statement of the relation between symbols, indices, legisigns, and proper names. I wish to thank Columbia University Library for granting me permission to use this letter.

4. For a more extensive account, see *DiLeo 1990*.

5. I wish to thank the Indiana University College of Arts and Sciences and the Graduate Advisory Committee for awarding me a Graduate Student Travel Award which enabled me to present this paper at the Peirce Congress. Also, I am indebted to the editors of the Peirce Edition Project at Indiana University–Purdue University at Indianapolis for their generous assistance.

REFERENCES

Abbott, Edwin A.
1952 *Flatland: A Romance of Many Dimensions.* Dover: New York.
Achinstein, Peter
1970 Inference to Scientific Laws. In *Minnesota Studies in Philosophy of Science,* vol. 5, ed. Roger H. Stuewer. University of Minnesota Press: Minneapolis, pp. 87–104, 109–111.
1980 Discovery and Rule Books. In *Nickles, ed. 1980,* pp. 117–137.
1987 Scientific Discovery and Maxwell's Kinetic Theory. *Philosophy of Science* 54: 409–34.
Adler, Irving
1967 *Groups in the New Mathematics.* John Day: New York.
1968 *Mathematics and Mental Growth.* John Day: New York.
Alexander, Peter
1965 On the Logic of Discovery. *Ratio* 7: 219–233.
Anderson, Alan Ross and Nuel D. Belnap, Jr.
1975 *Entailment: The Logic of Relevance and Necessity.* Princeton University Press.
Anderson, Douglas R.
1986 The Evolution of Peirce's Concept of Abduction. *Transactions of the Charles S. Peirce Society* 22: 145–164.
Anderson, Myrdene and Floyd Merrell, eds.
1991 *On Semiotic Modeling.* Walter De Gruyter: Berlin.
Andréka, Hajnal
1988 Building *n*-ary Relations from Binary Ones. In C. Bergman, ed. *Abstracts, Iowa State University Conference on Algebraic Logic and Universal Algebra in Computer Science,* 1–4 June 1988, Ames, Iowa, p. 24.
1990 Weakly Representable but not Representable Relation Algebras (abst. # 90T-03-223). *Abstracts Presented to the American Mathematical Society* 11: 505.
Andréka, Hajnal, James Donald Monk, and István Németi, eds.
1991 *Algebraic Logic, Proceedings of the Conference in Budapest, 8–14 August, 1988.* Colloquia Mathematica Societatis János Bolyai. Vol. 54. North Holland: Amsterdam.
Anellis, Irving H.
1986–87 Notes on Conversations with Roger Maddux on Tarski and Algebraic Logic, 9 September 1986–22 April 1987. Unpublished.
1987a Notes on a lecture by R. Maddux entitled "Tarski's Formalization of Set Theory without Variables." Iowa State University Mathematics Colloquium, 20 April 1987. Unpublished.
1987b Notes on a lecture by István Németi entitled "On Logic in Relation with

Other Fields." Iowa State University Mathematics Colloquium, 10 November 1987. Unpublished.

1989 Notes on a Talk by George McNulty entitled "An Overview of Term Rewriting Systems." Presented to the Fifth Southeastern Logic Symposium, University of North Carolina-Charlotte, 11 March 1989. Unpublished.

1995 Peirce Rustled, Russell Pierced: How Charles Peirce and Bertrand Russell Viewed Each Other's Work in Logic, and an Assessment of Russell's Accuracy and Rôle in the Historiography of Logic. *Modern Logic* 5, pp 270–328.

Anellis, Irving H. and Nathan Houser
1991 Nineteenth Century Roots of Algebraic Logic and Universal Algebra. In *Andréka et al., eds. 1991,* pp. 1–36.

Archibald, Raymond Clare
1925 *Benjamin Peirce 1809–1880: Biographical Sketch and Bibliography.* The Mathematical Association of America: Oberlin, Ohio. Repr. in *Cohen, ed. 1980.*

Asquith, Peter and Henry Kyberg, eds.
1979 *Current Research in Philosophy of Science.* Philosophy of Science Association: East Lansing.

Austin, John Langshaw
1962 *Sense and Sensibilia.* Reconstructed from the manuscript notes by G. J. Warnock. Oxford University Press.

Audi, Robert, ed.
1995 *The Cambridge Dictionary of Philosophy.* Cambridge University Press.

Axelrod, Robert.
1984 *The Evolution of Cooperation.* Basic Books: New York.

Bach, Emmon and Robert T. Harms, eds.
1968 *Universals in Linguistic Theory.* Holt, Rinehart and Winston: New York.

Bachtijarov, K. I.
1979 On the Group Properties of Operations of the Algebra of Logic (in Russian). In *Collection of Scientific Works: Automatization of Agricultural Production* 15: 117–121. Gorjachkin Institute of Engineers for Agricultural Production: Moscow.

Bacon, John
1989 Review of *Tarski and Givant 1987. History and Philosophy of Logic* 10: 244–246.

Baldwin, James Mark
1906, 1908, 1911 *Thoughts and Things,* 3 vols. Swan Sonnenschein and George Allen: London.

1915 *Genetic Theory of Reality.* Putnam: New York.

Baldwin, James Mark, ed.
1901, 1905 *Dictionary of Philosophy and Psychology.* 3 vols. Macmillan: New York.

Banchoff, Thomas F.
1990 *Beyond the Third Dimension: Geometry, Computer Graphics, and Higher Dimensions.* Scientific American Library: New York.

Barone, Francisco
1966 Peirce e Schroeder. *Filosofia* 17: 181–224.

Barwise, Jon and Robin Cooper
1981 Generalized Quantifiers and Natural Language. *Linguistics and Philosophy* 4: 159–219.

Beatty, Richard
1969 Peirce's Development of Quantifiers and of Predicate Logic. *Notre Dame Journal of Formal Logic* 10: 64–76.

Behnke, Heinrich, F. Bachmann, K. Fladt, and W. Suss, eds.
　1974　*Fundamentals of Mathematics*, 3 vols. Trans. S. H. Gould. MIT Press: Cambridge. (*Analysis,* Vol. 3.)
Bell, Eric Temple
　1945　*The Development of Mathematics.* McGraw-Hill: New York.
Belnap, Nuel D. Jr. and Thomas B. Steel, Jr.
　1976　*The Logic of Questions and Answers.* Yale University Press: New Haven.
Benacerraf, Paul
　1983　Mathematical Truth. In *Benacerraf and Putnam, eds. 1983,* pp. 403–420.
Benacerraf, Paul and Hilary Putnam, eds.
　1983　*Philosophy of Mathematics: Selected Readings,* 2nd ed. Cambridge University Press.
Bentham, George
　1827　*Outline of a New System of Logic: With a Critical Examination of Dr. Whately's "Elements of Logic."* Hunt and Clarke: London.
Bergson, Henri
　1907　*Creative Evolution.* Trans. Arthur Mitchell, 1911. Holt: New York.
Berlin, Isaiah
　1978　The Hedgehog and the Fox. In Isaiah Berlin, *Russian Thinkers,* ed. Henry Harcy and Aileen Kelly, with an introduction by Aileen Kelly. Hogarth Press: London.
Bernays, Paul
　1959　Über eine natürliche Erweiterung des Relationenkalküls. In *Heyting 1959,* pp. 1–14.
　1975　Review of *Schröder 1890–1895,* vol. 1. *Journal of Symbolic Logic* 40: 610–611.
Beth, Evert William
　1959　*The Foundations of Mathematics.* North-Holland: Amsterdam. Repr. 1966, Harper and Row: New York.
Bird, Otto
　1962　What Peirce Means by Leading Principles. *Notre Dame Journal of Formal Logic* 3: 175–178.
Birkhoff, Garrett
　1948　*Lattice theory.* American Mathematical Society: Providence.
Black, Max
　1971　The Elusiveness of Sets. *Review of Metaphysics* 24: 614–636.
Blanché, Robert
　1957　Sur la structuration du tableau des connectifs interpropositionnels binaires. *Journal of Symbolic Logic* 22: 17–18.
Blok, Willem J.
　1989　*Algebraizable Logics.* American Mathematical Society: Providence.
Bocheński, Innocentius Marie
　1951　*Ancient Formal Logic.* North-Holland: Amsterdam.
　1961　*A History of Formal Logic.* Trans. and ed. Ivo Thomas. University of Notre Dame Press: Notre Dame. Repr. 1970 with corrections, Chelsea: New York.
Bôcher, Maxime
　1904　The Fundamental Conceptions and Methods of Mathematics. *Bulletin of the American Mathematical Society* 11: 115–135.
Boisen, Monte B. and G. V. Gibbs
　1990　*Mathematical Crystallography: An Introduction to the Mathematical Foundations of Crystallography.* Rev. ed. *Reviews in Mineralogy* 15. Mineralogical Society of America: Washington, D.C.

Boler, John F.
  1963 *Charles Peirce and Scholastic Realism.* University of Washington Press: Seattle.
Boll, Marcel
  1948 *Manuel de Logique Scientifique.* Dunod: Paris.
Boole, George
  1844 On a General Method in Analysis. *Philosophical Transactions of the Royal Society of London* 134: 225–282.
  1847 *The Mathematical Analysis of Logic, Being an Essay Towards a Calculus of Deductive Reasoning.* Macmillan, Barclay, & Macmillan: Cambridge; George Bell: London. Repr. 1948, Blackwell: Oxford. Also 1952 with later notes and additions in *Boole 1952,* pp. 45–124.
  1848 Notes on Quaternions. *Philosophical Magazine* 33: 278–280.
  1854 *An Investigation of the Laws of Thought, on Which are Founded the Mathematical Theories of Logic and Probabilities.* Walton and Maberly: Cambridge and London. Repr. 1958, Dover: New York. New ed. 1916, ed. P. E. B. Jourdain, Open Court: Chicago.
  1952 *Studies in Logic and Probability.* Preface by A. E. Heath; note in editing by R. Rhees. Watts: London.
Borges, Jorge Luis
  1962 *Ficciones.* Grove Press: New York.
Bottazzini, Umberto
  1985 Dall'analisi matematica al calcolo geometrico: origini delle prime ricerche di logica di Peano. *History and Philosophy of Logic* 6: 25–52.
  1986 *The Higher Calculus: A History of Real and Complex Analysis from Euler to Weierstrass.* Trans. Warren Van Egmond. Springer-Verlag: Berlin.
Brent, Joseph
  1993 *Charles Sanders Peirce: A Life.* Indiana University Press: Bloomington.
Brink, Chris
  1978 On Peirce's Notation for the Logic of Relatives. *Transactions of the Charles S. Peirce Society* 14: 285–304.
  1987 *Some background on Multisets.* Technical report (TR-ARP-2187), Research School of Social Sciences, Australian National University.
Brock, William Hodson, ed.
  1967 *The Atomic Debates: Brodie and the Rejection of Atomic Theory; Three Studies.* Leicester University Press: Leicester.
Bromberger, Silvain
  1960 Why Questions. In *Colodny, ed. 1960,* pp. 86–111.
Brown, W. M.
  1983 The Economy of Peirce's Abduction. *Transactions of the Charles S. Peirce Society* 19: 397–411.
Brumbaugh, Robert S.
  1989 *Platonic Studies of Greek Philosophy: Form, Arts, Gadgets, and Hemlock.* State University of New York Press: Albany.
Brunning, Jacqueline
  1981 *Peirce's Development of the Algebra of Relations.* Ph.D. dissertation, University of Toronto.
  1991 Peirce's Relative Product. *Modern Logic* 2: 33–49.
  1993 Degenerate Triads. In *Creativity and Logical Form: Studies in Peirce's Evolutionary Semiotics,* ed. Helmut Pape. Surkamp: Frankfurt. Forthcoming.

Buchler, Justus
1939 Peirce's Theory of Logic. *Journal of Philosophy* 36: 197–215.
Budden, F. J.
1972 *The Fascination of Groups.* Cambridge University Press.
Buerger, Martin J.
1963 *Elementary Crystallography.* Rev. printing. Wiley & Sons: New York.
Burch, Robert W.
1991 *A Peircean Reduction Thesis.* Texas Tech University Press: Lubbock.
1996 Peirce on the Application of Relations to Relations. This volume, pp. 206–233.
Burch, Robert W. and Herman J. Saatkamp, Jr., eds.
1992 *Frontiers in American Philosophy*, vol. 1. Texas A&M University Press: College Station.
Burge, Tyler
1983 Russell's Problem and Intentional Identity. In *Tomberlin, ed. 1983*, pp. 79–110.
Burger, Dionys
1965 *Sphereland: A Fantasy About Curved Spaces and an Expanding Universe.* Trans. Cornelie J. Rheinboldt. Crowell: New York.
Burks, Alice and Arthur Burks
1988 *The First Electronic Computer: The Atanasoff Story.* The University of Michigan Press: Ann Arbor.
Burks, Arthur W.
1943 Peirce's Conception of Logic as a Normative Science. *The Philosophical Review* 52: 187–193.
1946 Peirce's Theory of Abduction. *Philosophy of Science* 13: 301–306.
1949 Icon, Index and Symbol. *Philosophy and Phenomenological Research* 9: 673–689.
1951 A Theory of Proper Names. *Philosophical Studies* 2: 36–45.
1970 Von Neumann's Self-Reproducing Automata. In *Burks, ed. 1970*, pp. 3–64. Repr. in *von Neumann 1986*, pp. 491–552.
1972 Logic, Computers, and Men. *Proceedings and Addresses of the American Philosophical Association* 46: 39–57.
1975 Models of Deterministic Systems. *Mathematical Systems Theory* 8: 295–308.
1977 *Chance, Cause, Reason—An Inquiry into the Nature of Human Reason.* University of Chicago Press.
1979 Computer Science and Philosophy. In *Asquith and Kyberg, eds. 1979*, pp. 399–420.
1980 Man: Sign or Algorithm? A Rhetorical Analysis of Peirce's Semiotics. *Transactions of the Charles S. Peirce Society* 16: 279–292.
1984 Computers, Control, and Intentionality. In *Kerr et al., eds. 1984*, pp. 29–55.
1986a An Architectural theory of Consciousness. In *Rescher, ed. 1986*, pp. 1–14.
1986b *Robots and Free Minds.* College of Literature, Science, and the Arts, The University of Michigan: Ann Arbor.
1986c A Radically Non-von Architecture for Learning and Discovery. In *Handler, ed. 1986*, pp. 1–17.
1988a The Logic of Evolution, and the Reduction of Coherent-Holistic Systems to Hierarchical-Feedback Systems. In *Harper and Skyrms, eds. 1988*, pp. 135–191.

1988b Teleology and Logical Mechanism. *Synthese* 76: 333–370.
1990 Replies. In *Salmon, ed. 1990:* pp. 349–524.
1996 Peirce's Evolutionary Pragmatic Idealism. *Synthese* 106: 323–372.
Burks, Arthur, ed.
1970 *Essays on Cellular Automata.* University of Illinois Press: Urbana.
Byrne, Robert, ed.
1988 *1,911 Best Things Anybody Ever Said.* Fawcett Columbine: New York.
Cajori, Florian
1929 *A History of Mathematical Notations.* Open Court: La Salle.
Cantor, Georg
1895 Beiträge zur Begründung der transfiniten Mengenlehre. *Mathematische Annalen* 46: 481–512. Repr. in *Cantor 1932.*
1932 *Gesammelte Abhandlungen mathematischen und philosophischen Inhalts,* ed. Ernst Zermelo. Springer: Berlin. Repr. 1962 George Olms Verlagsbuchhandlung: Hildesheim.
Carnap, Rudolf
1950 *Logical Foundations of Probability.* University of Chicago Press.
1959 *The Logical Syntax of Language.* Littlefield, Adams: Paterson.
1960 Semiotic; Theory of Signs. In *Runes, ed. 1960,* pp. 288–289. Repr. in *Runes, ed. 1983,* p. 305.
1963 Intellectual Autobiography. In *Schilpp, ed. 1963,* p. 5.
Carroll, Lewis
1894 A Logical Paradox. *Mind* (n.s.) 3: 436–438.
1958 *Symbolic Logic and the Game of Logic.* Dover: New York.
1977 *Symbolic Logic.* Part I, Elementary, repr. of 1896 5th ed. Part II, Advanced, never previously published, ed. William Warren Bartley, III. Clarkson N. Potter: New York.
Castañeda, Hector-Neri
1975 *Thinking and Doing.* D. Reidel: Dordrecht.
1989 *Thinking, Language, and Experience.* University of Minnesota Press: Minneapolis.
Castonguay, Charles Ernest
1972 *Meaning and Existence in Mathematics.* Springer-Verlag: Berlin.
Cavaillès, Jean
1938 *Méthode Axiomatique et Formalisme,* 3 nos. Hermann: Paris.
Chin, Louise H. and Alfred Tarski
1951 Distributive and Modular Laws in the Arithmetic of Relation Algebras, *University of California Publications in Mathematics* (n.s.) 1: 341–384.
Chomsky, Noam
1957 *Syntactic Structures.* Mouton: The Hague.
Church, Alonzo
1939 Schroeder's Anticipation of the Simple Theory of Types. *Erkenntnis* 9: 149–153.
1956 *Introduction to Mathematical Logic, I.* Princeton University Press.
1960 Logistic System. In *Runes, ed.1960,* pp. 182–183. Repr. in *Runes, ed.1983,* p. 199.
Cicero, Marcus Tullius
1923 *Di Senectute, De Amicitia, De Divinatione.* Loeb Classical Library.
Clark, William Glenn and Shea Zellweger
1993 Let the Mirrors Do the Thinking. *Mount Union Magazine* 93: 2–5.
Claus, V., H. Ehrig, and G. Rozenberg, eds.

1979 *Graph Grammars and Their Application to Computer Science and Biology.* Springer-Verlag: Berlin.

Clifford, William Kingdon

1879 On the Types of Compound Statement Involving Four Classes. In *Lectures and Essays by the Late William Kingdon Clifford*, 2 vols., ed. Leslie Stephen and Frederick Pollock. Vol. 2, pp. 89–105. Macmillan: London.

Cohen, I. Bernard, ed.

1980 *Benjamin Peirce: "Father of Pure Mathematics" in America.* Arno Press: New York.

Colodny, Robert Garland, ed.

1960 *Mind and Cosmos: Essays in Contemporary Science and Philosophy.* University of Pittsburgh Press: Pittsburgh.

Comer, Stephen D.

1983 A New Foundation for the Theory of Relations. *Notre Dame Journal of Formal Logic* 24: 181–187.

Comer, Stephen D., ed.

1985 *Universal Algebra and Lattice Theory: Proceedings of a Conference at Charleston, 11–14 July, 1984.* Lecture Notes in Mathematics 1149. Springer Verlag: Berlin.

Commons, Michael L., Francis A. Richards, and Cheryl Armon, eds.

1984 *Beyond Formal Operations: Late Adolescent and Adult Cognitive Development.* Praeger: New York.

Copi (Copilowish), Irving M.

1948 Matrix Development of the Calculus of Relations. *Journal of Symbolic Logic* 13: 193–203.

1967 *Symbolic Logic,* 3rd ed. Macmillan: New York; Collier-Macmillan: London.

Coulmas, Florian

1989 *The Writing Systems of the World.* Blackwell: Cambridge, Massachusetts.

Coxeter, Harold Scott Macdonald

1948 *Regular Polytopes.* 2nd ed. 1963, Macmillan: New York. 3rd ed. 1973, Dover: New York.

1988 Regular and Semi-regular Polyhedra. In *Shaping Space: A Polyhedral Approach*, ed. Marjorie Senechal and George Fleck. Birkhauser: Boston, pp. 67–79.

1991 *Regular Complex Polytopes.* 2nd ed. Cambridge University Press.

1995 Two Aspects of the Regular 24-Cell in Four Dimensions. In *Kaleidoscopes: Selected Writings of H. S. M. Coxeter,* eds. F. Arthur Sherk, Peter McMullen, Anthony C. Thompson, and Asia Ivic Weiss. Wiley: New York, pp. 25–34.

Craig, William

1974 *Algebraic Logic.* North Holland: Amsterdam.

Crombie, Edward James

1970a Peirce, Cognition and the Modes of Inference. M. A. thesis, University of Waterloo: Waterloo, Ontario.

1970b Philosophy and Evolution. Unpublished.

Curd, Martin

1980 The Logic of Discovery: An Analysis of Three Approaches. In *Nickles, ed. 1980*, pp. 201–219.

Cushing, Steven

1987 Some Quantifiers Require Two-Predicate Scopes. *Artificial Intelligence* 32: 259–267.

Dantzig, Tobias
  1967 *Number: The Language of Science.* 4th ed. Free Press: New York.
Dauben, Joseph W.
  1979 *Georg Cantor: His Mathematics and Philosophy of the Infinite.* Harvard University Press.
  1982 Peirce's Place in Mathematics. *Historia Mathematica* 9: 311–325.
  1988 Review of *Hallett 1984. British Journal for the Philosophy of Science* 39: 541–550.
Davis, Ellery W.
  1903 Some Groups in Logic. *Bulletin of the American Mathematical Society* 9: 346–348.
Dawkins, Richard
  1976 *The Selfish Gene.* Oxford University Press.
  1987 *The Blind Watchmaker.* Norton: New York.
Dedekind, Richard
  1888 *Was sind und was sollen die Zahlen?* Vieweg: Braunschweig. Trans. W. W. Beman as "The Nature and Meaning of Numbers," in *Essays on the Theory of Numbers,* Open Court: Chicago, 1901. Repr. 1963 Dover: New York, pp. 31–115.
Deely, John and Terry Prewitt, eds.
  1993 *Semiotics 1991.* University Press of America: New York.
Deledalle, Gérard (gen. ed.), Michel Balat and Janice Deledalle-Rhodes, eds.
  1992 *Signs of Humanity. Proceedings of the Fourth Congress of the International Association for Semiotic Studies,* vol. 3, Barcelona and Perpignan, 1989. Mouton de Gruyter: Berlin.
DeLong, Howard
  1971 *A Profile of Mathematical Logic.* Addison-Wesley: Reading.
De Morgan, Augustus
  1831 *On the Study and Difficulties of Mathematics.* 4th repr. ed. 1943, Open Court: La Salle.
  1836 Calculus of Functions. *Encyclopædia Metropolitana* 2: 305–392.
  1842–1849 The Foundations of Algebra. *Transactions of the Cambridge Philosophical Society* 7: 173–187, 287–300; 8: 139–142, 241–254.
  1846 On the Syllogism: I. On the Structure of the Syllogism. *Transactions of the Cambridge Philosophical Society* 8: 379–408. Repr. in *De Morgan 1966,* pp. 1–21.
  1847 *Formal Logic, or the Calculus of Inference, Necessary and Probable.* Taylor and Walton: London. Repr. 1926 with notes by A. E. Taylor, Open Court: London.
  1850 On the Syllogism: III; and on Logic in General. *Transactions of the Cambridge Philosophical Society* 10: 173–230. Repr. in *De Morgan 1966,* pp. 74–146.
  1860a On the Syllogism: IV; and on the Logic of Relations. *Transactions of the Cambridge Philosophical Society* 10: 331–358. Repr. in *De Morgan 1966,* pp. 208–246.
  1860b Logic. *English Cyclopaedia,* vol. 5. Repr. in *De Morgan 1966:* pp. 247–270.
  1868 Book review. *The Atheneum* 2: 71–73.
  1966 *On the Syllogism and Other Logical Writings,* ed. Peter Heath. Routledge & Kegan Paul: London; Yale University Press: New Haven.
Descartes, René
  1984, 1985 *The Philosophical Writings of Descartes.* 2 vols. (Vol. 1 1985, Vol. 2 1984.) Trans. John Cottingham, Robert Stoothoff, and Dugald Murdoch. Cambridge University Press.

Dhombres, Jean G.
1986 Quelques aspects de l'histoire des équations fonctionnelles lié à l'évolution du concept de fonction. *Archive for History of Exact Sciences* 36: 91–181.
DiLeo, Jeffrey
1989 A Semiotic Classification of Proper Names. In *Prewitt et al., eds. 1989,* pp. 143–149.
1991 Peirce's Haecceitism. *Transactions of the Charles S. Peirce Society* 27: 79–109.
Dipert, Randall R.
1978 *Development and Crisis in Late Boolean Logic: The Deductive Logics of Peirce, Jevons, and Schröder.* Ph.D. dissertation, Indiana University: Bloomington.
1980 Ein Karlsruher Pionier der Logik. Ernst Schröder's Beitrag zur Logik und der Grundlagen der Mathematik. *Fridericiana* (Karlsruhe) no. 27: 23–44.
1981 Peirce's Propositional Logic. *Review of Metaphysics* 34: 569–595.
1982 Set-Theoretical Representations of Ordered Pairs and Their Adequacy for the Logic of Relations. *Canadian Journal of Philosophy* 12: 353–374.
1984 Peirce, Frege, the Logic of Relations, and Church's Theorem. *History and Philosophy of Logic* 5: 49–66.
1990 Individuals and Extensional Logic in Schroeder's *Vorlesungen uber die Logik der Algebra. Modern Logic* 1: 22–42.
1991 The Purposes of Logic. Paper delivered to the Department of Philosophy, University of Bologna, Italy, 1990. Unpublished.
1994 The Life and Logical Contributions of O. H. Mitchell: Peirce's Gifted Student. *Transactions of the Charles S. Peirce Society* 30: 515–542.
1995 Peirce's Underestimated Place in the History of Logic. In *Ketner, ed. 1995,* pp. 32–58.
Diringer, David
1968 *The Alphabet: A Key to the History of Mankind.* 2 vols., 3rd ed. Funk and Wagnalls: New York.
Driesch, Hans
1908 *The Science and Philosophy of the Organism,* 2 vols. A. and C. Black: London.
Dunn, Michael
1986 Relevance Logic and Entailment. In *Handbook of Philosophical Logic,* vol. 3, ed. Dov Gabbay and F. Guenthner. D. Reidel: Dordrecht, pp. 117–224.
Edmondson, Amy C.
1987 *A Fuller Explanation: The Synergetic Geometry of R. Buckminster Fuller.* Birkhauser: Boston.
Eisele, Carolyn
1957 The Charles S. Peirce-Simon Newcomb Correspondence. *Proceedings of the American Philosophical Society* 101: 416–421.
1975 C. S. Peirce's Search for a Method in Mathematics and the History of Science. *Transactions of the Charles S. Peirce Society* 11: 149–158.
1982 Mathematical Methodology in the Thought of Charles S. Peirce. *Historia Mathematica* 9: 333–341.
Engel, Kenneth
1980 Shadows of the 4th Dimension. *Science 80* 5: 68–73.
Enros, Philip Charles
1979 *The Analytical Society: Mathematics at Cambridge University in the Early Nineteenth Century.* Ph.D. dissertation, University of Toronto.
Euclid
1956 *The Elements,* Trans. T. Heath. Dover: New York.

Evans, Martha Walton, ed.
    1988 *Relational Models of the Lexicon: Representing Knowledge in Semantic Networks.* Cambridge University Press.

Everett, C. J. and Stanislaw M. Ulam
    1946 Projective Algebra, I. *American Journal of Mathematics* 68: 77–88. Repr. in *Ulam 1974,* pp. 231–242.

Fann, Kuang Tih
    1970 *Peirce's Theory of Abduction.* Martinus Nijhoff: The Hague.

Fargues, Jean, Marie Claude Landau, Anne Dugourd, and Laurent Catach
    1986 Conceptual Graphs for Semantics and Knowledge Processing. *IBM Journal of Research and Development* 30: 70–79.

Feibleman, James K.
    1946 *An Introduction to Peirce's Philosophy Interpreted as a System.* Harper: New York.

Feynman, Richard P.
    1985 *"Surely You're Joking, Mr. Feynman!" Adventures of a Curious Character.* As told to Ralph Leighton, ed. Edward Hutchings. W. W. Norton & Company: New York.

Fillmore, Charles J.
    1968 The Case for Case. In *Bach and Harms, eds. 1968,* pp. 1–88.

Findler, Nicholas V., ed.
    1979 *Associative Networks: Representation and Use of Knowledge by Computers.* Academic Press: New York.

Fisch, Max H.
    1967 Peirce's Progress from Nominalism toward Realism. *The Monist* 51: 159–178. Repr. in *Fisch 1986,* pp. 184–200.
    1972 Peirce and Leibniz. *Journal of the History of Ideas* 33: 485–496.
    1974 Supplements to the Peirce Bibliographies. *Transactions of the Charles S. Peirce Society* 10: 94–129.
    1978 Peirce's General Theory of Signs. In *Sebeok, ed. 1978,* pp. 31–70. Repr. in *Fisch 1986,* pp. 321–355.
    1980, 1982a The Range of Peirce's Relevance. *The Monist* 63 (1980): 269–276 and 65 (1982): 123–141. Repr. in *Freeman, ed. 1983,* pp. 11–37 and in *Fisch 1986,* pp. 422–448.
    1982b Introduction. W1: xv–xxxv.
    1986 *Peirce, Semeiotic, and Pragmatism.,* eds. Kenneth L. Ketner and Christian J. W. Kloesel. Indiana University Press: Bloomington.

Fisch, Max H. and Jackson I. Cope
    1952 Peirce at the Johns Hopkins University. In *Wiener and Young, eds. 1952,* pp. 277–311. Repr. in *Fisch 1986,* pp. 35–78.

Fisch, Max H. and Atwell Turquette
    1966 Peirce's Triadic Logic. *Transactions of the Charles S. Peirce Society* 2: 71–85. Repr. in *Fisch 1986,* pp. 171–183.

Fisher, Ronald Aylmer
    1929 *The Genetical Theory of Natural Selection.* 2nd ed. rev. 1958, Dover: New York.

Flower, Elizabeth and Murray G. Murphey
    1977 *A History of Philosophy in America.* Putnam: New York.

Foss, Jeff
    1984 Reflections on Peirce's Concepts of Testability and the Economy of Research. *Philosophy of Science Association* 1: 28–39.

Fraenkel, Abraham
  1968 *Abstract Set Theory.* North Holland: Amsterdam.
Fraenkel, Abraham and Yohoshua Bar-Hillel
  1958 *Foundations of Set Theory.* North-Holland: Amsterdam.
Frankfurt, Harry G.
  1958 Peirce's Notion of Abduction. *The Journal of Philosophy* 55: 593–597.
Frazee, Jerome
  1988 A New Symbolic Representation of the Basic Truth-Functions of the Propositional Calculus. *History and Philosophy of Logic* 9: 87–91.
  1990 A New Symbolic Representation for the Algebra of Sets. *History and Philosophy of Logic* 11: 67–75.
Freeman, Eugene, ed.
  1983 *The Relevance of Charles Peirce.* The Hegeler Institute: La Salle, Ill.
Frege, Gottlob
  1879 *Begriffsschrift, eine der arithmetischen nachgebildete Formelsprache des reinen Denkens.* Verlag von Louis Nebert: Halle. Repr. in *Frege 1964b.* Trans. Stefan Bauer-Mengelberg as *Begriffsschrift, a Formula Language, Modeled upon that of Arithmetic, for Pure Thought,* in *van Heijenoort,* ed. *1967,* pp. 1–82.
  1882–1883 Über den Zweck der Begriffsschrift. *Jenaische Zeitschrift für Naturwissenschaft (Supplementar-Heft)* 16: 1–10. Trans. in *Frege 1968.*
  1884 *Die Grundlagen der Arithmetic.* Koebner: Breslau. Repr. 1950, trans. J. L. Austin, as The *Foundations of Arithmetic.* Basil Blackwell: Oxford; Philosophical Library: New York. Also repr. in *Frege 1964.*
  1892 Über Sinn und Bedeutung. *Zeitschrift für Philosophie und philosophische Kritik* 100: 25–50.
  1893 *Grundgesetze der Arithmetik, begriffsschriftlich abgeleitet.* Vol. 1. Verlag Hermann Pohle: Jena. Repr. 1962, Georg Olms Verlagsbuchhandlung: Hildesheim. Trans. in *Frege 1964a.*
  1903 *Grundgesetze der Arithmetik, begriffsschriftlich abgeleitet.* Vol. 2. Verlag Hermann Pohle: Jena. Repr. 1962, Georg Olms Verlagsbuchhandlung: Hildesheim. Trans. in *Frege 1964a.*
  1964a *The Basic Laws of Arithmetic. Exposition of the System.* Trans. and ed. Montgomery Furth. University of California Press: Berkeley.
  1964b *Begriffsschrift und andere Aufsätze,* ed. I. Angelelli. Georg Olms Verlagsbuchhandlung: Hildesheim.
  1967 *Kleine Schriften,* ed. I. Angelelli. Georg Olms Verlagsbuchhandlung: Hildesheim.
  1968 On the Purpose of the Begriffsschrift. Trans. Victor Dudman. *Australasian Journal of Philosophy* 46: pp. 89–97.
  1969 *Nachgelassene Schriften,* ed. Hans Hermes, Friedrich Kambartel, and Friedrich Kaulbach. Felix Meiner Verlag: Hamburg.
Freytag Löringhoff, Baron v., Bruno
  1967 *Logik II: Definitionstheorie und Methodologie des Kalkülwechsels.* W. Kholhammer Verlag: Berlin.
  1985 *Neues System der Logik: Symbolisch-symmetrische Rekonstruktion und operative Anwendung des aristotelischen Ansatzes.* Felix Meiner Verlag: Hamburg.
Fumerton, Richard A.
  1980 Induction and Reasoning to the Best Explanation. *Philosophy of Science* 47: 589–600.
Gardner, Martin
  1958 *Logic Machines and Diagrams.* McGraw-Hill: New York.

Garey, M. R. and D. S. Johnson
1979   *Computers and Intractability: A Guide to the Theory of NP-Completeness.* Freeman: San Francisco.
Gärdenfors, Peter, ed.
1987   *Generalized Quantifiers.* D. Reidel: Dordrecht.
Geach, Peter T.
1958–1959   Russell on Meaning and Denoting. *Analysis* 19: 69–72.
Gelb, Ignace J.
1963   *A Study of Writing.* Rev. 2nd ed. University of Chicago Press.
Gilman, Benjamin Ives
1892   On the Properties of a One-Dimensional Manifold. *Mind* (n.s.) 1: 518–526.
1923   The Paradox of the Syllogism Solved by Spatial Construction. *Mind* (n.s.) 32: 38–49.
Gindinkin, Semen Grigor'evich
1985   *Algebraic Logic.* Trans. Robert H. Silverman. Springer Verlag: New York.
Givant, Steven
1988   Recent Progress and Open Problems in Algebraic Logic Based on the Calculus of Relations. Abstract in *Németi, ed. 1988.*
1991   Tarski's Development of Logic and Mathematics Based on the Calculus of Relations. In *Andréka et al., eds. 1991,* pp. 189–215.
Gleick, James.
1987   *Chaos – Making a New Science.* Viking-Penguin: New York.
Gödel, Kurt
1930   Die Vollständigkeit der Axiome des logischen Funktionenkalküls, *Monatshefte für Mathematik und Physik* 37: 349–360. Trans. as "The Completeness of the Axioms of the Functional Calculus of Logic" by Stefan Bauer-Mengelberg, in *van Heijenoort, ed. 1967,* pp.583–591.
Goldfarb, Warren D.
1979   Logic in the Twenties: The Nature of the Quantifier. *Journal of Symbolic Logic* 44: 351–368.
Gonseth, Ferdinand
1937   *Qu'est-ce que la logique?* Hermann: Paris.
Goodman, Nelson
1966   *The Structure of Appearance.* Bobbs-Merrill: New York.
Gottshalk, W. H.
1953   The Theory of Quaternality. *Journal of Symbolic Logic* 18: 193–196.
Goudge, Thomas A.
1950   *The Thought of C. S. Peirce.* University of Toronto Press.
Grassmann, Hermann
1861   *Lehrbuch der Arithmetik für höhere Lehranstalten.* Adolf Eslin: Berlin.
Grattan-Guinness, Ivor
1975   Wiener on the Logics of Russell and Schröder. An Account of His Doctoral Thesis, and of His Subsequent Discussion of it with Russell. *Annals of Science* 32: 103–132.
1977   *Dear Russell—Dear Jourdain. A Commentary on Russell's Logic, Based on His Correspondence with Philip Jourdain.* Duckworth: London; Columbia University Press: New York.
1986   From Weierstrass to Russell: a Peano Medley, in *Celebrazioni in memoria di Giuseppe Peano nel cinquantenario della morte.* University of Turin (Depart-

ment of Mathematics): Turin, pp. 17–31. Also in *Revista di storia della scienza* 2 (1985, publ. 1987): 1–16.

1988 Living Together and Living Apart: On the Interactions Between Mathematics and Logics from the French Revolution to the First World War. *South African Journal of Philosophy* 7: 73–82.

1990a *Convolutions in French Mathematics, 1800–1840. From the Calculus and Mechanics to Mathematical Analysis and Mathematical Physics,* 3 vols. Birkhäuser: Basel; Deutscher Verlag der Wissenschaften: Berlin.

1990b Charles Babbage: Production by Numbers. *Annals of science* 47: 81–87.

1990c Bertrand Russell (1872–1970), after Twenty Years. *Notes and Records of the Royal Society* 44: 280–306.

1991 The Correspondence between George Boole and Stanley Jevons, 1863–1864. *History and Philosophy of Logic* 12: 15–35.

1992 Charles Babbage as an Algorithmic Thinker. *Annals of the History of Computing* 14: 34–48.

Grattan-Guinness, Ivor, ed.

1987 *History in Mathematics Education.* Belin: Paris.

Green, Thomas M. and Charles L. Hamberg

1968 A Study of N-Dimensional Equations, Modulo Two. *Mathematics Teacher* 61: 741–748.

Grefenstette, John, ed.

1985 *Proceedings of an International Conference on Genetic Algorithms and Their Applications, 24–26 July.* Carnegie Mellon University: Pittsburgh.

Griffin, Nicholas

1980 Russell on the Nature of Logic. *Synthese* 45: 117–188.

Grmek, Mirko Drazen, Robert S. Cohen, and Guido Cimono, eds.

1980 *On Scientific Discovery: The Erice Lectures 1977.* D. Reidel: Dordrecht.

Groenendijk, Jeroen A. G., Theo M. V. Janssen, and Martin B. J. Stokhof, eds.

1981 *Formal Methods in the Study of Language.* Mathematical Centre Tracts: Amsterdam.

Gruber, Jeffrey

1965 *Studies in Lexical Relations.* Ph.D. dissertation, MIT: Cambridge.

Haack, Susan

1993 Peirce and Logicism: Notes Towards an Exposition. *Transactions of the Charles S. Peirce Society* 29: 33–56.

Hacker, Edward A.

1975 The Octagon of Opposition. *Notre Dame Journal of Formal Logic* 16: 352–353.

Hailperin, Theodore

1976 *Boole's Logic and Probability: A Critical Exposition from the Standpoint of Contemporary Algebra, Logic, and Probability Theory.* Vol. 85, *Studies in Logic and the Foundation of Mathematics,* ed. Jon Barwise, D. Kaplan, H. J. Keisler, Patrick Suppes, and A. S. Troelstra. 2nd ed. 1986. North Holland: Amsterdam.

1988 The Development of Probability Logic from Leibniz to MacColl. *History and Philosophy of Logic* 9: 131–191.

Hallett, Michael.

1984 *Cantorian set Theory and Limitation of Size.* Oxford University Press.

Halmos, Paul

1960 *Naive Set Theory.* Springer-Verlag: Berlin.

Hamilton, William D.
> 1982 Pathogens as Causes of Genetic Diversity in Their Host Populations. In *Anderson and May, eds. 1982*, pp. 269–296.

Hamilton, William D., Robert Axelrod, and Reiko Tanese
> 1990 Sexual Reproduction as an Adaptation to Resist Parasites (A Review). *Proceedings of the National Academy of Sciences (US)* 87: 3566-3573.

Hamilton, William D. and Marlene Zuck
> 1982 Heritable Fitness and Bright Birds: A Role for Parasites? *Science* 218: 384–386.

Hamilton, Sir William
> 1852 *Discussions on Philosophy and Literature, Education and University Reform. Chiefly from the Edinburgh Review, Corrected, Vindicated, Enlarged, in Notes and Appendices.* Edinburgh. 2nd ed. enlr. 1853. 3rd ed. 1866. Edinburgh. 2nd ed. enlr. 1853. 3rd ed. 1866. With an Introductory Essay by Robert Turnbull, Harper: New York, 1853, 1855, 1860.

Handler, Wolfgang, ed.
> 1986 *CONPAR 86: Conference on Algorithms and Hardware for Parallel Processing. Aachen, 17–19 September, Proceedings.* Springer-Verlag: Berlin.

Hanson, Norwood Russell
> 1958 *Patterns of Discovery.* Cambridge University Press.
> 1961 Retroductive Inference. *Philosophy of Science: The Delaware Seminar* 1: 21–37.
> 1965 The Idea of a Logic of Discovery. *Dialogue* 4: 48–61. Repr. in *Hanson 1971*, pp. 288–300.
> 1971 *What I Do Not Believe and Other Essays*, ed. Stephen Toulmin and Harry Woolf. D. Reidel: Dordrecht.

Harman, Gilbert
> 1965 The Inference to the Best Explanation. *The Philosophical Review* 74: 88–95.
> 1968 Enumerative Induction as Inference to the Best Explanation. *The Journal of Philosophy* 65: 529–533.
> 1986 *Change in View.* MIT Press: Cambridge.

Harman, Gilbert and Donald Davidson, eds.
> 1972 *Semantics of Natural Language.* D. Reidel: Dordrecht.

Harper, William and Bryan Skyrms, eds.
> 1988 *Causation in Decision, Belief Change in Statistics.* Kluwer Academic: Dordrecht.

Harris, James F. and Kevin Hoover
> 1982 Abduction and the New Riddle of Induction. In *Freeman, ed. 1983*, pp. 132–144.

Hausdorff, Felix
> 1914 *Grundzüge der Mengenlehre.* Veit: Leipzig. Repr. 1965, Chelsea: New York. 3rd German ed. of 1935 trans. John R. Aumann *et al.* as *Set Theory*, 2nd ed. Chelsea, 1962.

Hawkes, Herbert Edwin
> 1902 Estimate of Peirce's Linear Associative Algebra. *American Journal of Mathematics* 24: 87–95.

Hawkins, Benjamin S. Jr.
> 1975a A Compendium of C. S. Peirce's 1866–1885 Work. *Notre Dame Journal of Formal Logic* 16: 109–115.

1975b Review of *Roberts 1973*. *Transactions of the Charles S. Peirce Society* 11: 128–139.
1979 A Reassessment of Augustus De Morgan's Logic of Relations: A Documentary Reconstruction. *International Logic Review* 10: 32–61.
1981 Peirce's and Frege's Systems of Notation. In *Ketner et al.,1981*, pp. 381–389.
1986 J. W. Dauben on C. S. Peirce's Place in Mathematics: Some Reflections. *International Logic Review* 17: 62–69.

Hempel, Carl
1966 *Philosophy of Natural Science*. Prentice Hall: Englewood Cliffs.
1970 *Aspects of Scientific Explanation*. Free Press: New York.

Hendrix, Gary G.
1975 Expanding the Utility of Semantic Networks through Partitioning. *Proc. IJCAI-75:* 115–121.
1979 Encoding Knowledge in Partitioned Networks. In *Findler, ed. 1979*, pp. 51–92.

Henkin, Leon
1953 Some Notes on Nominalism. *Journal of Symbolic Logic* 18: 19–29.
1956 Two Concepts from the Theory of Models. *Journal of Symbolic Logic* 21: 28–32.
1962 Are Logic and Mathematics Identical? *Science* 138: 788–794.
1971 The Representation Theorem for Cylindrical Algebras. In *Heyting 1971*, pp. 85–97.
1973 Universal Semantics and Algebraic Logic. In *Leblanc, ed. 1973*, pp. 111–127.

Henkin, Leon and James Donald Monk
1974 Cylindric Algebras. *Proceedings of Symposia in Pure Mathematics* (Providence: American Mathematical Society) 25: pp. 105–121.
1974 Cylindric Algebras and Related Structures. In *Henkin et al., eds. 1974*, pp. 105–121.

Henkin, Leon, James Donald Monk, and Alfred Tarski
1971 *Cylindric Algebras, Part I*. North-Holland: Amsterdam.
1985 *Cylindric Algebras, Part II*. North-Holland: Amsterdam.

Henkin, Leon, James Donald Monk, Alfred Tarski, H. Andréka, and István Németi, eds.
1981 *Cylindric Set Algebras*. Springer Lecture Notes in Mathematics 883. Springer Verlag: Berlin.

Henkin, Leon, John Addison, C. C. Chang, William Craig, Dana Scott, and Robert Vaught, eds.
1974 *Proceedings of the Tarski Symposium, University of California at Berkeley, 1971*. American Mathematical Society: Providence.

Henkin, Leon, Patrick Suppes, and Alfred Tarski, eds.
1959 *The Axiomatic Method, with Special Reference to Geometry and Physics*. North-Holland: Amsterdam.

Herbrand, Jacques
1930 Investigations in Proof Theory: The Properties of True Propositions. Chapter 5 of *Recherches sur la théorie de la démonstration*. Ph.D. dissertation, University of Paris. Trans. Burton Dreben and Jean van Heijenoort, in *van Heijenoort, ed. 1967*, pp. 529–581.

Herzberger, Hans G.
1981 Peirce's Remarkable Theorem. In *Sumner et al., eds. 1981*, pp. 41–58; 297–301.

Heyting, Arendt, ed.
  1959 *Constructivity in Mathematics.* North-Holland: Amsterdam.
  1971 *Metamathematical Interpretation of Formal Systems.* North-Holland: Amsterdam.
Hilbert, David
  1899 *Grundlagen der Geometrie.* B. G. Teubner: Stuttgart. Repr. 1971 as *Foundations of Geometry,* trans. Leo Unger/rev. Paul Bernays. Open Court: La Salle.
  1927 The Foundations of Mathematics. In *van Heijenoort, ed. 1967,* pp. 464–479.
Hilpinen, Risto
  1982 On Peirce's Theory of Proper Names. Unpublished.
  1989 Peirce on Language and Reference. Forthcoming in *Ketner, ed. 1993.*
Hintikka, Jaako
  1980 C. S. Peirce's "First Real Discovery," and its Contemporary Relevance. *The Monist* 63: 304–315.
Hintikka, Jaako, J. Moravcsik, and Patrick Suppes, eds.
  1973 *Approaches to Natural Language: Proceedings of the 1970 Stanford Workshop on Grammar and Semantics.* D. Reidel: Dordrecht.
Hintikka, Jaako and F. Vandamme, eds.
  1985 *Logic of Discovery and Logic of Discourse.* Plenum Press: New York.
Hiż, Henry
  1946 Remarque sur le degré de complétude. *Comptes Rendus des Séances de l'Academie des Sciences* 223: 973–974.
  1973 A Completeness Proof for C-Calculus. *Notre Dame Journal of Formal Logic* 14: 253–258.
Hobbes, Thomas
  1947 *Leviathan,* ed. Michael Oakeshott. Oxford.
Hocutt, M. O.
  1972 Is Epistemic Logic Possible? *Notre Dame Journal of Formal Logic* 13: 433–453.
Holland, John
  1975 *Adaptation in Natural and Artificial Systems—An Introductory Analysis with Applications to Biology, Control, and Artificial Intelligence.* University of Michigan Press: Ann Arbor.
  1985 Properties of the Bucket-Brigade Algorithm. In *Grefenstette, ed. 1985,* pp. 1–7.
  1986 Escaping Brittleness: The Possibilities of General-Purpose Learning Algorithms Applied to Parallel Rule-Based Systems. In *Michalski et al., eds. 1986,* chap. 20.
  1989 Using Classifier Systems to Study Adaptive Nonlinear Networks. In *Stein, ed. 1989,* pp. 1–36.
  1990a Concerning the Emergence of Tag-mediated Lookahead in Classifier Systems. *Physica D* 42: 188–201.
  1990b Emergent Models. In *Scott, ed. 1990,* pp. 107–125.
Holland, John and Arthur Burks
  1987 Adaptive Computing System Capable of Learning and Discovery. U.S. Patent 4,697,242; issued 29 September 1987.
  1989 Method of Controlling a Classifier System. U.S. Patent 4,881,178; issued 11 November 1989.
Holland, John, Keith Holyoak, Richard Nisbett, and Paul Thagard
  1986 *Induction: Processes of Inference, Learning, and Discovery.* MIT Press: Cambridge.

Houser, Nathan
 1985 *Peirce's Algebra of Logic and the Law of Distribution.* Ph.D. dissertation, University of Waterloo: Waterloo, Ontario.
 1989a The Fortunes and Misfortunes of the Peirce Papers. In *Deledalle, ed. 1992,* pp. 1259–1268.
 1989b Introduction. W4: xix–lxx.
 1990 The Schröder-Peirce Correspondence. *Modern Logic* 1: 206–236.
 1993 On "Peirce and Logicism": a Response to Haack. *Transaction of the Charles S. Peirce Society* 29: 57–67.
Huntington, Edward Vermilye
 1904 Sets of Independent Postulates for the Algebra of Logic. *Transactions of the American Mathematical Society* 5: 288–309.
Huntington, Edward Vermilye and Christine Ladd-Franklin
 1905 Symbolic Logic. In *The Americana. A Universal Reference Library* 9: 6pp. (unpaginated). Repr. in 1934 ed., 17: 568–573.
Hutchins, Robert M.
 1952 *The Great Conversation, The Substance of a Liberal Education.* Vol. 1 of *Great Books of the Western World,* Robert Maynard Hutchins, Ed. in Chief; Mortimer J. Adler, Associate Ed. William Benton, Publisher; Encyclopædia Britannica, Inc.
Ifrah, Georges
 1985 *From Zero to One: A Universal History of Number.* Viking: New York.
Iliff, Alan J.
 1994 The Role of the Matrix Representation in Peirce's Development of the Quantifiers. This volume, pp. 193–205.
Jackendoff, Ray
 1983 *Semantics and Cognition.* MIT Press: Cambridge.
Jevons, W. Stanley
 1880 *Studies in Deductive Logic: A Manual for Students.* Macmillan: London. Repr. 1908.
Johnson, James S.
 1969 Nonfinitizability of Classes of Representable Polyadic Algebras. *Journal of Symbolic Logic* 34: 344–352.
Johnson, William Ernest
 1922 *Logic,* pt. 2. Cambridge University Press.
Jones, E. E. Constance
 1890 *Elements of Logic as a Science of Propositions.* T. & T. Clark: Edinburgh.
 1898 The Paradox of Logical Inference. *Mind* 23: 205–218.
Jónsson, Bjarni
 1951 Boolean Algebras with Operators. *American Journal of Mathematics* 73: 891–939.
 1959 Representation of Modular Lattices and of Relation Algebras. *Transactions of the American Mathematical Society* 92: 449–464.
Kamp, Hans
 1981a Events, Discourse Representations, and Temporal References. *Languages* 64: 39–64.
 1981b A Theory of Truth and Semantic Representation. In *Groenendijk et.al., eds. 1981,* pp. 277–322.
Kant, Immanuel
 1781 *Kritik der Reinen Vernunft.* Verlegts Johann Friedrich Hartknoch: Riga. 2nd ed. 1787. Repr. in 1902–1942, *Gesammelte Schriften,* Berlin. Trans.

1929 as *Immanuel Kant's Critique of Pure Reason,* by Norman K. Smith. Macmillan: London; 2nd impr. with corrections 1933; repr. 1958, St. Martin's: New York.

Kapitan, Tomis

1984 Castañeda's Dystopia. *Philosophical Studies* 46: 263–270.

1990 In What Way is Abductive Inference Creative? *Transactions of the Charles S. Peirce Society* 26: 499–512.

1994 Peirce and the Structure of Abductive Inference. This volume, pp. 477–496.

Katzan, Jr., Harry

1975 *Introduction to Computer Science.* Petrocelli/Charter: New York.

Kauffman, Louis H.

1988 New Invariants in the Theory of Knots. *American Mathematical Monthly* 95: 195–242.

1991 *Knots and Physics.* World Scientific: New Jersey.

Keenan, Edward L.

1972 On Semantically Based Grammar. *Linguistic Inquiry* 3: 413–462.

Keenan, Edward L., ed.

1975 *Formal Semantics of Natural Language.* Cambridge University Press.

Kelly, Kevin T.

1987 The Logic of Discovery. *Philosophy of Science* 54: 435–452.

Kempe, Alfred Bray

1885 A Memoir Introductory to a General Theory of Mathematical Form. *Proceedings of the Royal Society* 38: 393–401.

1886 A Memoir on the Theory of Mathematical Form. *Philosophical Transactions of the Royal Society of London* 177: 1–70.

1887 Note to A Memoir on the Theory of Mathematical Form. *Proceedings of the Royal Society of London* 42: 193–196.

1889–1890 On the Relation between the Logical Theory of Classes and the Geometrical Theory of Points. *Proceedings of the London Mathematical Society* 21: 147–182.

1890 The Subject Matter of Exact Thought. *Nature* 43: 156–162.

1894 Mathematics. *Proceedings of the London Mathematical Society* 26: 5–15.

1897 The Theory of Mathematical Form. A Correction and Explanation. *The Monist* 7: 453–458.

Kent, Beverley E.

1987 *Charles S. Peirce: Logic and the Classification of the Sciences.* McGill-Queen's University Press: Kingston and Montreal.

Kerr, Donald, Karl Braithwaite, N. Metropolis, David Sharp, and Gian-Carlo Rota, eds.

1984 *Science, Computers, and the Information Onslaught.* Academic Press: New York.

Ketner, Kenneth L.

1985 How Hintikka Misunderstood Peirce's Account of Theorematic Reasoning. *Transactions of the Charles S. Peirce Society* 21: 407–418.

1986 Peirce's "Most Lucid and Interesting Paper": An Introduction to Cenopythagoreanism. *International Philosophical Quarterly* 26: 375–392.

1987 Identifying Peirce's "Most Lucid and Interesting Paper." *Transactions of the Charles S. Peirce Society* 23: 539–555.

1991 *Elements of Logic.* Texas Tech Press: Lubbock.

Ketner, Kenneth L., ed.

1995 *Peirce and Contemporary Thought.* Plenary papers from Charles S. Peirce Sesquicentennial Congress. Fordham University Press: New York.

Ketner, Kenneth L., Joseph Ransdell, Carolyn Eisele, Max H. Fisch, and Charles Hardwick, eds.
1981 *Proceedings of the C. S. Peirce Bicentennial International Congress.* Texas Tech Press: Lubbock.
Keynes, John Neville
1884 *Studies and Exercises in Formal Logic.* Cambridge University Press. 3rd ed. 1894, Macmillan: London.
King-Hele, Desmond
1974–1975 A Discussion with Bertrand Russell at Plas Penrhyn, 4 August 1968. *Russell: The Journal of the Bertrand Russell Archives* 16: 21–25.
Kiss, Stephen A.
1947 *Transformations on Lattices and Structures of Logic.* Stephen A. Kiss: New York.
Kleene, Stephen C.
1952 *Introduction to Metamathematics.* D. Van Nostrand: Princeton. 6th Repr. 1971, American Elsevier: New York.
1967 *Mathematical Logic.* Wiley & Sons: New York.
Kleiner, Scott A.
1983 A New Look at Kepler and Abductive Argument. *Studies in the History and Philosophy of Science* 14: 279–313.
1988 Erotetic Logic and Scientific Inquiry. *Synthese* 74: 19–46.
Kneale, William and Martha
1962 *The Development of Logic.* Oxford University Press.
Koestler, Arthur
1960 *The Watershed: A Bibliography of Johannes Kepler.* Doubleday: Garden City, N.Y.
1963 *The Sleepwalkers.* Grossett and Dunlap: New York.
König, Julius
1905 Zur Kontinuum-Problem. *Verhandlungen des Dritten Internationalen Mathematiker-Kongresses in Heidelberg (1904).* B. G. Tuebner: Leipzig.
Koppelman, Elaine
1971 The Calculus of Operations and the Rise of Abstract Algebra. *Archive for History of Exact Sciences* 8: 155–242.
Korzybski, Alfred
1933 *Science and Sanity: An Introduction to Non-Aristotelian Systems and General Semantics.* 4th ed. 1958. The Non-Aristotelian Library Publishing Company: Lakeville, Connecticut.
Kripke, Saul A.
1959 A Completeness Theorem in Modal Logic. *The Journal of Symbolic Logic* 24: 1–15.
1972 *Naming and Necessity.* Harvard University Press.
Kruse, Felicia E.
1986a Indexicality and the Abductive Link. *Transactions of the Charles S. Peirce Society* 22: 435–447.
1986b Toward an Archaeology of Abduction. *American Journal of Semiotics* 4: 157–167.
Kurepa, Djuro and B. Schön
1974 Real Functions. Chap. 12 in vol. 3 of *Behnke et al., eds. 1974,* pp. 446–478.
Ladd-Franklin, Christine
1880 On De Morgan's Extension of the Algebraic Process. *American Journal of Mathematics* 3: 210–225.
1883 On the Algebra of Logic. In *Peirce, ed. 1883,* pp. 17–71.

1890 Some Proposed Reforms in Common Logic. *Mind* 15: 75–88.

1892a Review of *Schröder 1890*. *Mind* (n.s.) 1: 126–132.

1892b Dr. Hillebrand's Syllogistic Scheme. *Mind* (n.s.) 1: 527–530.

1912 Implication and Existence in Logic. *Philosophical Review* 21: 641–665.

Laing, Richard

1975 Some Alternative Reproductive Strategies in Artificial Molecular Machines. *Journal of Theoretical Biology* 64: 63–84.

1977 Automaton Models of Reproduction by Self-Inspection. *Journal of Theoretical Biology* 66: 437–456.

Lakoff, George

1972 Linguistics and Natural Logic. In *Harman and Davidson, eds. 1972*, pp. 545–665.

Lalvani, Haresh

1989 Structures and Meta-Structures. In *Symmetry of Structure,* Interdisciplinary Symmetry Symposia 1 (Budapest, 13–19 August), pp. 302–306.

Langer, Susanne K.

1967 *Mind: An Essay on Human Feeling.* Vol. I. The Johns Hopkins Press: Baltimore.

Langley, Pat, Herbert A. Simon, Gary L. Bradshaw, and Jan M. Zytkow

1987 *Scientific Discovery: Computational Explorations of the Creative Processes.* MIT Press: Cambridge.

Langton, Christopher, ed.

1989 *Artificial Life.* Addison-Wesley: New York.

Laplace, Pierre

1814 *A Philosophical Essay on Probabilities.* Trans. F. W. Truscott and F. L. Emory. Dover: New York, 1951.

Laszlo, Ervin

1972 *Introduction to Systems Philosophy.* Harper Torchbooks: New York.

Lebesgue, Henri

1905 Sur les fonctions représentables analytiquement. *J. de math.* 6: 139–216.

Leblanc, Hugues, ed.

1973 *Truth, Syntax and Modality.* North-Holland: Amsterdam.

Lehmann, Frederick W. IV, ed.

1992 *Semantic Networks in Artificial Intelligence.* Pergamon Press: New York.

Leibniz, Gottfried Wilhelm

1690 A Study in the Logical Calculus. In *Philosophical Papers and Letters,* trans. and ed. Leroy E. Loemker. D. Reidel: Dordrecht, 1956. 2nd ed. 1969, pp. 371–382. (Date is approximate; Loemker gives "Early 1690's.")

Lenzen, Victor F.

1965 Reminiscences of a Mission to Milford, Pennsylvania. *Transactions of the Charles S. Peirce Society* 1: 3–11.

1969 An Unpublished Scientific Monograph by C. S. Peirce. *Transactions of the Charles S. Peirce Society* 5: 5–24.

1973 The Contributions of Charles S. Peirce to Linear Algebra. In *Riepe, ed. 1973,* pp. 239–254.

1975 Charles S. Peirce as Mathematical Physicist. *Transactions of the Charles S. Peirce Society* 11: 159–166, 225–226.

Leśniewski, Stanisław

1929 Grundzüge eines neuen Systems der Grundlagen der Mathematik. *Fundamenta Mathematicae* 14: 1–81.

1931 Über Definitionen in der sogenannten Theorie der Deduktion. *Comptes Rendus des Séances de la Société des Sciences et des Lettres de Varsovie* 24: 289–

309. Trans. E. C. Luschei as "On Definition in the So-called Theory of Deduction" in *McCall 1967*, pp. 170–187.

1939 Einleitende Bemerkungen zur Fortsetzung meiner Mitteilung u.d.T. "Grundzüge eines neuen Systems der Grundlagen der Mathematik." *Collectanea Logica* 1: 1–60. Trans. W. Teichmann and S. McCall as "Introductory Remarks to the Continuation of My Article 'Grundzüge etc.' " in *McCall 1967*, pp. 116–169.

Levy, Stephen H.

1982a *A Comparative Analysis of Charles S. Peirce's Philosophy of Mathematics*, Ph.D. dissertation, Fordham University: New York.

1982b The Significance of Charles Peirce's Philosophy of Mathematics. *Proceedings of the Semiotic Society of America*. Indiana University Press: Bloomington.

1983 Peirce's Concept of Collection or Set. Unpublished.

1986 Peirce's Ordinal Concept of Number. *Transactions of the Charles S. Peirce Society* 22: 23–42.

1991 Peirce's Theory of Infinitesimals. *International Philosophical Quarterly* 31: 127–140.

Lewis, Clarence Irving

1918 *A Survey of Symbolic Logic*. University of California Press: Berkeley. Repr. with corrections and with the omission of Chs. 5 and 6, 1960, Dover: New York.

Lewis, David

1975 Adverbs of Quantification. In *Keenan, ed. 1975*, pp. 3–15.

Lieb, Irwin C.

1988 Pragmatism and the Normative Sciences. In *Burch and Saatkamp, eds. 1992*, pp. 273–284.

Lieb, Irwin C., ed.

1953 *Charles S. Peirce's Letters to Lady Welby*. Whitlock's: New York.

Link, Godehard

1987 Generalized Quantifiers and Plurals. In *Gärdenfors, ed. 1987*, pp. 151–180.

Listing, Johann Benedict

1862 Der Census räehmlicher Complexe, oder Verallgemeinerung des Euler'schen Satzes von den Polyëdern. *Abhandlungen der Königlichen Gesellschaft der Wissenschaften zu Göttingen* 10: 97–182.

Loeb, Arthur L.

1976 *Space Structures: Their Harmony and Counterpoint*. Addison-Wesley: Reading, Massachusetts.

1988 Polyhedra: Surfaces or Solids? In *Shaping Space: A Polyhedral Approach*, ed. Marjorie Senechal and George Fleck. Birkhauser: Boston, pp. 106–117.

Lowe, Victor

1985, 1990 *Alfred North Whitehead. The Man and His Work*, 2 vols. Vol. 1, 1985. Vol. 2, ed. Jerome B. Schneewind, 1990. Johns Hopkins University Press: Baltimore.

Löwenheim, Leopold

1915 Über Möglichkeiten im Relativkalkül. *Mathematische Annalen* 76: 447–470. Trans. Stefan Bauer-Mengelberg as "On Possibilities in the Calculus of Relatives" in *van Heijenoort, ed. 1967*, pp. 228–251.

1940 Einkleidung der Mathematik in Schröderschen Relativkalkul (The Adaptation of Mathematics to Schröder's Relative Calculus). *The Journal of Symbolic Logic* 5: 1–15.

Łukasiewicz, Jan
   1934 Z Historii Logiki Zdań. *Przegląd Filozoficzny 37: 417–437. Repr. as "On the History of the Logic of Propositions" in Łukasiewicz 1970*, pp. 197–217.
   1950 On the System of Axioms of the Implicational Propositional Calculus. Publ. as a Supplement to vol. 22 of *Rocznik Polskiego Towarzystwa Matematycznego.* Repr. In *Łukasiewicz 1970*, pp. 306–310.
   1961 O Determiniźmie, rev. version of speech delivered by Łukasiewicz as Rector of the University of Warsaw at the opening of the academic year 1922–1923. In *Z zagadnień logiki i filozofii*, an anthology of Łukasiewicz's works, ed. Jerzy Słupecki. Warsaw. Trans. Z. Jordan as "On Determinism" in *McCall 1967*, pp. 19–39, and in *Łukasiewicz 1970*, pp. 110–128.
   1970 *Selected Works*, ed. L. Borkowski. North Holland: Amsterdam.
Łukasiewicz, Jan and Alfred Tarski
   1930 Untersuchungen über den Aussagenkalkül. *Comptes Rendus des Séances de la Société des Sciences et des Lettres de Varsovie* 23: 30–50. Trans. J. H. Woodger in *Tarski 1956*, pp. 38–59, and in *Łukasiewicz 1970*, pp. 131–152.
Luschei, Eugene C.
   1962 *The Logical Systems of Leśniewski*. North-Holland: Amsterdam.
Lyndon, Roger C.
   1950 The Representation of Relational Algebras. *Annals of Mathematics* 51: 707–729.
   1956 The Representation of Relation Algebras, II. *Annals of Mathematics* 63: 294–307.
   1961 Relation Algebras and Projective Geometries. *Michigan Mathematics Journal* 8: 21–28.
MacColl, Hugh
   1877a The Calculus of Equivalent Statements (First Paper). *Proceedings of the London Mathematical Society* 9: 9–20.
   1877b The Calculus of Equivalent Statements (Second Paper). *Proceedings of the London Mathematical Society* 9: 177–186.
   1878 The Calculus of Equivalent Statements (Third Paper). *Proceedings of the London Mathematical Society* 10: 16–28.
   1880a The Calculus of Equivalent Statements (Fourth Paper). *Proceedings of the London Mathematical Society* 11: 113–121.
   1880b Symbolical Reasoning. *Mind* 5: 45–60.
   1896 The Calculus of Equivalent Statements (Fifth Paper). *Proceedings of the London Mathematical Society* 28: 172, 182–183.
   1897a The Calculus of Equivalent Statements (Sixth Paper). *Proceedings of the London Mathematical Society* 28: 555–579.
   1897b The Calculus of Equivalent Statements (Seventh Paper). *Proceedings of the London Mathematical Society* 29: 98–109.
   1897c Symbolic Reasoning (II). *Mind* (n.s.) 6: 496–510.
   1900 Symbolic Reasoning (III). *Mind* (n.s.) 9: 75–84.
MacHale, Desmond
   1985 *George Boole: His Life and Work*. Boole Press: Dublin.
Maddux, Roger D.
   1978 Some Sufficient Conditions for the Representability of Relation Algebras. *Algebra Universalis* 8: 162–172.
   1982 Some Varieties Containing Relation Algebras. *Transactions of the American Mathematical Society* 272: 501–526.

1983 A Sequent Calculus for Relation Algebras. *Annals of Pure and Applied Logic* 25: 73–101.

1985 Finite Integral Relation Algebras. In *Comer, ed. 1985*, pp. 175–197.

1991a Introductory Course on Relation Algebras, Finite-Dimensional Cylindric Algebras, and Their Interconnections. In *Andréka et al., eds. 1991*, pp. 361–392.

1991b The Origin of Relation Algebras in the Development and Axiomatization of the Calculus of Relations. In *Blok and Pigozzi, eds. 1991*, pp. 438–449.

Mannoury, Gerrit

1909 *Methodologisches und Philosophisches zur Elementar-Mathematik*. P. Visser: Haarlem.

Marquand, Allan

1881 On Logical Diagrams for *n* Terms. *The London, Edinburgh, and Dublin Philosophical Magazine and Journal of Science* 12: 266–270.

Martin, Richard M.

1969 On the Peirce Representation-Relation. *Transactions of the Charles S. Peirce Society* 5: 143–157. Repr. as "The Relation of Representation" in *Martin 1980*, pp. 67–79.

1976a Some Comments on De Morgan, Peirce, and the Logic of Relations. *Transactions of the Charles S. Peirce Society* 12: 223–230. Repr. as "De Morgan and the Logic of Relations" in *Martin 1980*, pp. 46–53.

1976b On Individuality and Quantification in Peirce's Published Logic Papers, 1867–1885. *Transactions of the Charles S. Peirce Society* 12: 231–245. Repr. as "Individuality and Quantification" in *Martin 1980*, pp. 11–24.

1978 Of Servants, Lovers, and Benefactors: Peirce's Algebra of Relatives in 1870. *Journal of Philosophical Logic* 7: 27–48. Repr. in *Martin 1980* as "Of Lovers, Servants, and Benefactors,*"* pp. 25–45.

1980 *Peirce's Logic of Relations and Other Studies*. Foris: Dordrecht.

Mautner, Felix I.

1946 An Extension of Klein's Erlanger Program: Logic as Invariant-Theory. *American Journal of Mathematics* 68: 345–384.

Maynard Smith, John

1978 *The Evolution of Sex*. Cambridge University Press.

1982 *Evolution and the Theory of Games*. Cambridge University Press.

McCall, Storrs, ed.

1967 *Polish Logic 1920–1939*. Oxford University Press.

McCarthy, John

1980 Circumscription. A Form of Nonmonotonic Reasoning. *Artificial Intelligence* 13: 27–39.

1986 Application of Circumscription to Formalizing Common-Sense Knowledge. *Artificial Intelligence* 28: 89–116.

McCord, Michael

1987 Natural Language Processing in Prolog. In *Walker et al., eds. 1987*, pp. 291–402.

McCulloch, Warren S.

1965 *Embodiments of Mind*. MIT Press: Cambridge.

McKenzie, R.

1970 Representations of Integral Relation Algebras. *Michigan Mathematics Journal* 17: 279–287.

McKinsey, J. C. C.
1940 Postulates for the Calculus of Binary Relations. *Journal of Symbolic Logic* 5: 85–97.
1948 On the Representation of Projective Algebras. *American Journal of Mathematics* 70: 375–384.

McNulty, George F.
1976a Undecidable Properties of Finite Sets of Equations. *Journal of Symbolic Logic* 41: 589–604.
1976b The Decision Problem for Equational Bases of Algebras. *Annals of Mathematical Logic* 11: 193–259.

Mendelson, Elliott
1964 *Introduction to Mathematical Logic*. Van Nostrand: New York. Repr. 1977.

Menger, Karl
1937 The New Logic. *Philosophy of Science* 4: 299–336.
1962 A Group in the Substitutive Algebra of the Calculus of Propositions. *Archive der Mathematik* 13: 471–478.

Menninger, Karl
1969 *Number Words and Number Symbols*. MIT Press: Cambridge.

Merrill, Daniel D.
1978 De Morgan, Peirce, and the Logic of Relations. *Transactions of the Charles S. Peirce Society* 14: 247–284.
1984 The 1870 Logic of Relatives Memoir. Introduction, part 3, W2, pp. xlii–xlvii.
1990 *Augustus De Morgan and the Logic of Relations*. Dordrecht: Kluwer Academic Publishers.

Mertz, Donald W.
1979 Peirce: Logic, Categories, and Triads. *Transactions of the Charles S. Peirce Society* 15: 158–175.

Messenger, Theodore
1986 Expressing and Using N-adic Operators in the Propositional Calculus (abstract). *Journal of Symbolic Logic* 51: 1086.

Meyer, Robert K. and Michael Dunn
1969 E, R, and Γ. *The Journal of Symbolic Logic* 34: 460–474.

Michael, Emily
1974 Peirce's Early Study of the Logic of Relations. *Transactions of the Charles S. Peirce Society* 10: 63–75.
1975 Peirce's Paradoxical Solution to the Liar's Paradox. *Notre Dame Journal of Formal Logic* 16: 369–374.
1976 Peirce's Earliest Contact with Scholastic Logic. *Transactions of the Charles S. Peirce Society* 12: 46–55.
1979 An Examination of the Influence of Boole's Algebra on Peirce's Development in Logic. *Notre Dame Journal of Formal Logic* 20: 801–806.

Michalski, Ryszard Stanislaw, Jaime Guillermo Carbonell, and Tom M. Mitchell, eds.
1986 *Machine Learning II*. Morgan Kaufmann: Los Altos.

Mill, John Stuart
1843 *A System of Logic, Ratiocinative and Inductive*, 2 vols. John Parker: London. Repr. 1973–1974 in a critical edition by J. M. Robson, as vols. 7 and 8 of *Collected Works of John Stuart Mill*. University of Toronto Press.

Miller, George Abram
1904 On the Definition of Infinite Number. *The Monist* 14: pp: 469–472.

Mitchell, Oscar Howard
1883 On a New Algebra of Logic. In *Peirce, ed. 1883*, pp. 72–106.

Miyazaki, Koji
1986 *An Adventure in Multidimensional Space: the Art and Geometry of Polygons, Polyhedra, and Polytopes.* John Wiley and Sons: New York.

Monk, James Donald
1965 Model-Theoretical Methods and Results in the Theory of Cylindric Algebras. In *The Theory of Models, Proceedings of the 1963 International Symposium at Berkeley*, eds. J. W. Addison, Leon Henkin, and Alfred Tarski. North-Holland: Amsterdam, pp. 238–250.
1969 Nonfinitizability of Classes of Representable Cylindric Algebras. *Journal of Symbolic Logic* 34: 331–343.
1989 Review of *Tarski and Givant 1987*. *American Mathematical Society Bulletin* (n.s.) 20: 236–239.

Montague, Richard
1973 The Proper Treatment of Quantification in Ordinary English. In *Hintikka, Moravcsik, and Suppes 1973*, pp. 221–242.

Moore, Gregory H.
1982 *Zermelo's Axiom of Choice: Its Origins, Development, and Influence.* Springer Verlag: Berlin.

Morris, Charles
1937 *Logical Positivism, Pragmatism, and Scientific Empiricism.* Hermann: Paris.
1955 Foundations in the Theory of Signs. In *International Encyclopedia of Unified Science*. 2 vols., eds. O. Neurath, R. Carnap, and C. Morris. Vol. 1, pp. 77–137. University of Chicago Press.
1970 *The Pragmatic Movement in American Philosophy.* George Braziller: New York.

Mostowski, Andrzej
1957 On a Generalization of Quantifiers. *Fundamenta Mathematicae* 44: 12–36.

Müller, Andreas
1991 *Signal-Systems as Emergent Properties of Interacting Processes: Signs and Meaning in Artificial Intelligence.* John Benjamins: Amsterdam.

Murphey, Murray G.
1961 *The Development of Peirce's Philosophy.* Harvard University Press. Repr. 1993, Hackett: Indianapolis.

Naimark, Mark Aronovich and Aleksandr Isaakovich Stern
1982 *Theory of Group Representations.* Trans. Elizabeth Hewitt. Springer-Verlag: New York.

Németi, István
1985 Exactly Which Varieties of Cylindric Algebras are Decidable? *Mathematical Institute of the Hungarian Academy of Sciences, 29 June, 1985, Report No.34*, preprint.
1991 Algebraization of Quantified Logics, an Introductory Overview. In *Blok and Pigozzi, eds. 1991*, pp. 485–569.

Németi, István, ed.
1988 *Abstracts of the Conference on Algebraic Logic, August 8–14, 1988, Budapest, Hungary.* J. Bolyai Mathematical Society: Budapest.

Nessarian, Nancy, ed.
1987 *The Process of Science.* Martinus Nijhoff: Dordrecht.

Newman, James R.
1956 *The World of Mathematics.* Simon and Schuster: New York.

Nickles, Thomas
1984 Positive Science and Discoverability. In *Proceedings of the Philosophy of Sci-*

ence Association, vol. 1, ed. Peter D. Asquith and P. Kitcher. Philosophy of Science Association: East Lansing, pp. 13–27.

1985 Beyond Divorce: Current Status of the Discovery Debate. Philosophy of Science 52: 177–206.

1987 'Twixt Method and Madness. In *Nessarian, ed. 1987,* pp. 41–68.

Nickles, Thomas, ed.

1980 *Scientific Discovery, Logic, and Rationality.* D. Reidel: Dordrecht.

Novy, Lubos

1973 *Origins of Modern Algebra.* Trans. Jaroslav Tauer. Noordhoff International: Leyden.

Nozick, Robert

1981 *Philosophical Explanations.* Harvard University Press.

Ogg, Oscar

1961 *The 26 Letters.* Crowell: New York.

Pagels, Heinz

1982 *The Cosmic Code.* Bantam: New York.

Panteki, M.

1992 *Relationships between Algebra, Differential Equations, and Logics in England, 1800–1860.* Ph.D. dissertation, NAA London.

Parker, Francis D.

1964 Boolean Matrices and Logic. *Mathematics Magazine* 37: 33–38.

Parry, William Tuthill

1954 A New Symbolism for the Propositional Calculus. *Journal of Symbolic Logic* 19: 161–168.

Patin, H. A.

1957 Pragmatism, Intuitionism, and Formalism. *Philosophy of Science* 24: 243–252.

Peacock, George

1830 *Treatise of Algebra.* J. and J. J. Deighton: Cambridge.

Peano, Giuseppe

1889 *Arithmetices Principia Nova Methodo Exposita.* Bocca: Turin. Trans. Hubert Kennedy in *Peano 1973,* pp. 101–134.

1895 Review of *Frege 1893,* vol. 1. *Revista di Mathematica* 5: 122–128.

1973 *Selected Works of Giuseppe Peano.* Trans. and ed. Hubert Kennedy. University of Toronto Press.

Peirce, Benjamin

1870 *Linear Associative Algebra.* Private lithograph ed.: Washington D.C. Edited by C. S. Peirce and repr. 1881 in *The American Journal of Mathematics* 4: 97–229. 2nd ed. repr. in *Cohen, ed. 1980* (unpaginated).

Peirce, Charles Sanders

1849–1914 The Charles S. Peirce Papers. Manuscript collection in the Houghton Library, Harvard University.

1867a An Improvement in Boole's Calculus of Logic. *Proceedings of the American Academy of Arts and Sciences* 7: 250–261. Repr. in 3.1–19 and in W2: 12–23.

1867b Upon the Logic of Mathematics. *Proceedings of the American Academy of Arts and Sciences* 7: 402–12. Repr. in 3.20–44 and in W2: 59–69.

1870 Description of a Notation for the Logic of Relatives, Resulting from an Amplification of the Conceptions of Boole's Calculus of Logic. *Memoirs of the American Academy of Arts And Sciences* ns 9: 317–378. Repr. in 3.45–149 and W2: 359–429.

1875 On the Application of Logical Analysis to Multiple Algebra. *Proceedings of the American Academy of Arts and Sciences* ns 10: 392–394. Repr. in W3: 177–179.

1877 Note on Grassmann's 'Calculus of Extension.' *Proceedings of the American Academy of Arts and Sciences* ns 10: 115–116. Repr. in W3: 238–239.

1880 On the Algebra of Logic. *American Journal of Mathematics* 3:15–57. Repr. in 3.154–251 and in W4: 163–209.

1881 On the Logic of Number. *American Journal of Mathematics* 4:85–95. Repr. in 3.252–288 and in W4: 299–309.

1882a *A Brief Description of the Algebra of Relatives* (7 January 1882). Privately printed: Baltimore. Repr. in 3.306–322 and in W4: 328–333.

1882b On the Relative Forms of Quaternions (abstract). *Johns Hopkins University Circular,* February 1882: 179. Repr. in 3.323 and in W4: 334–335.

1883a A Theory of Probable Inference. In *Peirce, ed. 1883,* 126–181. Repr. in 2.694–754 and in W4: 408–450.

1883b On a Limited Universe of Marks. Included as " 'Note A' " in *Peirce, ed. 1883,* pp. 182–186. Repr. in 2.517–531 as rewritten (making up the first two sections of "Extension of the Aristotelian Syllogistic") in 1893 for *How to Reason* (called "Grand Logic" in CP).

1883c The Logic of Relatives. Included as " 'Note B" in *Peirce, ed. 1883,* pp. 187–203. Repr. in 3.328–358 and in W4: 453–466.

1885 On the Algebra of Logic: A Contribution to the Philosophy of Notation. *American Journal of Mathematics* 7: 180–202. Repr. in 3:359–403 and in W5: 162–190.

1892 The Critic of Arguments. *The Open Court* 6:3391–3394. Repr. in 3.404–424.

1896 The Regenerated Logic. *The Monist* 7: 19–40. Repr. in 3.425–455.

1897 The Logic of Relatives. *The Monist* 7:161–217. Repr. in 3.456–552.

1901 Insolubilia. In *Baldwin, ed. 1901* I, p. 554.

1902 Logic (exact). In *Baldwin, ed. 1901* II, pp. 23–27. Repr. in part in 3.616–625.

1903 Review of Welby and Russell. *The Nation* 77: 308–309. Repr. 8.171–175; N 3: 143–145.

1931–1958 *Collected Papers of Charles Sanders Peirce,* vols. 1–6 eds. Charles Hartshorne and Paul Weiss; vols. 7–8 ed. Arthur W. Burks. Harvard University Press.

1976 *The New Elements of Mathematics* (NEM), ed. Carolyn Eisele. Mouton: The Hague. 4 vols. in 5 books.

1977 *Semiotic and Significs The Correspondence between Charles S. Peirce and Victorial lady Welby* (SS). Ed. Charles S. Hardwick. Indiana University Press: Bloomington.

1982– *Writings of Charles S. Peirce: A Chronological Edition* (W). Ed. Max H. Fisch, Edward C. Moore, *et al.* Indiana University Press: Bloomington.

1992 *Reasoning and the Logic of Things. The Cambridge Conferences Lectures of 1898.* Ed. Kenneth Laine Ketner. Harvard University Press.

Peirce, Charles Sanders, ed.

1883 *Studies in Logic, By Members of the Johns Hopkins University.* Introduction by C. S. Peirce. Little, Brown: Boston. Repr. 1983 in the series *Foundations of Semiotics* with an introduction by Max H. Fisch and a preface by Achim Eschbach. John Benjamins: Amsterdam.

Peirce, Charles Sanders and Christine Ladd-Franklin

1902a Logic. In *Baldwin, ed. 1901* II, pp. 21–23.

1902b Symbolic Logic. In *Baldwin, ed. 1901* II, pp. 645–650.

Pera, Marcello
  1980 Inductive Method and Scientific Discovery. In *Grmek et. at., eds. 1980*, pp. 141–165.
  1987 The Rationality of Discovery: Galvani's Animal Electricity. In *Pitt and Pera, eds. 1987*, pp. 117–202.
Piaget, Jean
  1957 *Logic and Psychology.* Basic Books: New York.
Pietarin, Juham
  1972 *Lawlikeness, Analogy and Inductive Logic.* North Holland: Amsterdam.
Pitt, Joseph C. and Marcello Pera, eds.
  1987 *Rational Changes in Science: Essays on Scientific Reasoning.* D. Reidel: Dordrecht.
Popper, Karl
  1975 *Objective Knowledge.* Oxford University Press.
Potter, Vincent and Paul Shields
  1977 Peirce's Definitions of Continuity. *Transactions of the Charles S. Peirce Society* 13: 20–23.
Prewitt, Terry, John Deely, and Karen Haworth, eds.
  1989 *Semiotics 1988.* University Press of America: New York.
Prior, Arthur N.
  1955 Diodoran Modalities. *The Philosophical Quarterly* 5: 205–213.
  1957 *Time and Modality.* Oxford University Press.
  1958 Peirce's Axioms for Propositional Calculus. *The Journal of Symbolic Logic* 23: 135–136.
  1962 *Formal Logic.* Oxford University Press.
Putnam, Hilary
  1971 *Philosophy of Logic.* Harper Torchbooks: New York.
  1982 Peirce the Logician. *Historia Mathematica* 9: 290–301.
  1983 Mathematics Without Foundations. In *Benacerraf and Putnam, eds. 1983*, pp. *295–311.*
  1995 Peirce's Continuum. In *Ketner, ed. 1995*, pp. 1–22.
Pycior, Helena M.
  1979 Benjamin Peirce's *Linear Associative Algebra. Isis* 70: 537–551.
  1983 Augustus De Morgan's Algebraic Work: The Three Stages. *Isis* 74: 211–226.
  1987 British Abstract Algebra: Development and Early Reception, 1750–1850. In *Grattan-Guinness, ed. 1987*, pp. 152–169.
Quine, Willard V.
  1935 Review of the *Collected Papers of Charles Sanders Peirce* Vol. 3, *Isis* 22: 285–297.
  1950 *Methods of Logic.* Holt, Rinehart and Winston: New York. Rev. 1959, 1972.
  1960 *Word and Object.* Wiley & Sons: New York.
  1961 *Mathematical Logic.* Harvard University Press.
  1963 On What There Is. *From a Logical Point of View.* Harper & Row: New York.
  1966a Russell's Ontological Development. *Journal of Philosophy* 63: 657–667. Repr. with omissions and other changes in *Quine 1981*, pp. 73–85.
  1966b *Selected Logical Papers.* Random House: New York.
  1966c *The Ways of Paradox.* Harvard University Press. Rev. and enlarged ed. 1976.
  1969 *Set Theory and Its Logic.* Harvard University Press.
  1970 *Philosophy of Logic.* Prentice-Hall: Englewood Cliffs. 2nd ed., 1986, Harvard University Press.
  1981 *Theories and Things.* Harvard University Press.
  1985 In the Logical Vestibule. *Times Literary Supplement* 12: 767.

1995 Peirce's Logic. In *Ketner, ed. 1995*, pp. 23–31.

Quinton, Anthony
1973 *The Nature of Things*. Routledge & Kegan Paul: London.

Reichenbach, Hans
1947 *Elements of Symbolic Logic*. Macmillan: New York.
1949 *The Theory of Probability*. University of California Press: Berkeley.

Reiter, R.
1980 A Logic for Default Reasoning. *Artificial Intelligence* 13: 81–132.

Renton, William
1887 *The Analytic Theory of Logic*. James Thin: Edinburgh.

Rescher, Nicholas
1978 *Peirce's Philosophy of Science*. University of Notre Dame Press: Notre Dame.

Rescher, Nicholas, ed.
1986 *Current Issues in Teleology*. University Press of America: New York.

Rescher, Nicholas and R. Brandom
1979 *The Logic of Inconsistency*. Rowan and Littlefield: Totowa, NJ.

Resnik, Michael D.
1985 Logic: Normative or Descriptive? The Ethics of Belief or a Branch of Psychology. *Philosophy of Science* 52: 221–238.

Riepe, Dale Maurice, ed.
1973 *Phenomenology and Natural Existence: Essays in Honor of Marvin Farber*. State University of New York Press: Albany.

Riolo, Rick
1989a The Emergence of Coupled Sequences of Classifiers. In *Proceedings of the Third International Conference on Genetic Algorithms*. Morgan Kaufmann: San Mateo, pp. 256–264.
1989b The Emergence of Default Hierarchies in Learning Classifier Systems. In *Proceedings of the Third International Conference on Genetic Algorithms*. Morgan Kaufmann: San Mateo, pp. 322–327.

Robbin, Tony
1992 *Fourfield: Computers, Art, & the 4th Dimension*. Boston: Little, Brown and Company.

Roberts, Don D.
1973 *The Existential Graphs of Charles S. Peirce*. Mouton: The Hague.
1992 The Existential Graphs. In *Computers & Mathematics with Applications* 23: 639–663. Repr. in *Lehmann, ed. 1992*.

Roberts, Don D. and Henry H. Crapo
1969 Peirce Algebras and the Distributivity Scandal (abstract). *Journal of Symbolic Logic* 34: 153–154.

Robin, Richard S.
1967 *Annotated Catalogue of the Papers of Charles S. Peirce*. University of Massachusetts Press: Amherst.
1971 The Peirce Papers: A Supplementary Catalogue. *Transactions of the Charles S. Peirce Society* 7: 37–57.

Robinson, H.
1975 The Class as One and as Many. *International Logic Review* 6: 172–182.

Robinson, Raphael M.
1959 Binary Relations as Primitive Notions in Elementary Geometry. In *Henkin, Suppes, and Tarski, eds. 1959*, pp. 68–85.

Rodríguez-Consuegra, Francisco A.
1991 *Bertrand Russell's Mathematical Philosophy: Origins and Development*. Birkhaüser: Basel.

Root-Bernstein, Robert S.
  1988 Setting the Stage for Discovery. *The Sciences* 28: 26–35.
Roth, Robert J.
  1988 Anderson on Peirce's Concept of Abduction: Further Reflections. *Transactions of the Charles S. Peirce Society* 24: 131–139.
Rotman, Brian
  1987 *Signifying Nothing: The Semiotics of Zero.* St. Martin's: New York.
Royce, Josiah
  1905 The Relation of the Principles of Logic to the Foundations of Geometry. *Transactions of the American Mathematical Society* 24: 353–415. Repr. in *Royce 1951*, pp. 379–441.
  1913 An Extension of the Algebra of Logic. *Journal of Philosophy* 10: 617–633. Repr. in *Royce 1951*, pp. 293–309.
  1951 *Royce's Logical Essays: Collected Logical Essays of Josiah Royce,* ed. Daniel S. Robinson. Wm. C. Brown: Dubuque.
Royce, Josiah and Fergus Kernan
  1916 Charles Sanders Peirce. *Journal of Philosophy* 13: 701–709.
Rumelhart, David E., James L. McClelland, and the PDP Research Group, eds.
  1988 *Parallel Distributed Processing,* vol. 1. MIT Press: Cambridge.
Runes, Dagobert D., ed.
  1960 *Dictionary of Philosophy.* Rev. and enlarged ed., 1983. Philosophical Library: New York.
Russell, Bertrand
  1901a On the Notion of Order. *Mind* 10: 30–51.
  1901b Recent Work on the Principles of Mathematics. *The International Monthly* 4: 83–101. Repr. as "Mathematics and the Metaphysicians" in *Russell 1918* and in *The Collected Papers of Bertrand Russell,* vol. 3 (Routledge 1993), pp. 366–379.
  1903 *Principles of Mathematics.* Cambridge University Press. 2nd ed. 1937, Allen & Unwin: London. Repr. 1956.
  1906a Les Paradoxes de la logique. *Revue de métaphysique et de morale* 14: 627–650.
  1906b The Theory of Implication. *American Journal of Mathematics* 28: 159–202.
  1918 *Mysticism and Logic.* Longmans, Green: New York. Repr. 1929, 1949 Allen & Unwin: London.
  1919 *Introduction to Mathematical Philosophy.* Allen & Unwin: London. Repr. 1960.
  1944 My Mental Development. In *Schilpp, ed. 1944,* pp. 3–20.
  1946 Foreword to *Feibleman 1946,* pp. xv–xvi.
  1956 *Logic and Knowledge: Essays, 1901–1950,* ed. Robert Charles Marsh. Allen & Unwin: London.
  1959 *My Philosophical Development.* Allen & Unwin: London.
  1967 *The Autobiography of Bertrand Russell 1872–1914.* Little, Brown: Boston.
Saccheri, Giralamo
  1733 *Euclides ab Omni Naevo Vindicatus.* Milan. First part trans. George Bruce Halsted. Open Court: Chicago, 1920. Repr. 1968 Chelsea: New York.
Sagan, Carl
  1977 *The Dragons of Eden—Speculations on the Evolution of Human Intelligence.* Random House: New York.
Sáin, Ildiko
  1988 Concerning Some Cylindric Algebra Versions of the Downward Löwenheim-Skolem Theorem. *Notre Dame Journal of Formal Logic* 29: 332–344.

Saliĭ, Viacheslav Nicolaevich
  1988 *Lattices with Unique Complements.* Trans. G. A. Kandall, ed. B. Silver. American Mathematical Society: Providence. Orig. pub. Moscow 1884.
Salmon, Merrilee, ed.
  1990 *The Philosophy of Logical Mechanism: Essays in Honor of Arthur W. Burks, with his Responses.* Kluwer Academic: Dordrecht.
Sampson, Geoffrey
  1985 *Writing Systems.* Stanford University Press.
Savan, David
  1976 *An Introduction to C. S. Peirce's Completed System of Semiotics.* Enlarged 2nd ed. 1988. Toronto Semiotic Circle Monograph, Victoria College: Toronto.
  1988 Peirce and the Trivium. *Cruzeiro Semiotico* 8: 50–56.
Scharle, Thomas W.
  1962 A Diagram of the Functors of the Two-Valued Propositional Calculus. *Notre Dame Journal of Formal Logic* 3: 243–255.
Schein, Boris Moiscevich
  1991 Representation of Subdirects of Tarski Relation Algebras. In *Andréka et al., eds., 1991,* pp. 621–635.
Schilpp, Paul Arthur, ed.
  1944 *The Philosophy of Bertrand Russell.* Open Court: La Salle. Repr. 1951, 4th ed. 1971.
  1963 *The Philosophy of Rudolf Carnap.* Open Court: La Salle.
Schmeikal-Schuh, Bernd
  1993 Logic from Space: Logic Derived from the Original Concept of Orientation in Space. *Quality and Quantity* 27: 117–137.
Scholz, Heinrich
  1961 *A Concise History of Logic.* Philosophical Library: New York.
Schroeder, Marcin
  1991 Syllogistics Defined in Relation Algebras (abst. # 863-03-659). *Abstracts Presented to the American Mathematical Society* 12: 9.
Schröder, Ernst
  1877 *Der Operationskreis des Logikkalkuls.* Verlag von B. G. Teubner: Leipzig.
  1880 Recension of *Frege 1879. Zeitschrift für Mathematik und Physik* 25: 81–87, 90–94.
  1890, 1891, 1895, 1905 *Vorlesungen über die Algebra der Logik.* Teubner: Leipzig. (Vol. I 1890; Vol. II.1 1891; Vol. III.1 1895; Vol. II.2 1905.) Vol. II, 2nd ed. 1966. Chelsea: Bronx, NY.
  1892 Signs and Symbols. *The Open Court* 6: 3431–3434, 3441–3444, and 3463–3466.
  1898a On Pasigraphy. Its Present State and the Pasigraphic Movement in Italy. *The Monist* 9: 44–62 (corrections [1899] p. 320).
  1898b Über zwei Definitionen der Endlichkeit und G. Cantor'sche Sätze, *Nova Acta Academiae Caesareae Leopoldina-Carolinae Germanicae Naturae Curiosorum* 71: 303–362.
Schütte, Kurt
  1933 Über einen Teilbereich des Aussagenkalkuls. *Comptes Rendus des Séances de la Société des Sciences et des Lettres de Varsovie* 26, Cl. iii, pp. 30–32.
Scott, Andrew, ed.
  1990 *Frontiers of Science.* Blackwell: Oxford.
Sebeok, Thomas A., ed.
  1978 *Sight, Sound, and Sense.* Indiana University Press: Bloomington.

Servois, François Joseph
    1814 Essai sur un nouveau mode d'exposition des principes du calcul différen-
          tial. *Annales des mathématiques pures et appliquées* 5: 93–140.
Shanahan, Timothy
    1986 The First Moment of Scientific Inquiry: C. S. Peirce on the Logic of Ab-
          duction. *Transactions of the Charles S. Peirce Society* 22: 449–466.
Shapiro, S. C., ed.
    1992 *Encyclopedia of Artificial Intelligence*, 2 vols. Wiley & Sons: New York.
          2nd ed.
Sheffer, Henry Maurice
    1913 A Set of Five Independent Postulates for Boolean Algebras, with Appli-
          cation to Logical Constants. *Transactions of the American Mathematical So-
          ciety* 14: 481–488.
Shen, Eugene
    1927 The Ladd-Franklin Formula in Logic: The Antilogism. *Mind* 36: 54–60.
    1929 The "Complete Scheme" of Propositions. *Psyche* 9: 48–59.
Shephard, G. C.
    1974 Combinatorial Properties of Associated Zonotopes. *Canadian Journal of
          Mathematics* 26: 302–321.
Shields, Paul
    1981 Charles S. Peirce on the Logic of Number. Ph. D. dissertation, Fordham
          University: New York.
Simon, Herbert A.
    1973 Does Scientific Discovery have a Logic. *Philosophy of Science* 4: 471–480.
    1977 *Models of Discovery*. D. Reidel: Dordrecht.
Simons, Peter
    1987 Frege's Theory of Real Numbers. *History and Philosophy of Logic* 8: 25–44.
Sinisi, V. F.
    1969 Leśniewski and Frege on Collective Classes. *Notre Dame Journal of Formal
          Logic* 10: 239–246.
Skemp, Richard R. ed.
    1982 Communicating Mathematics: Surface Structures and Deep Structures.
          Visible Language 16: 204, 281–288.
Skidmore, Arthur W.
    1971 Peirce and Triads. *Transactions of the Charles S. Peirce Society* 7: 3–23.
Skolem, Thoralf
    1920 Logisch-kombinatorische Untersuchungen Über die Erfüllbarkeit oder
          Beweis-barkeit mathematischer Sätze nebst einem Theoreme über dichte
          Mengen, *Videnskapsselskapets skrifter, I. Mathematisk-naturvidenskabelig
          klasse*, no. 4. Trans. Stefan Bauer-Mengelberg as "Logico-combinatorial
          Investigations in the Satisfiability or Provability of Mathematical Propo-
          sitions: A Simplified Proof of a Theorem by L. Löwenheim and Gener-
          alizations of the Theorem" in *van Heijenoort, ed. 1967*, pp. 254–263.
    1923 Einige Bemerkungen zur axiomatischen Begründung der Mengenlehre,
          *Matematikerkongressen i Helsingfors den 4–7 Juli 1922, Den femte skandinav-
          iska matematikerkongressen, Redogörelse*. Akademiska Bokhandeln: Hel-
          sinki, pp. 217–232. Trans. Stefan Bauer-Mengelberg as "Some Remarks
          on Axiomatized Set Theory" in *van Heijenoort, ed. 1967*, pp. 291–301.
    1928 Über die mathematische Logik, *Norsk matematisk tidsskrift* 10: 125–142.
          Trans. as "On Mathematical Logic" by Stefan Bauer-Mengelberg and
          Dagfinn Føllesdal, in *van Heijenoort, ed. 1967*, pp. 512–524.

Skolimowski, Henryk
    1967 *Polish Analytical Philosophy.* Humanities Press: New York.
Smith, David E. and Jekuthiel Ginsburg
    1934 *A History of Mathematics in America Before 1900.* The Mathematical Association of America/Open Court: Chicago.
Smolensky, P.
    1988 Information Processing in Dynamical Systems: Foundations of Harmony Theory. In *Rumelhart et al., eds. 1988,* pp. 194–281.
Sowa, John F.
    1976 Conceptual Graphs for a Data Base Interface. *IBM Journal of Research and Development* 20: 336–357.
    1979a Definitional Mechanisms for Conceptual Graphs. In *Claus et al., eds. 1979,* pp. 426–439.
    1979b Semantics of Conceptual Graphs. In *Proc. 17th Annual Meeting of the ACL,* pp. 39–44.
    1984 *Conceptual Structures: Information Processing in Mind and Machine.* Addison-Wesley: Reading, Mass.
    1988 Using a Lexicon of Canonical Graphs in a Semantic Interpreter. In *Evans, ed. 1988,* pp. 73–97.
    1991 Towards the Expressive Power of Natural Languages. In *Sowa, ed. 1991,* pp. 157–189.
    1992 Semantic Networks. In *Shapiro, ed. 1992,* pp. 1493–1511.
Sowa, John F., ed.
    1991 *Principles of Semantic Networks: Explorations in the Representation of Knowledge.* Morgan Kaufmann: San Mateo.
Spencer Brown, G.
    1969 *The Laws of Form.* Allen and Unwin: London.
Stein, D., ed.
    1989 *Complex Systems.* Addison-Wesley Longman: New York.
Stern, August
    1988 *Matrix Logic.* North-Holland: Amsterdam.
Stokes, George J.
    1900 The Theory of Mathematical Inference. *American Mathematical Monthly* 7: 1–8.
Stringham, William I.
    1880 Regular Figures in n-dimensional Space. *American Journal of Mathematics* 3: 1–14.
Styazhkin, Nikolai Ivanovich
    1959 *On the History of the Development of Mathematical Logic in the 19th Century* (in Russian). Ph.D. thesis. Moscow State University.
    1969 *History of Mathematical Logic from Leibniz to Peano.* MIT Press: Cambridge.
Sumner, Leonard Wayne, John Greer Slater, and Fred Wilson, eds.
    1981 *Pragmatism and Purpose: Essays Presented to Thomas A. Goudge.* University of Toronto Press.
Sylvester, James Joseph
    1884 Lectures on the Principles of Universal Algebra. *American Journal of Mathematics* 6: 270–286.
Taber, Henry
    1890 On the Theory of Matrices. *American Journal of Mathematics* 12: 337–395.
Tarski, Alfred
    1923 Sur le terme primitif de la logistique. *Fundamenta Mathematicae* 4: 196–

200. Trans. as "On the Primitive Term of Logistic" in *Tarski 1956*, pp. 1–23.

1924 Sur les ensembles finis. *Fundamenta Mathematicae* 6: 45–95.

1936 On the Concept of Logical Consequence. In *Tarski 1956*, pp. 409–420.

1941 On the Calculus of Relations. *The Journal of Symbolic Logic* 6: 73–89.

1952 A Representation Theorem for Cylindrical Algebras: Preliminary Report. *Bulletin of the American Mathematical Society* 58: 65–66.

1953a Some Metalogical Results Concerning the Calculus of Relations. *Journal of Symbolic Logic* 18: 188–189.

1953b A Formalization of Set Theory without Variables. *Journal of Symbolic Logic* 18: 189.

1954a A General Theorem Concerning the Reduction of Primitive Notions. *Journal of Symbolic Logic* 19: 158.

1954b On the Reduction of the Number of Generators in Relation Rings. *Journal of Symbolic Logic* 19: 158–159.

1954c An Undecidable System of Sentential Calculus. *Journal of Symbolic Logic* 19: 159.

1956 *Logic, Semantics, Metamathematics; Papers from 1923 to 1938*. Trans. J. H. Woodger. Oxford University Press. 2nd ed. 1983, ed. John Corcoran. Hackett: Indianapolis.

Tarski, Alfred and Steven Givant

1987 *A Formalization of Set Theory without Variables*. American Mathematical Society Colloquium Publications 41. American Mathematical Society: Providence.

Tarski, Alfred and F. B. Thompson

1952 Some General Properties of Cylindrical Algebras: Preliminary Report. *Bulletin of the American Mathematical Society* 58: 65.

Taylor, Alfred Edward

1926 *Plato: The Man and His Work*. Methuen: London. 6th ed. 1949. Repr. 1960, Meridian Books: New York.

Tesnière, Lucien

1959 *Eléments de Syntaxe Structurale*. 2nd ed. 1965, Librairie C. Klincksieck: Paris.

Thagard, Paul R.

1977 The Unity of Peirce's Theory of Hypothesis. *Transactions of the Charles S. Peirce Society* 13: 112–121.

1981 The Autonomy of a Logic of Discovery. In *Sumner et al., eds. 1981*, pp. 248–260.

1982 From the Descriptive to the Normative in Psychology and Logic. *Philosophy of Science* 49: 24–42.

1984 Conceptual Combination and Scientific Discovery. *Philosophy of Science Association* 1: 3–12.

1988 *Computational Philosophy of Science*. MIT Press: Cambridge.

Thompson, Manley H.

1949 The Logical Paradoxes and Peirce's Semiotic. *Journal of Philosophy* 46: 513–536.

1981 Peirce's Conception of an Individual. In *Sumner et al., eds. 1981*, pp. 133–148.

Thompson, Richard J.

1988 High Deeds in Hungary (abstract). In *Németi, ed. 1988*, p. 27.

Tomberlin, James E., ed.
1983 *Agent, Language and the Structure of the World: Essays Presented to Hector-Neri Castañeda, with His Replies*. Hackett: Indianapolis.
Turrisi, Patricia
1990 Peirce's Logic of Discovery: Abduction and the Universal Categories. *Transactions of the Charles S. Peirce Society* 26: 465–498.
Turquette, Atwell R.
1972 Dualism and Trimorphism in Peirce's Triadic Logic. *Transactions of the Charles S. Peirce Society* 8: 131–140.
Tursman, Richard
1987 *Peirce's Theory of Scientific Discovery*. Indiana University Press: Bloomington.
Ulam, Stanislaw M.
1974 *Sets, Numbers, and Universes, Selected Works*, eds. W. A. Beyer, J. Mycielski, and Gian-Carlo Rota. MIT Press: Cambridge.
van Benthem, Johan and Alice ter Meulen, eds.
1985 *Generalized Quantifiers in Natural Language*. Foris: Dordrecht.
van der Waerden, Bartels L.
1985 *A History of Algebra*. Springer-Verlag: Berlin.
Van Evra, James
1977 A Reassessment of George Boole's Theory of Logic. *Notre Dame Journal of Formal Logic* 8: 363–377.
van Heijenoort, Jean, ed.
1967 *From Frege to Gödel: A Source Book in Mathematical Logic, 1879–1931*. Harvard University Press.
Venn, John
1866 *The Logic of Chance*. Macmillan: London. 3rd ed. 1888.
1881 *Symbolic Logic*. London. 2nd ed. 1894. Macmillan: New York.
Vercelloni, L.
1989 *Filosofia delle strutture*. Nuovo Italia: Florence.
Verheyen, Hugo F.
1996 *Symmetry Orbits*. Birkhauser: Boston.
von Bertalanffy, Ludwig
1968 *General System Theory*. George Braziller: New York.
von Neumann, John
1966 *The Theory of Self-Reproducing Automata*, ed. and completed by Arthur W. Burks. University of Illinois Press: Urbana.
1986 *Papers of John von Neumann on Computers and Computer Theory*, eds. William Aspray and Arthur Burks. MIT Press: Cambridge.
Vuillemin, Jules
1968 *Leçons sur la première philosophie de Russell*. Colin: Paris.
Wajsberg, Mordchaj
1937 Metalogische Beitrage. *Wiadomości Matematyczne* 43: 1–38. Trans. S. McCall and P. Woodruff as "Contributions to Metalogic" in *McCall, ed. 1967*, pp. 285–318, and in *Wajsberg 1977*, pp. 172–200.
1977 *Logical Works*, ed. Stanisław J. Surma. Zakład Narodowy Imienia Ossolińskich, Wydawnictwo Polskiej Akademii Nauk: Wrocław, Warszawa.
Walker, A, Michael McCord, John F. Sowa, and W. G. Wilson
1987 *Knowledge Systems and Prolog*. Addison-Wesley: Reading.
Wallace, Doris B. and Howard E. Gruber, eds.

1989 *Creative People at Work: Twelve Cognitive Case Studies.* Oxford University Press.

Wang, Hao

1957 The Axiomatization of Arithmetic. *Journal of Symbolic Logic* 22: 145–158.

Water, Kenneth

1987 Relevance Logic Brings Hope to Hypothetico-Deductivism. *Philosophy of Science* 54: 453–464.

Weinberg, Julius Rudolf

1965 *Abstraction, Relation, and Induction. Three Essays in the History of Thought.* University of Wisconsin Press: Madison.

Wennerberg, Hjalmar

1962 *The Pragmatism of C. S. Peirce: An Analytical Study.* C.W.K. Gleerup: Lund.

Weyl, Hermann

1952 *Symmetry.* Princeton University Press.

1963 *Philosophy of Mathematics and Natural Science.* Trans. Olaf Helmer. Atheneum: New York.

Whately, Richard

1826 *Elements of Logic, Comprising the Substance of the Article in the Encyclopædia Metropolitana.* J. Mawman: London. Repr. with intro. by P. Dessì: 1987, Editrice CLUEB: Bologna.

Whitehead, Alfred North

1898 *A Treatise on Universal Algebra with Applications.* Vol. 1. Cambridge University Press. Repr. 1960, Hafner: New York.

1901 Memoir on the Algebra of Symbolic Logic. *American Journal of Mathematics* 23: 139–165, 297–316.

1902 On Cardinal Numbers. *American Journal of Mathematics* 24: 367–368, 378–382.

1948 *Essays in Science and Philosophy.* Philosophical Library: New York.

Whitehead, Alfred North and Bertrand Russell

1910–1913 *Principia Mathematica.* Cambridge University Press. 2nd ed. 1925–1927. Repr. 1959–1960.

Whitney, William Dwight, ed.

1889 *The Century Dictionary, An Encyclopedic Lexicon of the English Language.* The Century Company: New York. Repr. 1895. Rev. ed. 1911.

Wiener, Norbert

1913 *A Comparison Between the Treatment of the Algebra of Relatives by Schroeder and that by Whitehead and Russell.* Ph.D. dissertation, Harvard University.

Wiener, Philip P., ed.

1958 *Values in a Universe of Chance.* Doubleday: New York. Repr. 1966 as *Charles S. Peirce: Selected Writings (Values in a Universe of Chance).* Dover: New York.

Wiener, Philip P. and Frederic H. Young, eds.

1952 *Studies in the Philosophy of Charles Sanders Peirce.* Harvard University Press.

Wilder, Raymond L.

1968 *Evolution of Mathematical Concepts: An Elementary Study.* Wiley & Sons: New York.

Winchester, I. and K. M. Blackwell, eds.

1988 Antinomies and Paradoxes: Studies in Russell's Early Philosophy. *Russell* (n.s.) 8: 1–247.

Wiscamb, Margaret
    1969 Graphing True-false Statements. *Mathematics Teacher* 62: 553–556.
Wittgenstein, Ludwig
    1972 *Über Gewißheit/On Certainty*, eds. G. E. M. Anscombe and G. H. von
          Wright. Harper & Row: New York.
Wood, Robert E.
    1991 Plato's Line Revisited: The Pedagogy of Complete Reflection. *Review of
          Metaphysics* 44: 525–547.
Woods, W. A.
    1967 *Semantics for a Question-Answering System*. The Aiken Computation Labo-
          ratory, Harvard University, Report n. NSF-19, Mathematical Linguistics
          and Automatic Translation.
Yale, Paul B.
    1968 *Geometry and Symmetry*. Holden-Day: London.
Zahar, Elie
    1982 Logic of Discovery or Psychology of Invention? *British Journal of the Phi-
          losophy of Science* 34: 243–261.
Zellweger, Shea
    1979 A Logical Garnet as Both a 3-D and a 4-D Symmetry Model of the 16
          Binary Connectives. In *Program Abstracts, Sixth International Congress of
          Logic, Methodology, and Philosophy of Science, Hannover, Germany, 22–29 Au-
          gust*, pp. 205–209.
    1981 Devices for Displaying or Performing Operations in a Two-Valued
          System. Patents issued in the United States (1981, 1983, 1985), Canada
          (1983), and Japan (1991).
    1982 Sign-Creation and Man-Sign Engineering. *Semiotica* 38: 17–54.
    1987 Notation, Relational Iconicity, and Rethinking the Propositional Calcu-
          lus. In *Program Abstracts, Eighth International Congress of Logic, Methodology,
          and Philosophy of Science, 17–22 August*, Institute of the Academy of Sci-
          ences of the USSR: Moscow, 1: 376–379.
    1989 Symmetry and the Crystallography of Logic. In *Symmetry of Structure, In-
          terdisciplinary Symmetry Symposia, Budapest, Hungary, 13–19 August*, 1:
          641–644.
    1991 Peirce, Iconicity, and the Geometry of Logic. In *Anderson and Merrell,
          eds. 1991*, pp. 483–507.
    1992 Cards, Mirrors, and Hand-held Models that Lead Into Elementary Logic.
          Displayed at the Sixteenth Meeting of the North American Region of
          the International Group for the Psychology of Mathematics Education,
          6–11 August, University of New Hampshire: Durham.
    1993 Cognitive Frames, Cognitive Overload, and Mind-Held Diagrams in
          Logic. In *Deely and Prewitt, eds. 1993*, pp. 35–45.
    1994 On a Deep Correspondence Between Sign-creation in Logic and Sym-
          metry in Crystallography. Proceedings of the Fifth International Con-
          gress of the International Association for Semiotic Studies at Berkeley,
          June 12–18, 1994.
    1995 A Symmetry Notation that Coactivates the Common Ground Between
          Symbolic Logic and Crystallography. In *Symmetry: Culture and Science* 6:
          556–559.
Zeman, J. Jay
    1964 *The Graphical Logic of C. S. Peirce*. Ph.D. dissertation, University of Chicago.

1973 *Modal Logic: The Lewis Modal Systems.* Oxford University Press.
1986 Peirce's Philosophy of Logic. *Transactions of the Charles S. Peirce Society* 22: 1–22.

Zytkow, Jan M. and Herbert A. Simon
1988 Normative Systems of Discovery and Logic of Search. *Synthese* 74: 65–90.

BIOGRAPHICAL NOTES
ON CONTRIBUTORS

Irving H. Anellis is Executive Publisher and founder of Modern Logic Publishing, which publishes the journal *Modern Logic*. He has taught mathematics at various American universities and has served as a Research Associate at the Bertrand Russell Editorial Project. He has published several papers on the history of logic and mathematics and a book on Jean van Heijenoort. Currently, he is working on a book on the history of mathematical logic in Russia during the Soviet period.

Geraldine Brady is a Ph.D. student at the University of Groningen where she works with Professor E. M. Barth. She is Chief Manuscript Editor of *The Astrophysical Journal*, which is published for the American Astronomical Society by the University of Chicago Press.

Jarrett Brock is Professor of Philosophy at San Jose State University. His previous work on Peirce includes: "On Peirce's Conception of Semiotic," "Principal Themes in Peirce's Logic of Vagueness," "An Introduction to Peirce's Theory of Speech Acts," and "Peirce's Anticipation of Game Theoretic Logic and Semantics." He is currently considering the relation between the pragmatic maxim and its logical and semantical presuppositions.

Jacqueline Brunning is Associate Professor of Philosophy at the University of Toronto. Her areas of research interest are late 19th century and 20th century logic, American philosophy, especially that of Charles Sanders Peirce, and the philosophy of science. She is currently completing a book on Peirce's logic of relations.

Robert W. Burch received his Ph.D. from Rice University in 1969. He is currently Professor of Philosophy at Texas A&M University, College Station, Texas. His areas of specialization are American philosophy, logic, and the history and philosophy of logic. He is the author of *A Peircean Reduction Thesis: The Foundations of Topological Logic*.

Arthur W. Burks is Professor Emeritus of Philosophy and Electrical Engineering and Computer Science at the University of Michigan. He was a principal designer of the ENIAC, the first general-purpose electronic computer, and worked with John von Neumann on the logical design of the first electronic computer with random-access memory. His publications include *Chance, Cause, Reason: An Inquiry into the Nature of Scientific Evidence.*

Glenn Clark is former Chairman of the Mathematics Department at Mount Union College, Alliance, Ohio. In 1977 he joined Shea Zellweger's Logic Alphabet Project. He taught abstract algebra for many years and his knowledge of group theory added to a better presentation of the logic alphabet. He died in January 1993.

E. James Crombie (Ph.D., Waterloo, 1979) teaches philosophy at l'Université Sainte-Anne in Pointe-de-l'Église, Nova Scotia. His research interests include Peirce, Thomas Reid, 20th-century French philosophers, perception, philosophy of science, philosophy of education, and the question of minority cultures and languages. He is currently working on a comparative study of Charles Sanders Peirce and the French philosopher and historian of science, Gaston Bachelard.

Jeffrey R. DiLeo is completing a double-major doctorate in philosophy and comparative literature at Indiana University, Bloomington. He has published a number of articles on Peirce including "Peirce's Haecceitism," "The Semiotics of Indexical Experience," "A Semiotic Classification of Proper Names," and "Charles S. Peirce and the Rhetoric of Science." He is Editor-in-Chief of the journal *Symplokē*.

Randall R. Dipert is Professor of Philosophy at the United States Military Academy (West Point). He is a graduate of the University of Michigan and Indiana University. He is co-author of a logic text, *Logic: A Computer* (1985) and *Artifacts, Art Works, and Agency* (1993), and numerous articles on logic and its history, artificial intelligence, Peirce, the history of philosophy, and aesthetics. He is presently at work on a book on the history of logic, and a book on graph theory and the philosophy of mind.

Ivor Grattan-Guinness is Professor of History of Mathematics and Logic at Middlesex University, England. He earned his Ph.D. and D.Sc. in the history of science at London University. He was editor of the history of science journal *Annals of Science* from 1974 to 1981 and founder-editor of the journal *History and Philosophy of Logic* from 1979 to 1992. He has edited the *Companion Encyclopedia of the History and Philosophy of the Mathematical Sciences*, two volumes published in 1994. He has completed a general history of mathematics for the Fontana History of Science series, and is working on the relationships between mathematics, logic, and set theory between 1870 and 1930.

Benjamin S. Hawkins, Jr., received his Ph.D. from the University of Miami in 1971. He is the author of many articles and reviews and has held academic appointments at universities and colleges in Australia, Nigeria, and the United States.

Risto Hilpinen is Professor of Philosophy at the University of Turku (Turku, Finland), and a Research Professor at the University of Miami in Coral Gables. He has published articles on philosophical logic, epistemology, the philosophy of science, theoretical ethics, and the philosophy of Charles S. Peirce, and edited books in these areas. He is an editor of the journal, *Synthese,* and was Secretary General of the International Union of History and Philosophy of Science, Division of Logic, Methodology and Philosophy of Science from 1983–1991.

Jaakko Hintikka is Professor of Philosophy at Boston University. He is author or co-author of 29 books and about 300 scholarly papers. His most recent book is *The Principles of Mathematics Revisited* (1996). He is a former Guggenheim Fellow and officer of several international and national scholarly organizations. His research interests comprise philosophy of language and theoretical linguistics, philosophical and mathematical logic, epistemology, philosophy of science and philosophy of mathematics, history of philosophy and history of ideas, especially as they pertain to Aristotle, Descartes, Leibniz, Kant, Peirce, and Wittgenstein.

Henry Hiż is Professor Emeritus of Linguistics and Philosophy at the University of Pennsylvania. He was educated at the University of Warsaw, Université Libre de Bruxelles, and Harvard University where he received his Ph.D. in 1948. He taught at the University of Utah, Pennsylvania State University, New York University, Columbia University, Jagellonian University and the University of Warsaw. He has published articles on logic, semantics, semiotics, philosophy of language, linguistics, and ethics.

Nathan Houser is Director of the Peirce Edition Project and is Associate Professor of Philosophy at Indiana University, Indianapolis. He is general editor of *Writings of Charles S. Peirce: A Chronological Edition.* He is a co-editor of the *Essential Peirce* and a contributor to Ivor Grattan-Guinness's *Companion Encyclopedia of the History and Philosophy of the Mathematical Sciences.*

Alan Iliff is Associate Professor of Computer Science at North Park College in Chicago. He took graduate degrees in mathematics from the University of Illinois, Chicago, where his doctoral thesis concerned Peirce's contributions to mathematical logic. He teaches computer sciences as well as mathematics and philosophy.

Tomis Kapitan is Associate Professor of Philosophy at Northern Illinois University. He has authored essays on metaphysics, the philosophy of logic, and the philosophy of religion that have appeared in *Analysis, Logique et Analyse, History and Philosophy of Logic, American Philosophical Quarterly, Nous, Philosophical Studies, Erkenntnis,* and *Philosophical Perspectives.*

Beverley Kent has taught at universities in Australia, New Zealand, Canada, and the United States and is currently Associate Professor of Philosophy at Lakehead University, Thunder Bay, Ontario. Her publications include *Charles S. Peirce: Logic and the Classification of the Sciences* (1987), "Peirce's Esthetics: A New Look," "Objective Logic in Peirce's Thought," and "Fuming Against Smoking; Some Ethical Issues" (forthcoming).

Angus Kerr-Lawson is Professor in Pure Mathematics and Philosophy at the University of Waterloo, Ontario, Canada. He has written on American philosophers including Charles Sanders Peirce, and George Santayana. His other interests are logic and the foundations of mathematics.

Stephen H. Levy is a philosopher, logician, and computer scientist working with Calculemics, Inc. in Westfield, N.J. He holds a Ph.D. from Fordham University and has written widely on Peirce's logic and philosophy of mathematics. His publications have appeared in the *Journal of Philosophy, Modern Logic,* and the *International Philosophical Quarterly,* among others. He has taught at the Pennsylvania State University and Pace University.

Ana H. Maróstica is a Professor of Logic for Expert Systems at the School of Economics in the University of Buenos Aires. She is also the logician of a European group of investigators in artificial intelligence with headquarters in the University of Göttingen. Her research interests include the logic and epistemology of Peirce, cognitive science, and nonmonotonic logics. She holds her Ph.D. from Indiana University and is the author of many journal articles both in the United States and Europe.

Daniel D. Merrill is Professor of Philosophy at Oberlin College, Oberlin, Ohio. He has taught at Oberlin since receiving his Ph.D. from the University of Minnesota in 1962. His publications include *Augustus De Morgan and the Logic of Relations* (1990) and among other articles, "De Morgan, Peirce, and the Logic of Relations" (1978).

Don D. Roberts is a graduate of the University of Illinois (Ph.D., 1963). His work has appeared in *Studies in the Philosophy of Charles Sanders Peirce* and *Transactions of the Charles S. Peirce Society.* He is author of *Logical Fragments* and *The Existential Graphs of Charles S. Peirce.* He has been associate editor of the Peirce Edition Project and is currently Chairman of the Board of Advisors to the Project.

Paul Shields received his Ph.D. from Fordham University in 1981, where he worked with the late Vincent Potter, S.J. He is currently Associate Professor of Philosophy and Computer Science at St. Ambrose University.

John F. Sowa worked at IBM for thirty years on a variety of research and development projects, took early retirement in 1992, and is teaching in the program on Philosophy and Computers and Cognitive Science at Binghamton University. He is a Fellow of the American Association for Artificial Intelligence and an active participant in the ANSI and ISO standards efforts on conceptual schema modeling facilities. His theory of conceptual graphs, based on the existential graphs of C. S. Peirce, has been adopted as a basis for a number of research projects around the world. Sowa's graphs are being proposed as a standard display language for representing conceptual schemas by the ANSI X3T2 Committee on Information Interchange and Interpretation.

James Van Evra received his Ph.D. in 1966 from Michigan State University. His writing has appeared in *Inquiry, Theory and Decision, Analysis, Notre Dame Journal of Formal Logic, History and Philosophy of Logic, Journal of Theoretical Medicine,* and anthologies. He is co-author of a volume of essays on the philosophy of science.

Shea Zellweger, former Chairman of the Psychology Department, retired after 24 years (1969–1993) at Mount Union College, Alliance, Ohio. His longstanding Logic Alphabet Project began as offhand insights in 1953, he devised his notation in 1961–62, and it has evolved into patents in the United States, Canada, and Japan. During two full-year sabbaticals at the Peirce Edition Project in Indianapolis, his prior practice with the logic alphabet easily became an ideal preparation for what he found in Peirce's manuscripts, especially in the "Simplest Mathematics" (1902). He continues to develop and clarify the use of the logic alphabet.

Jay Zeman is a Professor of Philosophy at the University of Florida, Gaines-ville. He works with modal logic and with Peirce's thought, most recently with Existential Graphs in a computer environment. Recent papers include "Existential Graphs and Thirdness," to appear in *Semiotica,* and "The Tinc-tures and Implicit Quantification Over Worlds," to appear in a collection edited by Jacqueline Brunning and Paul Forster for the University of Toronto Press.

INDEX

Abbott, Edwin A., 372
Abduction, 10, 19, 106, 460–61, 465–66, 493n1, 556n5; action as purpose of, 485–86; in Artificial Intelligence, 536–41; autonomy of, 478, 482, 488, 490–93, 493n3; comprehensibility of for scientific theory, 477, 481–84; deductive reasoning in, 483–84; evolutionary, 512, 514–15, 524; form of, 480–81, 495n10; inferential process in, 477, 478–80; justification of, 489–90; and plausibility, 481, 484–88, 494n4, 495n11; as the production of hypotheses, 106, 469, 477, 481–84, 493n2, 495n13, 496n15; and sign production, 473; thought processes of, 470–71; tychism as, 501–2; in tychist logic, 536–41, 557n6; validity in, 490
Absolute terms, 159, 169, 177, 178, 209–10, 213, 224
Absolute zero, 272
Abstract reasoning, 29, 102–3, 456
Achinstein, Peter, 480, 495n11
Action theory, 473; and abduction, 485–86
Addition, 49, 96; and disjunction, 201; logical, 175; logical "or" linked to, 148; relative, 114, 159
Adicity, 212–13, 216, 234, 242
Aesthetics, 16, 456, 458
Agapism, 471, 498, 500, 502–3, 525, 531, 532; and creativity, 506–8, 512, 518; and deduction, 502
Alexander, Irving, 495n13
Algebra, 272; integral relation, 300; primitive, 285; projective, 302; proper relation, 273; quadrate, 285; relation of logic to, 152–56; representable relation, 273, 301
Algebraic logic, 12–13, 15, 24–30, 39n3; Boolean, 23, 25, 26–27, 147, 148–49, 156n2; extensionality of collection theories, 30–31; and form, 35–36; links with mathematics, 28–29, 147–52; Mitchell's,

180–84; and names, 32; part-whole theory of collections, 29, 30; popularity of, 32; and quantity, 35
Algebraic logic of relations (ALR), 158–60, 163–72, 176–79, 189–90
Alphabets, 339; Jevons' logical alphabet, 334–35; Łukasiewicz's, 368; symmetry of letters in, 336, 337, 353, 359–60, 377
—Zellweger's logical alphabet, 331, 333n9, 353–60, 362, 364–65, 380n1; and the cognitive ladder, 367–79; diagonal test, 356; flips of, 359–60, 361, 368–69; generalization to the 4th Series, 369–75; size test, 356; symmetry rules, 359–60; system of stems in, 354–59; uses for, 377–79
Analogy, 28, 147, 148–52, 155–56
Anancasm, 471
Anderson, Alan Ross, 535
Anderson, Douglas R., 495n13
Andréka, Hajnal, 301
Anticipations, 387, 399
Application of a relation, 206–7; associativity of, 216–19, 223; binary operation of, 214–20, 222–23; generalization of, 219–24; graphical syntax of, 214–23; marks of reference in, 219–20, 221–22, 223; notation for, 213–14; and the reduction thesis, 223, 231; subjacent numbers used in, 219–21, 223–24, 233n9; teridentity terms in, 225–32; unary operations of, 220–21, 228–29. *See also* Relations, logic of
Applied mathematics, 78–79
Arabic numbers, transition to use of, 339, 359, 368
Archibald, Raymond Clare, 152
Aristotelian syllogistics, 275
Aristotle, 106, 112, 137, 174, 179, 182
Arithmetic, 17, 44, 49
Arraying operation, 241
Artificial Intelligence (AI), 418, 426, 428, 436, 444; abduction in, 536–41; and

nonmonotonic logic, 535, 554; and qualitative induction, 557–63

Associativity: in algebraic logic, 28; in the application of a relation, 216–19, 223; in arithmetic, 49; and corollarial reasoning, 96

Austin, John Langshaw, 474n2

Axelrod, Robert, 518

Axiom system for natural numbers: of Dedekind, 46–48; and logicism, 50n2; of Peano, 46–48, 51n6; of Peirce, 3, 17, 43–49, 51n10

Axiomatics, 29, 40n11, 86

Ayim, Maryann, 16

Babbage, Charles, 39n4

Bacon, John, 302

Bar-Hillel, Yohoshua, 49, 51n11

Beatty, Richard, 210

Belief, 478, 487, 494n6

Bell, Eric Temple, 6, 112

Belnap, Nuel D., Jr., 535, 536, 537

Benacerraf, Paul, 79, 80–81

Bergson, Henri, 532

Berkeley, George, 132, 446

Bernays, Paul, 92, 113, 236, 279

Beth, Evert William, 51n10, 121

Binary functions, 292–93

Binary relations: algebra of, 176–79, 272; Mitchell's theory of, 182–83; as primitive for geometries, 286; in Schröder, 190

Binyon, F., 337, 370, 374

Black, Max, 55, 56

Blok, Willem J., 302

Bôcher, Maxime, 13

Boll, Marcel, 353

Bolyai, Farkas, 101

Bolzano, Bernhard, 66–67, 70

Boole, George, 2, 7, 194; algebraic logic, 23, 25, 26–27, 147, 148–49, 156n2; and classes, 59–60, 189; differential operator methods, 26–27; education of, 156n1; and existentials, 173; and Hamilton, 157n6; influence on Peirce, 37, 42n21, 194; mathematics applied to logic, 33; on the meaning of algebraic referents, 72n1; and particular propositions, 164, 174–75; on probability, 35, 40n15, 201; on the relation between algebra and logic, 153; universal propositions in, 163; universe of discourse in, 203

Boolean algebras, 15, 273; for chemistry, 42n20; cylindric algebra as, 277–78; Mitchell's quantification for, 186; and syllogistics, 275; truth functions in, 186–87

Boolean product, 255, 256, 258

Boolos, George, 57

Borel, Émile, 53

Borges, Jorge Luis, 399

Bourbaki, Nicolas, 296

Bowen, Francis, 193

Brandom, R., 535, 544, 554

Brink, Chris, 151, 161

Brodie, B., 42n20

Brown, W. M., 495n13

Brumbaugh, Robert S., 398

Brunning, Jacqueline, 156n3, 281

Burch, Robert W., 14, 172n13, 283, 297

Burge, Tyler, 574

Burger, Dionys, 372

C-calculus, axiomatization of, 264–65

Cantor, Georg, 23, 25, 40n7, 53, 124; on cardinality, 36, 66, 91–92; on the continuum, 30; Peirce's correspondence with, 69–69, 75n30

—set theory, 28–30, 36–37, 65–71; conceptual difficulties in, 55–56, 75n37, 89; continuity in, 449; individuals in, 60–61; theoremic reasoning in, 97–98

Cardinality: Cantor on, 36; of a set, 49, 66, 67, 71; transfinite, 67

Carnap, Rudolf, 11, 559n22, 593n1

Carroll, Lewis, 127, 335, 353

Cartesian Product, 236, 241, 242, 255, 256, 285, 374

Carus, Paul, 4, 104

Castañeda, Hector-Neri, 75n33, 491, 496n19, 574, 593

Categories, Peirce's theory of, 252–53, 448; and existential graphs, 253, 256–63. *See also* Dyadic relations; Monads; Triads

Cattell, J. McKeen, 125

Cauchy, A. L., 25, 27–28, 50n1, 88

Causality, 79–81

Cavaillès, Jean, 29

Cayley, Arthur, 2, 33, 41n17, 199, 280

Census Theorem, 235

Certainty: and deductive reasoning, 460–62, 475n4

Chaos, 528, 530–31

Chemistry: Boolean algebra for, 42n20

Chin, Louise H., 273, 274

Choice, Axiom of, 50, 52n14, 64, 65, 74n27; independence results for, 54–55; and theoremic reasoning, 101; Russell and, 40n11

Church, Alonzo, 61, 108, 387, 388

Clark, William Glenn, 333n10, 338, 365, 374–77, 382n7, 385n24

Classes, 58–59, 373; Boolean conceptions of, 59–60, 189; empty, 197; equational,